BIO-MEDICAL TELEMETRY

Sensing and Transmitting Biological Information from Animals and Man

R. STUART MACKAY

Biology Department
Boston University

Department of Surgery
Boston University Medical School

Second Edition

JOHN WILEY & SONS, INC.
New York / London / Sydney / Toronto

Library of Congress Catalogue Card Number: 74–121909
ISBN 0 471 56030 8

Printed in the United States of America

10 9 8 7 6 5 4 3 2

To T.O.M.

Preface to Second Edition

Interest in the telemetry course has remained, and thus it has been continued and expanded. Following the Boston presentation, it was given at the Royal College of Surgeons in London, England from June 26–28, 1968. It was next given for the University of California at San Diego from June 25–28, 1969. As a final activity that generated considerable enthusiasm and seemed to bring the whole topic "down to reality," a laboratory session was instituted. Each participant was given a set of small components, and each succeeded in building a physiological monitoring transmitter in a short length of soda straw. With this start, it seemed easier for the participants to proceed with the immediate use of these techniques in their own areas of research. The lectures, exhibit, and laboratory were presented in Canberra, Australia in February, 1970, under the joint sponsorship of the Australian National University and the University of New England (New South Wales). Since the demand persists, it is presently scheduled to be given next in September 1970, at the American Museum of Natural History in New York.

In the first edition, many of the circuits employed transistor types whose number was preceded with the letters FK, denoting a standard transistor in a small package. The manufacturer has since discontinued this packaging, but none of the circuits are thus rendered obsolete. At any time equivalent transistors are available in small packages, and the reader is advised to inquire of manufacturers what equivalent types are available at the time of use. Indeed, transistors are being continuously improved at reduced cost, and thus it is expected that the original circuits can serve not only as examples of useful configurations, but may actually function under more difficult conditions.

As indicated previously, a steadily increasing number of investigators are using these methods and writing about them. A supplementary reading list may prove valuable to some, and thus a list of approximately 1,400 items has been added. This list is not to be considered as complete, but rather a random sampling of the literature. Some of the items were suggested by I. Ball, R. Barwick, P. Fullagar, C. Henssge, H. Hornicke, and J. Schladweiler.

The recent studies on the fetus are being aided by a grant from the John A. Hartford Foundation, New York. The satellite oriented studies are aided by National Aeronautics and Space Administration grants, especially NGr 22-004-024.

R. Stuart Mackay

Boston, Massachusetts
February, 1970

Preface to First Edition

Though some early research had been done by using procedures that might be termed bio-medical telemetry, the radio version has received the most attention during the last decade, especially since junction transistors have become available. Despite a 1962 conference at the American Museum of Natural History in New York devoted exclusively to these matters, many scientists were not using these methods, not because they had rejected them but because they were unaware of their existence or potential. On the other hand, in certain groups a sudden interest in biotelemetry brought about great confusion, and two government agencies complained that some investigators who had requested research funds did not understand the methods they were proposing to adopt.

Thus it seemed essential that someone produce a course of instruction to explain to the scientific community the advantages, procedures, and limitations of these methods. It seemed clear that reports at the usual scientific symposia, although important, were somewhat overlapping and would not serve this particular function of instruction.

To meet this need an intensive course for biologists, physicians, and engineers was organized. My lectures lasted for three full days and the associated exhibit gave the participants some experience with equipment. The interactions between participants of various disciplines seemed to stimulate the engineers to learn biological and medical needs and procedures, whereas the biologists and physicians learned something of the most modern techniques and scientific approaches to the analysis or interpretation and design of experiments. Radio techniques were emphasized, for I felt that it was a simplification to go to systems that employed telephone lines.

The first presentation in May 1965 in San Francisco was sponsored by the University of California; approximately 150 scientists, engineers, and physicians participated. The next presentation in March 1966 was sponsored by the University of California in cooperation with the American Institute of Biological Sciences. The enrollment and the diversity of the participants was significantly larger. Again, special interest was shown in the exhibit in which both apparatus and live animals demonstrated telemetry in progress. It seemed essential that the third presentation be extended to four days and this took place in August 1966 in Washington, D.C., under the sponsorship of the Smithsonian Institution, also in cooperation with the American Institute of Biological Sciences. Attendance numbered 250. I gave the lectures and the exhibit was manned by several of my students and staff—in particular, Barbara Dengler, George Rubissow, Ernest Woods, Sam McGinnis, and Charles Brown. A new feature was a simultaneous display of large-scale telemetry at the National Zoo. The next presentation is scheduled at Boston University, September 27–30, 1967, and a European presentation has been planned, probably in London in cooperation with the International Institute for Medical Electronics and Biological Engineering.

This book contains, essentially, this last set of lectures and should prove useful in future courses. Introductory electronics material is included for the benefit of the biologist, and some elementary biomedical material will give engineers an impression of the problems. Some biologists may find those aspects of transducers and electronics summarized in the early chapters helpful, even if they decide not to telemeter their information. The more detailed knowledge of each group should allow them to use this book as a point of departure.

Metric units are employed except for some lengths for which English units give a more convenient impression of the tolerances to which a system was built, or where standard materials or data tables are not so specified.

In the last few years interest in this field has burgeoned. It is no longer practical to prepare a complete list of references. In many cases it has proved convenient to take examples from work at Berkeley, which often involved my students. Thus much of the material is original; however, many excellent research workers throughout the world have contributed to this field.

In connection with the original research in this book several acknowledgments should be made. One relatively early development took place while I was on sabbatical in Sweden on a Guggenheim Fellowship. Both before and after this period parts of the research were supported by the Berkeley and San Francisco campuses of the University of California. Recently, many advances have been made, thanks to grant NSG-600 of the National Aeronautics and Space Administration.

Many people have assisted in the studies reported here. Special mention should be made of Barbara Dengler, George Rubissow, Sam Toy, Ernest Woods, Fred Jenkinson, and Harvey Fishman. Others are mentioned in the text, but I should like also to acknowledge the early help of John Carbone, Raymond Watten, Don Buckla, Elaine Ross, and Mark Bohrod.

R. Stuart Mackay

Berkeley, California and
Boston, Massachusetts
February 1967

Contents

BIO-MEDICAL TELEMETRY

1

Introduction

This section introduces some of the potentialities, limitations, and ideas of bio-medical telemetry by a series of specific examples of some of the work that has already been accomplished. Certain of these experiments are discussed in detail in later sections. For the reader who wants a survey introduction it is hoped that this section alone will suffice.

At the outset it should be stated that, to some experiments, the use of telemetry methods adds nothing but extra complication, and sometimes unreliability. But certainly there are many situations in which the use of telemetry either makes an otherwise impossible experiment feasible or results in more productive results being obtained in a given time with significantly less statistical averaging being required. These methods prove important particularly in those situations where it is desirable to leave the subject in a relatively normal physiological and psychological state by interfering with his normal pattern of activities as little as possible. With uncooperative animals, this can take on an added importance. These methods not only allow freedom of activity but also provide for the production of signals from relatively inaccessible regions. Thus a radio transmitter can be swallowed to pass into regions that would otherwise be difficult to explore, while leaving the subject totally unaware of its presence. The absence of electrical leads penetrating or extending from the skin can be important to human subjects, who might otherwise tangle or pull them loose while showering or sleeping or romancing or carrying on any other activity. In the case of animals these same problems are avoided, with the added advantage that an internally placed transmitter is out of reach of active attempts to pull loose equipment or measuring devices that otherwise would be placed on the surface of the body.

Although these situations are the most generally important ones (and, indeed, telemetry from a completely restrained animal is seldom advantageous), in a number of static situations these methods can be important. Thus it is actually found useful to work in experimental forests by shooting a series of transmitters into the upper reaches of trees. The investigator can then walk through the forest simply retuning his receiver to record the signal from various transmitters monitoring different variables at different positions. Similarly, the inside of chemical-reaction vessels can be monitored without physical penetrations that might prove troublesome at high pressures. The sterilization of food in a can, the curing of concrete, the temperature of centrifuge rotors, and the like can similarly be studied by suitably placed temperature transmitters. The main purpose here, however, is to consider the telemetry and perhaps the reverse process, telestimulation, of unrestrained animals.

It might be noted that many things could be included under the literal meaning of the term telemetry. For example (Fig. 1.1), of interest from a century ago are the experiments of the Frenchman Marey (1869) where a flying bird was studied by having it trail wires and a hose leading to a smoked drum recorder. This sample has the spirit of some of the present topics, but will not be discussed further, nor other history given from before the recent radio methods which seem to provide the greatest flexibility and the greatest diversity of experiments.

If one wishes to do more than locate or track the position of an animal, it is necessary especially to consider the sensor of the information to be transmitted. It is this input transducer that will often prove the most uncertain or difficult part of the overall system. Thus attention is given to a number of combinations of sensors and transmitters. Special emphasis is given to methods in which radio or other wireless transmission is employed, although in many cases it can prove valuable to communicate signals over wires or telephone lines. In general, however, such methods will prove to be simplified cases of the radio case, with the initial sensor proving to be the most critical element. Although the receivers for these signals are given some attention, the greatest emphasis is placed on the transmitter system, since it is this element that often must be designed to have special properties of life, size, or stability in the presence of body fluids or other unusual engineering characteristics.

To make these ideas as definite and graphic as possible we shall first discuss a specific transmitter type in some detail. It is based on a rather simple but effective circuit that was early discussed for this purpose in the general literature (Mackay and Jacobson, 1957). Figure 1.2 shows an oscillator circuit that has come to be known to engineers as a Hartley oscillator. In the lower part of the drawing a coil and capacitor are connected in

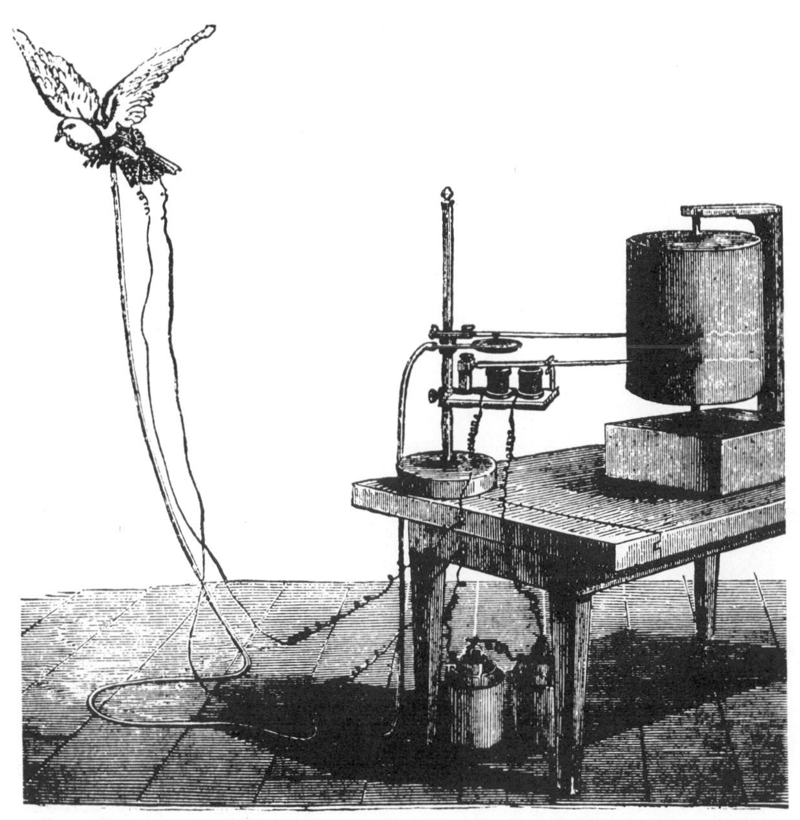

Fig. 1.1 Conveying remotely-sensed information without the use of radio in a century-old telemetry experiment. A rubber bulb detects the shortening of the pectoral muscle of a pigeon by its thickening, the pneumatic signal traversing a rubber tube to a bulb pushing a stylus on a smoked drum. A flapping vane at the wingtip opens and closes an electric contact to indicate the relative duration of the period of elevation and depression of the wing.

parallel; this combination has a natural frequency of oscillation in the same sense that a swing has a basic frequency at which it tends to oscillate. The coil is tapped approximately at its center and this connection goes to a transistor whose action will maintain oscillatory currents in the tuned circuit. The source of power for these oscillations is shown as a single cell battery, although experiments have demonstrated the possibility of using biological power sources here or even inwardly induced radio power from outside the body of the subject. The second capacitor is required in the circuit to feed the signal that actually causes the maintenance of oscillations into the transistor. In the vicinity of the coil can be placed an ironlike material, which

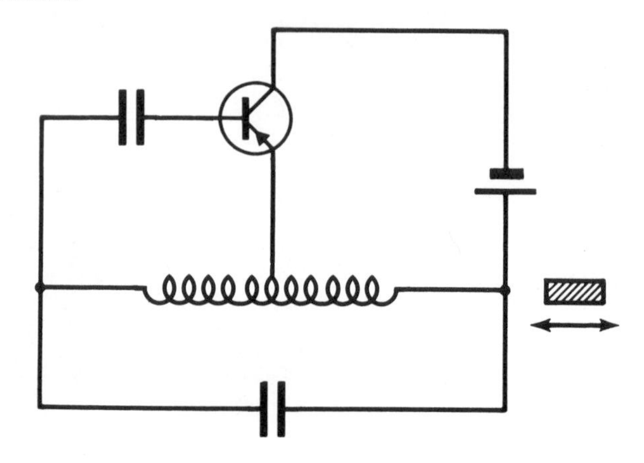

Fig. 1.2 Early circuit used to telemeter internal information, which still has wide applicability. Pressure changes on a diaphragm move the core to cause changes in the radio frequency of transmission. Oscillations periodically turn themselves off at a rate that depends on temperature if the transistor is of germanium or at a rate controlled by an associated resistance if the transistor is of silicon.

can be suspended from a diaphragm. Then pressure changes on the diaphragm will cause motion of this ironlike substance towards and away from the coil, which will modify the frequency of oscillation. An increase in pressure will lower the frequency of oscillation. If a coil of wire is placed in the vicinity of this oscillator, it will pick up the signal being generated, and this voltage can be applied to a radio receiver for amplification and interpretation. Thus this combination can be considered as a frequency-modulated pressure-sensing radio transmitter.

The use of changes of frequency to signify changes in pressure can be seen to have an importance extending beyond the usual advantages of frequency modulation (FM). If simple amplitude modulation were being used, the change in the physiological variable of interest would give rise to a change in the amplitude of the outgoing signal. But if the transmitter were to move behind a different thickness of intervening absorbing tissue, for example, after having been swallowed, or should it change its orientation so that the strength of the signal reaching the receiving antenna were to change, this would also produce a change in the magnitude of the received signal. It would be impossible, in general, to say whether a sudden increase in intensity were due to a change in the variable of interest, or to a change in the position of the transmitter. However, if a given number of cycles per second starts out from inside the body, then this same number must arrive at the receiving antenna independently of the previously mentioned factors to give an unambiguous indication. This can be summed up by saying that frequency is invariant under changes in attenuation and orientation.

This simple circuit has another interesting property. Oscillations in the tuned circuit produce an alternating voltage between the two ends of the coil, and thus a similar alternating voltage will be found between the center tap and the left end of the coil. Whatever else a transistor may be, it does consist of a pair of rectifying junctions, and this alternating voltage can result in the charging of the left-hand capacitor through the lower junction in such a way that a backward voltage is built up which actually turns the transistor off. Thus after some cycles of oscillation activity stops. This condition of no signal will persist until the charge on the left-hand capacitor can leak off, which in the original circuit took place mostly through the other rectifying junction of the transistor. In germanium transistors this resistance is relatively low and quite sensitive to temperature, and thus the higher the temperature the more quickly do oscillations again start. Thus periodic bursts of radio frequency energy are generated, with the frequency of these pulses transmitting temperature. It is interesting that such a simple circuit can simultaneously transmit temperature and pressure information, both in unambiguous frequency modulated form.

Some representative wave forms are shown in Fig. 1.3. At the top is the

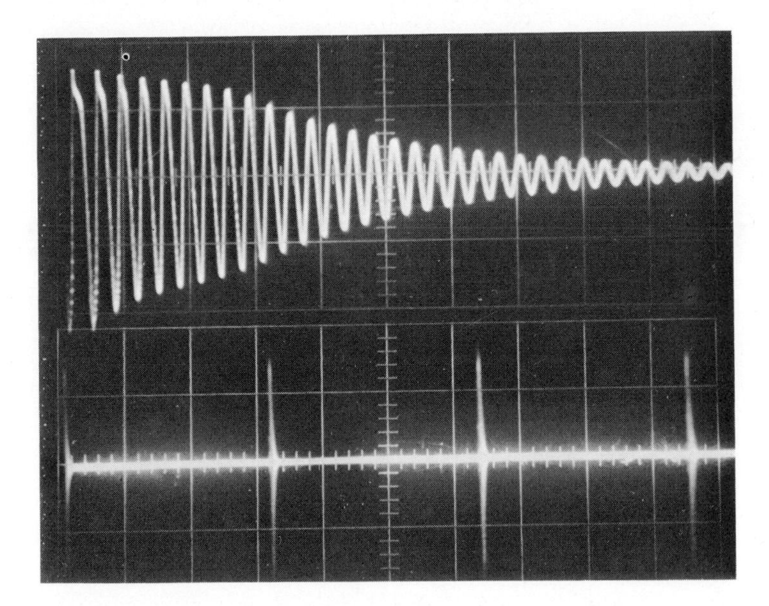

Fig. 1.3 Representative wave forms associated with the circuit in Fig. 1.2. At the top are the radio-frequency oscillations within a single burst. A change of parameters can give many radio-frequency cycles of approximately the same amplitude, as in Fig. 3.5. Below is the pattern of such bursts displayed by greatly reducing the oscilloscope sweep speed.

pattern of cycles during the active period while oscillations are taking place. The frequency here is in the radio range, perhaps between 100,000 cycles per second or 100 kilohertz (sometimes abbreviated 100 kc or 100 kHz) to 100,000,000 cycles per second (100 Mc or 100 MHz). Below that is seen the repetition pattern of these bursts of energy, perhaps taking place anywhere between 1 per second to 10,000 per second. If an ordinary amplitude modulation (AM) receiver is placed in the vicinity of such a circuit, a clicking or buzzing will be heard with a tone of the latter frequency. This transmitter type, because it is actually inactive or turned off much of the time, does tend to prolong battery life.

A receiver configuration that is capable of accepting this compound signal and decoding it into two separate graphs, one of pressure and the other of temperature, is described in the chapter on receivers. A receiver for one or the other of the two variables alone, however, is more standard and considerably easier to construct, and this type of equipment was used in most of the early experiments. Thus, if pressure indications were to be transmitted, a resistor was introduced into the circuit so that the charge on the capacitor that gave the periodic action tended to leak off, thus leading to continuous oscillations. The receiver then need only decode or follow changes in frequency of this continuous oscillation to tell about pressure changes. Similarly, if temperature alone were of interest, the diaphragm assembly would be omitted so that there would be no changes in the basic radio frequency, but merely changes in the rate of turning on and off. This action is sometimes termed squegging or blocking oscillation.

Some of the first studies on human subjects in which transmitters of this general type were used were initiated by having the subject swallow the transmitter, after which patterns of pressure fluctuation associated with muscular activity along the gastrointestinal tract could be followed. Such a transmitter left the subject in a completely comfortable state, since he would be quite unaware of its presence. The transmitter passes in the usual manner, still transmitting, after a day or two in a typical case. The early transmitters of this type were about the size of a large vitamin capsule, but with the reduction in size of normally available commercial components, it is now quite possible for anyone to construct a considerably smaller and more convenient transmitter of this type. A specific layout is indicated in Fig. 1.4. This transmitter of pressure fluctuations is inserted into a tied off finger cut from a rubber glove before swallowing to simplify reuse of the same transmitter by another subject.

From Fig. 1.4 the functioning of a number of the components can be understood. Near the coil is placed the core whose motion causes frequency changes. If the core is of ferrite (a ferromagnetic ceramic), an increase in pressure will cause a decrease in frequency. If the core is instead composed

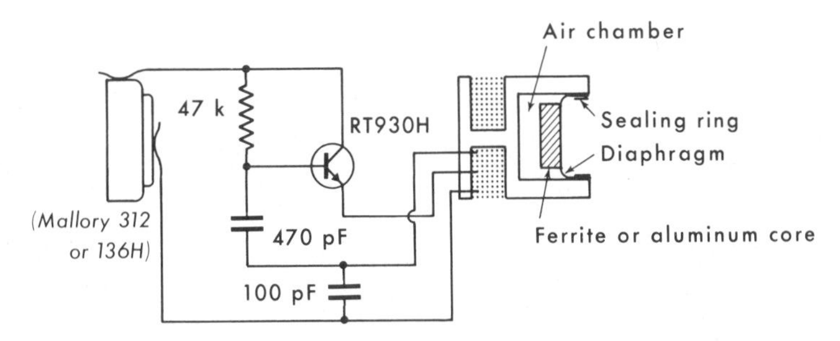

Fig. 1.4 Components in a capsule to be swallowed in order to telemeter the pattern of pressure fluctuations associated with muscle activity along the gastrointestinal tract. The presence of a resistor sufficiently low in value ensures continuous oscillation, thus simplifying the receiver if temperature information is not also required.

of a conducting material, such as a thin piece of aluminum, then an increase in pressure will cause an increase in frequency. The transmitter responds to changes in position of the core, and pressure is sensed because of a springy force being applied to restore the core to its initial position. This restoring force should not be supplied principally by the elasticity of the diaphragm because this changes upon contact with the various body fluids of the gastrointestinal tract. It is the compressibility of the air trapped behind the diaphragm which supplies this spring force, and if there are no leaks this will be constant, resulting in a similar calibration before ingestion and after passage from the body. Most of the signal radiated from the transmitter comes from the coil. The coil thus serves not only as part of the resonant circuit but also as the modulator and the antenna.

This configuration of components is essentially that previously described in some detail (Mackay, 1959), and for experiments on peristalsis there seems to have been no necessity to do research on improvements. More sophisticated configurations for more demanding purposes however, are given in Chapter 4. The resistor is included in the circuit to maintain continuous oscillations. The transistor can be the one indicated, or almost any readily available small one. The battery shown energizing the transmitter will typically give continuous operation for about three days, with a sufficiently strong signal that there should be no interruption in reception at a loop antenna placed near the abdomen of the subject, perhaps looped over his shoulder or around the waist. This particular battery, based on the chemistry of mercury, is sometimes used in hearing aids and electric watches, and it gives a very constant voltage during discharge.

Such a capsule is easily swallowed by most adults by simply placing it in the mouth and drinking a glass of water. It will drop directly to the

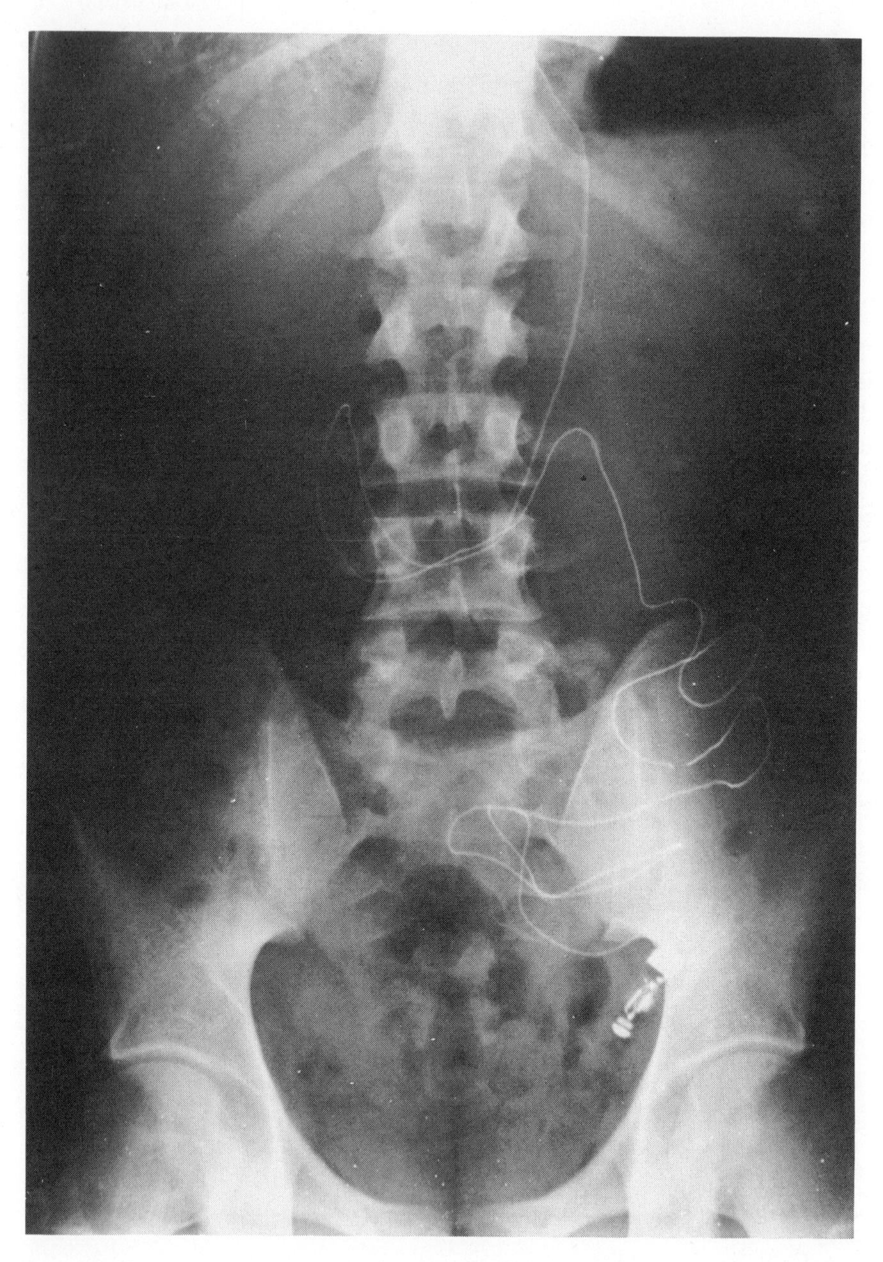

Fig. 1.5 Radiograph of a male human after swallowing a pressure-transmitting endoradiosonde. The transmitter is restrained in the small intestine from moving onward by an opaque thread which is seen looping upward through the gastrointestinal tract and which has been tied to the subject's tooth.

stomach, from which passage will take place after some hours. One can tell if the transmitter has passed into the small intestine by having the subject take a sip of cool water. Because of a little residual temperature sensitivity left in the transmitter, there will be a shift in the recorded base line if the transmitter is still in the stomach, but if it has left the cool water will not be able to alter the temperature. Exact positioning further along in the gastrointestinal tract generally requires an X-ray image, perhaps in conjunction with a meal of barium. In some cases, when we wish to prolong an observation in a given region of the small intestine, we attach the transmitter to a thread that is tied around a tooth. In most cases this procedure is surprisingly comfortable, and we often extend observations for several days in this manner. Figure 1.5 shows a radiograph of a young adult male in such a situation. The thread, which has been rendered radio-opaque, is seen extending upward through the small intestine and stomach and out toward the mouth. The components in the transmitter are shown also.

One of the applications of this type of transmitter is to study the effects of drugs, in some cases as they relate to the proper dosage for a particular individual. There are spasmolytic drugs whose purpose is to reduce activity along the gastrointestinal tract, but in many cases they have been difficult to study without interfering with the normal pattern of activity. Drugs such as atropine usually produce some side effects, for instance, dryness of the mouth, and there are continuing searches going on for drugs with lesser extra effects. The important thing is to determine what constitutes an adequate dose, and Fig. 1.6 depicts a test on an experimental anticholinergic

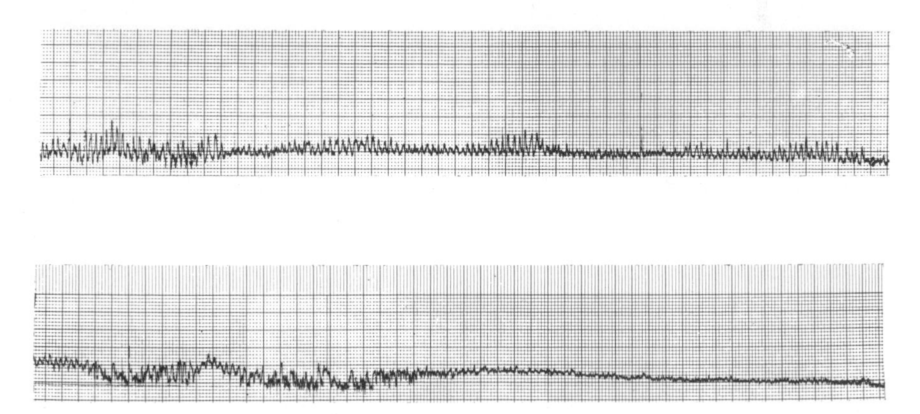

Fig. 1.6 A representative recording from the small intestine of a human. At the top an inadequate dose of a drug meant to stop activity had been administered. In the lower tracing the effect of an adequate dose is shown. In testing this drug it was found that when the dose was adequate there were side effects.

drug. In the top tracing is seen the characteristic pattern of pressure changes in the small intestine of a human subject. About a half hour before the start of this tracing a dose of the new drug was given, which proved inadequate, for it did not alter the pattern of activity. On another occasion the second tracing was produced, and with this adequate dose is seen the reduction in activity. Unfortunately, when the dose was adequate, the drug elicited just as many side effects.

Similar units can transmit pressure fluctuations from within the bladder or uterus. Transmitters of this same general form can serve other functions also. One early application was to the measurement of standing steadiness, in which case three transmitters with stiff springs were placed in the sole of a shoe to measure rocking pressures to the side and forward in connection with studies on certain alcohol-induced and diseased states. Such a transmitter can give a continuous record of digestive enzyme activity dissolving a piece of meat or an egg albumen substrate (rather than the discontinuous indication mentioned in Chapter 13). In another of the earliest papers such a unit was used to follow expansion and contraction of a special polymer with pH changes (Chapter 7).

It was mentioned that this early circuit could also transmit temperature, and in many biological and medical experiments that variable is of extreme interest. A specific version of such a circuit is diagrammed in Fig. 1.7. This combination of components can be incorporated into a sphere approximately a centimeter in diameter, and will transmit temperature information

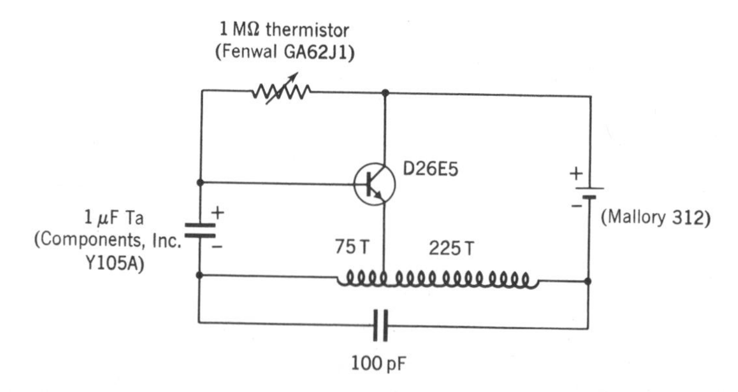

Fig. 1.7 A circuit in which blocking action is retained to transmit temperature, but no pressure information is sensed to modify the radio frequency. This circuit is quite reliable and supplies a good introductory exercise for those wishing to undertake biomedical telemetry. The indicated components are small, but others as well can be employed. The thermistor can be omitted if a germanium transistor is substituted. (See also the modified version of Appendix 4.)

Fig. 1.8 Telemetering deep body temperature from within a 170-kg tortoise on the Galapagos Islands. A circuit, as shown in Fig. 1.7, has been ingested by the animal and the click rate timed by ear with a portable receiver and stop watch.

continuously for months. The range of transmission is not great, but if an AM pocket radio is brought within a foot or two a click will be heard approximately every second, with a rate depending on temperature. These clicks can be timed with a stop watch, and in conjunction with a preliminary calibration at a series of known temperatures this provides a very rugged, simple, and reliable overall system for remotely monitoring the temperature at an inaccessible spot. In the circuit shown, rather than employing the temperature sensitivity of a germanium transistor, as in the original circuit, a silicon transistor is used in conjunction with a thermistor. This type of transistor is more readily available in small sizes at this time, and thermistors (resistors that change with temperature) are available with a variety of properties.

Such units have some applicability in field work with wild animals. Thus Fig. 1.8 is a photograph of the monitoring of the deep-body temperature of a 375-lb tortoise on one of the Galapagos Islands. Since it was desired to approach the animal anyhow to make periodic local measurements of the environmental temperatures, the short range of transmission did not cause significant additional problems. Further mention of this work is made in a later chapter.

It is strongly advised that anyone wishing to interest himself in this general field should construct this circuit before proceeding. The circuit is quite reliable and always seems to work if the connections are correct. The coil can be wound by hand on a pencil, bunching the turns into a small space and taking a loop of wire to the side for the tap. Success with such a simple configuration generally proves encouraging to a new worker during the temporary failures that inevitably occur in the later fabrication of more complicated and specialized equipment. The expense is low enough that such a project can be undertaken routinely by classes in high school, in which the receiving equipment can be simply an inexpensive transistorized pocket receiver. It will be found that the radio frequency generated covers a broad range, and thus the tuning of the receiver is not especially critical in the present case. This can be an advantage if the investigator is attempting continuous recording, because if there is any drift in the receiver the signal will still be properly received. It does make it difficult to work with several transmitters in the same vicinity, although their pulsing repetition rates can be sufficiently varied so that they are individually recognizable though appearing simultaneously at the loud speaker.

The rather slow clicking rate of this transmitter, which allows the signal to be interpreted by the human ear, is determined by the relatively high capacitance but physically small tantalum capacitor in the base connection of the transistor. We have found such components to be adequately stable to allow temperature measurements to about 0.1°C. Related circuits have been used also for environmental studies (Cole, 1962), and to follow the temperature of an incubating penguin egg in Antarctica (Eklund, 1959). If this capacitor is replaced by one of lesser capacitance, for example 0.01 microfarad (μF), then coming from the receiver will be heard a tone whose pitch varies with temperature. A pair of wires from the loudspeaker connection can communicate this signal through simple circuits (to be discussed in the chapter on receivers) which convert this changing frequency into a variable voltage for direct recording by a penwriter. The pulses can be directly recorded by a magnetic tape recorder for later analysis. In a hospital situation, a spare electrocardiograph machine can be used either to record the individual pulses, thereby determining their rate, or it can be used to record the decoded voltage from the previously mentioned circuit.

We named such transmitters as these "endoradiosondes," but they have also been called radio pills, gutniks, transensors, and the like. Extended applications in animal studies are suggested by the fact that the range of transmission is almost as great through fresh water or ice or desert sand as through air. In later sections various transmitters are discussed that are either bigger or smaller, that have longer and shorter lives, and that are capable of transmitting a variety of other physiological variables. From these two examples the reader should gain some impression of just what is meant when mention is made of a radio transmitter.

Implanted Transmitters

In many cases, principally in animal studies, a transmitter is placed within the body of an animal by a surgical procedure rather than through a normal body opening. In these cases special emphasis is often placed upon long life, both of the power source and of the stability of the sensing transducer in its interaction with the biological system. Several days after the surgery, it is often found that the animal has fully recovered and presumably feels no discomfort from the placement of the transmitter. Whether in a cage or in a wild environment, the animal is then free to go about his normal routine without the tendency to tangle or to pick and pull at items that would be exposed if the transmitter were simply placed external to the body. Similar transmitters however, have been used in a number of significant experiments with external placement on various collars or harnesses, often with no noticeable effect on the animal. In that case there is often less of a premium on size and weight of the transmitter, and the decision as to placement depends on the particular experiment. In some cases the placement of the transmitter will be dictated further by the nature of the physiological variable that is to be sensed.

In Fig. 1.9 is shown a radiograph of a rhesus monkey into which surgical procedures have placed four separate transmitters, each simultaneously active on a different frequency.

The upper central transmitter senses the voltage due to the beating of the heart; that is, it transmits the electrocardiogram. Extending from this transmitter can be seen a pair of wires which, in this case, were simply sutured near the top and bottom of the sternum where voltages generated by the functioning of other muscles would be minimized. To prevent breakage due to flexing, these leads are actually very narrow helices of steel wire which are encased in silicone rubber. To the right of this transmitter is another implanted transmitter that senses and telemeters temperature. Below this pair is a transmitter of acceleration. Actually it might better be termed a motion sensor for measuring activity, and functioning can be proper even in the weightless state. The actual circuit is quite like the electrocardiogram transmitter, except that the input electrical signal comes from a small piece of piezoelectric ceramic fastened at one end and with a small mass of solder at the other. At the bottom is a transmitter of blood pressure. A transducer (which is discussed in some detail in Chapter 5) is placed on the outside of the abdominal aorta; it is able to sense the absolute pressure within the

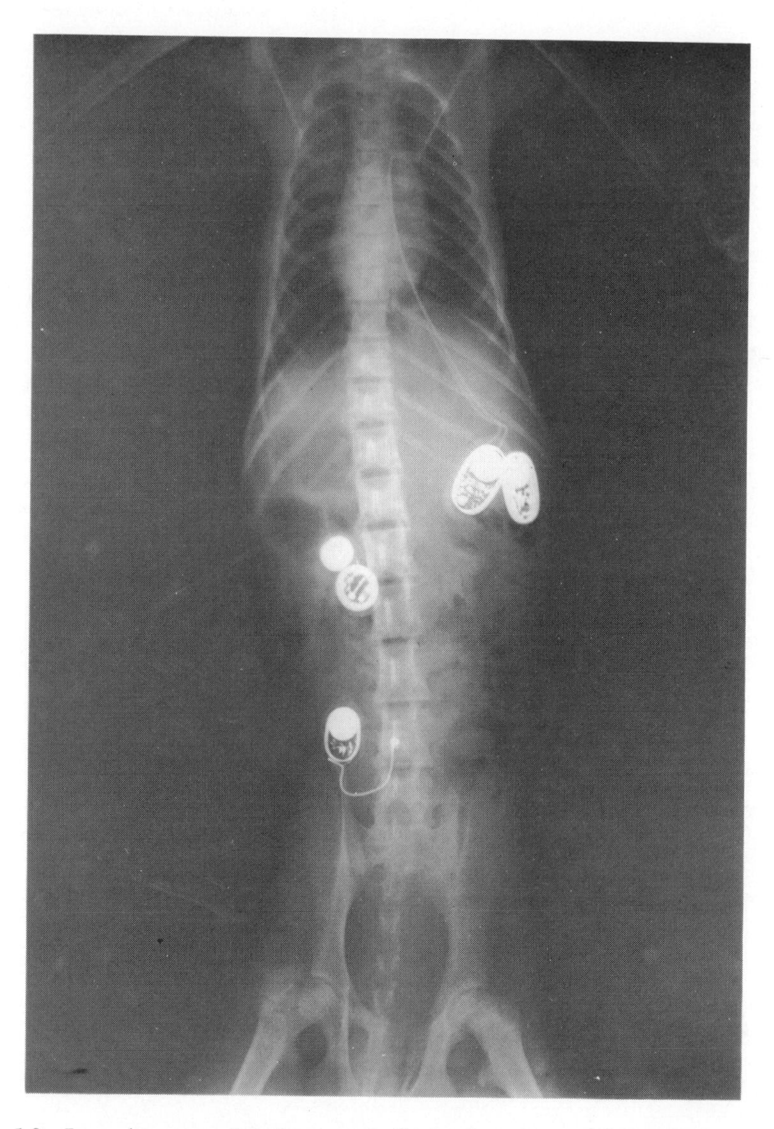

Fig. 1.9 In a rhesus monkey four surgically implanted transmitters were simultaneously active on different frequencies. The top two transmitters are for temperature and electrocardiogram transmission, whereas the transmitter below them is an implanted accelerometer for indicating activity. At the bottom is shown a transmitter of the signal from a sensor attached to the outer surface of the abdominal aorta and which is capable of sensing instantaneous blood pressure independent of changes in the properties of the intervening vessel wall.

vessel independently of such things as changes in the elastic properties of the intervening vessel wall. The signal from this sensor is then communicated to the transmitter shown.

The number of transmitters employed in a particular experiment is largely determined by the demands of the experiment. In some cases it is most convenient to have a single transmitter simultaneously telemeter several variables. In other cases it is more convenient to place several independent standardized transmitter types at appropriate locations in an experimental animal. Figure 1.10 shows a section of an eight-channel recording of data from a rabbit. In this case six transmitters were simultaneously operative

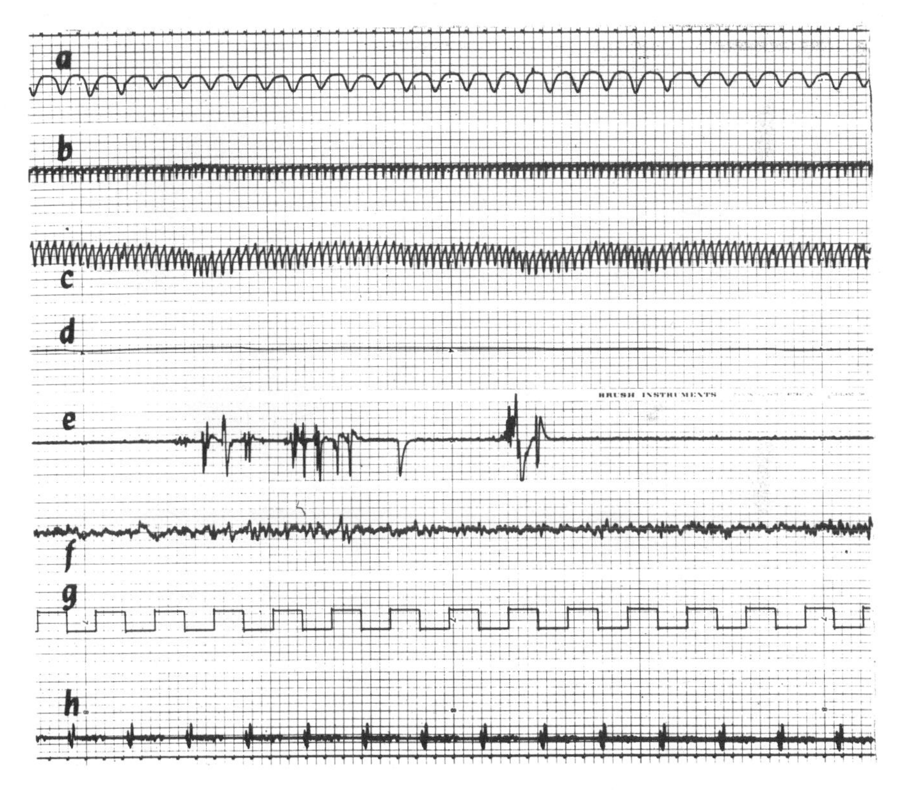

Fig. 1.10 Simultaneous recordings from six transmitters, through an omnidirectional receiver system, plus two channels of auxiliary information, on a section of paper from an eight-channel recorder. At *a* the pattern of breathing is shown monitored by pressure fluctuations; *b* is the electrocardiogram and *c* the instantaneous blood pressure; *d* shows temperature and *e* indications of activity; *f* is the electrical activity of the brain in the vicinity of the region involved in seeing. With each downstroke of the pen at *g* there was a flash of light, and the part of the tracing at *f* correlated with this light flash was continuously computed and recorded at *h*.

from different locations in the body of the animal. At the top of Fig. 1.10 is seen the tracing of the pressure changes due to breathing, and from this tracing not only can breathing rate be inferred but also its pattern. Thus it is possible to tell if respiration involved choking, or a fast inspiration with slow expiration, and so on. Below this tracing is an electrocardiogram tracing from which the various waves in the pattern can be studied. Below that is a recording of instantaneous blood pressure, and below that a recording of temperature. On the next line is seen the signal recorded from an activity-monitoring implanted accelerometer, with the voluntary activity shown having produced little change in temperature. Next is a brain-wave pattern from a transmitter mounted outside the skull but beneath the skin. One electrode was placed over a sinus (an air-filled cavity where there is little

Fig. 1.11 Radiograph of a snake after swallowing a mouse containing a transmitter of temperature and pressure. Intensity changes in addition indicate activity.

activity) and the other electrode was placed in the vicinity of that part of the brain involved in vision. The last pair of lines are not the signals from transmitters, but represent simultaneous recordings of ancillary information. Thus on each downstroke of the square wave there was a flash of light in the room with the rabbit. There is no obvious change in the brain-wave pattern to be seen on the previous line. If, however, the electroencephalographic signal is run through a computer of evoked responses, which continuously calculates the part of a varying signal that is correlated with a repeated stimulus, then the tracing on the bottom section appears. This is the part of the brain-wave response that can be related to the flashing of the light, and it represents an average over a large number of events, which average is continuously slowly changing as new information comes in.

Especially when several variables can be sensed in one region, it may be more convenient to employ a single transmitter to send the several pieces of information. For example, peristalsis in cold-blooded animals can be studied by a single transmitter of pressure and temperature if it is swallowed by the subject. Such a transmitter in a mouse has been ingested by the snake (a boa) in Fig. 1.11. The transmitter was passed in the usual way 22 days later. A single transmitter in this case is convenient, and also dispenses with the use of two antenna coils and two batteries. Such studies will be mentioned further in Chapter 5 and more information given on multichannel transmission in Chapter 8.

Antennas

From the transmitter, whether it is inside or outside of the subject, the signal is then propagated to some sort of a receiving antenna. In many cases this is simply a few turns of wire connected to a receiver. For observations within a laboratory, that is, at ranges of a few meters or less, one can think of the transmission as taking place rather in the fashion of the transfer of energy from the primary winding of a transformer into the secondary winding. For longer range transmissions, up to many kilometers, the more usual concepts of radio apply, and indeed the receiving antennas may well resemble the familiar antennas used with home television sets.

In making short-range physiological observations, there are possibilities for the signal to become useless. Thus, if the transmitter strength is marginal and the subject moves away from the receiving antenna, the signal can become very weak and be lost in any background noise. Changes in orientation can also cause a loss of signal which might otherwise be quite strong. It will be clear to most readers that, if two coils of wire are arranged in a perpendicular fashion, there will be no energy coupled from the first coil into the second. Thus a signal from an ingested transmitter can vanish as the capsule moves along the turns of the gastrointestinal tract. Similarly, the signal from an implanted transmitter can disappear at a fixed receiving loop if the animal reorients himself during play. If the signal is basically strong, this contingency may be unlikely and not prove annoying, but in other situations it may cause the loss of important information. There are omnidirectional systems in which three perpendicular antennas are arranged in the vicinity of the subject so that there will always be a signal in one of them. This type of antenna, suitable for placement under a cage, is shown in Fig. 1.12. The output from the three coils can then either be scanned cyclically or else suitably combined into a single signal that does not vanish for any orientation of the transmitter. These matters are taken up in one of the later chapters.

On the other hand, in some cases it may be desirable to allow the signal to fluctuate in strength and to vanish occasionally. Thus the motion of a transmitter within the gastrointestinal tract can be monitored by changes in the magnitude of the received signal; this is not a precise indication but is a measure of activity in some sense.

Figure 1.13 shows another aspect of this concept. Here is recorded a temperature tracing from one of the large lizards, *Ctenosaura pectinata* which has a small temperature transmitter implanted in its abdomen. A loop of wire encircling the general area to which the animal was confined was enlarged until the signal tended to be marginal. Then random motions on

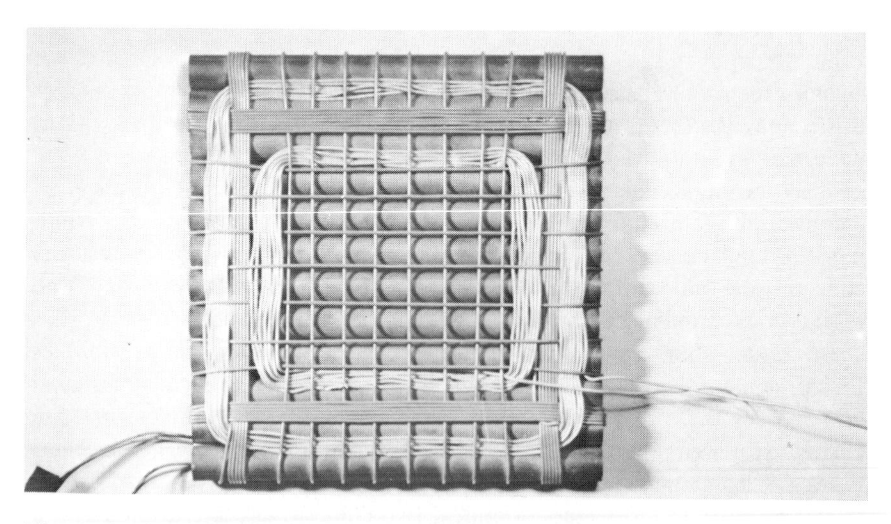

Fig. 1.12 Omnidirectional receiving antenna for relatively low frequency applications. Three perpendicular coils are wound on a block of ferrite (glued up from rods). No matter how the resulting radio field reorients itself, a signal will be induced into at least one of the coils from a nearby transmitter.

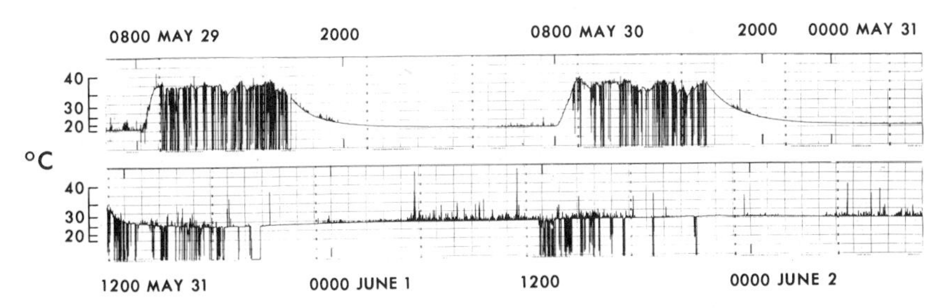

Fig. 1.13 Temperature recording from a transmitter implanted in the abdomen of a lizard *Ctenosaura pectinata*. Cooling at night and cycling within a small temperature range during the day is shown in the top tracing. (Heating and cooling rates are generally not the same in nonlinear and nonfixed systems.) By making the receiving antenna large and limiting the transmitter strength, it is assured that random activity will cause a periodic loss of signal and a resulting momentary deflection to zero. The frequency of such interruptions measures activity. In the lower trace, with constant ambient temperature and illumination, the persistence of the activity pattern is seen. Sharp upward deflections represent added signals due to radio interference.

the part of the animal would cause a periodic loss in signal, with the recorder dropping back to zero as is seen. Thus the frequency of these "drop outs" is a measure of activity. In this single recording two kinds of information are then readily available to the eye of the observer. First of all it can be seen that, as day arrives with a source of heat, the animal quickly warms, and by moving about in his area he is able to hold his temperature within a small range. There are small cyclic fluctuations in temperature which are readily observed from the recording. Then, at the end of the day, when the light and heat source are removed, the lizard cools off in an expected fashion and remains relatively inactive during the night. On the second line the persistence of the circadian activity rhythm is seen when the overall temperature and illumination are kept constant. Small upward deviations of the trace merely represent the appearance of radio noise and are not confused with the "drop-outs" associated with activity. Of course, a simple circuit could simultaneously record the rate of appearance of these drop-outs if a graph of activity were required.

In the longer range experiments, in which the more usual concepts of radio apply, there are inevitable circumstances under which a signal may disappear; these matters are discussed further in a later chapter.

The Scale of Experiments

The scope of observations that can be made is too broad to more than hint at with a few examples. Transmitters introduced through normal body openings in the human can sense pH in the stomach, the site of bleeding along the gastrointestinal tract, radiation intensity, the pressure changes in the bladder due to micturition, the pressure of teeth grinding together during sleep, vaginal temperature, and the like. Human and subhuman species have been studied, as have aquatic and terrestrial animals, cold- and warm-blooded animals, and so on. Some transmitters need only send their signal for an hour, whereas others are expected to transmit continuously for a year or more.

Certain transmitters need extend their signals for only a few centimeters, whereas others are useful only at ranges of kilometers. In these laboratories there has been a continuing interest in the problem of glaucoma and thus continuous observations of pressure in the eye of an animal have been of importance. A number of students have considered the problem of producing a transmitter small enough to put into the eye. In connection with his recent doctoral research, Carter Collins has produced small transmitters, ranging in size down to 2 mm in diameter by 1 mm thick, which absorb inwardly directed radio energy, and reradiate it as a modulated signal carrying information about instantaneous pressure. These transmitters can be placed in the eye of an animal, the eye remaining functional, and Collins has found that all five sensory modalities produce a transient increase in pressure within the eye of the rabbit.

All of this is discussed further in the chapter on passive transmission. A photo of the experiment from his thesis is seen in Fig. 1.14. A pair of coils close to the head of the animal picks up the signals from the two transmitters, one in each eye, for continuous pressure recording. A useful signal here need only pass from inside the eye to outside. It is clear that transmitters can be valuable in carrying a signal through a sensitive closed region even though the transmission range may be limited to millimeters or centimeters.

The collection of useful data may require a longer range of transmission; for example, a study was made of the temperature changes in the marine iguana which lives on the Galapagos Islands and is the only lizard in the world to enter the ocean regularly to feed. The water in that area is cooled by the Humboldt current. In this case the animal was sometimes on land and sometimes in the ocean, and the investigator could not get very close to the animal on land. It was possible (Mackay, 1963) to construct a transmitter that was strapped to the base of a lizard's tail and to have a thermistor probe sensing his temperature. The transmitter had to be neutrally buoyant so as not to sink the animal when swimming. It had to be out of the way so that he could crawl into crevices in a usual fashion and have a range of at least 100 ft. A radio frequency lower than otherwise desirable was selected so that there could be some penetration through the shielding effect of the ocean water (which is a partial conductor of

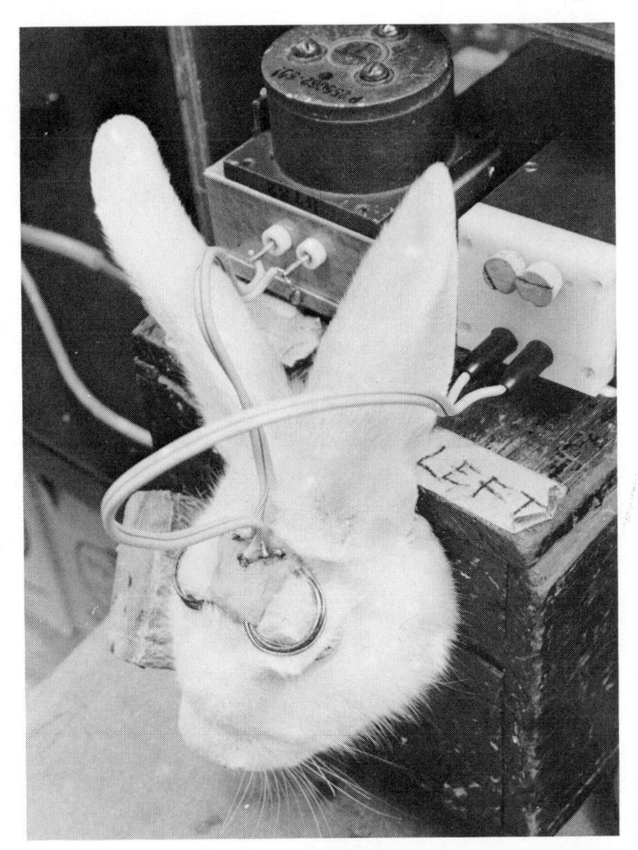

Fig. 1.14 In both eyes of a rabbit a tiny passive transmitter has been placed into which power is induced from outside and from which a pressure signal is reradiated. The two receiving antenna coils appear above the rabbit's eyes. It was found that all sensory modalities, including unexpected sights, produced a transient rise in pressure. (Photo from thesis by Dr. Carter Collins.)

electricity). A photograph of the animal with the externally mounted transmitter is shown in Fig. 1.15. By using this apparatus it was possible to demonstrate that the lizard, on entering the water, did remain active and vigorous although it cooled to the temperature of the surrounding water rather quickly. Thus it adjusted to large fluctuations in temperature, quite unlike the tortoises in the same area which tended to maintain a relatively constant temperature throughout the 24-hr cycle.

On a still larger scale there are the experiments in which it may simply be desired to know the location of an animal and this tracking may have to be done from ranges of kilometers. In the well-known experiments

Fig. 1.15 A male marine iguana on the Galapagos Islands with an externally placed temperature transmitter taped to the base of its tail. The transmitting antenna loop lies across its back, down between the spines. Core temperatures were obtained at satisfactory ranges from the unrestrained animal, whether on land or in the ocean.

by the Craighead brothers methods were used in which a radio transmitter was incorporated into a collar placed around the animal's neck. The collar itself also served as a transmitting loop antenna, thus contributing to the substantial ranges observed. A photograph from one of their experiments is shown in Fig. 1.16. These matters also are discussed in a later chapter.

Serious consideration is being given to each of several possibilities for automatically following movements of animal tracking transmitters from artificial satellites in the sky (see Fig. 14.8). All regions are accessible to them, and plans are being made for their employment both in tracking and in relaying physiological information.

Perhaps the most extensive application of these techniques will be in hospital monitoring, first in intensive care units and then in general. It is this application that may be the stimulus for setting aside specified radio frequencies not interfered with by others. In the health sciences another wide application may prove to be the electromyographic control of prostheses, such as artificial hands or arms, by passive transmitters implanted for the life of the host.

Aquatic animals pose special problems because of the difficulty of transmitting electromagnetic signals through conducting media such as ocean water. It has been possible, however, to build both radio transmitters and sound-wave transmitters to carry signals from an animal in the water to a remote observer. In Fig. 1.17 is shown the feeding of a radio transmitter, placed in the gill of a dead fish, to a dolphin of the type *Tursiops truncatus*. This animal tended not to chew her fish, which resulted in successful telemetry of temperature from deep in the body while swimming and diving

Fig. 1.16 An immobilized cow elk being fitted with a transmitter collar for later tracking from a distance. The entire circumference of the collar also serves as the transmitting loop antenna, which contributes to the transmission range. (Photograph by Drs. Frank and John Craighead.)

Fig. 1.17 Feeding a dolphin (*Tursiops truncatus*) a dead fish, into the gill of which has been stuffed a temperature transmitter. The fish is swallowed whole, and the temperature deep in the body can then be monitored continuously as the untethered animal swims freely in a large pool of ocean water.

in a normal fashion. In experiments that are mentioned later it was found that the temperature regulator of a dolphin is quite good in providing a temperature essentially as constant as that of other warm-blooded animals, but that if the dolphin were removed into the air the regulator became ineffective.

A greater range of transmission in water is achieved by using sound waves to carry the information, but we have employed those procedures largely with fish and invertebrates so far because of the extended range of hearing of many of the cetaceans. It might be noted that some of the methods suitable for telemetry from aquatic animals are also the most desirable for communicating information from inside metallic vessels such as high-pressure reaction chambers or diving chambers or from within other shielded regions.

In all cases it is possible to build a booster transmitter which is capable of being carried by a subject and receiving the weak signals from an internal transmitter and reradiating them a great distance to a central receiving point. Examples are cited later. Other transmitters turn themselves on and off periodically in order to conserve battery life while achieving great range. Others, called transponders, return a signal only when activated by a radio signal from the investigator; these units give signals only when needed and by the delay in the returned response also allow estimation of the distance to the subject.

There are many related methods to be considered. Thus, if only the position or motion of a subject at short range is wanted, it can often be quite as effective to affix a small permanent magnet or a radioactive source to the animal instead of energizing a transmitter with an electric power source. In one series of experiments a small magnetized needle was affixed to the wing of a hummingbird to monitor the rate of wing beat. The needle could instead have been sutured to the wall of the ventricle of the heart to monitor heart beat.

In many experiments in which the information is not instantly needed and in which it is certain that the subject will again be seen it is sufficient to employ a recorder of any sort whose record can later be run off by the investigator. This would clearly not be suitable if the actions or welfare of a man in the ocean or an astronaut were being monitored, but in a surprising number of studies where it is really desired to collect data, this method is a simple alternative to the continuous transmission of information.

In some cases a reverse procedure can be employed. Thus power can be induced into an animal for purposes of stimulation of various sorts. One may then wish to monitor the resulting response by telemetry, in order to modify the next stimulating impulse. In this case the use of telemetry and telestimulation comprises almost a dialogue between the subject and the experimenter.

Other examples of the inward inducement of power are mentioned in a later section. It may eventually prove possible to do minor forms of surgery without actually going through a body wall into the region in question. We have worked with some success already on a blood-vessel clamp which, in response to external signals, would either shut off or restore the blood supply to a particular organ, thus allowing physiological observations on the result of this change in conditions. Examples of other precise ways of disturbing biological systems in order to understand their normal functioning and component interactions are given in Chapters 15 and 16, such methods often being especially effective when combined with telemetry.

It is hoped that these few preliminary words will give a feeling for the scope of this activity. The possibilities are limited only by the imagination of the investigator.

2

Electronics

The material in this chapter is not intended for people normally working in the field of electronics, but rather for those who may wish to make use of some of the techniques to be discussed later and who do not have a background in this field. Thus this is intended to be largely a summary in one place of kinds of material that a biologist might otherwise have to look through several sources to obtain. In some cases it is desirable for a biological scientist to reread the section on electricity of some familiar elementary physics text before continuing with this section. Some will find it useful to refer to the introductory sections of the *Radio Amateurs Handbook* (Goodman, 1966) or a similar book (Orr, 1966).

At the time of this writing a substantial number of scientists have had experience with vacuum tube circuits but had ceased to work with electronics before the introduction of the transistor. Thus in several places reference is made to circuits based on vacuum tubes, and then the analogous configuration using transistors will be indicated. Not only is this analogy a convenient help in understanding certain circuit types, but the more recently developed field-effect transistors have properties much like some vacuum tubes, and an analysis of circuits based on these can be almost identical to analyses of vacuum-tube circuits.

Emphasis is placed on semiconductor circuits. In most transmitters this is essential because of considerations of size and power drain. In receivers, these circuits are often advantageous on the basis of size, power drain, portability, ruggedness, and safety (due to low voltages). These factors are especially important in field work.

1. SEMICONDUCTORS

Materials have traditionally been classified either as good insulators or as good conductors of electricity. In recent years, materials that conduct at an intermediate degree, and called semiconductors, have become extremely important in the technology of electronics. All physics and electronics texts discuss this subject and only a few words are included to introduce its language and ideas.

Semiconductors are formed from elements by introducing minute amounts of other elements as impurities. From the standpoint of this book, at the present time the two most important semiconductor materials are germanium and silicon. If certain impurities are added, then some electrons are free to move through the crystal. Since electric conduction is then largely by negative charges, this material is referred to as *N*-type semiconductor. Other added elements instead effectively produce a deficiency of electrons with electrons moving from place to place to fill in vacancies. The overall effect is somewhat as if the vacancies, called "holes," were an entity moving in a direction opposite to the actual motion of electrons. In this case the majority of conduction is, in effect, as if by positive-charge carriers, and these materials are referred to as *P*-type. Holes and electrons drift in an electric field with a velocity proportional to the field; the holes go in a direction opposite to that of the electrons and with about half their velocity.

The actual situation is complex in many of its details, but a few practical points may suffice here. Structures based on silicon often work at a slightly higher voltage, are somewhat less temperature sensitive, and display less undesirable leakage current (are a better insulator) than those based on germanium. Although ordinary metals increase their resistance (conduct less current for a given voltage) for an increase in temperature, semiconductor devices generally show an increasing current with an increase in temperature. The above mentioned current-carrying processes can be slow enough to limit the frequency at which a semiconductor device can work. At this time, however, the technology of these devices has progressed to the point at which a telemetry experiment is seldom limited by this factor, although in some cases relatively expensive components must be employed.

2. DIODES

Two-element vacuum tubes, called diodes, incorporate a metal anode in a vacuum near a hot cathode which is able to emit electrons. When the anode is positive, a current, due to the attraction of the electrons, flows through the tube, but not when it is negative. Thus these devices have been used for converting alternating current to direct current and also for demodulating an amplitude-modulated radio signal in order to extract the information from it. Consistent with traditional electrical theory, the current flow is always considered as taking place from positive to negative (anode to cathode), which is opposite to the direction of any actual electron flow.

A similar rectifier or unidirectional flow action can take place at a junction between a piece of *P*-type and *N*-type semiconductor. In Fig. 2.1 the nature of this action is shown in simplified form. At the upper left, a battery is connected to this diode in such a way that the positive terminal is connected to the piece of *P*-type material. Both the positive and negative carriers are urged into the vicinity of the junction where a current can thus flow. With this connection of the battery, current flows in the direction indicated by the arrow. If the battery is reversed, as in the right-hand part of the figure, both the electrons and the holes are electrically attracted away from the junction region. Without charge carriers present, little current can flow across the junction, and no current flows in the overall circuit. The symbol for such a diode is a triangular mark (as shown in Fig. 2.1) with the direction in which current can flow being indicated by the point of the triangle or arrow. If an alternating voltage is applied to such a junction diode, current will flow in pulses in one direction only on every other half cycle. It is this rectifying action that provides one of the main uses for diodes, and the possible degree of perfection is discussed in Section 9 of this chapter, but there are other aspects to diode performance.

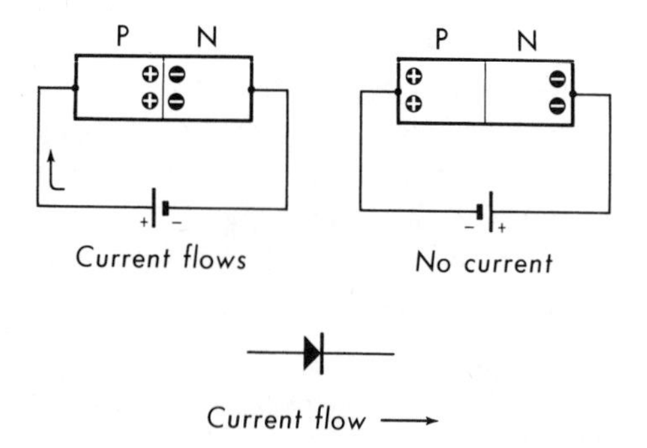

Fig. 2.1 A junction between a piece of *P*-type semiconductor and a piece of *N*-type semiconductor will pass current if voltage is applied in one direction but not in the other. The symbol for this diode rectifier, with the implied direction in which current can flow, is shown below.

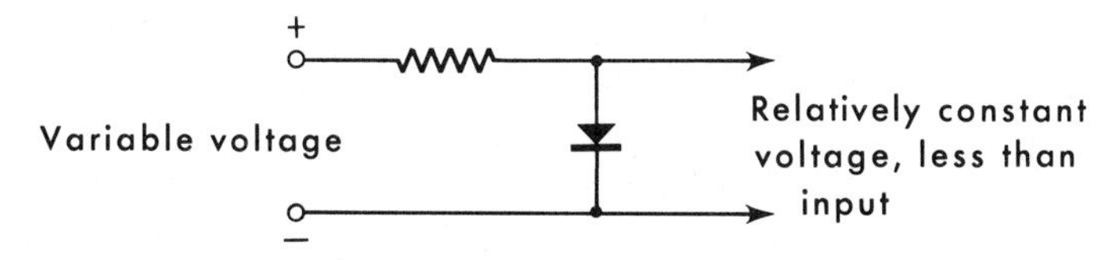

Fig. 2.2 The voltage across a conducting diode is relatively constant and can be used to somewhat regulate a voltage from a variable source.

When the diode is biased in the forward direction so that current is flowing, as in the upper left-hand part of Fig. 2.1, there will be a small voltage drop across the diode. (In Fig. 2.1 the battery is shown connected directly across the diode for the sake of simplicity of explanation, but in any real circuit there would be some other useful device also in the circuit, across which most of the battery voltage would actually appear.) This voltage drop is approximately 0.2 V for germanium and approximately 0.6 V for silicon. This property can actually be made use of when a small constant voltage is needed in an electronic circuit. Thus Fig. 2.2 shows a voltage-regulating configuration in which a potential difference of a few volts is applied through a resistance to a diode in such a direction as to insure conduction. The voltage taken from this configuration is simply the voltage across the diode, which will be relatively fixed independent of fluctuations in the applied voltage. Although much of the applied voltage appears across the diode, most of the *changes* in applied voltage appear across the series resistor. For example, some diodes act as if they have a resistance of 30 ohms (Ω) with regard to changes of voltage, and if the series resistor is 300 Ω, then only one tenth of the input-voltage fluctuations appear at the output. Use of a 30 Ω resistor in place of the diode would result not only in diminishing the fluctuations in applied voltage, but in only one tenth of the total input voltage appearing at the output, as well. Some standard diodes show an incremental resistance (resistance to changes in voltage) in ohms of 25 divided by the current being passed in milliamperes. Later in this book, specific circuits will incorporate this action of a diode to provide small fixed voltages or biases, where a separate battery would otherwise have to be incorporated.

If the diode of Fig. 2.1 is instead biased in the backward or nonconducting direction, then another effect is involved which can be either useful or detrimental, depending on the circuit. Remembering that a capacitor (condenser) consists of a pair of conducting electrodes separated by an insulator, then from the upper right part of Fig. 2.1 it can be seen that the diode will act like a capacitor. The higher the applied backward voltage, the larger the depletion region that is free of charge carriers, and thus the lower the capacitance (capacity) of this capacitor. In some operations it is extremely convenient to have a variable capacitor that can be controlled by an applied voltage. In other circuits the change in capacitance incidentally associated with semiconductor components can cause troublesome changes in frequency, and the like, with changes in battery voltage or other parameters.

Some diodes are sold specifically to serve as voltage-controlled variable capacitors. They are sometimes referred to as a "varicap" or "varactor." In many cases it is important to know the way in which capacitance varies with applied voltage. Many of these back-biased diodes will be found to have associated with them a capacitance that is inversely proportional to the square root of the reverse-bias voltage plus a small constant of about 0.7 V. Specially graded junctions (to which are applied the terms abrupt or hyperabrupt) have been constructed to give a more rapid variation of capacitance with voltage, in some cases in proportion to the -3 power (Lindner, 1962; Shimizu and Nishizawa, 1961; Chang et al., 1963; and Adam, 1963). Some of these diodes are less effective because, instead of acting simply as a capacitor, they also convert electricity to heat, and this resistive aspect can interfere with the sharp tuning or other functioning of a circuit. The capacitance of the usual voltage-sensitive capacitors also depends somewhat on temperature, with some typical units having a temperature coefficient of 300 parts per million per degree Centigrade.

It should be mentioned, at this point, that a capacitor formed by putting a metallic coating on two sides of a piece of barium titanate ceramic results in a capacitor that also has nonlinear properties; that is, capacitance depends on voltage since the current that flows is not proportional to the applied voltage. These materials were actually available before semiconductor diodes were, and their use will be mentioned in connection with special circuits in the chapter on passive transmitters. In some cases they still provide the best action of this sort, if it is noted that the electrode material also affects performance.

If the voltage applied in the backward direction to one of these junction diodes is increased, the current will remain extremely low over quite a range. This is especially true with silicon diodes where the leakage currents may be a very small fraction of a microampere. In some cases, however, at a particular backward voltage conduction will be observed to start again. This is a nondamaging type of action in which the backward current varies rapidly with small changes in voltage or, to state it in the reverse fashion, voltage is rather independent of the current that flows. Thus in this so-called Zener region there is again a voltage-regulating action. Some diodes are manufactured specifically for this purpose and provide stable output voltages in the range of a few volts upward. (Of course, if an excessive voltage is applied in the reverse direction to many rectifier types, there will be an irreparable puncture, this being quite a different effect.)

In Section 5 of this chapter another special diode type called a tunnel diode is discussed. Through the application of a special physical effect, it is able to respond extremely rapidly over a very wide range of temperatures, and it is directly able to cause oscillations (generate radio signals).

3. TRANSISTORS

The first transistors employed tiny wires placed on a base of semiconductor material. For a number of reasons, such point-contact transistors are no longer used. Instead, junction transistors consisting of a three-part sandwich of semiconductor materials are employed. Sections of *P*-type semiconductor can be placed on each side of a piece of *N*-type or pieces of *N*-type semiconductor can be placed on each side of a piece of *P*-type material. Thus there are two basic types of transistor, as shown in Fig. 2.3. At the left is a so-called *PNP* transistor with *P*-type materials on each side of a central piece of *N*-type material; on the right is the corresponding *NPN* transistor. The symbols and nomenclature derive from the older technology but are perhaps best understood from a simplified description of the actions taking place at the two junctions existing in an ordinary transistor.

First, consider the *PNP* transistor on the left. Notice the right-hand junction between the two pieces of material which are called the collector and the base. It is backward biased by the right-hand battery, and so the

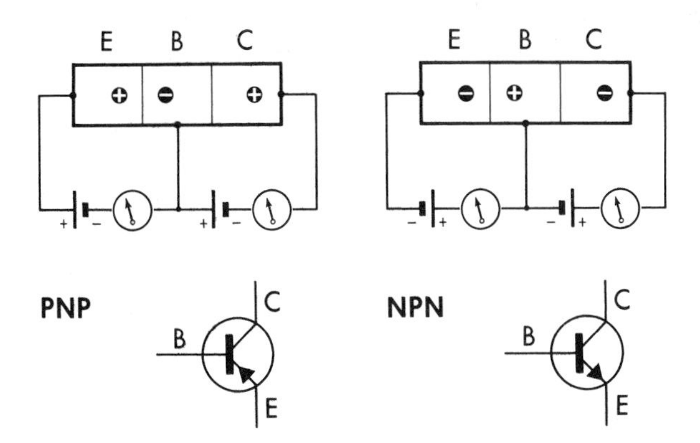

Fig. 2.3 A sandwich containing two junctions between *N*- and *P*- type semiconductors constitutes a transistor and is able to amplify changes in power. Two possible geometries lead to two transistor types with the symbolism and electrode designation shown.

meter in that part of the circuit would be expected to indicate the passage of practically no current. Now consider the left-hand junction between the two pieces of material that constitute the emitter and base. The left-hand battery biases this diode junction into conduction, and thus the left-hand meter would be expected to show the passage of current. Although there should be no current flowing in the collector, some of the charge carriers from the other junction do diffuse into this region, and thus there is a flow of current. The flow of current at the emitter junction thus can control the flow of current at the collector junction. In fact, suitable construction insures that very little current flows in the base circuit, and that most of it is involved in the collector circuit. Since the electrical condition of the base connection controls the flow of current from emitter to collector, there is a crude analogy to a vacuum-tube circuit, with the emitter corresponding to the cathode, the base corresponding to the grid, and the collector corresponding to the anode. This analogy is not perfect since currents in the base control the currents at the collector, whereas voltage at the grid controls the current to the anode of a vacuum tube. Many vacuum-tube circuits look quite similar to the corresponding transistor circuits through this analogy. The symbol for this transistor type places an arrow on the emitter with the point facing towards the base to indicate the sense of current flow.

The same discussion applies to the *NPN* transistor on the right. It will be noted that the two batteries have been reversed. In actual practice it is often possible to replace a *PNP* transistor in a given circuit with an *NPN* transistor, in which no other changes are required than the inversion of the battery (and any other voltage-sensitive elements such as diodes or electrolytic capacitors). The symbol for this transistor type is quite the same; the arrow is still shown on the emitter, but pointing away from the base to indicate the direction of current flow in that part of the circuit.

Although a transistor is composed of two diodes, it is able to complete operations that could not be accomplished by a pair of isolated diodes. It is possible to control large amounts of power in the collector circuit by the expenditure of relatively small amounts of power in the base circuit. Since there are two types of transistor, and only one type of vacuum tube (electrons can only flow from cathode to anode), it is not surprising to find that there are certain different kinds of transistor circuits that achieve special effects by employing both transistor types together. Although a general discussion of these circuits is beyond the scope of this book, specific examples of all of these types are given in later chapters.

Each junction in a transistor has not only the rectifying properties of any other junction but also the other properties previously mentioned.

Thus the junction from collector to base shows a capacitance that depends on the applied voltage. The higher the voltage, the lower this capacitance, and over a range, the higher the current, the lower this capacitance. Circuits can be constructed in which this effect is used profitably. Alternatively, resistive biasing networks can be designed in which a change in power supply (battery) voltage causes the change in collector voltage to compensate the change in collector current, leaving the capacitance unchanged.

The emitter junction is biased into a conducting condition, and thus an approximately constant voltage appears between emitter and base. As mentioned, for silicon this is approximately 0.6 V and for germanium approximately 0.2 V. In some of the circuits that are discussed, this small difference in voltage between the two types can be important because of the use of a single-cell battery that provides only a little voltage over and above this amount.

Some general statements can be made about transistors. At very low currents, at very high frequencies, or at very low temperatures performance tends to fall off. There are, however, relatively inexpensive units that maintain at room temperature a beta of 100 with a collector current of 1 μA. The parameter beta (β) is essentially the change in collector current produced by a given change in base current, and thus is a sort of measure of the amplification of the transistor. (Another designation for this parameter is H_{fe} or h_{fe}, depending on whether the ratio of currents or the ratio of small changes in currents is meant; the two are usually alike within a factor of 1 to 2, with the ac value being up to approximately three times as large at small currents.) These same units can function effectively at frequencies as high or higher than 100 MHz. Although there are many methods of fabrication for transistors, the so-called planar epitaxial units of silicon are especially appropriate for many of the circuits here. Transistors of the *PNP* and *NPN* types have often quite different performance specifications and price. Transistors at this time are often the smallest part of a circuit, and thus it is no longer necessary to minimize the number of this component type in small transmitters.

As the grid voltage is increased in a vacuum-tube circuit, the plate voltage (anode voltage) decreases, due to increasing currents flowing in the useful part of the rest of the circuit. The action does not remain effective when the anode voltage approximates or becomes less than the grid voltage. Transistor circuits can show useful gain and effective operation when the collector voltage is less than the base voltage. In some circuits the action is such that the collector voltage becomes essentially zero; that is, its potential drops very closely to that of the emitter, whereas the base voltage may be 0.6 V positive (in a silicon transistor). A number of the circuits discussed in this book use this mode of operation, although it may seem surprising to those readers who have had experience only with vacuum-tube circuits.

Biasing a transistor into a condition where it can amplify is an important aspect of the design of any given circuit. An increase in temperature can cause an increase in current, thus leading to a further increase in temperature or current. This regenerative process must be limited or the transistor may destroy itself, though it is rather unlikely when using a silicon transistor in a transmitter powered by a single cell battery. In Fig. 2.4 are shown several useful connections. They are by no means an exhaustive set and indeed apply only to the case in which the input signal is applied to the base of the transistor. (Standard texts take up the more general possibilities of applying the signal at any of the three electrodes, which procedures can be useful in certain cases.) The connection at *a* is rather like the usual vacuum tube amplifier stage in that the signal is applied between base and "ground" and is taken out between collector and ground, with a "load resistor" going to the positive side of the battery whose negative side is connected to ground. The resistor and capacitor connected between

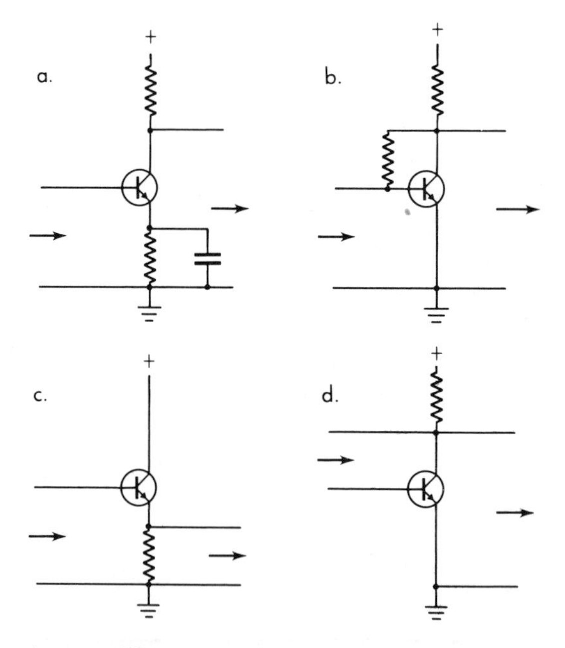

Fig. 2.4 A transistor amplified stage must also have its operating bias point stabilized. Several configurations, excluding details of the input circuit, are depicted. Specific examples are shown in later circuits. The two connections at the bottom display a relatively high input impedance.

emitter and ground ensure some feedback in a sense that will limit slow changes in current, because a slowly increasing current does cause a rise in emitter voltage. Through the input-signal source some intermediate bias is applied to the base connection. A voltage divider (two resistors from plus to ground, with the base attached to their common point) can fix the average base potential, and the useful signal can be coupled into the transistor through a capacitor. Specific arrangements of this aspect for the other configurations will be seen in the various actual circuits throughout the book.

Figure 2.4*b* shows a configuration that can place the average potential of the collector at that of the base. Thus any input signal will periodically cause the collector potential to drop below that of the base but, as has been mentioned, this is satisfactory in a transistor circuit when the signal amplitude is not large. If the resistor between collector and base has a high value, the average collector potential will be stably fixed at a voltage other than the average base potential.

Figure 2.4*c* shows a configuration which displays a relatively high resistance at its input and a relatively low resistance at its output. The gain in power is thus appreciable, although the magnitude of the output voltage swing is approximately that of the input voltage swing. This configuration is called an "emitter follower," and it corresponds to the cathode follower of vacuum-tube technology. At Fig. 2.4*d* is seen an unusual configuration which also has high input resistance and low output resistance properties. It can be shown that this unusual configuration has many of the properties of an emitter follower that has been improved, at least for higher frequencies, by "bootstrapping" or feeding back through a capacitor a signal from the emitter to the base resistor. The equations describing this configuration indicate a performance at low frequencies not otherwise readily achieved in some cases. Although its performance is superior in many cases and it was developed for use in some of the following circuits, it does suffer from the disadvantage that neither of the input connections remains fixed at ground potential.

There is a newer type of transistor called a field-effect transistor. The action of this device can be understood by visualizing a rod of *P*-type silicon, on the edge of which is placed a short piece of *N*-type silicon. If a voltage is applied to this junction in such a way that it is backward biased, then in that region, across part of the width of the bar, charge carriers will be depleted. Current flow from one end of the bar to the other will thus be impeded; that is, such a device appears as a variable resistance under the control of an applied voltage. The controlling voltage "sees" only a backward-biased junction and thus is required to supply little current. This type of device is rather like a vacuum tube, with the so-called "source," "gate," and "drain" corresponding respectively to the cathode, grid, and anode. In an ordinary transistor, although leakage currents can be very low and the amplification high, some current must be supplied by the signal to produce any useful output. In the field-effect transistor the only currents that need flow are the leakage currents. The material of the gate is actually placed on each side of the main conductor so that it can more effectively pinch the field existing between source and drain, and the main pathway can be of either of the two types of semiconductor, thus leading to "*P*-channel" and "*N*-channel" units.

Other devices that employ the control of fields such as the metal-oxide-semiconductor (MOS) transistor are also useful and well developed, but those presently available have not proved useful in the smallest telemetry transmitters because of their higher voltage requirements. However in receiving equipment or in externally placed transmitters where more than one cell can be employed, they can show real advantages. Their connections and biasing arrangements are often like those of vacuum tubes. Such devices have such a high input resistance and sensitivity they can be destroyed by induced voltages while soldering them into a circuit that is not even energized. Thus during installation it is well to wrap a bare wire around all the leads, thereby leaving them connected together until the circuit is completed. Such components are also becoming widely incorporated into integrated circuits that are designed to function at low current levels. An example of the application of the high input impedance property is seen in Fig. 17.3, where an inexpensive double MOS unit is employed. Zero bias could as well have been used there, but the tube-like bias shown reduces the current drain on the battery.

The field-effect transistor is important in circuits that are not to draw any current from their input devices, and several examples are discussed in the section on monitoring various chemicals with high-impedance electrodes (Figs. 3.7, 7.7, 7.8, 7.9, and 7.10). The input capacitance of a field-effect transistor is generally less than a transistor but more than a vacuum tube. The field-effect transistor is free of the conventional transistor noise due to carrier recombination in the transistor base region. All transistors share an advantage in being relatively nonmicrophonic and can function even under conditions of vigorous vibration.

Another transistor type that is suitable for generating periodic impulses is the unijunction transistor (e.g., 2N2840 or 2N2646). If the main electrodes are connected across a voltage source in series with a load resistor, cyclic conduction will take place if the control electrode is connected to the common point of a condenser in series with a resistor, both connected across the voltage source. Some aspects of the topology of this transistor type actually resemble the field effect transistor.

Although transistors are often the smallest component in an electronic circuit, some units do not come in small packages. If a special small unit with particular electrical properties is required, it is frequently possible to remove the case from around a transistor to reveal a small unit surrounded by a relatively large amount of empty space. This removal is often a satisfactory procedure with silicon units that are coated with an inert covering by a process called passivation, but it is not possible in general with germanium units. Illumination alters the electrical properties of transistors and thus if exposed, they must be shielded from light. The use of certain plastic-covered commercial transistors has resulted in strange performance variations with time of day due to changes in illumination penetrating the plastic, and has also led to the appearance of hum at the power frequency from nearby lights.

In closing this section it might again be noted that many of the original transmitter circuits had to place a premium on using few transistors because of the size of the cases of the early units. It is now true that one of the smallest of the available component types is the transistor and, neglecting factors of cost, it is sometimes expedient to include extra transistors in certain circuit configurations. Later mention will be made of integrated circuits in which an entire circuit including transistors, is formed by the same processes that would otherwise form single transistors. Although these procedures do not necessarily reduce the size of an overall telemetry transmitter, in some cases there are advantages.

4. NEGATIVE RESISTANCE AND IMPEDANCE

A very valuable concept is that of negative resistance and, as we shall see, it has special relevance to certain practical questions here. The drawings of Fig. 2.5 can be used to introduce this concept. In many electrical devices an increasing applied voltage results in a proportional increase in current. The ratio of voltage to current is called resistance, and ordinary resistors are characterized on a voltage versus current graph by a straight line such as is seen in the upper left-hand corner. Different resistances are characterized by different slopes to this line. If a device were to display a characteristic curve sloping downward to the right rather than upward to the right, we might say that it was representative of a negative resistance because an increasing voltage was accompanied by a decreasing current (negative change in current) or similarly, an increasing current was accompanied by a decreasing voltage. This is suggested at the upper right in Fig. 2.5. It might be more proper to speak of this as a negative incremental resistance rather than as a negative resistance, since it is the ratio of changes in voltage to changes in current which is negative, rather than actually

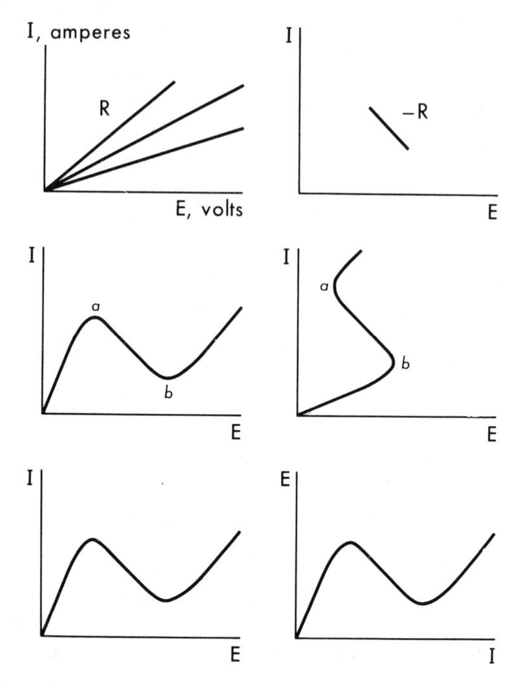

Fig. 2.5 A device with a negative sloping region on its current-voltage curve is often spoken of as displaying a negative resistance. There are two kinds of negative resistance, each with somewhat different properties. Such elements are capable of serving as the basis for an oscillator or switching circuit, and generally incorporate feedback.

representing a positive current accompanied by a reversed voltage, or a current flowing in the wrong direction accompanying an applied voltage. However, the simpler term negative resistance is usually applied to these cases.

Such a negative-sloping characteristic cannot extend indefinitely in either direction or the device associated with it would have to be capable of supplying infinite power. There are two possible ways in which the negative-sloping region can curve into adjacent positive-sloping characteristics, and these two are depicted in the center line of Fig. 2.5. In both cases the region *a-b* is the negative-resistance region, and in the left-hand case the curve bends in a counterclockwise direction, whereas in the right-hand case the bending is clockwise. Both curves are shown going through the origin to imply the flow of no current with no voltage applied.

In a device represented by the left-hand curve a steadily increasing voltage is accompanied first by an increasing current, then a decreasing current, and then again by an increasing current. In the device represented by the right-hand curve a steadily increasing voltage gives rise to some rather unexpected effects. First the current slowly increases to that value represented by point b and then there is a sudden discontinuous jump in current to the upper part of the curve. Reducing the voltage causes a slow drop in current to the value at a, whereupon there is a discontinuous drop in current to a rather low level, followed by a slow decrease in current back to zero. In this second case the negative-resistance region is not traced out, and there is hysteresis or backlash between the voltages at which upward switching and downward switching take place.

On the bottom line can be seen what must be done to trace out the negative-resistance characteristic in all cases. The left-hand curve is the same curve as the one above in which an N-shaped curve is traced out for a slowly increasing or decreasing voltage. At the right the curve immediately above is shown with its axes reversed; that is to say, if a steadily increasing *current* is applied to this device then it will be observed that the voltage across it will first increase, then decrease, and then increase again. For the sake of specificity it might be mentioned that thermistors show an electrical negative-resistance property. If a steadily increasing voltage is applied to them, the current will first slowly rise and then at some point the current will try to switch to a very high value, at which point the thermistor will blow up. The usual way for testing a thermistor is to apply a variable voltage to it through a large value of resistance which limits the current. The current through the thermistor is thus essentially the applied voltage divided by the resistance of the external series resistor, and the properties of the thermistor have little effect on the current that flows. Thus the current is the quantity that is controlled, and the resulting voltage across the thermistor can be observed or recorded.

A specific example of a physical device showing the second type of negative resistance has been given, and we might note that living nerve fibers apparently show an electrical negative-resistance property of the first type; it is by virtue of this property that they are able to generate nerve impulses. In general, it is true that devices capable of displaying an electrical negative-resistance property are able to convert dc to ac power by oscillation, and it is only these devices that are capable of switching or impulse production, or other kinds of discontinuous operation (for further discussion see Mackay, 1958). Many oscillatory systems, whether they be biological, mechanical, or the like, can be analyzed in these terms. In some cases of oscillators it is more convenient to consider a feedback process, but this always leads to the possibility of observing a characteristic curve with a region of negative slope. Some physical devices have the requisite feedback as an intrinsic property, and then it is most convenient to discuss negative resistance. The negative resistance can be considered as canceling positive resistance that would damp out any oscillations once started. Whenever such a property is noted, it is automatically true that the basis exists for the production of oscillations. In the next section we discuss tunnel diodes which display an electrical negative resistance and which thus are convenient for generating radio signals from a source of dc voltage.

To be used for producing sinusoidal electrical oscillations a dc source of power must be combined with a device that can display an electrical negative resistance and must then be coupled to a resonant circuit consisting of a capacitor and a coil of wire (an inductance). For the type of device characterized by the left-hand column of Fig. 2.5 a parallel connected resonant circuit should be used, whereas with devices of the type considered in the right-hand side of the figure a series-connected resonant circuit should be used. A specific example is given in the next section.

There is a separate application for the concept of negative resistance or negative impedance, and that is to the increasing of the overall sensitivity of a measuring operation. As an example, suppose that in response to a change in some physiological variable a transducer changed its impedance from 100 up to 101 Ω. The same considerations would apply if this were a change in resistance, a change in capacitance, a change in inductance, or a combination of these. This 1 percent change in the sensor could then be used to modulate some sort of radio transmitter. Unless there were considerable amplification, this small change might produce only a minimal useful signal. This might be overcome by noise in any part of the overall transmitter or receiver system. Consider combining a negative resistance of 90 Ω with the transducer, in order to subtract out much of the unchanging part of the sensor impedance. In that case the effective impedance change "seen" by the transmitter circuit would be from 10 to 12 Ω, which constitutes a 10 percent change. The same procedure can be carried even further by canceling almost all of the unchanging characteristic of a transducer, if both the negative and positive impedances are intrinsically quite stable. Here again it is necessary, in order to perform this operation, to use some feedback arrangement to generate a negative impedance. An example of a specific configuration for generating a negative inductance is shown in Fig. 2.6. It can be combined with a variable inductance sensor, for example, in a system in which pressure changes move a ferromagnetic armature in the vicinity of a coil, to give a large modulation in a simple telemetering transmitter. The second part of Fig. 2.6 shows a complete transmitter of simplified form that uses this principle. It also makes use of the fact that the feedback connection, which produces the negative impedance, has currents of various phases that can be used to produce oscillation. It should not be considered that the cancellation of some of the impedance

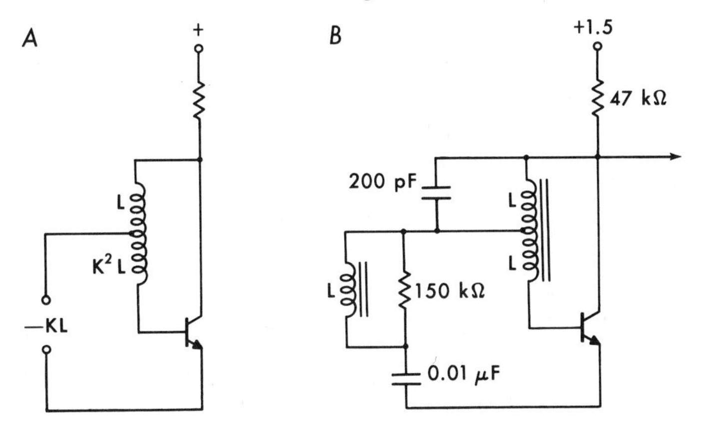

Fig. 2.6 Feedback which generates a negative impedance, as at the left, can be used to cancel some of the unchanging part of a variable impedance transducer to give higher sensitivity, and can be incorporated into an oscillator, as shown at the right.

of a variable-inductance element is merely a resonance effect such as would be obtained by inserting a capacitor, since it is clear that with a change of frequency, a negative inductance is not the same as a capacitance.

Figure 2.7 shows a general scheme for producing negative impedance. The desired impedance is placed in such a configuration, and then it is possible to dial over a decade of the negative of this impedance. If voltage is applied to the input terminals, a current will flow, but in the opposite direction to that which would exist if the impedance had merely been placed across the terminals themselves. Thus this is a true negative-impedance device over its range of operation. The same result can be obtained with a single amplifier, but three were shown for simplicity of explanation.

It might be noted that in setting up electrical analogues of biological systems, a non-constant impedance is sometimes required. For example, a nonlinear capacitance can represent the non-constant compliance of the globe of the eye or the lung as fluid or gas enters or leaves. A similar configuration employing amplifiers in which the gain depends upon the applied voltage will serve this function, the required amplification being less than unity.

The process of heterodyning, in which signals of various frequencies beat against each other, is described in other sections. If a frequency changes from 200 up to 201 kHz, the change may not be noticeable; but if this signal is first caused to beat with an oscillator of frequency 199 kHz the result is a 1000 Hz signal that increases up to 2000 Hz, and this is indeed noticeable. Thus in the frequency domain heterodyning or beating serves a function

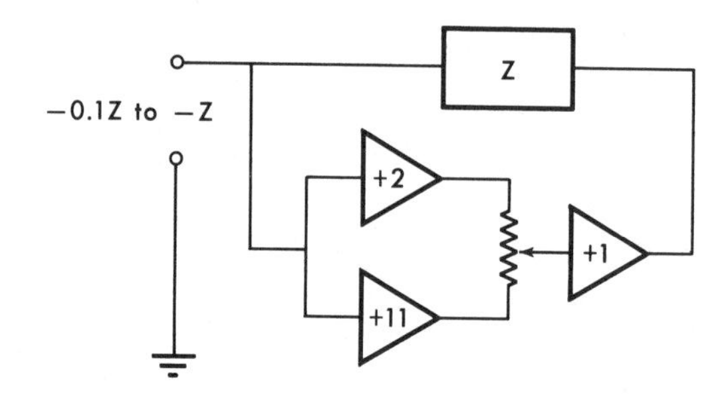

Fig. 2.7 With an ordinary impedance element and several stable amplifiers, a decade box can be set up from which the desired negative impedance can be dialed. The same result can be accomplished with a single amplifier, but this arrangement clarifies the action.

similar to the use here of a negative resistance to subtract off a fixed part of a variable quantity, so that small changes will be seen as large percentage changes in the remainder. This does not increase the intrinsic limiting sensitivity of a transducer by reducing interfering noise at the site of measurement, but it does reduce the effect of interfering signals in the later parts of the observational chain. It also results in large signals which may not need as much auxiliary amplification.

5. TUNNEL DIODES

A quantum mechanical effect known as tunneling leads certain diodes to show an electrical incremental negative resistance of the type depicted in the left-hand column of Fig. 2.5; that is, an increasing applied voltage results in a current that first increases, then decreases, and then increases again. It would thus be expected that a combination of this type of device with a parallel resonant electrical circuit would lead to oscillations. Indeed this is true, and very simple circuits can result. The tunneling is a rapid process which is not limited in speed to the extent that some other semiconductor effects are, and thus, tunnel diodes are capable of generating oscillations at very high frequencies, or of serving as switches and pulse generators for very rapid processes. The process also takes place over extremes in temperature.

One of the problems with the use of tunnel diodes in these applications is the inconvenience of the voltage that must be employed with all presently known units. The diode must be biased into the negative-sloping region, and for existing units this takes place at approximately 0.2 V. But the usual batteries supply a voltage in the neighborhood of 1.3 V. Thus most of the voltage (and power) supplied by a typical battery is inevitably thrown away by the usual tunnel-diode oscillator or radio transmitter. Batteries can be constructed using different elements for the electrodes and which, in combination, yield approximately the correct voltage, but these batteries are merely inefficient and do not deliver a correspondingly greater current. It is not possible to use several diodes in series for higher voltages. Thus at this time it is my feeling that tunnel diodes have a restricted range of experiments to which they are properly applied. Of course, certain energy sources (such as thermocouples) deliver power at a low voltage, and with these sources it is extremely appropriate to consider the use of tunnel diodes as the basic oscillator component. Tunnel-diode circuits for telemetry have been suggested (Ko et al., 1963; and Ko and Slater, 1966), and another type of circuit is described in Chapter 13.

An example for summarizing a number of applications of diodes is the workable telemetering transmitter shown in Fig. 2.8. The diode at the top of the transmitter is biased in the forward direction by the battery, and the constant voltage of the diode subtracts out a fixed amount from the battery voltage, thus leaving just the voltage needed to bias the tunnel diode into the negative-sloping portion of its characteristic. This bias current returns to the battery through the low dc resistance of the coil L. It is necessary that the bias circuit for the tunnel diode display an overall low impedance, and this is the case here. In the right-hand part of Fig. 2.8 it is seen that the battery backward biases a variable-capacitance diode through a variable resistance. This variable resistance, which, for example, might be a thermistor, is the sensor of this telemetry transmitter. An increase in resistance decreases the voltage applied to this diode, and results in an increase in capacity. A resistor could have been shown connected in parallel with the variable-capacitance diode to make it more evident how the bat-

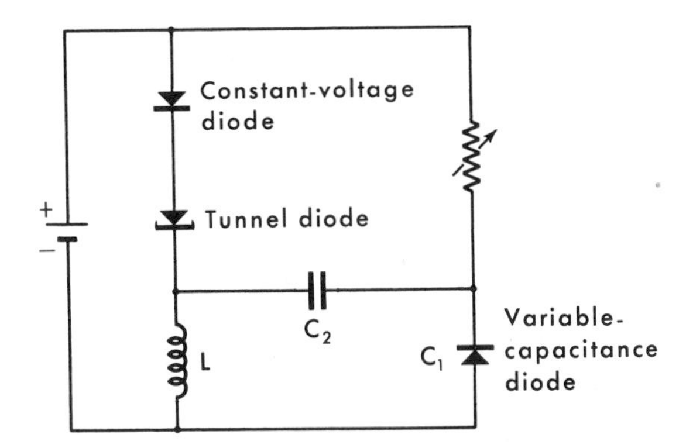

Fig. 2.8 A functional transmitter exemplifying three different diode actions. The transmitted frequency is modulated by changes in the resistor.

tery voltage divides between the diode and the sensor, but in any real case the leakage of the diode effectively provides the other part of the variable resistive voltage divider which applies the changing voltage to the capacitor. In order to maintain oscillations, a low-resistance power source must be connected in series with a tunnel diode and a parallel resonant circuit. In the present case, the resonant circuit is the inductance L connected in parallel with the capacitance C_1 of the diode. If the capacitance of the capacitor C_2 is considerably greater than C_1, the presence of C_2 does not have any effect on the resonance, and the diode and coil are effectively connected in parallel as far as alternating currents are concerned. Capacitor C_2 is present to prevent the steady voltage at the base of the left-hand pair of diodes from being applied directly to the right-hand diode, instead of the right-hand diode achieving its slowly changing signal voltages from the sensor. Oscillations take place at a frequency that can be determined from the usual formulas with the controlling factor being LC_1. A substantial radio signal is radiated from the coil. This circuit constitutes an effective frequency-modulated telemetering transmitter, in which changes in resistance bring about changes in the radio frequency of transmission. The use of the constant voltage drop across a forward-conducting diode to cancel out part of the battery voltage demonstrates the waste in any such system, but the present arrangement is more efficient than using a pair of resistors as a voltage divider across the battery to produce a fixed lower voltage.

Since a tunnel diode also shows a variable capacitance effect, a frequency-modulated voltage transmitter can be constructed by simply placing a tunnel diode with a resonant circuit in the emitter of an emitter-follower circuit, to whose base the variable voltage is applied.

6. MULTIVIBRATORS

Multivibrators are circuits generally characterized by two active elements, such as vacuum tubes or transistors, with the entire signal from the first being communicated to the input of the second, and the entire output of the second being communicated to the input of the first. At all times, except during brief intervals of transition, one of these elements is turned fully on and the other fully off; but periodically the element that is on goes off, at which time the element that is off turns on. Some readers may be familiar with the vacuum-tube configuration shown at the upper left in Fig. 2.9. When this circuit is switched on, one of the tubes which is slightly more active will tend to come on ahead of the other. Let us suppose that the left-hand tube starts to conduct current first. This will cause a reduction in anode voltage, which drop in potential will be communi-

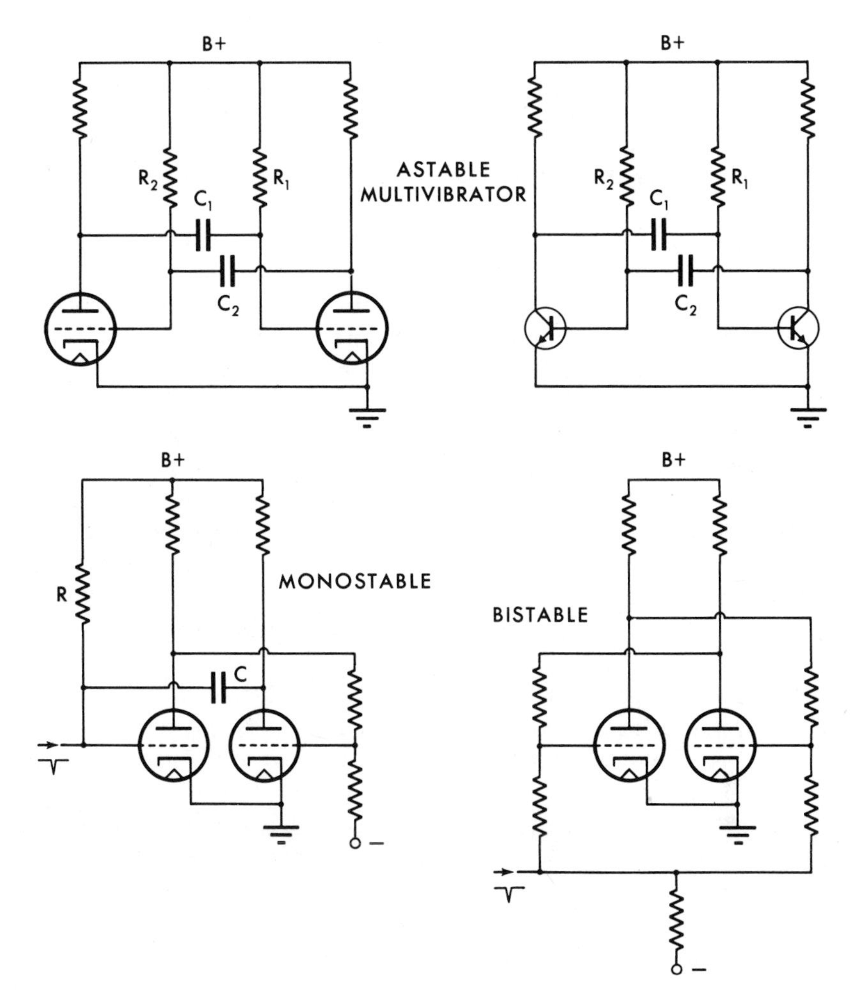

Fig. 2.9 Three types of multivibrator, one spontaneously switching back and forth continuously, another with a preferred state and running through a single cycle of operation when pulsed, and the third with two indefinitely stable states between which switching takes place only on arrival of an input impulse. For the benefit of those who have had experience with vacuum-tube circuits, but are less familiar with transistor circuits, these multivibrators are depicted in vacuum-tube form. At the upper right the similarity with the corresponding transistor circuit is seen; many specific examples are given in later circuits.

cated through C_1 to the grid of the right-hand tube, thus giving it less chance to conduct, with a consequent rise in plate potential of the right-hand tube. This increase in potential is communicated through C_2 to the grid of the first tube causing it to conduct even more heavily. As a result of this regenerative process, the first tube will come on fully and the second tube will be turned fully off. The grid of the second tube will be maintained negative for the length of time it takes C_1 to discharge through R_1, but during this period of rising potential a voltage will be reached at which the second tube will start to conduct. The start of conduction causes a drop in plate voltage of the second tube which is communicated through C_2 to start turning off the first tube, which causes a rising plate potential which is communicated through C_1 to more rapidly turn on the second tube. This regenerative effect causes an abrupt turning on of the second tube after a time roughly R_1C_1, at which time the first tube is abruptly turned off. The first tube will come on again after a time R_2C_2, which once more abruptly turns off the second tube. This rapid switching process repeats itself regularly at intervals, with abrupt transitions between well-defined states of one or the other of the tubes conducting. The grids, although returned through their resistors to a positive voltage, never actually go positive but instead are always either at a negative voltage or else "clamped" at ground potential by the diode action between grid and cathode when they attempt to go positive. The rapidity with which the grids increase in potential towards the critical switching voltage is determined by the voltage to which the labelled resistors are returned. Thus switching is soonest when the resistors are returned to the maximum positive voltage, least rapid if returned to ground potential, and the frequency is adjustable for various intermediate voltages. Thus such an astable multivibrator is not only an oscillator capable of producing rectangular waves at the anodes, but the frequency can be controlled by changing the voltage to which the grid is returned; that is, the action can be as a voltage-controlled oscillator (often abbreviated VCO).

To the top right of Fig. 2.9 is shown the transistorized version, which generally functions in quite the same way. Similar circuits can be constructed using field-effect transistors. At the lower left is shown a version of a multivibrator, which has a preferred state. There is feedback from the output of each tube to the input of the other, but one of the connections is through a resistor rather than a capacitor. Thus one characteristic time is eliminated and there is a normal state. In the present case, if the circuit is switched on, rather obviously the left-hand tube would tend to be fully on and the right-hand tube fully off. If, however, at any time a sharp impulse is injected into the circuit, then this multivibrator will run through a single cycle consisting of the right-hand tube going fully on and the left-hand tube going fully off. After a time of approximately RC the condition will revert to the original, and the duration of this single cycle is relatively independent of the properties of the input signal. There are other versions of this monostable multivibrator, some of which will be seen in detail in later parts of this book, but it can be said that the transistorized version is quite similar in appearance to the tube version.

The third type of multivibrator replaces both capacitor couplings by resistors, thus eliminating both characteristic times. In the lower right part of Fig. 2.9 is seen a bistable multivibrator which will remain indefinitely in whichever state it is placed. If, at any time, an impulse is injected, then the tube that was off will go on and the tube that was on will go off, thus reversing the original state until at some random later time another impulse comes in. These are the so-called binary counter stages found in large computers, or in the radioactivity-scaling circuits used by physicists. Here again the transistorized version is quite similar.

These multivibrators are seen in a number of the practical circuits, where they serve as oscillators or general purpose pulse generators. Between almost any pair of electrodes a negative resistance can be observed, as might be expected. Another way of regarding these circuits is that the output from one tube or transistor is fed into the other in order to receive a 180-degree phase shift, before being returned to the original tube or transistor. This phase shift is required for the regenerative feedback that is generally needed for oscillation or switching.

More theory could be given, but the detailed circuit diagrams that follow should provide sufficient insight into the functioning of multivibrators for the present purposes. We might note that occasionally an astable multivibrator will be observed to run at a considerably higher frequency than expected. In some cases this is due to the fact that when the base of the transistor is driven negative, Zener-like conduction can take place to limit the voltage excursion, thus bringing about a quicker return back to the switching voltage. This is more likely with high supply voltages, and in the circuits that follow this has not proved to be a problem. Another difficulty sometimes observed in transistor multivibrators is an occasional malfunction in which both transistors in an astable circuit somehow manage to go into conduction simultaneously, with a total cessation of oscillation. Although this has seldom produced any complication, in special circuits the insertion of two diodes (as in Fig. 2.10) can absolutely assure that both transistors will not simultaneously conduct. A switching action by the diodes is illustrated in that a positive voltage is applied to both bases only if one transistor is off.

There is a tendency for the frequency of an astable multivibrator to be independent of the magnitude of the applied voltage; that is, as batteries approach the end of their life, the circuit may not malfunction. This is because, although the negative voltage to which the bases are driven is less,

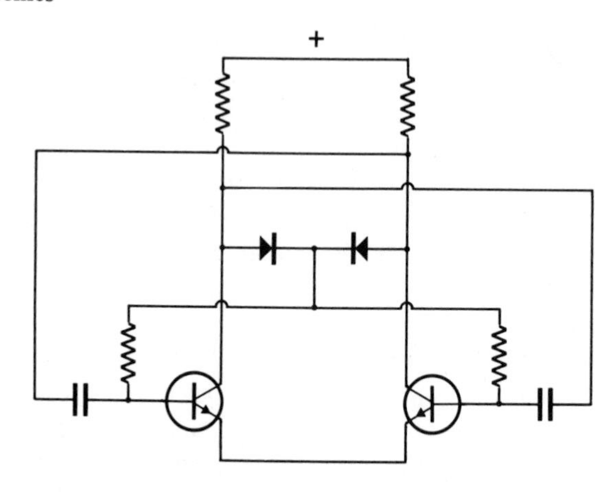

Fig. 2.10 Some astable multivibrators can go into a state in which both transistors are steadily conducting, but a pair of diodes inserted as above can introduce a switching action that removes base voltage and current if one of the transistors is not turned off.

the voltage toward which they are progressing is also less; thus the intermediate switching point is reached in approximately the same time independent of supply voltage. There are multivibrators that combine an *NPN* transistor and a *PNP* transistor to give certain useful properties, but such multivibrators do not necessarily have these compensating characteristics. Specific examples are given in later chapters.

Another aspect of the functioning of multivibrators is found whenever a highly asymmetric wave form is to be generated. Thus in a number of circuits (to be discussed later) a short impulse is to be generated once each second (a relatively long time). A high-capacitance capacitor is necessarily involved in determining the relatively long time interval, and it must be recharged to its original voltage level during the short time interval in order that it be ready to determine the next long time interval. Extra components are usually required to provide for this rapid recharging action, and three specific circuit arrangements for accomplishing this will be seen in later chapters (Figs. 10.3, 15.2, and 15.9).

Not only can the frequency of switching be controlled by an input voltage, but changes in either resistance or capacitance can equally well control the frequency. Thus in a number of the temperature-sensing transmitters to be described in later sections, one of the resistors (such as R_1 in the upper right part of Fig. 2.9) is replaced by a thermistor. An increase in temperature then causes the right-hand tube to more rapidly become conducting after having been switched off. Varactors can also control switching times. Nothing has been said about the collector resistors in these circuits, but they are merely set at some convenient value appropriate to the overall power limitations. These circuits are most stable if the active elements vigorously switch fully on or fully off rather than going only into partial conduction during the conducting phase. Thus a conducting transistor will have its collector voltage either at the full battery voltage with respect to "ground," or very close to zero volts with respect to ground. If a transistor is used as a switch, being either full on or full off except during a transition, then almost no power is dissipated in it (wasted in heating), since either the voltage or current is zero most of the time.

In the circuit shown in Fig. 2.9, if R_1 is a one megohm thermistor, then an approximate value of the collector resistor for the right-hand transistor can be calculated as follows. Current flowing through a one megohm thermistor into the base of a transistor in a multivibrator is able to turn that transistor fully on (the collector voltage going approximately to zero) if the collector resistor is 10 K ohms, or more, and if the transistor being used has a β of 100. In checking this transistor specification, it should be noted that the transistor must have this beta at the lowest expected temperature of operation.

Both resistors and both capacitors in the base circuits of the transistors can be made temperature sensitive, and if all resistances are made rather large, an extremely low current circuit results. Figure 2.11 shows a multi-

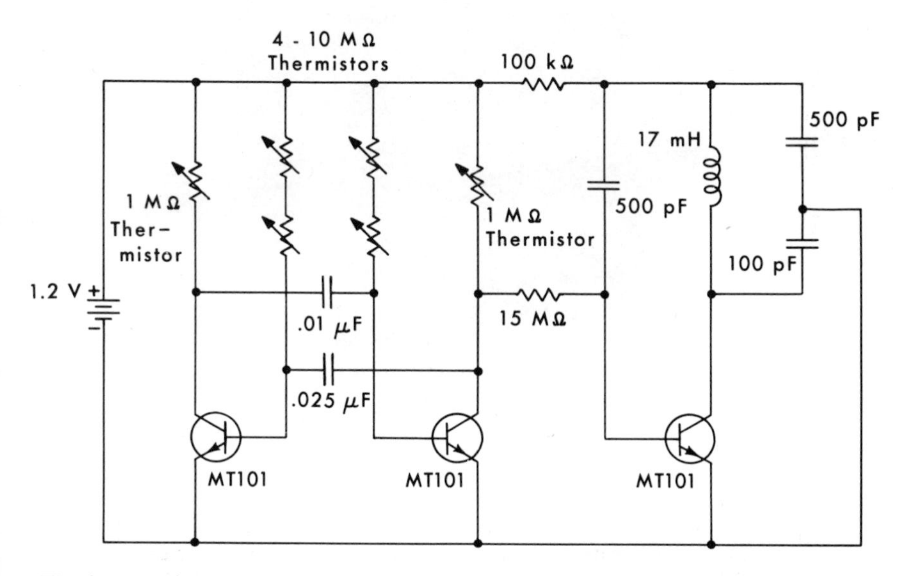

Fig. 2.11 A low-current temperature transmitter in which the switching of the astable multivibrator is controlled by thermistors resulting in the output oscillator being turned on periodically at a rate that carries information.

vibrator that periodically switches on a small oscillator whose output radio signal is the desired result. Such a unit ran continuously for approximately six years on a single small battery. Much of the total power drawn by this circuit is used by the output stage (radio transmitter) which is only turned on for a fraction of the total time. The workings of such an output oscillator are described in the next section.

The switching rate of one of these multivibrators also indirectly affects the current it draws. In one case it was found that reducing the frequency from 18 to 1.8 kHz allowed a savings of 15 μA. At low currents the transistor input capacitances are typically 30 pF from base to emitter, and the capacitors from the opposite collectors must be several times as large for effective driving. Recharging currents are thus relatively large so collector resistors must be somewhat reduced. At the lower frequency there is more time for recharging and the resistors need not be as small if waveshape is maintained with the same proportions.

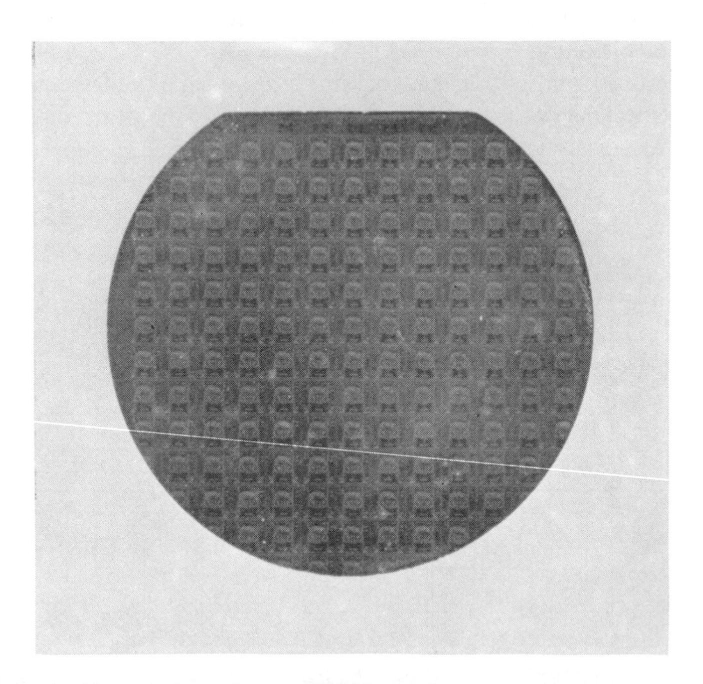

Fig. 2.12 A silicon wafer a few centimeters in diameter and containing more than 130 integrated circuits related to a multivibrator. The wafer is broken into parts for individual use of the elements. These components can be extremely convenient, but in many cases their use does not significantly reduce the overall size of a telemetry transmitter.

It is possible to form relatively complicated circuits (such as multivibrators) in a solid rugged chip by various photographic, etching, and deposition processes. Transistors and diodes are formed in the way they usually would be, and capacitors of limited capacitance can be deposited, as can resistors, within a certain range of values. Figure 2.12 shows a thin wafer of silicon approximately three centimeters in diameter upon which have been deposited 137 complete and identical circuits, each essentially comprising a multivibrator and a few extra components. This chip is broken into individual parts to which suitable leads are attached for use. Small size, ruggedness, and mass production are advantages of such techniques. It should not be assumed, however, that in all cases the size of telemetry transmitters is reduced by the use of these monolithic integrated circuits. These matters are discussed in Section 12.

7. SINUSOIDAL OSCILLATORS

Perhaps the most usual way of considering an oscillator is as an arrangement which takes the output from some device capable of amplifying power and feeding it back to the input in such a phase as to augment any change there. The resulting positive feedback leads to oscillations, and such oscillators are suitable for purposes including generation of radio signals from the steady voltage delivered by a battery. In the present case we shall be discussing systems that give smoothly changing voltage and current rather than the abruptly switching wave forms of the multivibrators. In such oscillators there is usually a tuned circuit that will tend to maintain the voltage sinusoidal by a sort of a "flywheel" effect even if currents are delivered in pulses, but it is often true that the more nearly sinusoidal the currents in the circuit the more stable the frequency of oscillation. In general, an oscillator adjusts its frequency to that value for which the phase shift between input and output is just right to maintain the proper feedback condition. The detailed theory of oscillators is covered in electronics texts, and in this section only two examples of specific kinds of circuits which have proved useful in bio-medical telemetry are cited.

Those who are not engineers may find it helpful in considering these circuits to remember a simple analogy. A tuned circuit is like a swing that can be kept in motion at its characteristic frequency by any impulse at the end of each cycle. In the case of the swing, the feedback timing is through the eye of the person pushing the swing, with the period of oscillation depending upon the properties of the oscillatory system rather than on the details of the periodic push, which are not especially critical. A transistor supplies regular impulses to a tuned circuit in a similar way to maintain regular oscillations.

Figure 1.2 shows an oscillator of the type being described here. The original vacuum-tube version of this circuit was known as a Hartley oscillator, and it is characterized by a coil that is tapped. In the transistor version the tap goes to the emitter junction with one side effectively being connected to the collector and the other side to the base. Thus an increase in collector current, because of transformer action, causes an increasing base current, which causes a further increase in collector current, and so on. If feedback is vigorous enough and the components are suitably chosen, the blocking action mentioned in Chapter 1 also will result, but that is a separate matter. This circuit has proved to be very reliable for short-range telemetry applications. In the specific circuits described in this book the coils should be wound in a closely packed fashion so that there is good coupling between the two halves. Actually, however, the circuit can be made to work with two independent coils between which there is no magnetic coupling because of the common current circulated through both via the parallel capacitor.

Rather than using a tapped coil or two coils with the common point going to the emitter junction of the transistor, an equivalent oscillator can be built using a single continuous coil and two capacitors (it being inconvenient to tap into a capacitor). Such a configuration with vacuum tubes was called a Colpitts oscillator, and a number of transistorized versions will be seen in the present volume.

An example is given in Fig. 2.13. This was a circuit that was used to telemeter the pattern of gastrointestinal muscular contractions from certain lizards. The inductance L_1 served not only as the transmitting antenna but also as the pressure-sensing transducer by virtue of the motion of a ferromagnetic core nearby in response to pressure changes. Rather than tapping this coil, two capacitors were connected across it, and they were in series. Their common point was connected to the emitter junction of the transistor. In the Hartley circuit the degree of feedback is determined by the point on the coil at which the tap is taken off, and in the present circuits the degree of feedback is determined by the ratio of the sizes of these two capacitors. In the Hartley circuit the resonant frequency (the frequency of oscillation) is determined by the total inductance of the coil and the capacity of the capacitor connected across it. In the present case the frequency of oscillation is determined by the inductance of the coil and the total capacitance of the two capacitors in series connected across it. If one capacitor is considerably larger than the other, however, the effective capacitance of the combination is just that of the smaller capacitor. Thus in Fig. 2.13 adjustment of C_1 is used to set the frequency of oscillation, and small changes do not upset the degree of feedback badly enough to

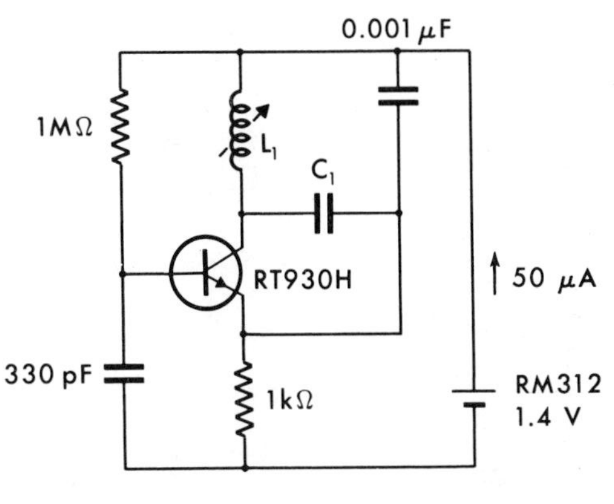

Note:

L_1 400 turns of #42 wire
Inductance approximately 750 μH

C_1 selected to resonate with L_1
at approximately 300 kc
Nominal value 500 pF

Fig. 2.13 An oscillator employing a pair of capacitors rather than a divided inductor, with frequency being modulated by changes in the inductance L_1. Changes in C_1 can also be used to frequency modulate the oscillator.

interfere with operation. The range of transmission of this circuit was required to be only a few feet, and the low frequency of operation was selected for reasons mentioned in a later chapter.

In some cases it is desirable to use an oscillator to control an output stage which then does the actual radiating of the signal. Circuits of this type are shown in Fig. 2.14, which shows two configurations that draw extremely low currents and thus give long life with a small battery. Both transmitters are meant to work from a single silver cell giving 1.5 V, and the range is relatively short. At the left is seen a configuration in which the signal from the Colpitts oscillator is injected into the base of a series-connected output stage in whose collector circuit an antenna coil is seen resonated with its own capacitor. Total current drain of this circuit is approximately 3 μA (which is not very different from that of a typical electric wristwatch). At the right is seen a similar oscillator stage driving a more powerful output stage connected in parallel. In both circuits an attempt was made to stabilize operation against voltage, temperature, or transistor changes by isolating the transistor capacitances from the capacitor in the oscillator resonant circuit. This was done by placing the 6.8 kΩ resistor in the emitter connection. Factors such as voltage changes which

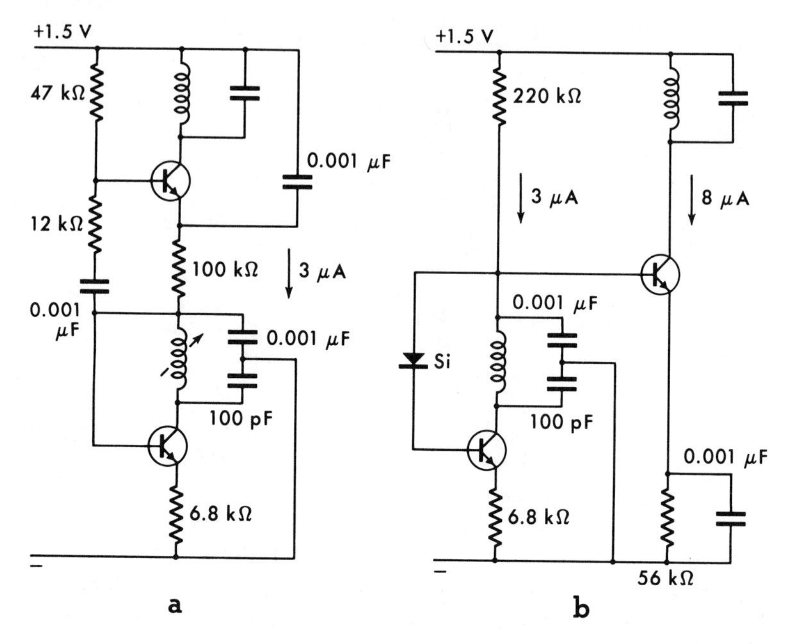

Fig. 2.14 Oscillator circuits with separate output stages. Changes in capacitance or inductance frequency modulate these low-current transmitters, with the stronger signal being generated by the parallel connected unit on the right. These circuits were originally studied by using a now outmoded transistor type MT101, but newer types are also effective. Stray capacitance across the emitter resistor of the oscillator transistor reduces overall stability.

lead to capacitance changes thus do not influence the oscillation frequency nearly as much, although the advantage is eliminated by any stray capacitances that might accidentally be placed in parallel with that resistor.

There are many other oscillator types that are valuable in these activities. The output from a transistor can be fed back to its input through a series of three resistors and capacitors to give sinusoidal oscillations whose frequency depends on changes in either the resistance or capacitance values (a "phase-shift" oscillator). If a separate resonant circuit is used in both the input and output circuits, even if there is no direct coupling between them, electrode capacitance will cause oscillations if the output circuit is tuned to a slightly higher frequency (tuned-plate tuned-grid circuit). A Clapp oscillator is very stable but somewhat less sure and simple than the Colpitts. For information about such oscillators, and many other types having advantages in special cases, the reader is referred to any standard electrical engineering text.

8. CRYSTAL OSCILLATORS

The frequency of oscillation of a vibrating mechanical system is often much sharper than the characteristic frequency of a resonant electrical combination consisting of a capacitor and a coil. Resonant systems are often described in terms of their quality, or Q. This is the ratio of the energy stored per cycle to that dissipated per cycle of oscillation. In a general way, it can be said that a high Q system has a very sharply defined resonant frequency at which response is great. Also, it takes approximately Q cycles for a resonant system once excited to have its vibrations die down significantly in amplitude. The Q of a good electrical circuit may be a few hundred, while that of a vibrating quartz crystal can be several hundred thousand. Thus, if the frequency of a radio transmitter is to be accurately controlled and sharply fixed, then it should be controlled in frequency by the vibrations of a quartz crystal. In some cases legal restrictions on frequency stability can only be met by control in this fashion. Aside from the convenience of knowing precisely at what frequency a telemetry transmitter will be operating there is the further practical aspect of being able to use, in conjunction with a crystal-controlled transmitter, a receiver that accepts only an extremely narrow band of frequencies and that thus more fully rejects radio noise at other frequencies.

Although various mechanical resonant systems can be used to impart their frequency to an electrical oscillator, the most common system is a vibrating quartz crystal. Such a crystal is piezoelectric, which means that an electrical signal applied to metal electrodes on the crystal will cause a mechanical deformation of the crystal, and conversely a mechanical deformation of the crystal will result in an electrical signal. To an input ac signal such a crystal appears as an electrical resonant circuit connected in parallel with the actual electrical capacity of the crystal and its mounting. The crystal can appear as either a series or a parallel resonant circuit. The series and parallel resonant frequencies are different by a fraction of a percent, and for many crystals the series mode has a higher Q and thus gives greater stability.

For the ultimate in stability the controlling crystal of an oscillator is mounted in a vacuum and placed in a small temperature-regulating oven. Great stability can be achieved, however, without going to this expedient. The actual response to temperature changes depends upon the way the section of the crystal is cut relative to the crystallographic axes of the original complete crystal. In Fig. 2.15 is shown the known effect on the frequency temperature coefficient of cutting one series of crystals from the parent crystal at different angles. It should be noted that certain cuts yield

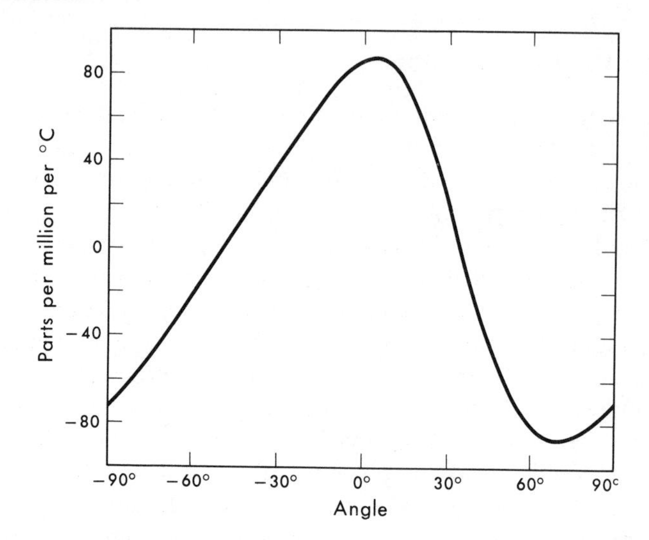

Fig. 2.15 The temperature coefficient of the frequency of oscillation of quartz crystals cut at different angles. A crystal can be cut to maximize or minimize a change in frequency with a change in temperature.

vibrations of frequency almost independent of temperature, and this is obviously desirable when stability is needed. (In Chapter 6 the measurement of temperature by the use of a crystal specifically cut to change its frequency is mentioned.)

There are many ways of constructing crystal oscillators. In a circuit in which a resonant electrical combination could take over control of the frequency, a crystal can be used; for example, a crystal can be placed in the feedback path of a Colpitts oscillator. The base capacitor in configurations like that of Fig. 2.13 can be replaced by a crystal to fix the frequency; or a crystal can be set in a Hartley oscillator in place of the capacitor feeding back to the base. In many cases the capacitances associated with transistors (or vacuum tubes) are important elements in the final circuit. Thus one of the resonant circuits in a tuned-plate tuned-grid oscillator can be replaced with a crystal. Above about 30 MHz a series resonant crystal is often preferred, for example, with a capacitive voltage divider or tapped coil from collector to battery and the crystal from the tap to the emitter (e.g., Fig. 17.3).

A crystal at series resonance may appear as approximately 70 Ω. Thus a circuit with a 70 Ω resistor in place of the crystal should oscillate at approximately the correct frequency and come to a full stop if the resistor is removed. Replacing the resistor with a crystal should then lead to oscillations at the crystal frequency. For work at high frequencies the capacitance in parallel with the crystal can be troublesome. It can be cancelled by connecting in parallel with the crystal a small inductance having a magnitude that will resonate with this capacitance. If the power in a circuit is increased, damage can result. It is not that the crystal will usually crack but rather that if dissipation is over approximately a milliwatt a small hole will melt from the crystal electrode (usually gold plating) and oscillation stops.

One crystal circuit has been widely applied by William Cochran (Cochran and Lord, 1963) to the tracking of freely moving animals in the field. A capacitor and resistor were later added to the circuit so that oscillations would periodically build up and then block themselves in order both to give a modulated signal and to conserve battery life. In all cases the output resonant circuit consisted of a single loop of wire in parallel with

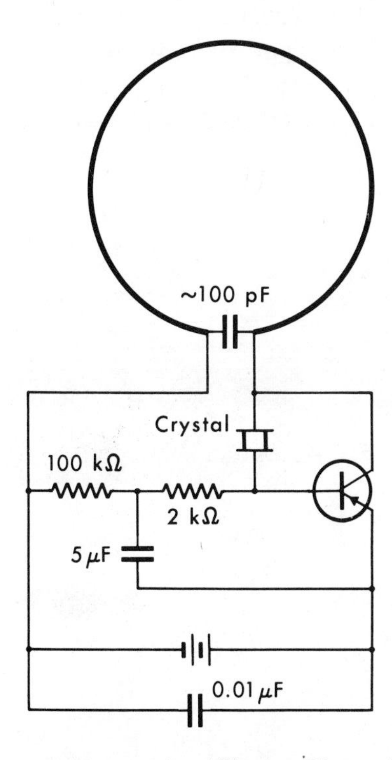

Fig. 2.16 One of many possible crystal-stabilized oscillator circuits, in this case fitted with a long *RC* time constant so that oscillations are periodically spontaneously interrupted. These circuits have been used by Cochran and others in tracking experiments in which the single-turn loop of the inductance serves as a collar around the neck of an animal.

a capacitor, which loop served as the transmitting antenna and was part of the harness fitting around the neck or body of the animal. It was found that tuning was somewhat altered by the act of placing the loop around the tissue, as was to be expected. Such a circuit, including the blocking feature, is shown in Fig. 2.16.

The higher the frequency of operation, the smaller and thinner becomes the crystal. It is generally not very successful to grind crystals for frequencies above roughly 30 MHz. For higher frequencies two approaches can be used. A crystal oscillator can be operated on a lower frequency and then Varactors or other nonlinear elements can be used to generate the higher frequency harmonics of this basic frequency. Such conversion to higher frequency is relatively efficient, but filters are required in the output to reject the lower harmonic frequencies. Thus it is useful to use a crystal that basically functions in an overtone mode, thus directly generating oscillations at the required frequency.

Building this type of oscillator can at first be tedious. It is best to build an oscillator that runs at approximately the correct frequency and that contains a small resistor in the feedback path, as mentioned. The resistor is then replaced by the crystal, and oscillations should commence at the

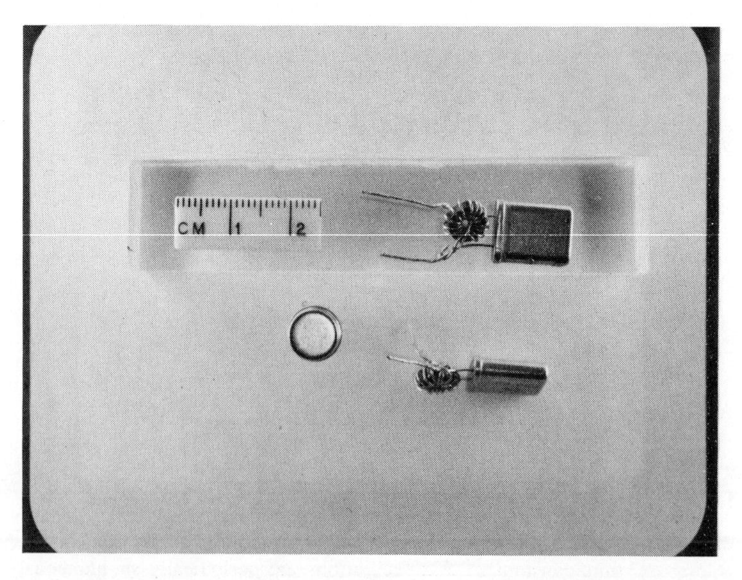

Fig. 2.17 Views of some standard small crystals in their containers. The can at the left is designated TO-5, while those at the right are HC-18U. The capacitance of the units at the right is cancelled by a toroidal coil.

proper frequency under the control of the crystal. As already mentioned, before the crystal is inserted into the circuit the effect of its actual electrical capacitance may be eliminated by a small inductance of just the right value to resonate out the unwanted capacitance. Thus at the base of the can containing the crystal can be placed a small powdered-iron torus upon which is wound a suitable few turns of wire, as in Fig. 2.17. A capacitor can be placed in series with this coil for use in any circuit in which a path for steady currents cannot be allowed where the crystal is connected. A given frequency crystal fitted into the smaller can in that figure may be the more rugged.

Another approach to this problem of the construction of a high-frequency crystal oscillator and several practical circuits are given in Fig. 2.18. In each of these cases a fifth overtone crystal is used to directly generate a frequency of approximately 90 MHz. Each of the circuits is shown to contain a 2.2 μF capacitor (of tantalum to reduce size) and a 330 kΩ resistor to give a periodic interruption of oscillations. Basically, the several circuits are all Hartley oscillators with the feedback to the base being through the crystal. To eliminate the effect of the capacity of the crystal a form of neutralization is used in which a reversed signal is fed through a small capacitor having the same capacitance as the crystal. This is perhaps most easily seen in the third line of the figure where the crystal is seen to be tapped up on the coil from the battery connection and a small capacitor is tapped down by the same amount. The actual layout of this circuit is shown at the bottom of Fig. 2.18, in which it is seen that the coil actually consists of a single large turn (approximately 4 cm in length) and a smaller fraction of a turn, with various taps. The size of the smaller turn determines what fraction of the energy in the larger turn is picked up and fed back to produce oscillation, and the size of the larger turn is used to control the tuning of the output. Both size adjustments are made by bending a thinner piece of copper wire permanently soldered into place across the heavier wire of the basic coil. The signal radiated from such a coil carries for hundreds of meters, and relatives of this circuit will be described in Figs. 14.4 and 18.4.

Without neutralization, a range of frequencies will be generated, and turning on a crystal too vigorously also tends to generate extra frequencies. If a receiver is tuned slightly off of a blocking signal and the neutralization is then connected, the signal should vanish due to the removal of any frequency sweep, especially during turning on. However, if observation is being done with an FM receiver, the single-frequency transmitter can seem to be generating two frequencies, since it can be noted when tuning is to either side of the discriminator characteristic (Fig. 11.6).

If a physiological variable is being monitored, then it is necessary to

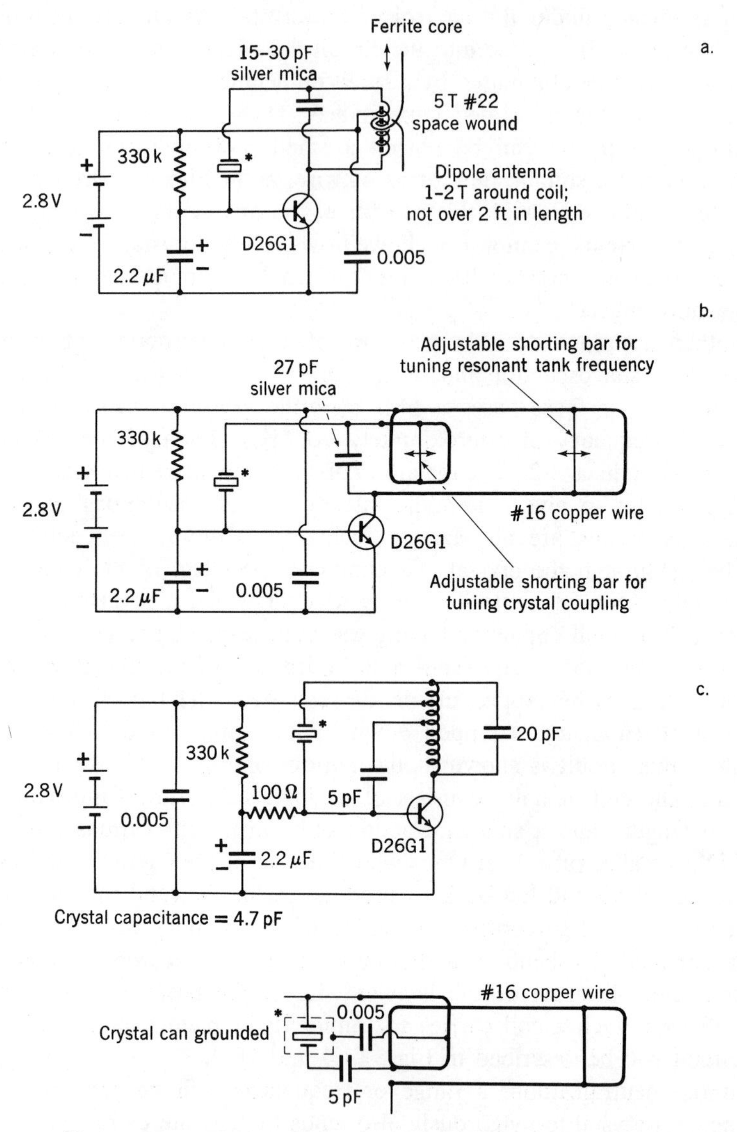

Fig. 2.18 Several high-frequency crystal-controlled blocking-oscillator circuits. A strong signal can be radiated from the single-turn loop antenna shown in the lower three sections, even though the loop is only 2 in. long by ¾ in. wide and employs a second turn 0.5 in. long. At these frequencies the layout of certain components is important, with the arrangement shown at the bottom of the figure being suggested.

superimpose some intelligence upon the oscillator signal. This modulation can be accomplished in several ways. One way is to periodically turn off the oscillator at a rate that will convey the information, thus giving an effect very much like the blocking oscillator temperature transmitter mentioned in Chapter 1. The intensity need not drop to zero, but there can be a periodic amplitude modulation at a rate proportional to the signal to be transmitted. (Such an FM-AM subcarrier modulation system is discussed in Chapter 3.) Modulation in amplitude can be achieved by changing the base voltage of the transistor, or if more than a single cell battery is employed, by placing a field-effect transistor in series with the oscillator to serve as a variable resistance.

It is tempting, however, to consider possibilities of frequency modulation (the implication being that the crystal would determine the average basic frequency around which small deviations would take place to communicate information). The very stability of the crystal oscillator makes direct frequency modulation somewhat difficult. However, it is known that the frequency of a quartz crystal oscillator operating in the parallel mode can be changed slightly by varying the capacitance shunting the crystal. Noble (1966) has continuously varied the frequency of operation of crystals over a part in a thousand without material degradation of frequency stability, given as 2 ppm/day for hours, and without temperature control. He has shown that the maximum possible frequency deviation will occur when the electrostatic capacitance of the crystal is a geometric mean of the maximum and minimum shunt capacitance imposed on it by the external circuit. Noble has also done work with inductive loading of the crystal, and found it possible to "pull" the crystal frequency as much as 2 percent, with some degradation of stability.

A more usual but cumbersome method for providing frequency modulation is to combine the signal from a relatively low-frequency oscillator whose frequency can be varied with the signal from the crystal oscillator. Beating one signal against the other gives a new frequency which can be varied by the variable-frequency oscillator, but yet whose major component is fixed by the crystal oscillator. The signal from a reference crystal oscillator may also be fed into an ordinary variable-frequency oscillator in such a way as to force control of the average frequency of the variable-frequency oscillator.

Another approach makes use of the fact that a few degrees of phase change in the output of a crystal oscillator gives a significant frequency shift since frequency is the rate of change of phase. Phase-shifting circuits are well known and can be electrically controlled. It should be noted that it is impossible to indefinitely decrease or increase the shift in phase, and so this system is acceptable only for the transmission of cyclically varying quantities with a relatively rapid rate of change (for ac quantities). With a subcarrier oscillator signal (a low-frequency oscillator whose rate is controlled by the quantity of interest) superimposed on the radio signal this method can also be used to transmit slowly varying or dc quantities. To get adequate deviations under these systems it is generally necessary to start with a lower radio-frequency oscillator and do the phase-shifting operation on it before multiplying the frequency up to the desired final value. In this way a phase shifter capable of acting only through a fraction of a cycle is able to have its effect act for an appreciable time (several cycles at the final frequency).

For the present applications it is probably most relevant to note that a variable capacitance, such as a varicap diode, in series with a crystal is able to "pull" the frequency sufficiently to produce useful frequency modulation in either transistorized or tunnel diode oscillators. However, difficulty of frequency modulation goes up with the number of the overtone, and thus in the example of Fig. 17.3 a harmonic of the frequency-modulated fundamental is employed. High crystal capacitance increases modulation ease, and this can be augmented by a greater-than-average metal coating.

9. POWER SOURCES

At the outset, it was realized and demonstrated that these telemetry transmitters could be powered by radio signals being induced into them, or by chemical reactions involving various body fluids (see Mackay and Jacobson, 1957), but for most purposes a battery remains the primary power source. Alternatives are discussed both in this section and in Chapter 13, but standard batteries are quite adequate and effective for many experiments. For short-range transmission, for instance, from an animal confined to a cage in a laboratory or from a human subject relatively fixed in position, a few microwatts of power are quite sufficient to communicate an adequate signal. The tiny batteries used to power electric watches or hearing aids can run such a transmitter continuously for one to two years. In long-range animal experiments the battery will often force a compromise between transmitter life, size and weight, and range. In those cases the alternatives to batteries may be unable to supply sufficient energy.

In the early experiments with Jacobson zinc-carbon flashlight cells were actually dismantled and reconstructed to make small batteries. Although these procedures were improved (Mackay, 1958), such LeClanché cells do have fundamental limitations. Their shelf life is long if they are stored in a frozen condition, and for low current drains they are relatively effective.

In such cells the cathode is actually manganese dioxide, into which current is conducted by a piece of carbon (or for more convenience, a platinum wire in reconstructed cells). Because of a variable valence by manganese (4, 3, or 2), these cells do not give a constant output voltage as they are run through their life. In addition, drying out and the formation of voids also causes a drop in voltage owing to increasing internal resistance. A dead battery can be detected by its inability to supply a momentary large current through the low resistance of an ammeter, but this test does not predict the life remaining in an apparently good cell. Since many transmitter circuits are most effective when supplied with a constant voltage, the variable voltage of a LeClanché cell is often a rather severe disadvantage.

Cells based on zinc and mercuric oxide give slightly less voltage than a fresh LeClanché cell, but they store better at higher temperatures. These mercury cells have become readily available in many sizes in recent years, and they maintain a relatively constant voltage as they are being discharged. Figure 2.19 shows a representative curve of voltage output as a function of time for many such cells. It should be noted that at the start, for a small percentage of the total discharge time, the voltage is somewhat high. For an extended time after that, the output voltage is quite constant. At the end of the life of the battery, the voltage suddenly and obviously drops to a small value. During discharge, intermediate products are not formed, and the

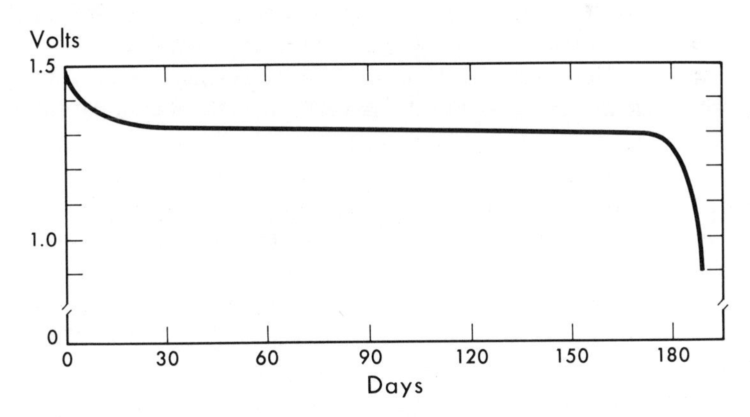

Fig. 2.19 Mercury batteries maintain a constant voltage over their lifetimes. A representative curve is shown. At the end of their lives the voltage suddenly drops to a negligible value (note the suppressed zero on the vertical scale). For a small percentage of the expected life at the start of an observation the voltage with many cell types will be a bit high, as shown. This curve for 21°C represents a Mallory W-1 battery loaded with a resistor of 150 kΩ. A type WH-1 cell does not show the initial high voltage because of its modified construction.

mercury resulting does not increase the internal resistance of the cell. The present curve shows the discharge of a small cell over a period of slightly more than six months. These cells normally contain some manganese dioxide for ancillary purposes. A different construction eliminates the initial extra voltage, such cells often being designated by the letter R at the end of their type number. Depending on whether long life and low current operation or lower temperature operation and higher current capability are most important, these cells can be supplied with a sodium hydroxide or a potassium hydroxide electrolyte. For example, a W-1 cell is a sodium hydroxide version of the 312 cell, while a WH-1 is a constant voltage W-1.

The mercury cells store considerable energy in a small volume, although they are rather heavy. A problem is always to obtain fresh cells from the manufacturer. It is said that the discharge of such a cell in use may be followed by X-ray images (Greatbatch, 1966) where the redistribution of opaque elements can be seen. The appearance of the initial high voltage, in those cells which show it, is a suggestion of freshness. Testing a few cells of a given batch to exhaustion at an accelerated rate indicates their energy content, but does not guarantee against failure by others of the same batch. Certain sized cells are produced with extra quality control for cardiac pacemakers, and these can be advantageous in some applications.

A Mallory cell, type RM312, is often rated at 36 mA hr, and indeed it will supply a current of 1 mA for about 1.5 days. If fresh so that it is not already partially self-discharged, the cell will supply a correspondingly lower current for months. But if a shorter experiment is attempted, in which the full capacity of that battery is to be withdrawn in five hours, the battery will go dead in a few minutes. If the convenient geometry of these batteries recommends them, several such cells can be connected in parallel to deliver a higher current for a short period, as suggested in some of the circuits of Chapter 15. Diodes in series with each battery can prevent a dead cell from discharging a good one, but this has generally not proven necessary. A summary of some available information on this and other cell types is given in Table 2.1.

Most battery chemists presently seem to be studying systems for producing large outputs for short times, rather than the long-life problem. There are, however, several approaches to the obtaining of a longer shelf life in a battery or to the production of better performance with low current drains. One is to take a system that is relatively stable thermodynamically, that is, one that is not vigorously attempting to react. A Weston standard cell is an example, but this system is not too appropriate for general use. Replacing the zinc in the above mercury cells by cadmium results in a more stable system which should supply the best battery of all for many purposes. It should provide an almost indefinite shelf life even at high temperatures, but there is a penalty in reduced output, the voltage being under 1 V.

Silver oxide cells produce a somewhat higher voltage than the usual mercury cells. Although the percentage difference is not great, the improvement in performance when an entire silicon transistor circuit is powered by a single cell can be significant because of the fact that it is voltages beyond 0.6 that are useful. These cells are also preferable at the low temperatures to which some external transmitters are exposed. Such cells come in an increasing variety of sizes, some convenient ones being the MS 212, 312, and 13 which are the silver oxide equivalents of the corresponding mercury cells.

A LeClanché cell with magnesium in place of zinc has a long "shelf life" because the metal becomes passive during storage, but it is not presently known how it reacts during low current drain, and considerable gas evolves during use.

Batteries having a solid state electrolyte (Owens and Argue, 1967) should provide a number of advantages with regard to life and freedom from drying out at high temperatures, though the present voltages and energy densities are somewhat low.

A few generalities about batteries can be given. A larger cell size is often more efficient and gives a longer shelf life than very tiny ones. A cylindrical geometry may give less uniform discharge characteristics than a planar geometry, although such a cell can be more difficult to seal properly. High-temperature storage of batteries in the field can ruin them. If a LeClanché cell spontaneously loses half of its stored energy at 21°C in a year, then at 45°C such a loss could take place in three months. Hot days are not fully compensated for by cool nights if the temperature cycles up and down. In field work, burial a few feet below the surface of the ground where it is generally cool can be extremely helpful. Gas released by cells during discharge has not caused problems in any of our transmitters, although this was a source of some concern in our earlier work.

In some cases a capacitor is connected in parallel with the battery to improve either circuit range or life. A capacitance of a few microfarads can store energy for sudden release in a very intense impulse, thus yielding a stronger signal in some circuits. In other circuits a total cessation of operation results when the internal resistance of the battery rises and prevents the passage of radio frequency currents. In these, operation is prolonged by bypassing the battery with a small capacitance, for example, a few hundred picofarads.

Another approach to long storage life before battery use is to assemble the battery as needed. Adding electrolyte to automobile storage batteries

Table 2.1

Sample	Type	Chemistry	Voltage	Service Capacity (ma·hour)	Weight (grams)	Height Diameter (cm)
	RM 212	Mercury	1.4	16	0.5	0.33/0.55
	RM 312 E 312 Hg 312	Mercury	1.4	36	0.64	0.36/0.79
	W-1	Mercury	1.4	36	0.64	0.36/0.79
	Mallory RM13GH	Mercury	1.4	60	0.98	0.535/0.79
	MS13H	Silver oxide	1.5	60	0.98	0.535/0.79
	RM 400 E 400 Hg 400	Mercury	1.4	80	1.15	0.345/1.16

	RM 575 HG 575	Mercury	1.4	100	1.49	0.335/1.16
	RM 675 E 675 Hg 675	Mercury	1.4	160	2.22	0.540/1.16
	RM 520 E 520 Hg 520	Mercury	1.4	130	1.98	0.730/1.26
	E 301	Silver oxide	1.5	100	1.68	0.415/1.160
	RM 640	Mercury	1.4	500	7.59	1.11/1.59
	Gould National 20B	Nickel- cadmium	1.2	20	1.50	0.55/1.15
	Gould National 50B	Nickel- cadmium	1.2	50	2.58	0.63/1.55

Letter prefixes in "type" description indicate the manufacturer as follows: RM, Ms, Mallory; E, Eveready; Hg, Burgess.

or to upper atmosphere radiosonde batteries are examples of this. The well known "seawater" battery with electrodes of magnesium and silver chloride is another example. The so-called Heidelberg capsule for measuring pH (to be mentioned in Chapter 7) is activated by adding saline and now uses such a system. These procedures allow the prediction of transmitter performance, which is advantageous, and the early attempts at the use of gastric-juice batteries (Mackay and Jacobson, 1957) were a step in this direction. These "reserve" cells are not usually the most efficient, at present.

Storage batteries help with the problems of initial energy content at the start of use, as well as conferring other advantages. Nickel-cadmium cells are especially appropriate in the present application. They supply the most recharge cycles of the sealed storage cells, and they have almost infinite shelf life when discharged and shorted. These cells often show a self-discharge rate at room temperature of 25 percent per month, which is not especially good. W. Scott has found that it is possible to heat-sterilize such a cell if discharged, although there is often some subsequent performance degradation and unreliability. It is also true that such cells can power receivers in field work at temperatures below freezing. These cells vary somewhat in their properties, but most must be periodically discharged if they are not to lose their ability to supply rated power.

Nickel-cadmium cells are available in small sealed packages. They have perhaps one third of the energy density of a LeClanché cell. A mercury cell has approximately the same weight but is smaller. By the inward induction of power in the form of radio-frequency energy, it is actually possible to recharge storage batteries inside an animal or human subject. Thus a recharging coil can be placed under the bed or nest of a subject in order to recharge his battery as he sleeps; the same automatic recharging could be done in connection with any other cyclically performed activity such as eating or bathing. It is difficult to seal batteries that are to be rapidly charged, since "over voltages" tend to liberate gases. Normally gas from the positive electrode is consumed at the negative one, with speed being limited by diffusion. Many of the small sealed cells can be recharged only a few times as rapidly as they are discharged. Thus the recharging cycle may have to be extended over hours in certain cases. There is some indication that a periodic momentary short current reversal allows more rapid recharge.

Reynolds (1964) noted that a stainless-steel electrode and a platinum-black electrode, when placed at various sites in the body of an animal, took on a potential difference of a few tenths of a volt and delivered a steady current of up to 480 μA. Whether this is some sort of fuel cell or oxygen electrode, or whether some other mechanism is involved is not known, but certainly such a power source can run a low-power transmitter. Whether there might be long-term damage either to the living system or to the electrodes is not entirely clear at this time. Under some conditions, the oxygen tension in tissues can become very low, and thus in general, power generating electrodes depending upon oxygen cannot be counted upon continuously. Since few experiments are required for more than a year and rather tiny batteries can supply such powers for a year, these methods would seem very interesting but will require further development before becoming important.

Energy in many other forms can be employed. The inward induction of radio energy is certainly of importance in special cases, and these will be taken up in Chapter 13. Light energy can also power these transmitters. Thus a set of solar cells mounted on an animal's head or back can either power transmitters directly or can be used for recharging storage batteries during the day or in lighted regions. Effective results can actually be obtained even if the solar cells are placed beneath the skin, although greater light intensities are then required. The transmitter power levels involved in such cases are usually not high. In an early experiment a radioactive light source and array of solar cells was used to activate an oscillator, but the arrangement was large and inefficient. Thermoelectric conversion of radioactive energy to electrical power can be achieved more practically for special applications.

Sound energy can be propagated through tissues to focus energy at a desired position. This has not seen much application to date in powering transmitters, although pulsed ultrasonic energy has been used for exploring body tissues in other ways, some of which will be mentioned. The inward induction of power and lower frequency mechanical motions will be mentioned in Chapter 16. In dealing with aquatic animals it might be thought that the ambient noise of the ocean or the vocal activities of the animals themselves might be used to activate piezoelectric transducers as power sources. In some cases this may be possible, but there is a fundamental limitation on the general use of the method. It can be seen that it is impossible to build a perfect rectifier diode because, if it were possible, the small voltages generated by thermal agitation in resistors could be rectified to charge capacitors. A number of these capacitors could be placed in series to give an appreciable dc voltage, which might be used to do work. This conversion of random thermal agitation to useful work disobeys the laws of thermodynamics, and thus such a rectifier of very low voltages is fundamentally impossible. (Phase-sensitive rectifiers can rectify very small voltages of predictable periodicity, but this is irrelevant here.) Thus a number of sound transducers could not have their signals combined to make use of very faint sounds in the ocean, although the louder noises might be used in some cases to supply energy. Motions in general are feasible sources of energy in some cases. Seismograph-like devices or accelerometers employing weights and voltage generators can convert gross body motions into small amounts of useful electricity. Similarly, the beating of the heart and the pulsing of the large blood vessels could activate piezo elements to supply power. Pressure changes in the thorax associated with breathing can also generate power. The swimming motion of aquatic animals could supply power, while the changes in pressure on diving and surfacing could be a further source of energy. In some cases, a relatively unimportant muscle might be detached at one end and sutured to a generator which would produce power and supply cyclic muscle stimulation.

It is well known that the flow of blood in a magnetic field generates a voltage, and this is one of the bases for certain flowmeters. It can be calculated that only the flow in the largest vessels could generate sufficient power to activate one of these small transmitters. It is tempting to consider placing several flowmeters in series, while using permanent magnets to set up the magnetic fields. However, this is not possible. Whenever electrodes are placed in a conducting system, the conductivity between sets of electrodes will be such that it is not possible to have several cells in series. This applies to magnetic flowmeters, to ocean-water batteries, and to batteries using gastric juices as the electrolyte; all are limited to a single cell and low voltage, although considerable current may be supplied.

Another source of energy is heat. Many of the animals of interest maintain their body temperature above that of their surroundings, and a thermopile having the body as a warm junction and the surroundings as the cold junction can use this temperature difference to produce power. It should be noted that the region of the body in contact with the hot junction will be cooled, not only by conduction, but by the abstraction of energy, and this could prove troublesome to the animal. Such systems generally generate power at a relatively low voltage and a corresponding current, which means that they can conveniently match the requirements of a tunnel-diode oscillator.

In certain cases special power sources can be visualized which might be appropriate. Thus a cylindrical rod passing through water is known to vibrate in a particular way that varies with speed. A transmitter powered by such a source on a swimming animal thus might not only require no external power source, but might automatically give a signal depending on velocity, which could be valuable. Other possibilities are left to the imagination of the reader.

10. TIMERS

There are a number of cases where one may wish to time or preset some function within the transmitter in or on an animal. For example, in order to conserve battery power during recovery from surgery or in some other preliminary period, it can be desirable to delay the turning on of the transmitter. If the subject cannot readily be approached, it is desirable that this be done automatically (though in some cases it can also be done by remote radio control). It may similarly be desirable that the transmitter function only a few minutes out of every hour or every day. In other cases it may be desirable that a transmitter function only at night. While a transmitter can readily be turned on and off by a photoelectric device, in some cases placed beneath the skin, the indications can become unreliable on dark or cloudy days, or if the animal rolls in mud or crawls into a crevice. In other cases the release of a transmitter or recorder package from the back of an animal is desired for recovery after a preset interval. In each of these cases, some sort of clock or interval timer is required which will give out one or more impulses at predetermined instants. In some cases little accuracy is required, as when a transmitter is to be released some time during a day, while in other cases rather great accuracy is desired, as when transmission is to take place only during the passage of a data collecting artificial satellite.

An electric watch constitutes a convenient timing mechanism, if a contact is run through the watch "glass." In some cases inexpensive watches and clocks function better when placed on an animal than if simply left sitting on a table, since the vibration of movement assures their running. The lubrication of a watch can "bind up" in such conditions as the cold of a desert at night. Rather good performance can be achieved with the so-called Accutron movement which is based upon a small tuning fork. Accuracy is usually a few percent. Accutron movements are somewhat altitude or pressure sensitive, a pressure increase causing slowing due to the increased mass of air being carried along by the mechanism. Typical sensitivities are 18 seconds per day per atmosphere, and balance-wheel watches show similar effects.

Electrolytic processes can also be used to determine a time interval. In these a fixed amount of some metal is electroplated from one electrode onto another. When the last of the metal is plated over, the voltage across the cell suddenly typically rises from a few millivolts (depending on temperature) to approximately a volt, and this can be used to activate a following circuit, that is, the current that was previously flowing through the cell will suddenly pass a current to burn through a wire and drop a recorder, or perform other similar functions. An example where electrolytic action instead gradually corrodes through a wire to directly produce a mechanical result will also be given. The time interval can be controlled over wide ranges by adjusting the current through the cell, which is done by changing the value of a series resistance attached to the fixed voltage of the power supply. These integrators or timers are produced by several companies, e.g., Bissett Berman, Curtis Instruments, Gould Ionics, Gibbs, and Sprague, and often have the appearance of a small capacitor. Such units are made with both liquid and solid electrolytes. If several intervals are to be timed, the activation of the first of several electrolytic cells can be used to start the energizing of the next one. It is also possible to use such cells repeatedly by plating the metal back on to the original electrode, or to integrate variables other than time. However, there is a certain art to fabricating cells which can be used repeatedly if undercutting and the dropping off of some of the metal prematurely or else the formation of bridging metal "whiskers" (dendritic growth) is to be avoided. The biological conditions of vibration and somewhat unsteady temperature seem to minimize the latter in liquid-filled units.

An application of these techniques is depicted in Fig. 2.20 which illustrates a bird harness that releases itself after an electrolytically timed interval. After all of metal #1 plates away, metal #2 disappears to suddenly

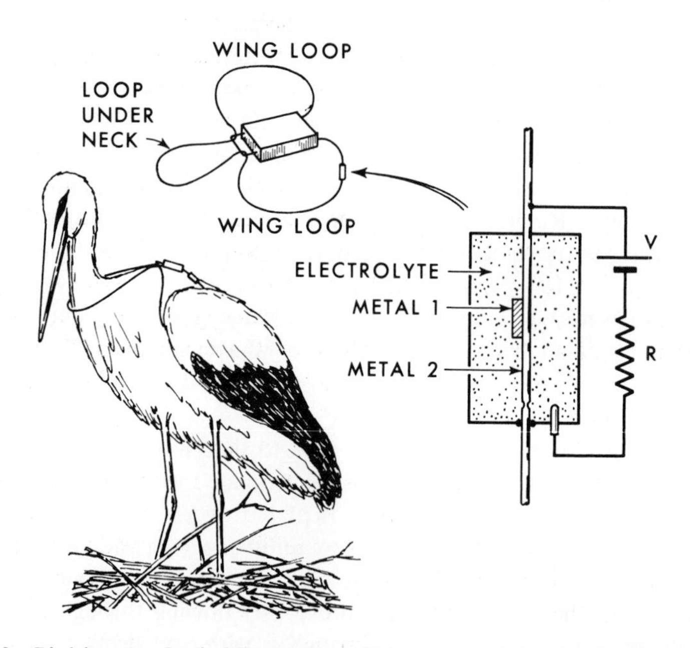

Fig. 2.20 Bird harness for holding a transmitter or recorder and releasing it after a preset interval. At the right are the details of the electrolytic fragmenting delay device. The use of two metals can reduce time variations due to irregularities in the deplating pattern of the base material. A representative combination of materials in the above is: lead on silver wire, lead nitrate solution and a silver cathode. Other strap configurations can be used on various animals, in some cases avoiding a neck loop. The cell can instead provide a trigger action to release a loop, and its activation can be initiated by cessation of transmitter power.

weaken the link. In some cases, the battery may be eliminated by selecting the cathode material. An analogy to this is the protection of boat propellers against electrolytic action for as long as a piece of zinc is present. In a similar way, the transmitter current could be run through the electrolytic cell to prevent self-corrosion until the battery "dies," and then the no longer protected wire would corrode through to automatically release the transmitter when it became inactive. No sudden surge of external power is needed late in the process nor is there trauma to the animal, as might be involved with an explosive release. For related considerations, see also Chapter 17, Sections 2 and 5.

An extremely accurate clock or timer can be made based upon the properties of a vibrating quartz crystal. However, vibrations or cycles perhaps occurring in the range of a million per second must be divided in number to give out impulses every few seconds or minutes or hours or days depending upon the application. Digital frequency dividing circuits are well known, but have always required an appreciable amount of space and power. Recent developments in the production of monolithic integrated circuits allow a very tiny cascaded series of counters to be constructed. Recent developments in MOS transistor circuit technology, especially in complementary symmetry units combining the two types, allow for the construction of extremely low current units in which charge is stored essentially without current flow during any interval, and then current flows only during the instant of switching; even then the currents required are only those to charge a few picofarads to a few volts. If such a unit were fabricated and combined with a vibrating crystal, the entire unit would be about the size of a finger and could run for about a year on a single battery. Furthermore, if turned on at a particular time of day, one would expect it to be inaccurate by less than a minute at the end of the year. Such a unit would supply the ultimate in flexibility since signals at various time intervals would be available from within the various parts of the dividing circuit. Since such a unit would have other applications as well, it is to be expected that they will eventually become available.

11. WIRING AND TESTING UNITS

In our work we have found that most transmitters can be constructed from selected small commercially available components. We have worked with devices for producing very tiny printed circuits, and we have worked with hybrid integrated circuits, but for most purposes to date, it has proven sufficient to wire circuits in the usual fashion. A small soldering iron is used with rosin-core solder and, for the smallest units, the actual assembly is done under a binocular dissecting microscope. We have not used spot welding to fabricate circuits because these procedures will occasionally produce circulating currents which are damaging to the components.

Soldering is done swiftly and it is found that heat-sensitive components such as transistors are generally not destroyed even though no "heat sink" is used to divert heat away from the sensitive regions. However, the first few attempts by a new worker can be discouraging.

There are several ways of attaching a battery into the circuit. In a unit that is to have long life or minimum size, wires are spot welded to the two sides of the battery or batteries. A standard spot welder is the best piece of apparatus for this purpose, although the discharge of a high-capacity low-voltage capacitor through a rapidly closed switch can also be used. Good material for welding to battery cases is to be found in the cut off leads to transistors or certain small capacitors. A number of tests were made in which leads were glued to batteries with conducting epoxy. Although this was fairly satisfactory, the slightest trace of moisture later caused erratic operation. Silver paint of certain kinds can also be used. If size is not of primary concern or the battery is to be changeable, then spring clips are used to hold it in place. Phosphor bronze has been used as the spring material rather than beryllium copper because of the potentially greater toxicity of the beryllium copper if the case should fracture. More recently, a spring silver has been used with extremely satisfactory results. If a battery is lightly rubbed with sandpaper, and a drop of 4N hydrochloric acid used as flux, solder will adhere. Ordinary copper wire can then be soldered on as a lead, but extreme care must be taken with small batteries to use speed and not cause overheating, which would rapidly remove an uncertain fraction of the total stored energy.

Components are tested before insertion with the usual instruments. A so-called universal test meter or volt-ohm-milliammeter is very valuable here. Such a unit is also valuable in trouble shooting a circuit that does not work.

A component that sometimes fails is the transistor in a circuit. It is possible to test the performance of a transistor using just an ohmmeter. Whatever else a transistor may be, it does consist of two rectifying junctions. Thus an ohmmeter connected between collector and base should show a high resistance for one attachment of the probes, and a low resistance if they are interchanged. The same should be true of connection to the base and emitter leads. Actually, transistor action can be checked by attaching the ohmmeter on its high-range scale from emitter to collector and touching the fingers from collector to base. The small changes in current through the hand and into the base give rise to changes in resistance if the transistor is functioning properly.

It is actually possible to estimate beta of a transistor, and to select preferable units with nothing but an ohmmeter. The ohmmeter is connected on the high-range scale from emitter to collector, being sure with an *NPN* unit that the positive lead is on the collector. A 100 kΩ resistor is placed from collector to base, and the lower the reading, the better the unit. If a 330 kΩ resistor connected from collector to base should give a resistance reading of 33 kΩ, then beta (H_{fe}) is approximately 10. The assumption here is that the meter voltage is considerably greater than 0.6 V, which on most universal test meters is certainly true on the high-resistance range.

In producing a number of these circuits, potting in plastic is common. But this is often done to the accompaniment of gentle heating. It might be noted that the maximum temperature to be allowed is limited not by the transistors in the circuit but by the batteries, and it is generally desirable that the temperature not rise much above 50°C. Ways of handling and repairing plastics are mentioned in Chapter 4.

Another component warranting comment is the transformer seen in several of the later circuit diagrams. In many cases this is formed by placing a few turns of wire into a ferrite cup core or pot core, whose other identical half is then slipped into place. It is extremely important that the two halves of such a core be pressed into tight contact and that no dirt particles remain in their junction. It is convenient to hold the two halves together with a spring clothespin while testing. If any air gap remains then reduced output signals result, and any frequency-sensitive aspect will be detuned. A small tube of ferrite can instead be used as a toroid core. Powdered-iron toroids are often preferable for the higher frequencies such as 100 MHz. The magnetic properties of all these materials tend to drop off at high frequencies, but in some cases a powdered-iron toroid is more reliable and less temperature sensitive than the corresponding ferrite unit. Ferrite generally does display higher permeability, but at high frequencies this can actually be a disadvantage since the proper number of turns for a given application might turn out to be a fraction of one, whereas with powdered iron the proper number of turns would be a convenient few.

In these transformers the turns ratio depends on the desired output relative to the available input. From the frequency and applied voltage and core properties, the required number of primary turns can be calculated. Some of the small cores will contain a magnetic flux appropriate to approximately one ampere turn before saturating. Examples of specific transformers will be seen in later diagrams.

Several comments on tantalum capacitors are warranted. These incorporate a large capacitance in a small space. The insulation in these is a thin film across which the voltage must be applied in the proper direction. Thus such capacitors are shown with a polarity of plus and minus. If voltage may reverse, then two can be used (as in Fig. 11.8), one holding off the reversed voltage from the other. A series of tests indicates the stability of these capacitors to be quite good. Thus in the small transmitters, which are to give out pulses slowly enough to count by ear, we sometimes use the high capacitance of such a capacitor to determine the relatively long interval (e.g., Figs. 1.7 and 15.9). In temperature transmission we then obtain extended observation with a precision of about 0.1 °C. This is better than might be expected from some of the older electrolytic capacitors.

Such a capacitor is also able to supply a bigger impulse of current for a short time than is a small battery. Thus in some of the aquatic transmitters (Figs. 15.2 and 15.3), one of these is shown connected across the battery to increase the peak current and the power in the outgoing pulse. In Fig. 15.3 the voltage rating on the battery capacitors is less than the battery voltage. Going to the next higher rating would have vastly increased overall size, and it was found that the voltage increase in the case shown merely increased the leakage current somewhat without causing a puncture or breakdown of the capacitor. Such capacitors as these always show some leakage current (a few microamperes in the larger capacitance ranges) and thus, when used for coupling "blocking capacitors" in ac amplifiers that must go to low frequencies (Fig. 7.4), the circuit bias must be stable in spite of changes in this quantity.

At low temperatures the capacitors in some transmitters cause problems. Some ceramic units work to about freezing, with cheap polystyrene units going somewhat further, and silvered mica capacitors going perhaps 30°C further. For low-temperature transmitters or receivers, care must be taken also with the transistors and their bias circuits.

If a small transmitter is sealed and found to have drifted to a wrong frequency (e.g., out of the frequency range of the receiver, or into interference with an overpowering commercial radio station) then external changes can sometimes be used to bring it to a more suitable range. A piece of ferrite placed near the antenna coil of a simple circuit will generally lower the radio frequency, while placing around the capsule a turn of wire soldered into a loop will often raise it. In circuits with a separate output stage this may not be effective, in which case provision for paralleling the tuning capacitor with an extra capacitor should be left in new or unfamiliar circuits.

12. INTEGRATED CIRCUITS

Complete electronic circuits can be prepared by evaporation of different materials through small photographically prepared masks, and by related processes, to provide a very tiny rugged and reliable finished unit that is referred to as a monolithic integrated circuit. Similarly, hybrid circuits can be constructed by combining various subassemblies along with standard small components. A simple example of a small assembly was depicted in Fig. 2.12. The production of such units is a rapidly developing field, with new components appearing regularly; the possibilities seem almost limitless. However, with the simpler transmitters the size is still limited by the battery and antenna or, if there is no battery, by the antenna which may then have to be extra large, thus yielding a negligible overall reduction in size or weight by employing an integrated construction.

It also should not be assumed that these small circuits will draw small current or work at low voltage. The power drawn by a microcircuit is no less than that drawn by one wired from similar discrete components. Indeed, in an integrated version of a specified circuit it may not be possible to incorporate resistors of as high a value in order to reduce current drain on the battery. Higher values of resistance can sometimes be prepared for use in the relatively constant temperature of the inside of "warm-blooded" animals.

The parts of an integrated circuit are very small and close together and thus a change in temperature affects all of them approximately equally at approximately the same time. Thus superior matching and balance can be expected under changing conditions. These comments are directed largely at circuits where the output is a continuously changing function of a changing input, but similar considerations apply to digital or counting circuits. Entire computers that are considerably more complicated than a standard desk calculator can be deposited on a single chip of relatively small size, and the possibility of a tiny crystal-controlled clock was mentioned in the section on Timers.

An integrated chip comprising a differential amplifier or a voltage-controlled oscillator can be very useful if designed to run at a suitable voltage. As an example of presently available components, the Westinghouse 4-stage (8 transistor) differential audio amplifier #WC183 gives over 60 db gain at a current drain of under 1 ma when energized with 1.5 volts, and is available either in a can or in a small flat package. Examples of other low power commercial amplifiers with noise and stability in the microvolt range are the Fairchild ADO-39, Analogue Devices No. 153, and Zeltex No. 170. When a monitoring transmitter is to be externally placed upon the subject, the requirements on size and power consumption are considerably relaxed, and there is a broader range of possibilities.

Hybrid construction seems indicated when a small number of identical units is required. Monolithic integration reduces size, and entire transmitters can thus be fabricated. In Fig. 2.21 is a multistage amplifier mounted in a standard hermetically sealed can with the top removed. The several connections into the circuit can be seen. The circuit is relatively complex, and yet there is room for more components. The advantage is perhaps greatest when a relatively complicated circuit function is involved.

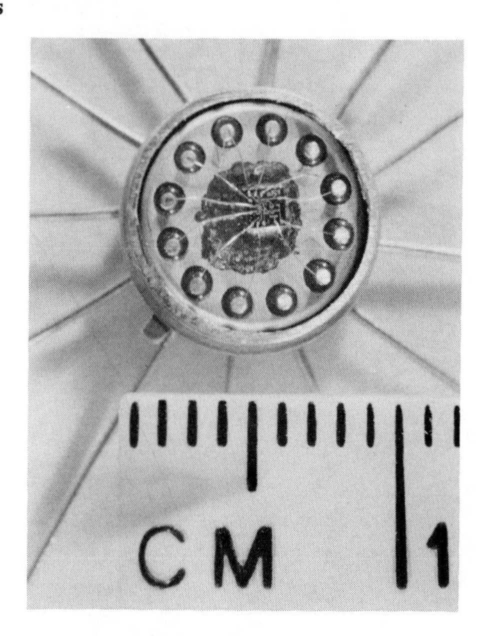

Fig. 2.21 An integrated circuit into which feed a number of connections, with the top of its hermetically-sealed can removed.

A specific example of the advantage to be gained from working with hybrid circuit components at this time may be helpful. The circuit of Fig. 8.1 was redesigned to eliminate the 15 μF capacitors which were the largest components. Use of a wire bonder plus the ability to prepare etched substrates allows construction of such a unit faster, somewhat smaller, and with a greater selection of semiconductor devices (for example, small field-effect transistors). In this particular case, the low frequency circuits were cut in size to approximately one-half, while the battery and radio frequency portion remained about the same.

Those who do not have such facilities need not be discouraged, since hand soldering with a small iron can be used with all the circuits in this book. However, a few comments on the reliable and routine packaging of transmitters might bear repetition. If funds are available for relatively expensive equipment, then it is useful to obtain a wire bonder for use with semiconductor chips. This allows the use of transistors in chip form, and other small integrated circuits. The emphasis here is more on systematic production than on ultimate size reduction. Both so-called thick-film and thin-film techniques are useful (e.g., Doyle, 1966). Small hermetically sealed metal cases are then also somewhat more easily employed. The latter can protect a circuit from penetration by body fluids, though lead wires can corrode in experiments to extend beyond a year. Special stainless steel or gold and glass parts seem needed for units to be left in place for the life of a subject.

Though these comments have related largely to transmitter construction, standard integrated circuits have applicability in all receiving systems. As an example, see Appendix 5.

3

Modulation Including Blocking Oscillation

1. MODULATION SYSTEMS, INCLUDING SUBCARRIERS

A steady unchanging signal can communicate only very limited kinds of information. Using direction-finding techniques at a receiver, such a transmitter can be used to indicate the location of a subject, which is valuable in studying such things as the migrations of animals. Similarly, the approximate location in a human subject of an internal transmitter can be ascertained by methods that resemble direction finding. Also, changes in received intensity from a transmitter of fixed strength indicate movement of the transmitter. Thus a transmitter of constant strength can be used in systems that are to monitor the temporal activity pattern of animals. The movement of an ingested capsule in the gastrointestinal tract of a subject can be monitored in the same way. It might be noted that battery power is conserved in these applications if the signal is periodically turned off, which in itself is a form of modulation. But aside from these kinds of data, a steady transmitter is quite unable to communicate intelligence.

Thus it is essential to consider various systems for superimposing useful information upon a radio signal. Perhaps the most familiar method of modulation is simple amplitude modulation (AM) such as is used by many commercial radio stations. In these systems a momentary increase in air pressure at the studio accompanying a compression half cycle of a sound wave gives rise to a momentary increase in amplitude or intensity in the outgoing radio signal. Cyclic fluctuations in pressure thus lead to cyclic fluctuations in radio intensity at the same frequency. Thus at the receiver the frequency of the sound being generated can be inferred by simply noting the frequency of intensity changes in the radio signal. We might note that the radio signal itself is typically generated in a frequency range between 500 kHz and something over one MHz, whereas the sound frequencies being transmitted extend perhaps up to 10 kHz. To retrieve the intelligence from the received signal it is common to rectify away the bottom halves of the radio frequency excursions, the result is an electric signal that will pull on a loud-speaker diaphragm in a varying way and at a frequency corresponding to the original sound frequency. It is important to note that the absolute amplitude of the received radio signal is not important and that the frequency of fluctuations alone carries the information; that is, there is no information at the receiver as to the absolute loudness of a singer in a studio, although relative changes in intensity that occur reasonably rapidly can be estimated. As an example of this last restriction, note that changes in atmospheric conditions which would cause an amplitude-modulated signal to fade might be interpreted instead as a signal coming from a less intense sound source.

It is generally true that simple amplitude modulation is a useful method for communicating information characterized by a frequency, while other methods of modulation, such as frequency modulation, are more appropriate for communicating amplitude information, or other information where absolute values of a quantity are to be transmitted. Thus simple amplitude modulation would be quite appropriate for transmitting a signal from which was to be determined heart rate. From such a signal it would also be possible to assess the general shape of an electrocardiogram, but it would be quite impossible to infer the magnitude of any of the voltages in a quantitative fashion. If the transmitter were arranged to alternately send the useful signal and then a standard signal, then at the receiver a comparison could be made in order to determine the absolute value of the quantity being transmitted, but this is hardly to be considered simple amplitude modulation.

It should be mentioned that simple amplitude modulation signals are somewhat more adversely affected by sources of radio noise and interfering radio stations than are many of the other forms of modulation, including frequency modulation. These matters are covered in most radio engineering texts, to which the reader is referred.

In some cases information is transmitted by using what is called a subcarrier frequency. In this case a subcarrier oscillator (sometimes referred to as SCO) is modulated by the variable of interest, and its signal is then superimposed as a modulation on the higher radio frequency; for example, changes in pressure might be used to change the frequency of a 100 Hz oscillator which would then be used to amplitude modulate a 500 kHz oscillator or transmitter. The frequency from the transmitter would be fixed at 500 kHz, and it is to this frequency that a radio receiver would be tuned. From the receiver would come a 100 Hz tone and changes in this frequency would indicate changes in pressure. Note that changes in amplitude would not cause ambiguity in the information and that a steady 100 Hz signal would indicate a steady pressure of the corresponding value; that is, very low frequency or dc information can be transmitted. Also, although small changes in stray capacitance might affect the high frequency of the radio oscillator, they would be much less likely to affect the lower frequency subcarrier oscillator; loading on the antenna also does not affect the information. This low frequency could instead be superimposed as frequency modulation, or in other ways, on the carrier frequency (the higher radio frequency) without changing the basic properties of this system. Large percentage changes in the subcarrier frequency require only small percentage changes in the radio frequency, which can lead to lower interference by noise while staying within necessary restrictions.

In the above example a second oscillator, perhaps at 1000 Hz, could be modulated by some other variable such as temperature, and these two lower audio frequencies could be added together before being used to modulate the radio frequency. The final radio signal would then carry both a 1000 Hz tone and a 100 Hz tone, and in the receiver, suitable filters could be used to concentrate attention on one or the other. Thus from one channel would come temperature information and from the other would come pressure information. Use of a number of subcarrier frequencies allows the simultaneous transmission of a number of variables from a single transmitter and antenna powered by a single battery. A similar result is sometimes achieved by cycling between variables at a high rate so different information is transmitted in successive intervals. In some cases of biological work such methods as these can be advantageous, although in others it is simpler or more practical to place a number of isolated transmitters simultaneously working on different frequencies at different locations in the body; an example of the latter was shown in Chapter 1. It might be mentioned that the blocking-oscillator circuit mentioned in Chapter 1 can be considered as using a signal like that normally generated in a subcarrier system. Thus it is the frequency of periodic amplitude change which carries the temperature information. If the basic radio frequency drifts somewhat, this does not cause a misinterpretation of temperature, because the signal remains unambiguous as long as it is picked up at all by the receiver.

In Fig. 3.1 are examples of the types of wave forms which are involved in several modes of modulation. It is assumed that in each case the signal is under the control of an applied voltage representing the changing variable of interest, and the examples show what happens to the wave in

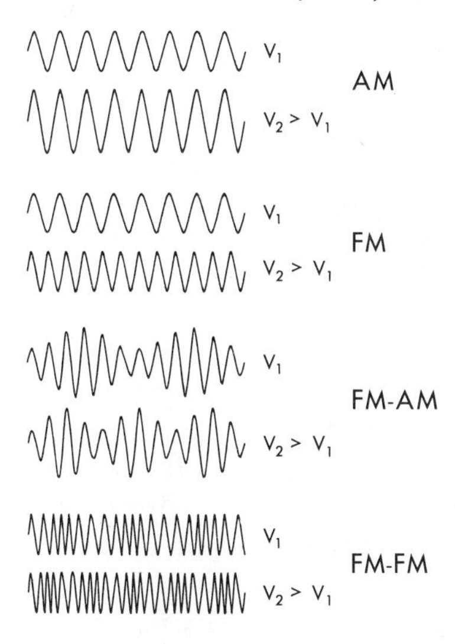

Fig. 3.1 Wave form changes under some common methods of modulation. It is assumed that a voltage is being monitored, and in each case the pair of lines shows what might be expected in response to an increase in voltage to a higher level.

response to an increase in voltage. At the top simple amplitude modulation (AM) is depicted in which an increase in voltage gives rise to a sinusoidal wave of the same frequency but having a greater amplitude. The second example shows an increase in voltage producing a wave of the same amplitude but of increased frequency, that is, this is frequency modulation (FM). The remaining two examples depict the use of a subcarrier frequency to carry the actual information. If the audio oscillator amplitude-modulates the radio frequency, then the condition in the third example prevails. Here an increase in voltage causes an increase in the frequency of the subcarrier oscillator, thus increasing the rate of appearance of high amplitude portions in the final wave. The information oscillator rate is frequency modulated and the final signal is amplitude modulated; this combination is thus referred to as FM-AM. In the final example the subcarrier oscillator (SCO) frequency modulates the output oscillator, and so an increase in voltage gives rise to an increase in the rapidity of the appearance of regions of higher than average radio frequency. The amplitude of the signal is unchanging, and this is referred to as FM-FM. There are many other systems of modulation, both simple and composite, but the foregoing examples should suffice to clarify a few methods. If the voltage in any of these examples periodically changes between the values V_1 and V_2, for example, at a frequency corresponding to that of a generated sound, the wave will change between the two conditions of a particular example in a periodic fashion and at a frequency corresponding to that of the sound. An amplitude modulated signal responding to a sound wave of approximately one sixth the frequency of the carrier frequency would yield a wave of approximately the same appearance as the V_1 wave in the FM-AM example.

In one application it was useful to take the SCO frequency to transmit one variable, and the frequency-deviation amplitude as proportional to a second variable. The latter was picked up with a simple peak detector from the demodulated SCO frequency signal and was invariant under orientation or carrier-frequency changes.

It is often convenient to make use of model airplane radio-control equipment in various telemetry experiments. Some of these activate various functions in the plane as a result of playing different audible tones into the transmitter. At the receiver is a small device termed a resonant-reed relay in which various lengths of magnetic metal are placed above an electromagnet fed by the receiver. A particular tone will set only one length of armature into vibration, thus closing only one set of contacts, from among perhaps a half dozen, each of which produces a different action of the plane. Such a system of transmitting different kinds of information at a single radio frequency can be considered as an example of FM-AM modulation.

An example of a circuit employing FM-AM modulation is shown in Fig. 3.2. Here an astable multivibrator cyclically increases and decreases the amplitude of oscillation of the transmitter oscillator at a rate that depends on temperature. The right-hand pair of transistors constitutes the active elements in the multivibrator, one half the period of which is determined by the thermistor. Instead of directly communicating a signal from the multivibrator into the radio-frequency oscillator, a different system is used. The collector resistors in the two halves of the multivibrator are quite different, and thus this part of the circuit alternately draws more and less current. The multivibrator is connected across the power source to the oscillator, and they are both supplied through a common impedance. Thus the supply voltage to the oscillator is cyclically increased and decreased by the switching of the multivibrator to give output amplitude modulation. By making the common impedance a pair of diodes in series as shown, it is possible to stabilize the radio frequency of the oscillator against temperature changes. Thus a receiver can be tuned to this transmitter and can be expected to record a continuous signal for days at a time. Any small residual drifting of either the transmitter or the receiver is compensated for by the automatic frequency control (AFC) built into many receivers. Unlike the blocking oscillators, this transmitter never goes

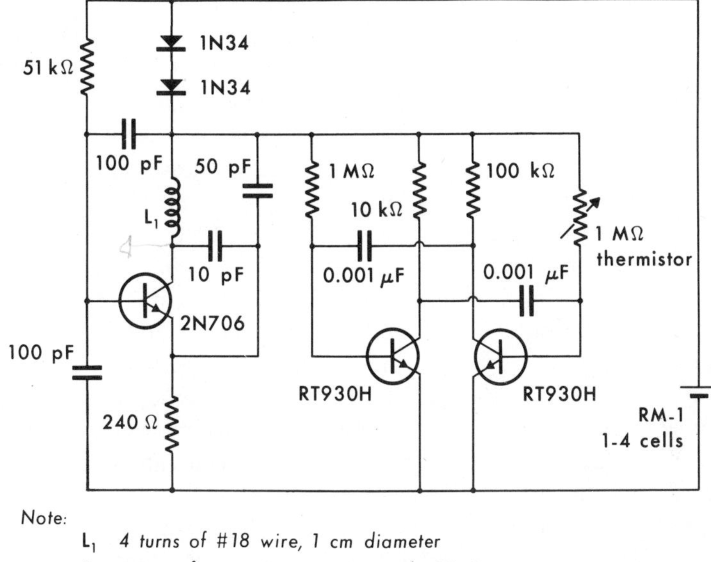

Note:
L_1 *4 turns of #18 wire, 1 cm diameter*
Frequency of operation approximately 88 Mc

Fig. 3.2 A small transmitter with a range up to approximately 100 m, with a sub-carrier frequency being determined by temperature. The diodes compensate the frequency of the output oscillator against major temperature changes so that simple standard receivers will be able to lock to the transmitter frequency in spite of any residual changes, thus ensuring continuity of reception.

off and thus, although the drain on the battery is higher, the automatic frequency-tracking circuit of receivers is able to function effectively. This circuit has proved to be quite valuable, and although small it gives ranges up to about 100 meters from inside an animal. It is the circuit, for example, that McGinnis and Brown (1966) chose for some of their later experiments with lizards. A short length of wire attached to the coil can significantly increase the transmission range, whether inside or outside the animal.

Not only is the transmission radio frequency of this circuit effective for reasons that will be discussed in Chapter 10, but this frequency is chosen to fall in the range of commercial FM stations. Thus it is possible to use readily available inexpensive transistorized pocket FM receivers with this circuit. We might note that not only is the carrier-frequency amplitude modulated, but due to the effect of the changing voltage on the capacitances of the oscillator transistor, there are also substantial amounts of frequency modulation; that is, this system is both FM-AM and FM-FM.

There is another type of modulation that warrants comment, and that is what might be called pulse-ratio modulation or duty-cycle modulation. This can be explained by an example. Suppose that a multivibrator is set to periodically generate pulses. The rate of arrival of such pulses then is used to carry information about some variable, such as pressure, which is controlling the signal to the multivibrator. It is possible that changes in other parameters such as temperature or voltage may also affect the frequency of the multivibrator, thus introducing small errors into the signal. Changes in these unwanted variables will affect both parts of the multivibrator's cycle similarly, whereas the variable of interest can be set to affect only the spacing between pulses. Taking the ratio of the duration of the pulse to the duration of the "off" time (which can be done automatically) gives a signal that is independent of superfluous changes, yet displays the true information. Somewhat higher accuracy in such a transmitter is achieved by effectively transmitting a standard signal periodically, which interval could otherwise be used to transmit information about a second variable of interest.

A specific useful circuit for accomplishing this type of modulation is shown in Fig. 3.3. At the center is an astable multivibrator; the two halves of its period are controlled by the voltage to which the base resistors

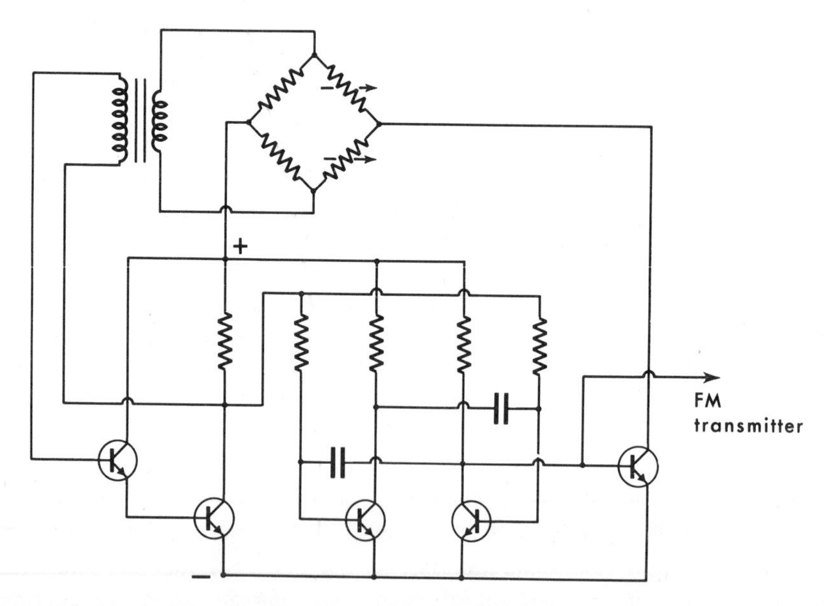

Fig. 3.3 A subcarrier oscillator that provides a variable fraction of "on" time in response to resistance changes. Good stability results, and the circuit is useful when working with the strain gauges that might be used in pressure or related measurements. Specific circuit values for some of the components can be found in Fig. 15.16.

are returned. The ac voltage from the multivibrator is applied through the transistor on the right to the bridge circuit at the top. This Wheatstone bridge gives out no ac voltage through the transformer if the resistance complies with the usual balance condition. The signal from the transformer goes through a special two-stage amplifier, which is discussed later, to control the base voltages in the multivibrator. If one of the resistances in the bridge is changed slightly, then an ac unbalance voltage results, and as this is generated by the multivibrator, it is in synchronism with switching therein. Unbalance of the bridge results in the base of one transistor of the multivibrator being a bit high during one part of the cycle and the base of the other transistor being a bit low for the other part of the cycle. Thus one half of the cycle is made longer and the other half shorter by an unbalance in the bridge. Therefore this circuit can constitute a subcarrier oscillator to modulate a transmitter, and has been found to be quite effective when used, for example, with one or a pair of strain gauges in the bridge circuit. Detailed possible values of the components in this circuit are given in a related circuit discussed in Fig. 15.16.

In the preceding circuit information is contained in the fraction of the time during which it remains in one particular state (duty cycle), and this is unaffected by parameters that lengthen or shorten both parts of the cycle equally. Thus the transmitter can be quite stable. Other variables can similarly be transmitted; for example, an electronic switch cycling between two levels will be forced to take on a duty cycle varying with applied voltage if placed in an amplifier feedback path so that the average output matches the input. But any such pulsatile signal can become altered on reception if the cyclic changes take place in times that are comparable to the time that it takes the intermediate frequency (IF) filter incorporated into most receivers to respond. It is not necessarily true that the overall system is quite accurate if the subcarrier oscillator period is comparable with the IF strip rise-time of the receiver. However, much of the basic accuracy of such a method can be restored in the receiver by setting the demodulator threshold at half the recent average signal amplitude to correct for finite rise-time or bandwidth. This can be done with a simple rectifier and capacitor shunted by a voltage divider having enough resistance to respond only after several cycles.

There are other methods of modulation that are more complex and reliable, but for the average scientist doing biological research the forms mentioned above suffice. In some satellite experiments greater subtlety is evoked not only for increased reliability, but for consistency between various kinds of signals emanating to the central receivers.

If a receiver can be sharply tuned to the transmission frequency, while passing only a very narrow adjacent band of frequencies, much noise can be rejected. However, the act of modulating a sinusoidal signal introduces nearby frequencies, which must be passed by the receiver if the information is to be recovered. This generation of extra frequencies or side bands by any modulation procedure is discussed in the next section.

2. BLOCKING OR SQUEGGING OSCILLATORS

When a long time constant (large resistance and capacitance) is incorporated into an oscillator circuit, then the normal action of the oscillator to stabilize its amplitude of oscillation can be accompanied by periodic changes in amplitude. In some cases oscillations actually totally cease and then, after a time, resume. As indicated in Chapter 1, these circuits are not only of historical interest in this field but also retain present utility in many cases because of their simplicity, low current drain, and general effectiveness.

An example of this type of circuit is shown in Fig. 3.4. If the resistance from collector to base is too high, then oscillations will not start spontaneously, whereas, if resistance is too low, then continuous oscillations result. Over an intermediate range of resistances, the blocking frequency is a continuous function of resistance. In the present case the variable resistance of a thermistor is used to provide temperature transmission. This particular transmitter was designed to transmit two variables simultaneously while drawing a minimum current, and thus a movable diaphragm and core were incorporated in the proximity of the coil to provide changes in the radio frequency in response to pressure changes. In the top line of Fig. 3.4 are given the circuit details that yielded the desired time sequence of events indicated on the second line. A resistor was incorporated into the emitter connection, where it somewhat limited variations in frequency during each period of oscillation. When the components were connected up on a small "breadboard," the performance was as indicated on the second line, but when they were all regrouped and reconnected to form a tiny finished transmitter it was observed that there were some changes in the timing of the various waves. It was found difficult to alter parameters to give the precise design requirements, and so the configuration shown at the bottom of Fig. 3.4 was employed. A tunnel diode was incorporated to limit the current during the half cycle when current flows from the coil into the base of the transistor. For most purposes the circuit shown in part A would have been quite adequate, but in the present case precise matching to a relatively complex receiver system (Fig. 11.8) was required. The circuit of Fig. 1.7 is closely related, except that it is set for a lower repetition rate and omits the pressure-change signal.

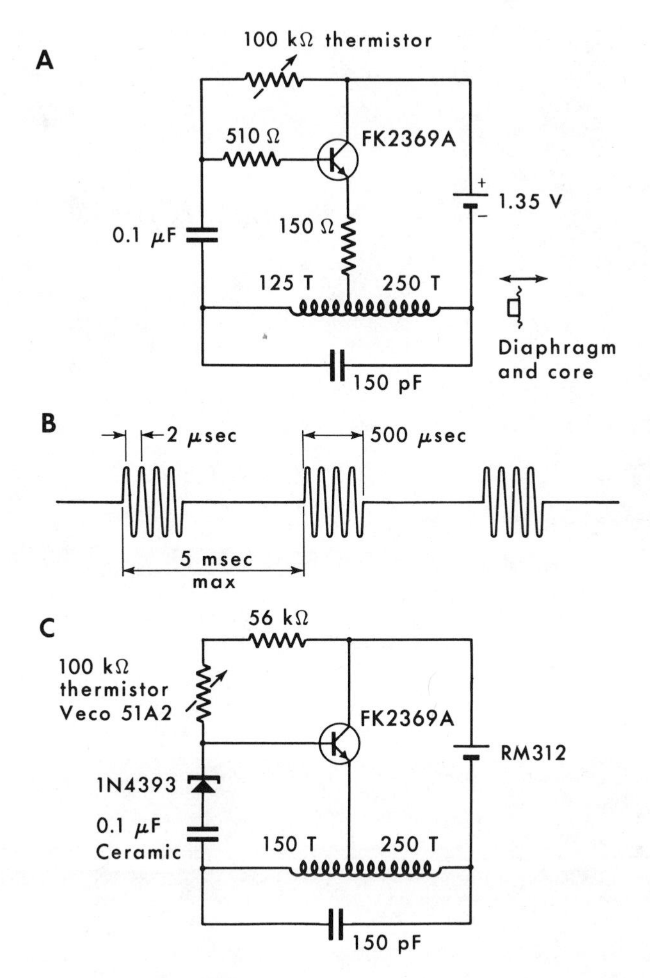

Fig. 3.4 A squegging or blocking oscillator designed to give the particular wave form shown at the center. The circuit was improved by the insertion of a tunnel diode, as shown at the bottom.

The general type of wave form produced by these circuits is shown at the top of Fig. 3.5 where the pattern of bursts is seen. Below that, with a higher oscilloscope sweep speed, is shown in more detail the envelope of the radio-frequency oscillations in a single burst. In this case oscillations are taking place during only approximately one tenth of the time, and thus the drain upon the battery is only approximately one tenth of that of an oscillator running continuously at a similar amplitude. In spite

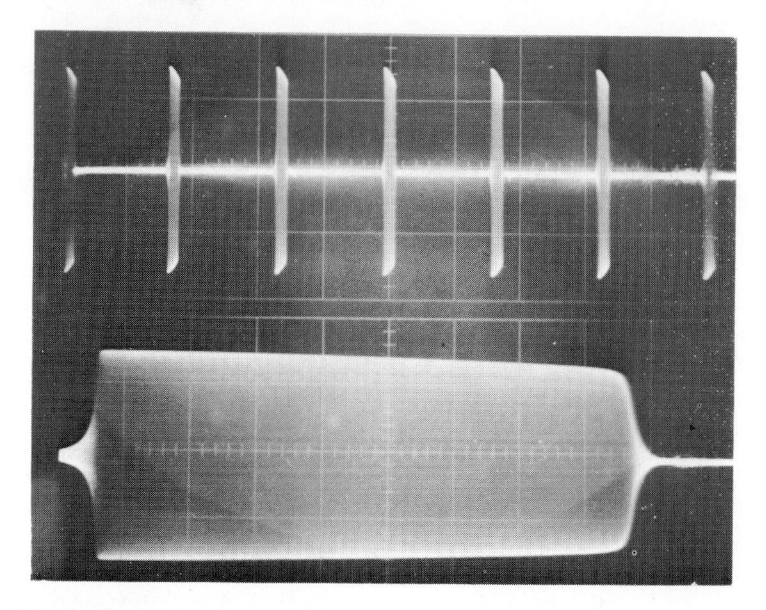

Fig. 3.5 At the top a pattern of bursts from a squegging oscillator; below, the envelope of many cycles of radio frequency in a single burst.

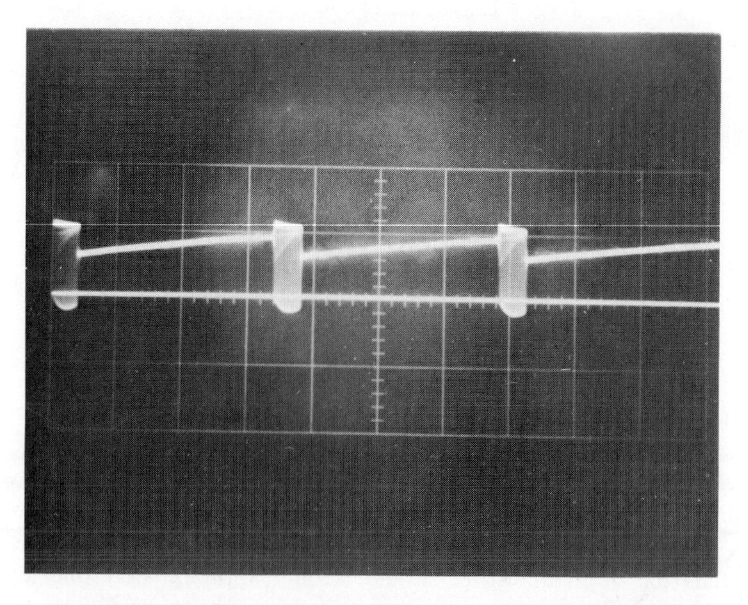

Fig. 3.6 Voltage from base to emitter in a squegging oscillator, with the line for zero volts also shown. The thickened vertical portions occur during oscillation.

of the low average current, the transistor is handling a relatively large current during oscillation, and thus maintains a high beta. These wave forms are those that appear if a small coil of wire is placed near the transmitter and connected into the oscilloscope, or if the oscilloscope is connected directly across the tuned circuit. Figure 3.6 shows the wave form of voltage of the transistor base with respect to emitter; the line of zero voltage is also displayed. A gradually increasing voltage prevails in the interval between oscillations. When the loop-gain of the system reaches unity, oscillations start, and they persist until there is a drop in average voltage to less than the starting value. An exact description of all of these effects requires highly nonlinear equations, but it is useful to note the implications that for certain values of base bias it would be expected that oscillations, once started, would continue steadily and unceasingly or if stopped would remain so. The possibility of switching on and off by remote signals a transmitter imbedded within the body of an animal is mentioned in Chapter 17. Useful comments on cyclically interrupted oscillating circuits are to be found in writings on superregenerative receivers (Chow, 1957, and Whitehead, 1950).

The term "squegging" is often applied to oscillations that run through a number of cycles and then stop periodically, whereas the term "blocking oscillator" is often used to apply to a circuit in which a single vigorous cycle periodically takes place. The oscillations in the present circuits, however, might be said to periodically block. The repetition rate of such an action does depend upon the supply voltage, and thus mercury cells with their constant voltage are often preferred as power sources. In that case in some circuits, for example, a drift in temperature indication of less than 0.1°C per month can be expected. In some cases it may be desired to transmit a variable signal in response to changes in voltage. Such an arrangement is shown in Fig. 3.7, in which a change in input voltage causes a change in the frequency at which periodic bursts are produced. In this particular case a field-effect transistor is combined with a second transistor to give a very high input resistance to the entire transmitter circuit. (The applications of these circuits to the measurement of such parameters as pH are mentioned in Chapter 7.) By applying the variable input voltage to the base of the oscillator transistor, but not to the collector, the amplitude of the useful signal is maintained relatively constant. The average current drain of this circuit, since it depends on the rate of the generation of impulses, is a function of the input voltage being monitored. Similarly, in the temperature-transmitting circuits, a long life between uses is assured by storing the transmitter in a refrigerator, not only to slow the self-discharge processes of certain batteries, but also to slow impulse production.

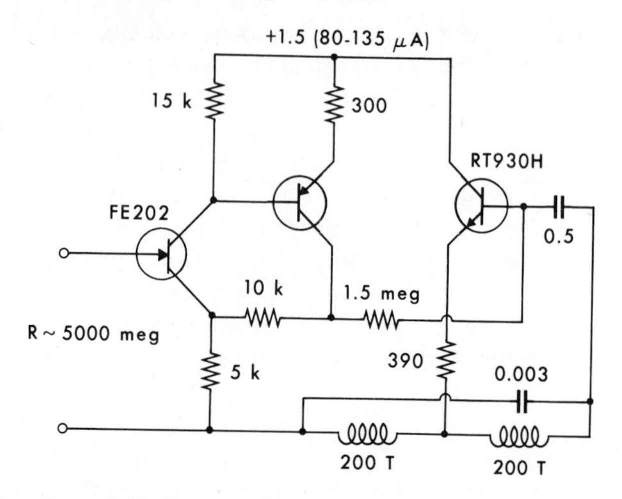

Fig. 3.7 By using a field-effect transistor it is possible to obtain a very high input resistance to a circuit which can then deliver a variable voltage to control the frequency of a squegging oscillator. The center transistor is associated with the input circuit, and its properties are relatively unimportant. Changes in input voltage cause pulse-frequency changes with little change in amplitude.

These same circuits can be used to transmit other variables when there is a transducer for converting the change in the parameter of interest into a change in resistance. Thus a cadmium sulfide cell placed in the position of the thermistor of Fig. 1.7 or Fig. 3.4 will allow the transmission of changes in light intensity. If needed, an increased range of intensities can be monitored if a fixed resistance is placed in series with the cell. Similarly, humidity (e.g., in the vicinity of an animal or in an environmental study) can be sensed by certain commercial variable resistance elements which change their properties in response to this variable. These elements can be connected directly in the position shown for the thermistor.

These circuits display a blocking rate that is dependent on their surroundings. As was early noted (Mackay and Jacobson, 1957), motion of a ferromagnetic core in the vicinity of the coil will produce small changes in blocking frequency. The magnitude of some of these interactions is suggested by Fig. 3.8, in which the effect of the motion of various materials is indicated. By using these effects it is possible to make blocking frequency responsive to pressure changes. These circuits can serve as a subcarrier oscillator of low frequency, followed by a radio-frequency stage, if it is desired to avoid such things as the effect with movement of a variable antenna loading on the transmission of information.

Similarly, placement of a ferrite-cored receiving antenna against one of the temperature transmitters can lead to an erroneous indication. To avoid uncertainty this type of antenna should not be placed nearer to the transmitter than a distance approximately equal to the diameter of the transmitting antenna coil.

The conductivity of the surrounding medium also affects these blocking circuits. Thus a temperature transmitter should be calibrated in physiological saline for ultimate precision, if the final transmitter is to be implanted in or ingested by an animal. The blocking frequency of the crystal-controlled oscillator circuit of Fig. 2.18 is quite sensitive to the motion of pieces of metal in its vicinity. This can either be useful or troublesome, depending upon the application.

In these blocking circuits changes in temperature somewhat alter the transistor beta and also the voltage drop corresponding to the base-emitter

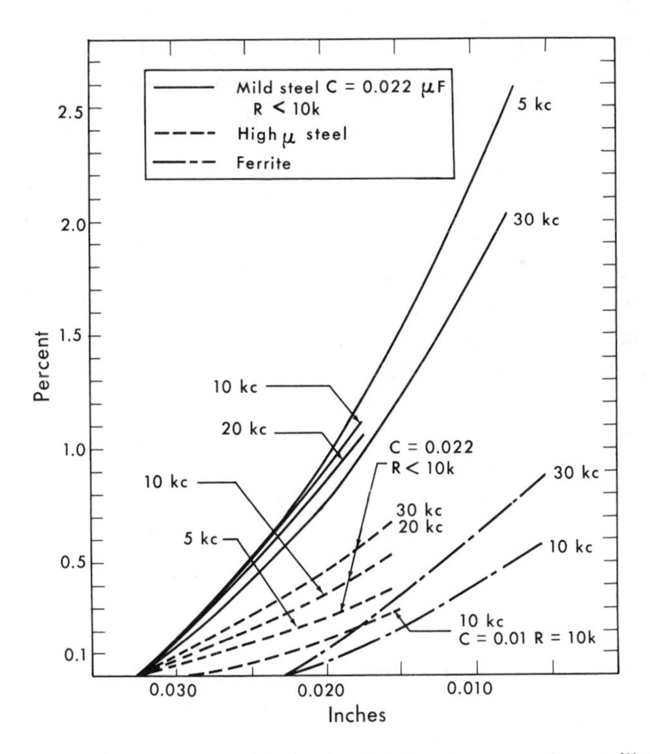

Fig. 3.8 Motion of various materials in the vicinity of a squegging oscillator causes a change in pulse frequency that can be useful or troublesome, depending on the application. In a representative circuit which has a resistance from collector to base of R and a base capacitor C these frequencies and sensitivities were observed.

diode. Thus temperature affects the condition at which the loop power gain becomes unity and the instant when oscillations resume.

Many modifications of the basic circuit are possible. Thus the resistance and capacitance that determine the basic time interval between bursts can be shifted from the base connection to one of the others, and the battery can be moved from the collector connection to one of the others. The capacitor which is connected across the coil to form a tuned circuit, can be reduced in size or left out entirely and functioning will continue, somewhat modified.

Generally, if the base resistor is below about 40 kΩ, oscillations will become continuous, and if above a few megohms, oscillations will not start. A change in turns ratio, a low base capacitance, or a high battery voltage (several cells in series) also tend to result in steady continuous oscillations. If blocking oscillations are taking place, an increase in base resistance will tend to decrease their rate of occurrence, but will leave the duration of each burst relatively unchanged. The result is a decrease in average power drawn from the battery and radiated. An increase in base-capacitor capacitance tends to increase the interpulse interval and also the duration of each pulse in the same way, thus leaving the average power drain approximately unchanged. By changing both R and C other adjustments such as changed pulse width with unchanged spacing can be obtained.

It should be noted that pulses of less than about 300-μsec duration begin to be filtered out by the intermediate-frequency filter's characteristic response time in many ordinary superheterodyne AM receivers. Also, blocking rates greater than a few kilocycles are not passed by the audio amplifiers of many inexpensive receivers. These last two factors are mentioned because they can alter the apparent performance of one of these transmitters when used in connection with typical receivers.

These circuits have an interesting and useful property because the transistor can function not only as an oscillator but can rectify due to diode action at the junctions. Thus it is found that if an oscillator is connected to a coil in the vicinity of one of these transmitters, the radio energy induced into the tuned circuit of the transmitter will be rectified and cause a reverse current to flow through the battery. Thus with no further components in the transmitter it is possible to recharge the battery, if it is a storage battery, even while it is in an animal. Several groups have commented on this aspect of these circuits (England and Pasamanick, 1961 and Nagumo et al., 1962). Nagumo and his associates have replaced the battery by a capacitor that is charged by an impulse to later momentarily serve as a power source. In this case, after the usual interpulse interval, a burst of oscillation is generated to return an impulse to the receiver. For each input pulse there is returned a pulse with a delay that carries information, thus leading to a passive system of transmission called an "echo capsule."

The circuits under discussion are generally used with a rate of burst generation in the range of a few cycles per second to a few thousand cycles per second. Two transmitters working simultaneously can be set on widely different radio frequencies so that they will not interfere with each other. Two transmitters sending on the same radio frequency can have their signals individually identified by transmitting at a different blocking rate. The human ear is quite good at separating out unrelated signals, and if a five-pulse-per-second transmitter and a one-pulse-per-second transmitter are placed beside a receiver there will be little trouble in timing either one with a stop watch. Nor will there be any confusion if a 100 Hz unit is also placed beside the receiver. Suitable electronic circuits can be set to separately count the signals from these three transmitters. Direct recording of the signal on magnetic tape for later decoding is also very convenient.

An AM receiver placed in the vicinity of one of these transmitters will pick up its signal over a broad range of tuning, in general; that is, a typical transmitter of this type generates a signal over a broad range of radio frequencies. This can be either a disadvantage or an advantage. Because the energy in the signal is spread over a wide range of frequencies, a receiver tuned to a narrow band of frequencies does not record a large amount of energy. The passband of a receiver could be adjusted to match the frequency range of transmission but such a receiver would also accept noise over a large range. For this reason, and because they are normally small low-powered units, the transmitters under discussion usually have a transmission distance limited to a few feet. If several of these transmitters are simultaneously active from within a single animal, or from nearby animals, then their signals may overlap in a receiver. For the reasons mentioned above, however, this may not cause problems. Because a transmitter of this type can be picked up over a range of settings of the tuning dial of the receiver, any drift in either the receiver or the transmitter need not result in a loss of a signal that is to be recorded continuously. Thus this effect actually aids unattended operation, especially where automatic recording from a single transmitter is being done. In the case of field work, in which the transmitter is adjusted in the laboratory and then taken to a distant place for use, there may be unexpected interference from commercial radio stations, especially at night when distant stations suddenly clutter a receiver. In that case the receiver can be retuned to an empty region, while still picking up the signal from one of these transmitters.

Three factors largely contribute to this spread in frequencies generated (places at which a receiver can be tuned to pick up the signal). Merely turning on and off a sine wave introduces some extra frequencies. These are the frequencies that combine with the basic signal to give periods of cancellation, and they are referred to as the added Fourier components. The reason that these added frequencies that are caused by modulation are called "side bands" can be visualized by thinking of a graph of frequency content extending from right to left, with these new frequencies appearing beside the basic one. A second reason for the extra frequencies is that interrupting the signal from an oscillator is not completely equivalent to turning the oscillator on and off periodically. If an oscillator is switched on a moment after it has been switched off, then the voltage of the next radio-frequency cycle may not happen to rise just at the time when the voltage would have been rising if the oscillator had not been turned off at the end of the previous burst of cycles. This lack of phase coherence between successive bursts introduces further side bands or extra frequencies.

A third reason for the spread in transmitted frequency is that the steadily changing conditions in the transistor during the pulse (which finally results in pulse termination) cause a sweep in frequency during each pulse. This can be seen on an oscilloscope by injecting both the pulse and the output of a steady oscillator. The time in the pulse of zero beat frequency can be seen to shift through the pulse as the oscillator frequency is changed. This sweep in frequency can be either minimized or maximized by the selection of components. Thus the sweep in frequency can be minimized by the use of a high capacitance in the tuned circuit, and a corresponding lower value of inductance to arrive at the desired radio frequency for transmission. The changes in effective transistor capacity that accompany this action are strongly dependent on the transistor type chosen, with some transistors giving a much smaller sweep in frequencies, at a specified frequency in a given circuit.

The crystal-controlled blocking oscillators which have been mentioned cover a much narrower range of frequencies, especially if the crystal is not turned on or off too abruptly. (Generally, turning on is the greater problem.) Usually, if an oscillator is turned on and off 100 times/sec, for example, then extra frequencies will be spaced around the basic frequency at intervals of 100 Hz. Thus an oscillator that is not turned on and off frequently will, when on, have all of its frequencies concentrated in a very narrow range. A representative AM receiver will accept or pass a range of frequencies of a few kHz for any given setting of the tuning dial, which means it would pass several side bands from one of these blocking transmitters for any given setting.

Figures 3.9, 3.10, and 3.11 give specific examples of these matters. Use was made of a radio-frequency analyzer which automatically graphs the

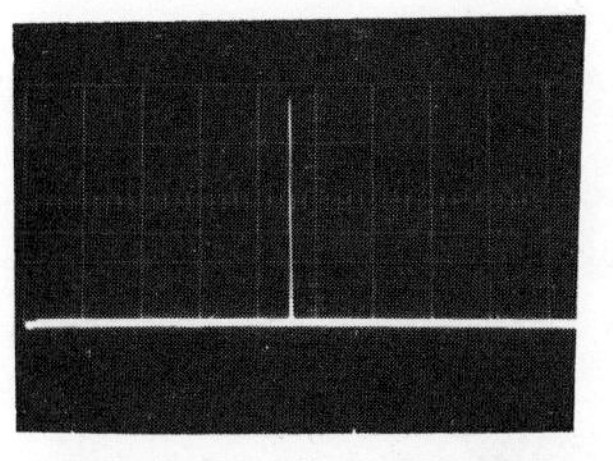

Continuously running oscillator
Continuously connected
Frequency: 1.001 Mc
Display: 2 kc/division
0.2-kc resolution

1010 kc 990 kc

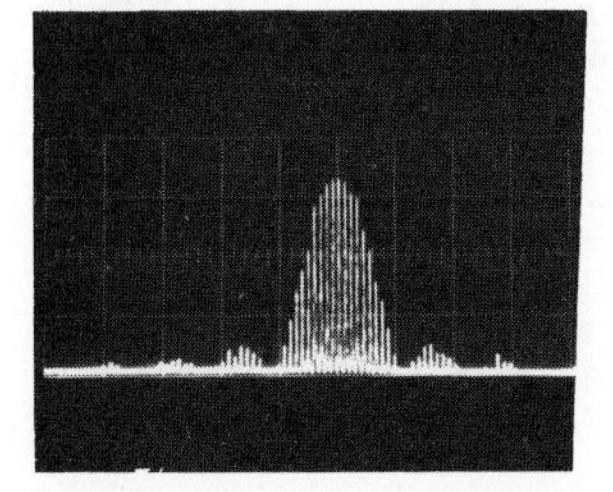

Continuously running
Periodically connected

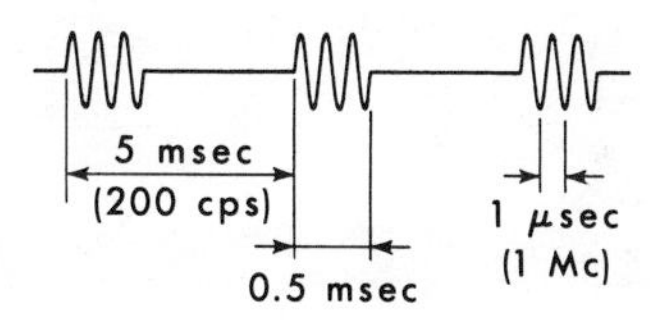

Fig. 3.9 Frequency analysis of various signals. A steady oscillation contains a single frequency as shown at the top, whereas a periodically interrupted oscillation contains the distribution of frequencies below.

relative magnitude at each frequency of an injected complex signal. Frequency is depicted as increasing from right to left, with degree of spreading being indicated. A continuously active sinusoidal oscillator concentrates all its energy at a single frequency. Thus, at the top of Fig. 3.9 a standard oscillator yields a single vertical line suitably positioned along the horizontal axis. Periodic interruption of the signal spreads the energy into a series of side bands whose relative intensities are predictable. Thus shown on the lower line of Fig. 3.9 is the result of periodically connecting a continuously running oscillator to the analyzing circuit. The wave shape delivered to the analyzer is as indicated and consists of pulses arriving at a rate of 200 Hz. Thus the side bands are spaced at regular intervals of 200 Hz. Their relative intensities (the envelope of the pattern) has a simple mathematical description which is mentioned in most textbooks on this subject.

Figure 3.10 depicts more complicated patterns arising from three different methods of generating approximately the wave shape shown at the bottom right of the figure. The top three sections depict the output from the corresponding self-blocking or squegging oscillator, in which there is some sweep in oscillator frequency during the pulse. It can be seen, for example, that a FK709 transistor in a given circuit spreads the output energy over a con-

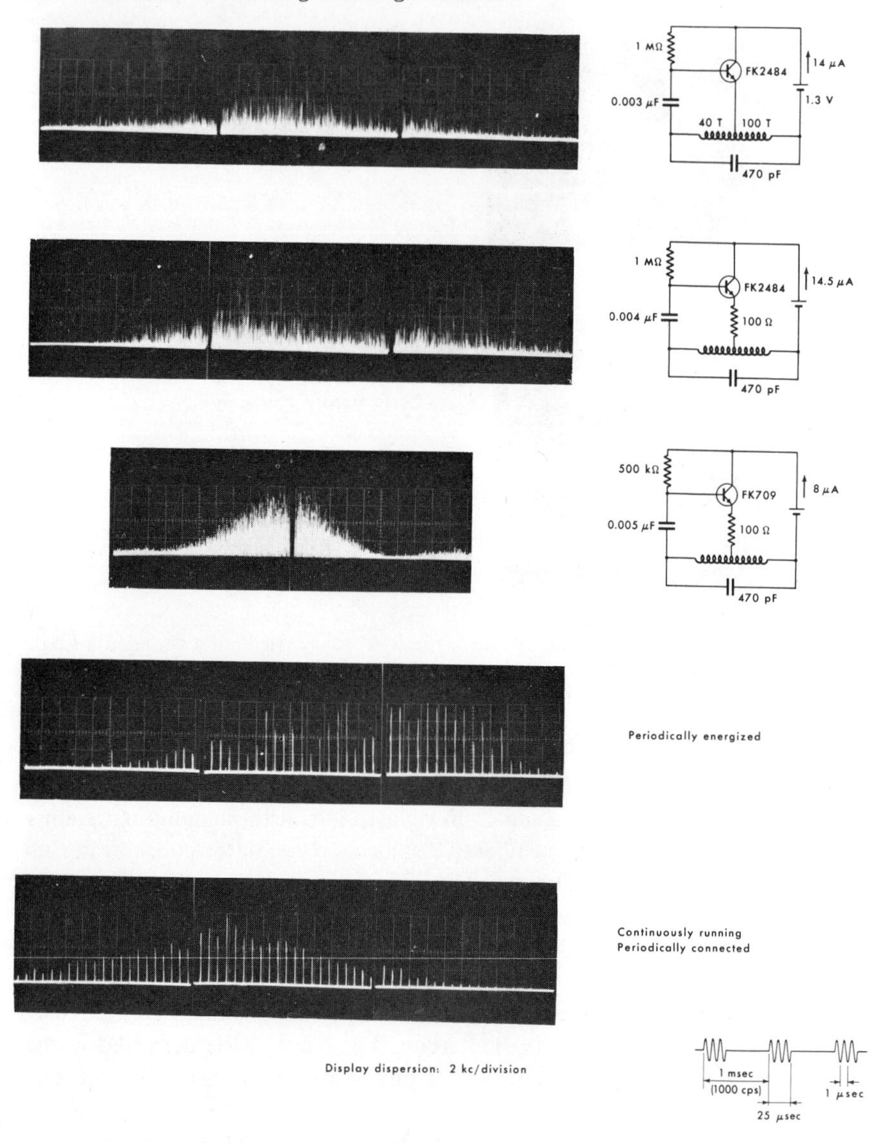

Fig. 3.10 The approximate indicated wave form can be achieved in a number of ways and there results several somewhat different frequency distributions.

siderably narrower range than does an FK2484. On the bottom two lines of the figure there are traces for comparison from a continuously running oscillator that is periodically connected to the output circuit, and from an oscillator that is periodically energized. The periodically energized oscillator generates all identical pulses, while the continuously running oscillator maintains radio frequency phase coherence in the overall wave train. At the top of Fig. 3.11 further properties of one of the previous circuits are depicted. Changing the base capacitor changes both pulse width and separation. Following that are given two other examples of this type of analysis, the first a comparison of two different pulse widths and the second the effect of an increase in display spread to bring out the expected fine structure. The details of these patterns change slightly with time when studying any real circuit and there are switching transients, but the generalities of the patterns are unchanging.

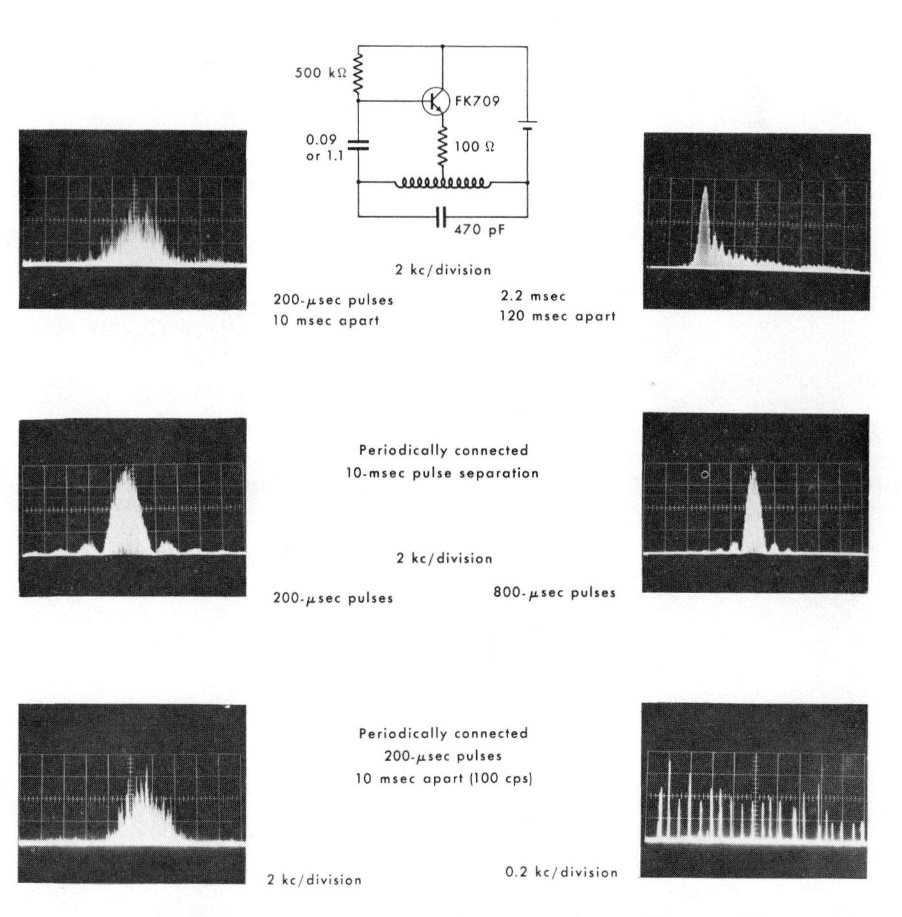

Fig. 3.11 The frequency spread produced by a squegging oscillator under two conditions of operation, as compared with a normal oscillator periodically connected to the analyzer as indicated.

Long pulses with many radio cycles tend to tune sharply and a great spacing between pulses tends to bring the extra frequencies in close to the central one. Thus the various tracking transmitters used with wild animals give a signal that tunes quite sharply in a receiver. Circuits of the type shown in Fig. 2.11 tend to tune sharply, whereas the self-interrupting or blocking circuits tend to spread their signal over a range of the tuning dial unless crystal control is involved. Without such control, choice of transistor and use of an emitter resistor affect the constancy of frequencies during each pulse, and hence the overall spread in transmitted frequency. Other forms of modulation produce similar effects, but this form of pulse generation has been covered because of its convenience and apparent simplicity.

4

Plastics and Other Materials

1. BODY FLUID PERMEABILITY

Plastics slowly pass moisture and thus special precautions must be taken with transmitters that are to remain within the body for more than a few weeks. To many investigators, this is the most surprising engineering problem to arise in connection with these telemetry transmitters. Even a circuit that is imbedded in a solid block of epoxy resin may begin to drift in its indication a few weeks after an experiment is started. Removal of the transmitter in order to dry out may slowly restore the original signal. A simple experiment can be used to demonstrate one aspect of the problem. If a one megohm glass-enclosed thermistor is coated with a layer of epoxy resin and then, after curing, allowed to soak in isotonic saline, after about a week it will be found that the resistance between the extending leads has dropped to perhaps 0.8 MΩ. Low-impedance circuits are less affected by these problems than those containing high resistances, but long-term experiments usually require the high-resistance circuits to minimize current drain on the battery. Some passive transmitters are able to work at rather low impedance levels for extended periods, because power for them is induced from outside, but they can still be caused to drift by the high dielectric constant of any water vapor which may diffuse into the capsule and perhaps condense.

Thus an important aspect of the construction of these transmitters is to maintain an effective moisture barrier. An attempt was made to evaluate various "potting" and coating materials by soaking them for extended periods in salt water. The change in weight in these materials upon soaking is not a perfect criterion of excellence in this connection, but in a general way ranks some of the materials as to expectation. Several epoxy resin imbedding materials are summarized as to increase in weight at two weeks, four weeks, eight weeks, and two years in Table 4.1. The epoxies in each case were cured for seven days at 55° to 60°C. Cylinders of the plastic approximately 1.5-cm diameter and 1-cm length were soaked in 0.9 percent saline at 39°C. In some cases the relative concentration of resin and hardener was also altered for parallel tests. Some of these epoxies, when "filled" with inert materials, are improved by 50 percent. It is said that some fluorinated resins under development should show water pickup of only about half that of other epoxy resins. In any case, for long-term observations none of these present materials is adequate alone.

Table 4.1 Water Absorption of Covering Materials

			Percent Change in Weight			
Resin	Hardener	Resin/ Hardener	Two Weeks	Four Weeks	Eight Weeks	Two Years
Epon 826 (1)	EM 308 (2)	1/1	2.3%	3.6%	4.5%	—
Epon 826	EM 308	2/1	1.0%	1.5%	1.9%	3.5%
Epon 826	Versamid 140 (3)	1/1	1.9%	3.3%	4.4%	—
Epon 826	Versamid 140	2/1	1.0%	1.6%	2.2%	4.1%
DER 332	EM 308	2/1	1.0%	1.3%	1.8%	3.2%
DER 332	Versamid	1/1	2.8%	4.6%	5.8%	—
DER 332	Versamid 140	2/1	0.9%	1.5%	2.0%	4.1%
EC-2216B (5)	EC-2216 A	2/3	2.0%	2.8%	3.2%	3.5%
Armstrong A-2	Activator A	100/4	0.9%	1.5%	2.0%	4.4%
Armstrong C-3	Activator E	100/12	3.9%	6.0%	7.6%	—
Armstrong C-3	Activator A	100/8	1.2%	1.7%	2.3%	5.2%
Epocast 202 (7)	D-40 (7)	100/15	2.0%	2.8%	3.3%	3.9%
Epocast 202	TETA (7)	10/1	0.9%	1.3%	1.7%	3.1%
Epocast 202	TETA	5/1	2.4%	3.5%	4.6%	—
Epocast 202	DTA (8)	10/1	1.0%	1.6%	2.1%	3.1%
Epocast 202	DTA	5/1	5.9%	8.8%	11.4%	—
Epocast 202	AEP (8)	10/1	1.0%	1.7%	1.9%	3.2%
Epocast 202	AEP	5/1	2.8%	3.9%	5.1%	—
DER 332	AEP	100/15	1.0%	1.6%	2.3%	6.0%
DER 332	DTA	10/1	2.1%	3.2%	4.3%	9.6%
DER 332	Versamid 140	3/1	0.9%	1.3%	1.7%	3.1%
Silastic 382 (10)			<0.1%	<0.1%	<0.1%	−0.2%
Type A adhesive (10)			0.2%	0.2%	0.2%	−0.1%

All epoxy resins were cured seven days at 55 to 60°C. Cylinders of the plastic approximately 1 1/2-cm diameter, 1-cm length, were soaked in 0.9% saline at 39° C (Numbers in parentheses refer to the manufacturer's name.)

1. Shell Chemical Co.
2. Thiokol Chemical Corp.
3. General Mills
4. Dow Chemical Co.

Table 4.1 Water Absorption of Covering Materials (*continued*)

Resin	Hardener	Resin/ Hardener	Percent Change in Weight Two Weeks	Four Weeks	Eight Weeks	Two Years
Acrylic			0.9%	1.3%	1.6%	—
Teflon (9)			0	0	0	—
Glass			0	0	0	—
Araldite 6005 (11)	131H (11) 136 (11)	10/3/3	0.5%	0.7%	0.9%	1.5%
Araldite 6005	131H	2/1	0.5%	0.7%	0.9%	1.5%
Versalon 1112 (3)			0.4%	0.5%	Decomposed	—
Versalon 1175 (3)			0.7%	0.9%	0.9%	1.0%
Epocast 202	Versamid 140	7/3	1.3%	1.9%	2.5%	4.0%
Epocast 202	Versamid 115	7/3	1.1%	1.5%	1.9%	2.7%
Epon 826	Versamid 140	7/3	1.0%	1.4%	1.8%	3.7%
Epon 826	Versamid 115	7/3	1.2%	1.6%	2.0%	2.9%
Araldite 6005	Versamid 140	7/3	1.0%	1.5%	1.9%	3.8%
Araldite 502	TETA	100/8	0.9%	1.2%	1.5%	3.4%
Epocast 202	TETA	100/8	0.9%	1.2%	1.6%	3.2%
Kadon dental resin (12)			1.3%	1.6%	—	1.7%
Dental acrylic (13)			1.1%	1.4%	—	1.4%
Paraffin embedding compound (14)			0	0	—	−0.3%
Epoxy coated with Silastic 382			2.3%	3.3%	4.1%	11%
Epoxy (uncoated control)			4.9%	6.4%	7.9%	13%
Epoxy coated with paraffin			—	0	—	0.7%
Epoxy (uncoated control)			—	1.2%	—	2.5%
Silastic 732 (10)			0%	0%	0.1%	0.1%

5. 3–M Company
6. Armstrong Products
7. Furane Plastics
8. E. V. Roberts and Assoc.
9. E. I. DuPont Co.
10. Dow Corning Co.
11. Ciba Products
12. L. D. Caulk Company
13. Hygienic Dental Mfg. Co.
14. Will Scientific, Inc.

A material of special importance is silicone rubber, because it is non-irritating to body tissues, as will be mentioned in a later section. Table 4.1 shows that soaking a sample of such a material produces a negligible change in weight. (More precisely, type 382 first loses a little weight, then back to original, then a small loss, while type A loses a little for a few days, and then proceeds as shown.) It is also true that soaking a piece of such a material for a year in salt water produces negligible changes in its electrical properties. Coating a transmitter with a layer of such an elastomer, however, only slightly slows the passage of moisture into the underlying material.

Thus, towards the bottom of the table, the change in weight of a rather strongly absorbing epoxy, both with and without a coating of Silastic 382, is summarized. Uptake is slowed by a factor of perhaps two, and the effect is less when a slower uptake base material is used.

We have investigated a number of substances that might be employed as coating materials. Among those materials that are readily formed, wax has proven best to date. Table 4.1 indicates the result of testing one of the more useful low-uptake epoxies when coated with paraffin. It is seen that even after two years the effect is small. Although it makes a good moisture barrier and melts at a convenient temperature, paraffin does not adhere especially well, and it is brittle and cracks. Thus we have made considerable use of a mixture of approximately half paraffin and half beeswax. Other waxes, such as those used for mounting strain gauges, seem to perform well but have not been tested in detail. The temperature of melted wax can be detrimental to batteries, and thus any coating process should not be prolonged.

In some cases a flexible covering is desired. We have not found materials such as Parafilm satisfactory in this demanding sense. Robert Goodman has developed some flexible paraffin derivatives that should be helpful in these cases. Rubbing silicone grease into rubber somewhat diminishes its permeability and results in an adequate combination for some purposes.

Teflon, and to a greater extent Kel-F, impede the passage of moisture if sufficiently thick, but their well-known permeability to certain gas molecules is indicative of what is to be expected from a thin covering.

Materials such as glass and certain ceramics make excellent transmitter coverings but are somewhat more difficult to form. A circuit placed in a glass test tube will be isolated from the effects of moisture if the tube can be satisfactorily closed off. Certain passive transmitters containing only coils and capacitors can be subjected to quite high temperatures, such as are involved in melting shut a piece of glass tubing. It is possible to obtain a lower melting temperature frit with a proper thermal coefficient of expansion to allow the sealing on of a coverplate in many cases. But the temperature involved is still too high to be tolerated by many electronic components.

Both glass and ceramic can be coated chemically from a liquid by metals, in some cases requiring baking to form a firm bond. The various DuPont silver preparations are often especially useful in this application. Once the container has been so treated at its open edge, a coverplate of metal, or of glass with a metalized ring, can be soldered into place without much warming of components that have been inserted into the case. This process is possible using low melting temperature solders or eutectic mixtures. If the transmitting antenna is a coil of wire, it can be important that the antenna coil be remote from this seal, or arranged perpendicular to it, so that the conducting ring does not act like a shorted turn on the secondary winding of a transformer to cancel out much of the radio field being generated.

The majority of a transmitter, including the parts sensitive to moisture, can be attached to the "header" of a standard transistor or integrated circuit mount, after which the cover can may be added and hermetically sealed. In some cases the components to be left outside would be the battery and the antenna which is to radiate the signal. The antenna can be wound on a ferrite tube into which is placed the battery and can.

With regard to metals that might be left exposed in the body, there are several stainless steels which are relatively inert, and thus suitable for electrodes, diaphragms, and the like. But with mixed metals (solders included), electrolytic action can lead to rapid decomposition, as can stray currents from associated circuits. A rubber covering can be indicated.

Imbedding a transmitter in a block of plastic gives it ruggedness and rigidity. Ruggedness is valuable when dealing with animal subjects, and rigidity is advantageous in minimizing frequency shifts caused by movement or vibration. Whether special precautions against moisture need be taken depends on the application. External transmitters generally pose no problem. A transmitter to be ingested by most animals (other than the ruminants) will probably reappear before drift caused by moisture can prove troublesome. On the other hand, most implanted transmitters will be expected to operate for a length of time following the recovery from surgery, so that this factor can become important. Also, a little drift may not prove troublesome in the transmission of a relative quantity such as the electrocardiogram, but if absolute values such as temperature or blood pressure are to be monitored extreme care must be taken.

2. ELECTRIC LOSS AND DIELECTRIC CONSTANT

Losses in the antenna coil result in a significantly reduced output signal for a given input power. These losses can be caused by the circulation of currents through fine wire having a high resistance. Perhaps a less obvious factor, however, is the loss introduced by plastics placed in the vicinity of such a coil. An alternating electric field heats certain plastics more than others and this energy must be supplied by the field's source. If an antenna coil is resonated with a suitable capacitor, then rather large circulating currents can exist, with an accompanying strong field in the vicinity of the transmitter. Losses that lower the Q of the system significantly damp these currents, with a direct reduction in the observable or useful field. The factor of "dielectric loss" is tabulated for most plastics by their maunfacturers, and this is a good indication of expected results. The effects are not drastic at relatively low frequencies such as 500 kHz but become extremely important at frequencies from about 30 MHz up. If transmitters that will work at frequencies near 100 MHz are being constructed, then it is extremely desirable that various components be checked on a "*Q* meter," of which several versions are commercially available.

In this same connection we might note that at these higher frequencies different capacitors perform quite differently. Thus some ceramic capacitors, which are quite acceptable at lower frequencies, display great losses at the higher frequencies. These must then be rejected for such purposes as tuning an output stage to resonance. The dielectrics in other ceramic capacitors may function quite acceptably. Similarly, the loss characteristic of various ferromagnetic materials such as ferrites and powdered-iron transformer cores must be selected to match the frequency range of operation if it is above about a dozen megahertz.

A specific example may be helpful. The following observations were made at 50 MHz on a useful antenna coil having a few turns. An initial *Q* of 150 was reduced to 100 by placing the transmitter circuit, without battery, inside of the coil. Although this would correspondingly reduce the useful current within the coil, the saving in volume can justify it in many cases. If the circuit is laid out in such a way that the wires form large low-impedance loops that couple into the main magnetic field, the *Q* can be further reduced. An initial coil *Q* of 180 was reduced to 150 by imbedding the coil in Epocast 202. This condition gradually changed after the first day, and 24 hours later the *Q* was up to 160. On the other hand, a similar coil potted in Epocast 212 had its *Q* reduced from 180 to 150, and the value was the same 24 hours later. Other plastics show similar effects, in some cases with greater losses being present. The *Q* of a coil can sometimes be calculated by multiplying 2 π times the frequency times the inductance and dividing by the resistance, the resistance being somewhat higher than the dc value at the frequency of interest. The best procedure is to make such measurements directly with a *Q* meter.

Introducing a dielectric other than air (or, more precisely, vacuum) into a capacitor raises the capacitance of the capacitor by a factor called the dielectric constant or relative permittivity. Similarly, imbedding a circuit in a block of plastic can alter some of its properties because of the appreciably greater than unity dielectric constant shown by many plastics. The dielectric constant of various imbedding materials is also usually tabulated by manufacturers. One common effect of potting a circuit in epoxy resin is to reduce the radio transmission frequency. This effect is small where an oscillator communicates its signal through a separate output stage, but can be significant in the simpler transmitters. There is no change in frequency when a crystal-controlled oscillator is imbedded in plastic, but the changing frequency of the output stage in some cases can cause sufficient detuning to significantly reduce the strength of the transmitted signal.

When a plastic is mixed with its catalyst to effect initial hardening, there remain some gradual changes over the next few days. It is often observed that a newly formed plastic solid will display 80 percent of its changed dielectric property in the first day, with full setting requiring perhaps a week. The imbedding process can shift the frequency of a high-frequency transmitter by 20 percent; that is, a 100 MHz oscillator can drop in frequency by 20 MHz upon potting. This can be enough to take the transmitter out of the range of its intended receiver. Lower frequency transmitters are less affected, if for no other reason than that the changes in capacitance are "swamped out" by the relatively large fixed capacitance of the tuning capacitor.

The electric field around a wire is strongest close to the wire, and thus the placement of any plastic there has the greatest effect. Similarly, the removal of undesirable plastic materials from the immediate vicinity of wires by a distance of one wire diameter significantly improves the performance of the system. From an electrical standpoint the best plastic seems to be polystyrene. It is possible to obtain this material dissolved in a solvent under the name "Q-dope." This can be painted on a coil before potting in order to move lossy high dielectric plastics out a way into the lower field region where they cause less loss and less lower the frequency. Thus, in the first example in this section, a 0.5-mm layer of Q-dope placed over the coil caused the Q of the coil to remain at 180, whether the coil was then potted in Epocast 202 or 212. The use of a Teflon-covered wire in the antenna is a helpful alternative.

3. TISSUE AND BLOOD REACTIONS

A foreign substance introduced into the body can cause a variety of reactions. Materials introduced into the blood stream result in clotting. This can be prevented by introducing the drug heparin into the blood stream, but then a small scratch or other wound can prove dangerous to the subject. Heparin solution can be contained in a thin silicone-rubber bag from which it will gradually diffuse to maintain a useful level at a particular site, but this procedure also is limited. Thus a nonthrombogenic surface is particularly desirable for certain studies. The inside of the vascular system is not considered as a foreign surface by the blood, and thus certain transducers are preferably fixed outside the vessels. All present plastics cause clotting.

A great advance in rendering surfaces nonthrombogenic came as a result of Gott's finding that plastics could be permanently heparinized. This was accomplished by coating plastic surfaces with graphite dispersion DAG 35 and then successively treating the graphite surfaces with Zephiran, a quaternary ammonium compound, and then with heparin. It has been shown that the quaternary ammonium compound is adsorbed strongly to the graphite surface and that the heparin will then react with the quaternary ammonium groups, thus binding the heparin to the surface (Gott et al., 1963 and 1964). Another method, applicable to certain other materials, is to attach quaternary ammonium groups to plastic surfaces by covalent bonds and then use these sites for ionic bonding to heparin. This may be through direct chemical reaction with the plastic surfaces to establish groups that can be quaternized and then heparinized, or through grafting of polymer chains (such as vinyl pyridine) which can be quaternized (Leininger et al., 1965). Another approach is given by Salzman et al. (1967).

It is known that the charge on an electrode placed in blood either helps or hinders clotting, a negative charge preventing clotting. However, a surface charge is not the same as an electric field, and the above investigations demonstrated that the zeta potential (derived from streaming potential measurements) and clotting time do not exactly go together, as was thought. Similarly, the heparinized surfaces are quite wettable compared with unmodified plastic surfaces. Thus the expectations of Lampert's rule that highly wettable surfaces clot fast do not hold completely. Heparin does not prevent the aggregation of platelets, so there can still be problems with these surfaces. Heparinized surfaces often are slightly more hemolytic than before treatment, but there seems not to be enough red-cell destruction with their use to cause anemia, even over extended periods.

A Dacron fabric mesh will display fibrin buildup with a tissue surface like a vessel forming and being presented to the blood; such a surface minimizes clotting, but buildup is not controlled. DeBakey and others have noted that Teflon is less satisfactory.

Some statically stable situations present clotting problems when there is flow, especially with considerable shear or vortex formation.

Foreign bodies placed in contact with tissues also cause reactions. Silicone rubber is excellent in this respect, with Teflon next. Polypropylene is generally somewhat better than polyethylene. The basic material of polyvinyl chloride is brittle, and plasticizers that are added leach out into the body. Some epoxies can prove toxic to tissues. Nylon degrades in the body, as does polyurethane. In all cases a pure polymer without additives is necessary for reliable results. Orlon loses less strength in the body than does Teflon, and nylon loses approximately 80 percent of its tensile strength in three years. Various enzymes attack various plastics, and breakdown products can appear in the urine of a subject a few weeks after placement (see Falb et al., 1966).

When working with implants the conservative procedure is to coat them with a layer of medical-grade silicone rubber. It is essential that the medical product be used, as others can prove quite irritating. The catalyst for some is quite toxic. We have had acceptable results with the direct placement of a wax coating, though some beeswax has caused irritation. Over wax a silicone rubber coating is probably generally preferable, and an outer layer of silicone at least prevents the exposure of unsterile regions in the wax due to handling during placement.

Various workers have successfully used canulas of Teflon, Silastic, polyvinyl chloride, polyethylene, and nylon. A summary of these various matters has been prepared by Falb et al. (1966), and the reader is referred to this source for further references.

One final comment upon the preparation of these materials before placement: an implant telemetry transmitter must be both clean and sterile before implantation. Any foreign material can lead to a tissue reaction if the cleaning is inadequate. With regard to sterilization, many units cannot be heat sterilized and thus must be subjected to chemical procedure. Plastics exposed to chemical vapors sometimes absorb enough gas to slowly evolve burning fumes over a period of several days. Thus we sometimes prefer to soak our transmitters in a Zepharin solution, which is discussed in Chapter 7.

4. FORMING AND REPAIR

In forming transmitters, we often make use of the fact that it is very difficult to glue to Teflon. Thus a small circuit to be potted into a block of epoxy is laid into a trough milled into a block of Teflon. The epoxy is poured in around the circuit, with any necessary leads extending up through the surface, and hardening is allowed to take place. The circuit can then be removed with confidence that sticking to the form will not take place. For ease of removal, however, it is important that the initial milling of the form be done with sides one angular degree off 90 degrees. This ensures that any irregularities will not impede smooth motion, and that once the circuit has "broken loose," it will come free easily.

Similarly, some of the very fine wire coils are wound on a Teflon form with a removable end piece. When working with wire sizes such as number 42, or smaller, extreme care must be taken not to nick the wire with a fingernail. The wire is pulled off the end of a spool, rather than unwound, and a suitable epoxy is smeared on with a wooden stick as random winding progresses. Here again a one degree taper is imposed on the Teflon form so the self-supporting coil will slide free readily after hardening.

In preparing the epoxy, it is necessary to mix in a proper fraction of hardener. Every effort is made in this process to stir in as few air bubbles as possible. However, we do spin the plastic in a simple centrifuge (an 8-in.

arm at 7000 rpm) and then somewhat evacuate a covering bell jar to minimize voids. The plastic is best prepared immediately before needed, although it can be stored in a freezer for a while. It is then poured into a warmed (40°C) mold and cured for two to three hours at approximately this same temperature. If a battery is not present in the circuit then a somewhat higher curing temperature such as 55°C can be used, and the curing time correspondingly reduced to perhaps an hour. The final transmitter can then be coated with a wax water barrier and/or silicone rubber.

It is extremely frustrating to discover a malfunction in a circuit that has been imbedded into a solid block of plastic. Such a unit, however, is not absolutely irreparable. A knife blade can be welded to the end of a soldering iron to give what might be called a hot knife. With this it is possible to cut into the potting material to repair mistakes. This knife is also convenient for removing plastic insulation from fine wires.

If a sample shape is to be duplicated, silicone rubber can be cast around it to yield a convenient mold into which various materials can be cast. If it proves necessary to glue to Teflon then etched films are available also.

There is another material commercially available which can be quite useful for special purposes. A polyolefin shrinkable tubing is available whose outside has been radiated with an electron beam. When subjected to a stream of hot air, the outside shrinks without melting at around 135°C, whereas the inside melts. The material shrinks 17 to 42 percent of its supplied diameter with less than a 10 percent longitudinal shrink. Such tubes have been slid over sound transducing cylinders to provide a watertight covering for certain aquatic studies, and shrink tubing has been used successfully as the entire covering for certain transmitters.

We have sometimes found it useful to machine the mountings for transmitter components, or covers for transmitters, from solid pieces of various commercial plastics. Rigidity of component mounting does minimize superposition of movement artifacts on the outgoing signal. In some cases a loose assembly is given rigidity with a thin coating of epoxy and then potted in the previously mentioned wax mixture. Access to a changeable battery is easiest if the transmitter case has a screw cover, but in certain minimum volume configurations it has been acceptable to melt away a wax covering. For resistance to abrasion, and possible chewing, milled plastic covers have been used, although in certain cases a specially configured case of stainless steel has been employed.

5

Pressure Sensing and Transmission

1. TRANSDUCERS

The sensor which does the initial detection of the variable of interest must transduce the parameter of interest into a suitable electrical form. Many of the comments that will be made about pressure transducers are applicable to transducers of other variables as well. Extensive discussions have been made of input transducers (Lion, 1959, and Neubert, 1963).

A mechanical change in position can be converted into a useful electrical signal in a number of ways. If an object such as a diaphragm is acted upon by a springy restoring force, then changes in pressure will cause displacements that can be sensed and transmitted; basically, this is the nature of a pressure transducer. One class of transducers converts some of the work done in deforming the transducer directly into electrical energy. Examples of these self-generating systems are the piezoelectric devices such as quartz crystals or pieces of barium titanate ceramic, and systems in which a magnet is forced to move in the vicinity of a coil, thereby inducing voltages into the winding. If the signal from the piezoelectric element was applied to a circuit of infinite resistance, and the element itself showed no leakage, then a signal once set up would remain indefinitely. In the moving coil case the circuit would have to display zero resistance for a steady response to a change in position. Since infinite or zero-resistance circuits can only be approximated, these units tend to give out signals in response to changes of applied conditions, with the signal dying away when position is fixed. Otherwise, energy would be coming from such a transducer continuously, even when no work was being done on the transducer, and this is impossible by the laws of conservation of energy. (Ways of obtaining a dc response from such transducers will, however, be described in Chapter 12.) It might be noted that the pyroelectric effect seen in some piezoelectric materials, and which can cause erroneous signals on occasion, is a generation of an electric signal accompanying a change in temperature, rather than a steady response to a given temperature, for similar reasons.

Other transducers take little energy from the system under study, but rather use changes in the system to modulate conditions in an observable fashion in an electrical circuit. The main parameters which might be modified in electric circuits are resistance, capacitance, inductance, and mutual inductance. Changes in one of these, when incorporated in a bridge circuit, an oscillator circuit, a voltage-sensing circuit, and the like, can yield a signal that modulates a transmitter. Each has a range of suitability in actual technology.

Changes in resistance can be used to control an oscillator signal in many ways, including those in Figs. 3.3, 3.2, and 1.7. Displacements are often converted into resistance changes by use of what is called a strain gauge. If a wire is pulled it becomes longer and thinner, and its resistance will increase, thus measuring the displacement of one end with respect to the other. Wire strain gauges often have resistance measured in tens of ohms, with some metal foil ones being higher. Semiconductor materials can be used to perform a similar function, and display resistances in the range of a few hundred to a few thousand ohms. Although the indication of one of the low-resistance gauges is little affected by moisture, such a low resistance does not lend itself to convenient circuit arrangements, and higher resistance gauges give out large changes for a given input, which is advantageous from a standpoint of necessary amplification. However, the ultimate stability of the overall system need not be greater because of a correspondingly greater temperature sensitivity. Two or four of these gauges are often used in a configuration where forces will produce deformations of opposite bridge arms in the same direction to augment each other, while temperature changes will cause resistance changes that cancel in the overall bridge. Strain gauges are very useful in many applications, especially when very small deflections of a rather stiff system are involved. However, it is also true that some workers choose this transducer out of force of habit in situations where it is not especially appropriate.

If direct current is used to energize a bridge circuit, an unbalance will produce a steady voltage that changes sign for a change in unbalance direction. Its use can lead to stability problems in the following circuits. This problem can be eliminated by "chopping" the output, that is, rapidly turning it on and off so that the following amplifiers can be ac coupled. Power is conserved, however, if instead the energizing signal to the bridge is periodically turned off. If an ac signal is applied to a bridge, the output will also be an ac signal, which changes phase as the sign of the unbalance changes and which becomes greater as the degree of unbalance increases in either direction.

One can make a pressure-sensitive electrical paint which changes its resistance in response to force changes by stirring powdered carbon into various kinds of liquid rubber. These transducers are small and can undergo a large percentage change in dimension. Although perpetually promising, at the time of this writing they still have not proven very reliable. A rubber tube filled with mercury, or even tap water, will show a variable resistance between electrodes at the ends, as the tube is stretched. These are also relatives of the strain gauge.

One can also go through intermediate steps to convert a displacement into a resistance change. Thus the motion in question can move a vane between a light of constant intensity and a cadmium sulphide cell. The resulting changes in resistance accompanying changes in light intensity can constitute a very sensitive detector of motion, even in the angstrom range of displacement.

There is a special class of what can be considered variable resistance transducers in which only two values are possible. For example, consider a limp diaphragm actuating a switch whose signal is then integrated to control the current to an electromagnet which restores the diaphragm towards its original position (drives the switch to its opposite state). There will be steady oscillation as in a buzzer, but a pressure on the diaphragm causes a change in percentage "on time" (duty cycle) of the magnet current, whose average is thus a continuous measure of pressure. As an example from another realm, the controllers of electric blankets usually follow temperature in a similar fashion, a heating element causing a bimetal strip to cyclically open and close a set of contacts.

The force can instead deform a capacitor in such a way as to give changes in capacitance in an electrical circuit. Thus a force can be applied to push a pair of capacitor plates close together or cause them to interleave further or force a dielectric between them. Little work is required to move such a system, although forces are involved which are necessarily greater than zero. In one pressure-sensitive capacitor unit we studied, anodized aluminum strips were interleaved, with changes in perpendicular force causing a change in capacitance. The configuration was unduly sensitive to moisture, and might have been more effective if enclosed in a bag of silicone fluid. As might be expected, some of the pressure-sensitive paints just mentioned show a large change in capacitance for an applied force, and in some cases this is a more stable effect than the changing resistance. At high frequencies, in a typical case, changes in capacitance tend to be more important (and effective) in altering a circuit than changes in inductance.

Forces can bring about a change in self inductance in a number of ways. If a ferromagnetic material is brought towards a coil, inductance increases. In order that the action may be sharp, it is desirable that the material be a nonconductor; hence the use of powdered iron or ferrite. If a conducting material is instead brought toward the coil, a decrease in inductance results. The properties of cores vary with temperature, magnetic history, and so on, and these factors can introduce irregularities into precise readings. At high frequencies, bringing a conducting core near a coil may give rise to changes not so much having to do with inductance changes, but instead associated with an increase in the stray capacitance between the turns of the coil. Another way of changing the self-inductance of a coil is to deform the coil, thus changing the relative position of the various turns, or the area of the entire configuration. Thus the signal from the loop antenna around the neck of an animal may contain a small amount of frequency modulation in synchronism with each cycle of the animal's movement because of small changes in the geometry of the situation. This may be either desirable or undesirable, and rigidity of mounting is sometimes important. At low frequencies a variable-inductance transducer can be quite effective and relatively unaffected by changes in moisture or movement of the leads, compared with the sensitivity to disturbance of a capacitive transducer.

The mutual inductance between two coils can be altered by either moving the two coils with respect to each other or by moving in the proximity of the two coils pieces of either magnetic or shielding (conducting) materials. Perhaps the most stable motion transducer consists of a suitably wound pair of coils on quartz forms which move with respect to each other. For most of the purposes we shall be discussing, however, this type of transducer is more cumbersome than needed. The familiar differential transformer is a variable mutual-inductance bridge, which can be miniaturized for telemetry use; an aluminum core is sometimes preferable to ferrite if the frequency is relatively high.

Other transducers work by changing the parameters of an amplifying system. Thus there are radio tubes with elements that move in response to applied forces. The negative resistance properties of a tunnel diode are changed in response to applied forces, as are transistor characteristics.

The above basic parameters of an electrical circuit can be combined to give oscillatory or time-delay effects, and some useful transducers have been built which directly affect such composite systems. For example, motion of the wall of a resonant cavity produces a predictable change in frequency, and such a system makes a convenient pressure transducer at high frequencies (see Fig. 13.8). It is known that a vibrating quartz crystal is somewhat altered in frequency by applied forces, and these can be especially effective if a given force is concentrated in a small area to give a high pressure region. Likewise, delay lines, either electrical or mechanical, can be distorted to yield a changing time delay in response to forces or displacements. These last considerations are included for the sake of generality, since it is largely the simpler circuit parameters which are used to modulate telemetry transmitters.

2. PRESSURE CHANGES

A number of groups have been interested in muscular activity along the gastrointestinal tract as monitored by pressure changes therein. Although absolute pressure in the gut could be of interest in assessing certain osmotic effects in connection with active transport studies, the studies to date have actually been concerned with changes of pressure representing activity and lasting only a few minutes. Thus slow baseline drift due to effects either in the transmitter or receiver were relatively unimportant in this application. With the availability of the junction transistor it became possible to build small low-powered active transmitters, and following this in quick succession three separate groups published articles reporting the successful transmission of pressure information from within human subjects (Mackay and Jacobson, 1957; Farrar et al., 1957; and VonArdenne and Sprung, 1958).

There are other parts of the body where transmission of pressure changes could also be informative. Thus the earlier attempts at passive transmission of pressure information were motivated by a desire to know of pressure changes within the human bladder during micturition. Similarly, the mechanism of the functioning of an intrauterine contraceptive might be elucidated by noting the activity pattern in the uterus with and without the device being present. To study bruxing (the grinding of teeth) during sleep does not require transmission of absolute pressure, which periodically returns to atmospheric, but merely the changes accompanying clenching. Sound transmission is also of this character, and will be dealt with in a separate section. In some cases the indication from a transmitter that drifts very slowly can be corrected each day by supplementary observation, if the body cavity in question is relatively exposed; the continuous measurement of the absolute pressure in the eye is of this character, and will be discussed further in Chapter 13.

The first chapter gives a description of one of these endoradiosondes, along with a sample tracing from a drug study. For the purposes of this section, it was felt that the early designs performed adequately (Mackay, 1959) and thus little development has been carried forward in these laboratories, other than the introduction of newer component types. A few special circuits, however, have been studied, and some have instructive

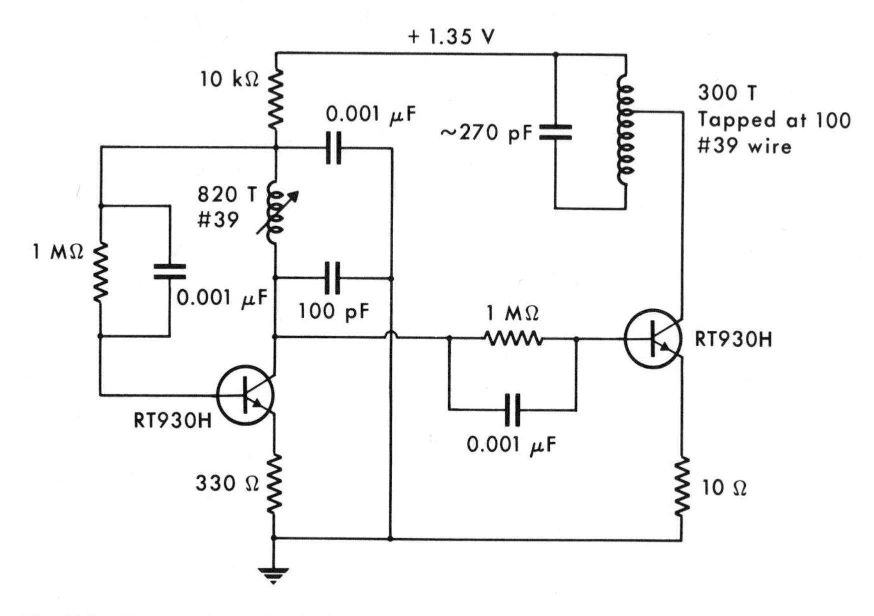

Fig. 5.1 A transmitter that is frequency modulated by inductance changes and has a separate, relatively powerful output stage. Tapping down into the antenna coil as shown increases the useful signal. The output transistor is biased off most of the time (Class C), so that pulses of current flow in the collector only when the voltage there is low, thus wasting little power in the transistor itself.

properties. Thus Fig. 5.1 shows a transmitter circuit which was built to give a strong signal from within extremely obese subjects. It employs a Colpitts oscillator stage which is frequency modulated by a ferrite core moving near a coil. The signal from this is communicated to an output stage for delivering a strong signal. The output stage also minimizes the effect of ambient conditions on the frequency of the basic oscillator. The output stage is so-called "class C," in that current flows to the tuned antenna circuit in short impulses, the transistor carrying significant current only when the collector voltage is low, thus dissipating little power there. Tapping down on the output coil results in an increased useful signal (see Appendix 1).

A performance only slightly inferior to that of the previous circuit can, however, be achieved with the simpler configuration of Fig. 5.2. The resistor from collector to base is small enough to prevent blocking oscillations, and the other resistor limits power drain without wasting too much power. The construction of such a unit is depicted in Fig. 5.3 where various parts are shown. From right to left along the top line are shown the two plastic rings that act like an embroidery hoop to hold the plastic diaphragm, then there is a resistor, a capacitor, a transistor, and the other capacitor. On the next line is shown one of the tapped coils, which is self-supporting because the wire was smeared with epoxy while winding on a Teflon form. Next is shown the transistor affixed in the center of the coil, then the resistors and capacitors are placed, and at the left is the ferrite core, which is merely a thin slab cut from a ferrite rod. From left to right along the bottom appears the completed diaphragm and core assembly, the coil and circuit assembly with battery clamp affixed, the core assembly in place over the coil, the battery, and finally the plastic cap which is pushed down over the back of the entire unit.

The previous circuits were meant to operate at frequencies just below the commercial AM broadcasting band, that is, around 400 kHz. Fig. 5.4 shows a unit for working at 100 MHz. The coil is simply a six-turn spiral or "pancake." The complete diaphragm assembly consists simply of a piece of standard aluminized plastic film. A short helix gives a better field as a transmitting antenna than does a flat spiral, and it is improved by the presence of ferrite. But, in the present case, this less desirable geometry is compensated for by the fact that the higher frequency gives increased effectiveness. The circuit was quite effective but somewhat difficult to stabilize against changes in temperature and other ambient conditions. It can be useful in areas not too crowded with FM stations.

Some representative traces are shown in Fig. 5.5 from transmitters of these sorts. At the top, tracings from the stomach show breathing and occasional large pressure fluctuations that indicate activity of the stomach. On the second line are traces from the small bowel. The first shows a period of maximum activity soon after the transmitter left the stomach, whereas

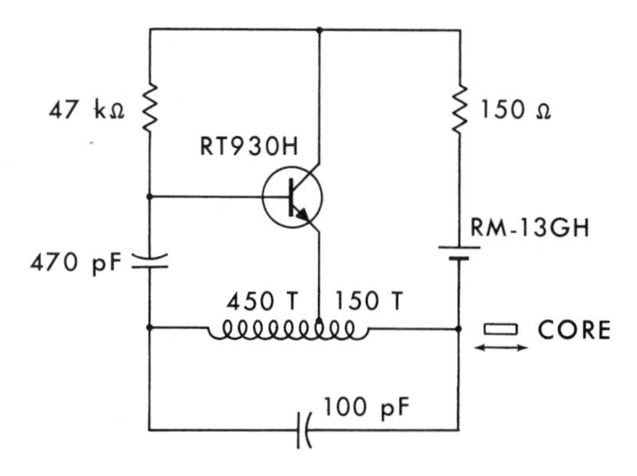

Fig. 5.2 A simple but relatively powerful gastrointestinal-pressure transmitter which has been used to give continuous recordings from within obese subjects.

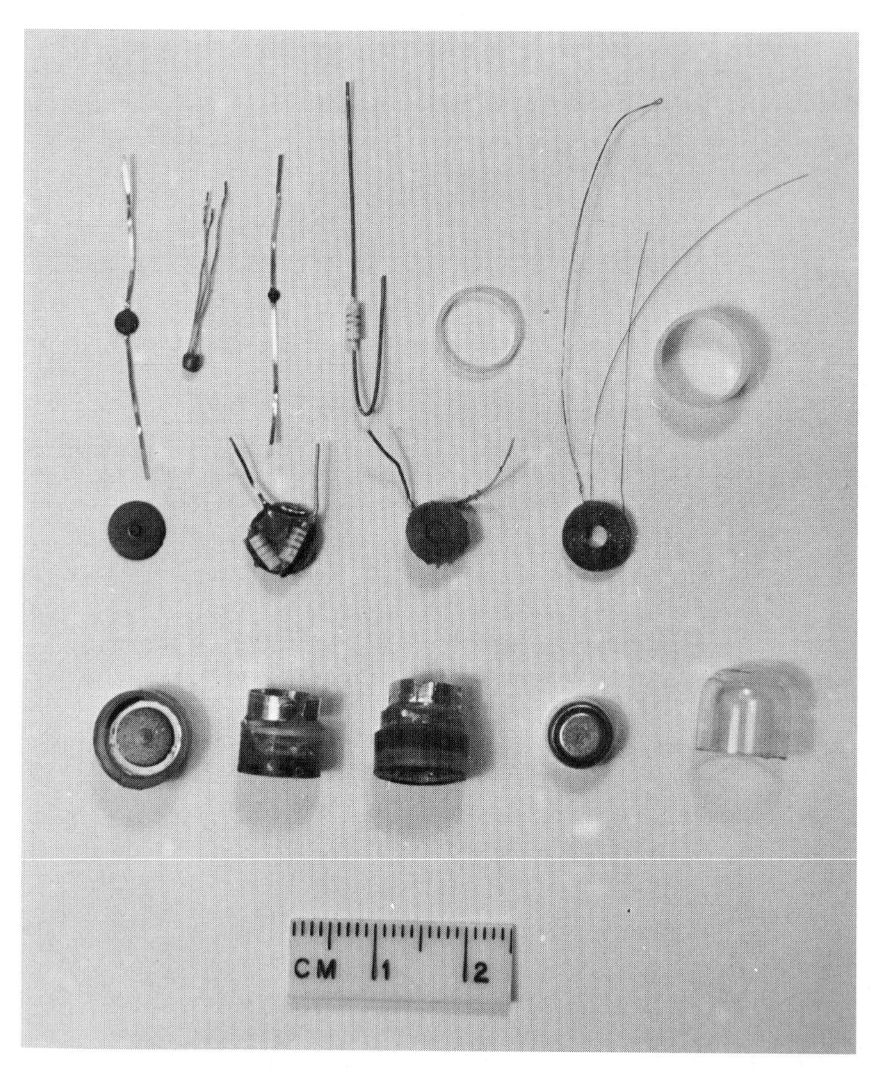

Fig. 5.3 Components and steps in assembly of the gastrointestinal-pressure transmitter. Along the bottom is shown the diaphragm assembly, completed units, and a battery. At the top are representative capacitors and a transistor and resistor. The intermediate components are described in the text.

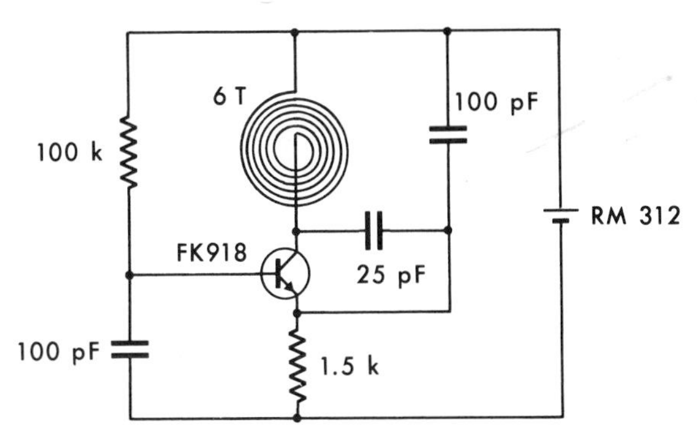

Fig. 5.4 In the frequency range of 100 MHz a simple "pancake" coil can be employed in a Colpitts oscillator circuit. Frequency modulation is produced by motion of an aluminized plastic diaphragm near the coil.

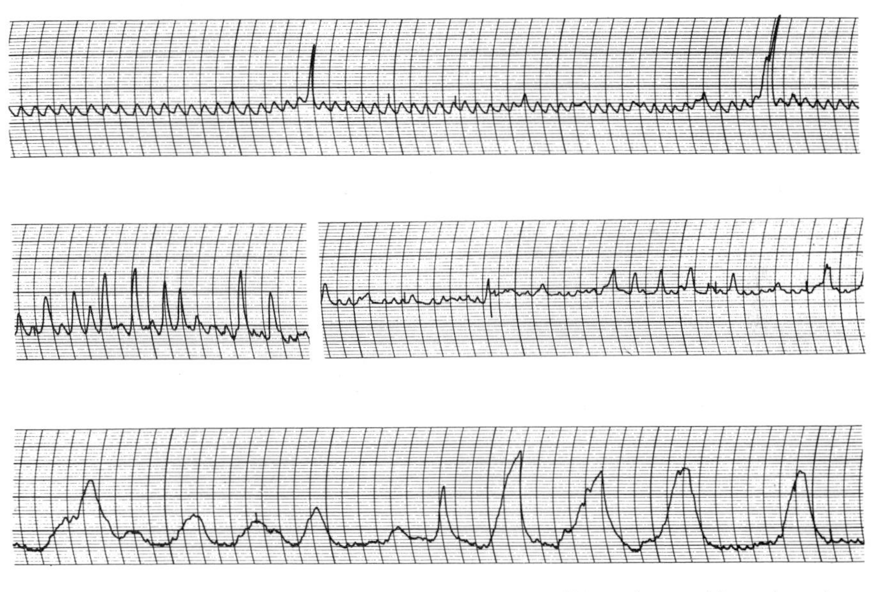

Fig. 5.5 Pressure tracings from a human stomach, small intestine, and large intestine. Chart speed in all cases was one major division in 5 sec. and the deflection sensitivity was one major division for 8 mm of Hg. Here the sensor was enclosed in a rubber balloon 1.2 cm in diameter. At the top are shown breathing and occasional large pressure fluctuations. In the center line at the left is a period of maximum activity soon after the transmitter left the stomach; to the right, intermittent activity several hours later; at the bottom, typical activity in the large bowel.

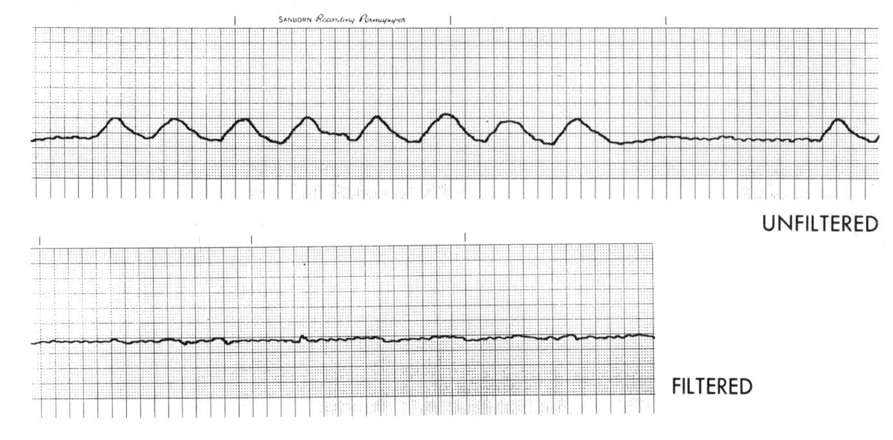

Fig. 5.6 Pressure fluctuations associated with breathing are recorded by an ingested gastrointestinal transmitter. Fluctuations due to the beating of the heart become more noticeable when the breath is held, as at the top, or when the signal is fed through a filter that emphasizes the higher frequencies (below). The tracing below is with no further amplification, but an estimate of heart rate can be made.

the second shows typical intermittent activity several hours later. At the bottom is seen typical activity in the large bowel. In each case the chart speed was 1 mm/sec, with a 6-mm deflection representing a pressure change of 10 mm of mercury.

Figure 5.6 depicts a tracing from the stomach in which the pressure fluctuations due to breathing are easily seen. There are also tiny fluctuations present which are associated with the beating of the heart. For each pulse there is a decrease in pressure and thus the transmitter is responding to a decrease in heart size rather than to a pulsation in some adjacent artery. An inward breath, of course, causes an increase in pressure in the stomach. Electrical filtering can somewhat emphasize the higher frequencies associated with the beating of the heart, and allow heart rate to be noted. For the second line of Fig. 5.6 there has been no increase in amplification but merely a partial rejection of the slow waves due to breathing.

In the small intestine slow waves seem more frequent in connection with pathology. Thus in Fig. 5.7, there are three tracings from the small intestine. The first line shows the pressure changes due to breathing, while the second line shows "Type I" activity. Slow waves, amounting to shifts in baseline from which other waves emerge, are referred to as "Type III" activity and seem increased in disease states. Thus Table 5.1, from a recent study with Wallen and Dengler, shows a significant increase in these slow waves in subjects suffering from peptic ulcer when compared with the situation in normal subjects. Ulcer patients sometimes complain of discomfort when a thread is used to restrain an endoradiosonde at a fixed position, and thus in this study freely moving capsules were used. In each case analysis of the data was accomplished by counting the appropriate waves by direct observation of the recordings, with the analysis being made for the first two hours after the capsule left the stomach. At a chart speed of 5 mm/sec the eye does not tend to notice slow waves that are of 20 to 40 sec duration and

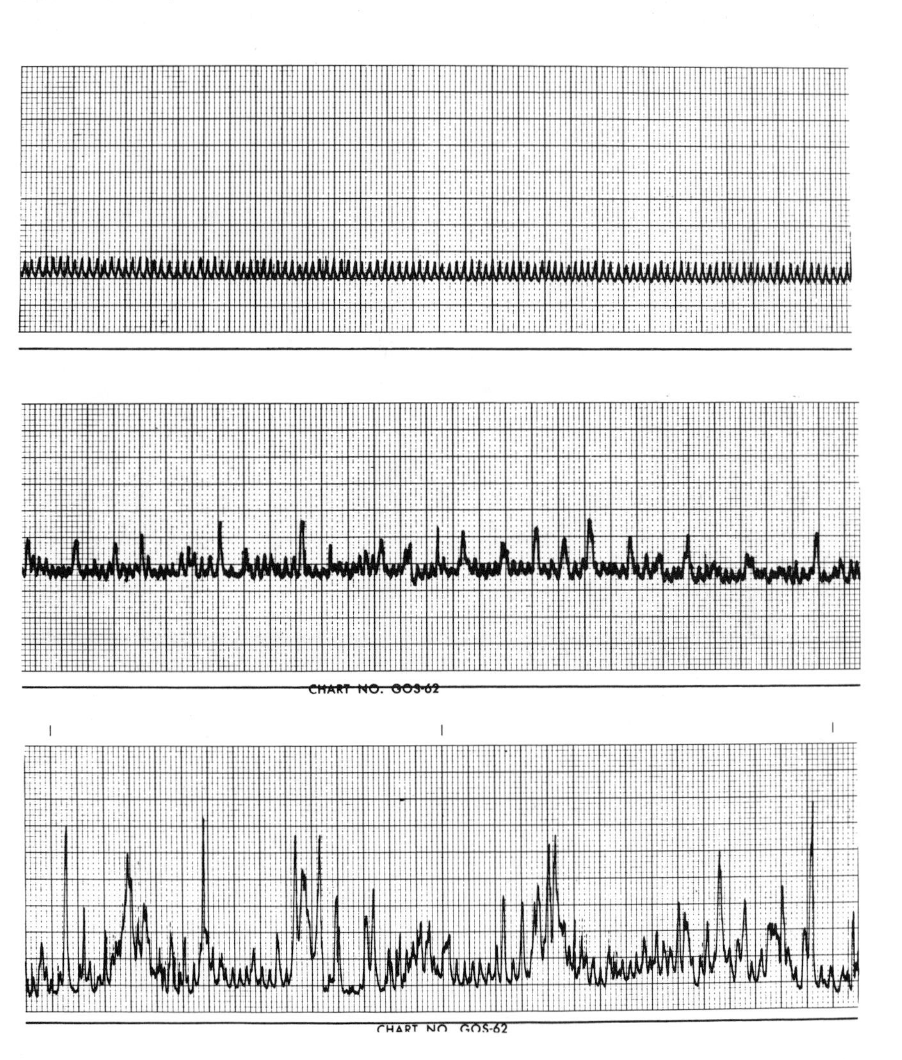

Fig. 5.7 At the top are the pressure fluctuations in the small intestine due to breathing, at the center Type-I activity, and at the bottom the slow waves called Type-III activity.

Table 5.1 Analysis of Intraluminal Pressure Waves in Ulcer Disease

Subject	Type I waves per five-minute period classified by amplitudes in centimeters of water.						Type III waves		
							Number	Amplitude	Amplitude Type I Superimposed
	0–5	6–10	11–15	16–20	21–25	> 25			
1 (ulcer)	4.4	7.9	1.3	0.4	0.4	0.0	11	5–20	10–25
2 (ulcer)	6.2	8.4	3.5	1.2	0.4	0.1	8	5–15	15–30
3 (ulcer)	3.8	11.2	7.1	2.9	0.5	0.1	18	5–15	10–25
4 (ulcer)	0.5	1.0	0.5	0.2	0.1	0.1	18	7–12	5–25
5 (ulcer)	5.2	7.1	2.6	0.4	0.5	0.2	10	10–20	20–30
6 (ulcer)	4.6	8.9	4.2	1.8	0.5	0.0	17	10–20	15–25
7 (ulcer)	7.2	4.9	1.9	1.0	0.6	0.9	21	15–30	15–30
8 (ulcer)	2.9	7.2	3.6	1.1	0.5	0.5	15	5–15	5–20
9 (normal)	1.8	2.7	0.7	0.0	0.0	0.0	0	—	—
10 (normal)	4.8	4.8	0.3	0.0	0.0	0.0	0	—	—
11 (normal)	1.8	6.4	1.0	0.2	0.0	0.0	5	5–10	5–20
12 (normal)	2.6	3.2	0.7	0.5	0.2	0.5	2	5–10	10–60
13 (normal)	3.3	2.1	0.4	0.2	0.1	0.1	0	—	—

of 1 to 2 mm amplitude, but they become quite obvious at a chart speed of 0.5 mm/sec. More large amplitude and "Type III" waves are seen in the ulcer subjects.

In all cases, as indicated in Chapter 1, the transmitter is tied into a rubber sack such as a finger cot or a finger cut from a rubber glove. Since the restoring force to the core does not much depend upon the elasticity of the diaphragm, but rather on the compressibility of the pocket of gas trapped behind the diaphragm, this does not change the sensitivity of the transmitter. It does, however, make the reuse of the capsule more convenient. This covering bag can be pulled tightly over the transmitter or it can be somewhat inflated with air or liquid. This obviously averages the applied pressures somewhat differently, and the resulting tracings do look somewhat different. In general, the use of an inflated bag results in larger, more frequent signals from a particular region. This raises the question as to whether the apparent difference in the pattern has to do with an increase in sensitivity to any existing activity by this configuration, or to a stimulating effect on the gut of the larger object. To check this a double transmitter was built having simultaneous transmission on two different frequencies and a diaphragm at one end, but with a small balloon at the other. Representative double-channel recordings are shown in Fig. 5.8. These tracings are from within a human subject, and in each case the lower trace represents

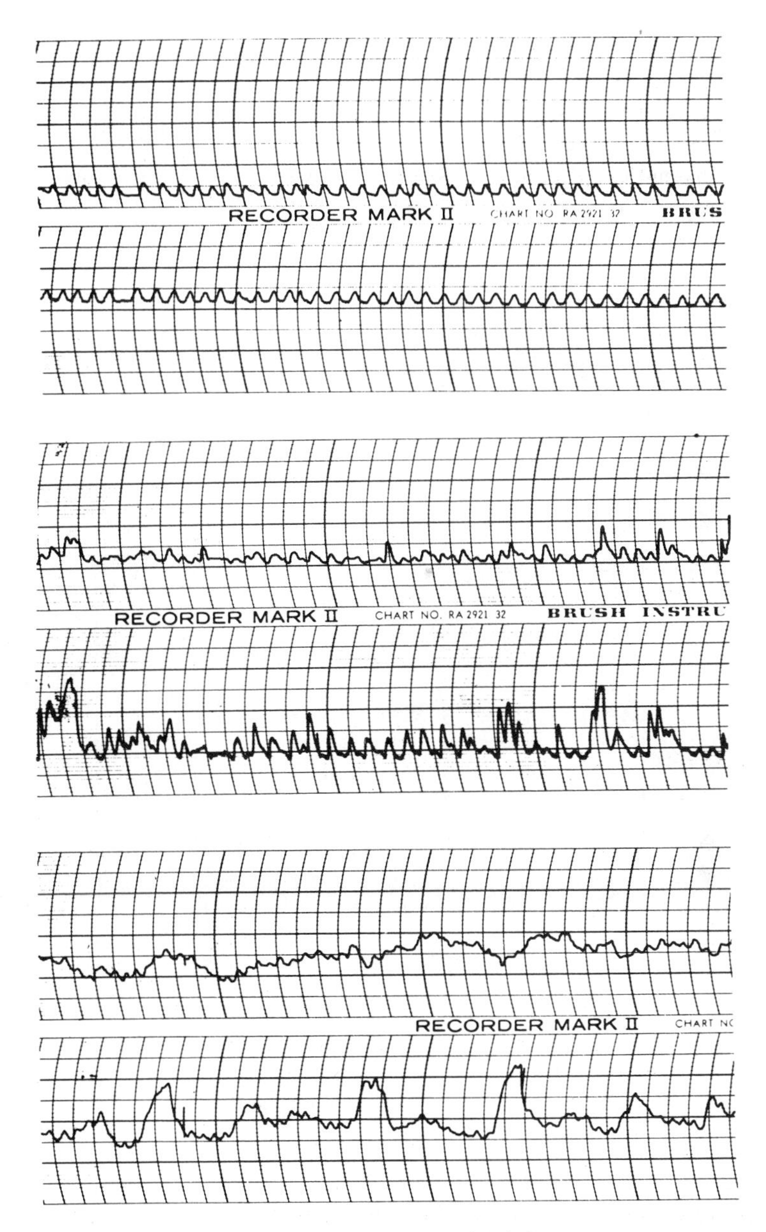

Fig. 5.8 Comparison of pairs of signals transmitted from a pressure sensor and another sensor in the small balloon distant by 2.5 cm. The stomach trace at the top shows the small regular fluctuations associated with breathing. The tracings at the center are from the small bowel and at the bottom, from the large bowel. In each pair the signal from the balloon-enclosed sensor is below.

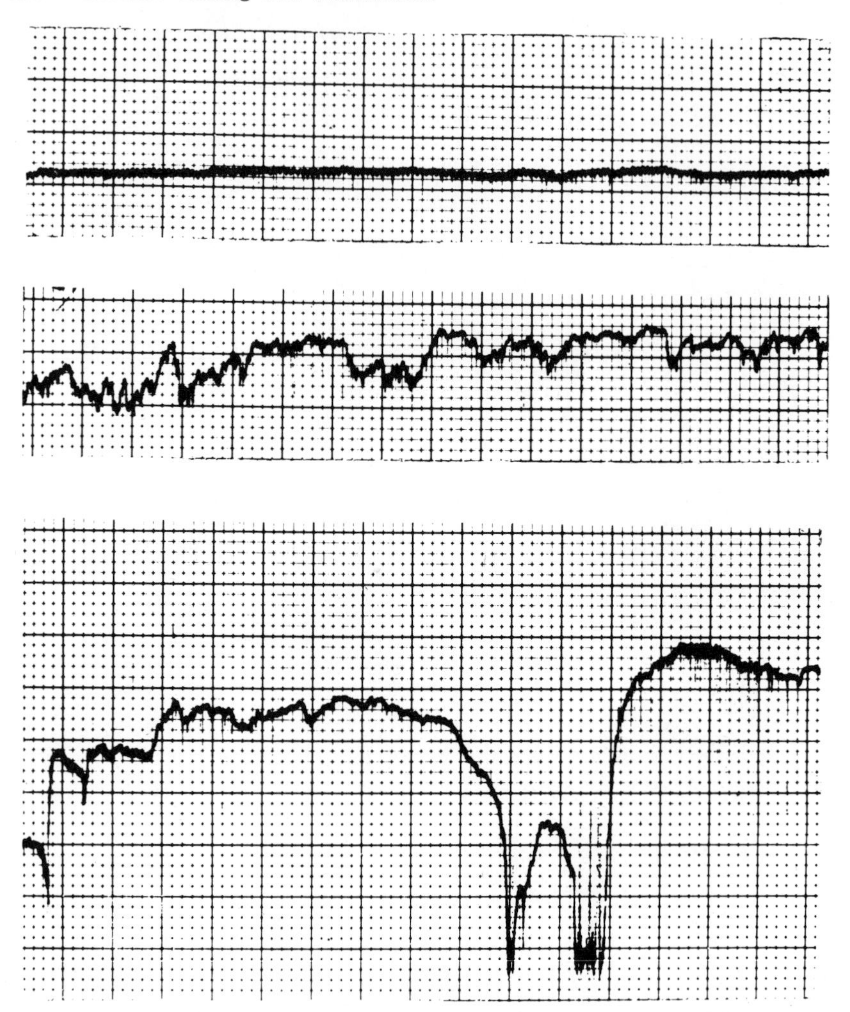

Fig. 5.9 Recovery of activity in the large bowel following the surgical trauma of gall-bladder removal. Small fluctuations at the top are associated with breathing. The center line shows the start of returning activity. Full amplitude is shown at the bottom.

the signal from the sensor in the rubber balloon. The three tracing pairs are successively from the stomach, the small bowel, and the large bowel. The small regular fluctuations seen most clearly at the top are associated with breathing. Such observations suggest that the larger pattern is mainly due to greater sensitivity rather than to a greater stimulating effect. Thus, if maximum sensitivity to activity, and discrimination against breathing fluctuations is desired, an inflated configuration is suggested. A true reflection of pressure, however, is undoubtedly more accurately conveyed from the diaphragm alone. In these double transmissions sometimes the signals are alike in shape, and sometimes rather different. Sometimes one tracing leads the other; this is emphasized by increasing separation, the diaphragm separation in the above case being 2.5 cm. The chart speed was 5-mm/sec, and the sensitivity set at 8-mm deflection for a 10-mm Hg change in pressure in the balloon, and at 6-mm deflection for a 10-mm Hg change in pressure on the diaphragm. The center pair of tracings from the small bowel gave alternately the pattern shown and periods of no activity resembling the stomach tracing; this particular tracing occurred two hours after the transmitter left the stomach. The colon tracing was recorded the following day.

The large amplitude waves in the colon are rather distinctive and, on occasion, we have observed changes in pressure that were larger than the blood pressure. Presumably, this leads to a momentary ischemic condition of the tissues. Figure 5.9 shows a set of tracings of the recovery of activity in this region following abdominal surgery. The transmitter can either be ingested the day before, or it can be introduced through the incision during surgery. As activity resumes the transmitter is then swept out in the usual way. It is useful to monitor abdominal sounds at the same time. In the present case recovery was rapid after gall bladder removal, although no special drugs were used to hasten recovery in the postoperative period.

We have used multiple transmitters operating simultaneously in other observations also. Figure 5.10 shows simultaneous tracings from the stomach

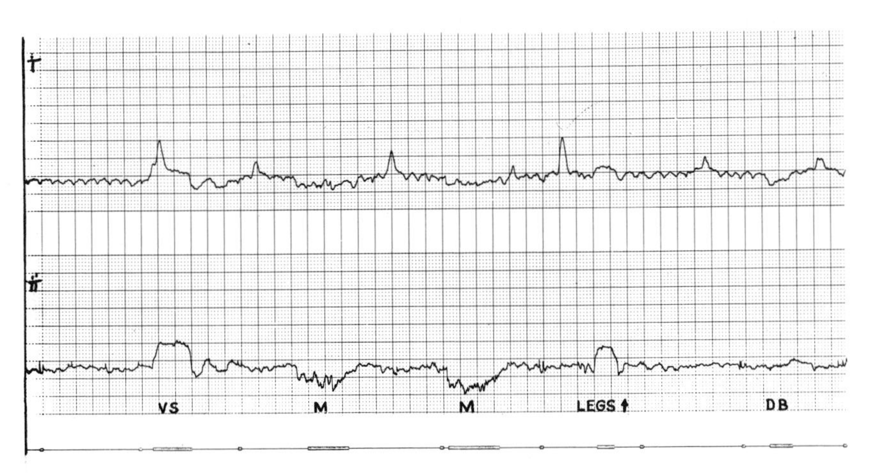

Fig. 5.10 Simultaneous recordings in the esophagus and (below) in the pouch above the diaphragm of a human subject suffering from hiatus hernia, as various maneuvers are performed. VS stands for expiration against resistance; M stands for inspiration against resistance, DB for deep breath. The time of raising of the legs from the bed is also indicated. The activity pattern at the top is normal for that region.

and the pouch of a patient suffering from hiatus hernia, as the patient goes through various maneuvers. The extension of the stomach up through the diaphragm causes distress in some subjects and not in others. In these observations with John Carbone, we were interested in recordings from the esophagus, pouch, and stomach in order to estimate the possibility of activity in subjects with certain body builds, possibly shifting the stomach contents back and forth. Two transmitters were also used to produce Fig. 5.11. In that case, the two were hung from threads in the esophagus, and each swallow-induced pressure wave was seen to arrive first at one transmitter and then at the other. By shifting the transmitters apart a known distance, it was possible to measure the velocity of wave propagation in this way.

We have also studied several subhuman species by these methods (Mackay, 1968). Figure 5.12 shows a radiograph of a monitor lizard, *Varanus flavescens*. A small transmitter was placed in a mouse whose outline is still visible, and the mouse was eaten by the lizard. The objective was to assess the effect of temperature on peristaltic activity in such a cold-blooded animal. It was found that activity increased in frequency and am-

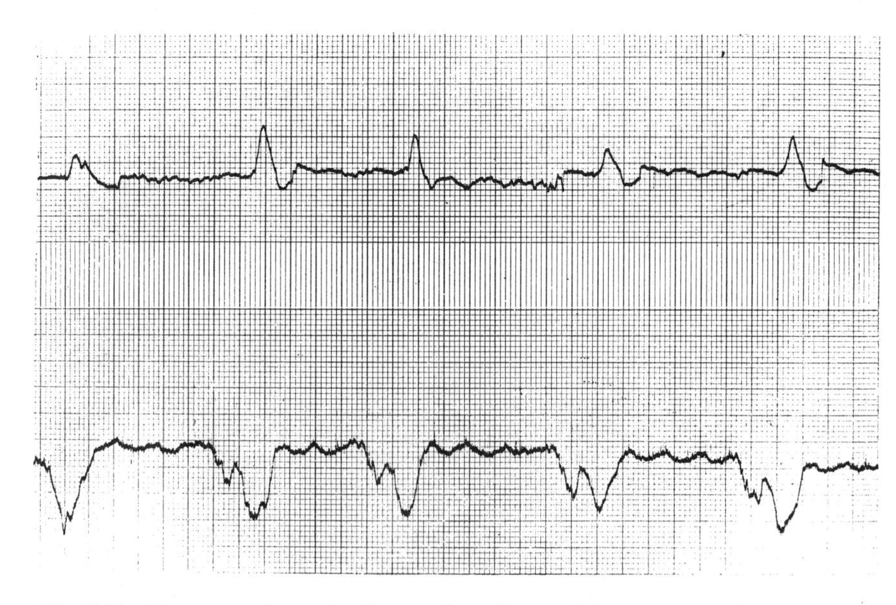

Fig. 5.11 Two transmitters simultaneously active on different frequencies were hung from a pair of threads in the esophagus of a human subject, the spacing being adjustable. Each swallow-induced wave activated first one transmitter and then the other, with the delay time allowing a velocity estimation of 16 cm/sec. The inverted lower trace is from the upper transmitter.

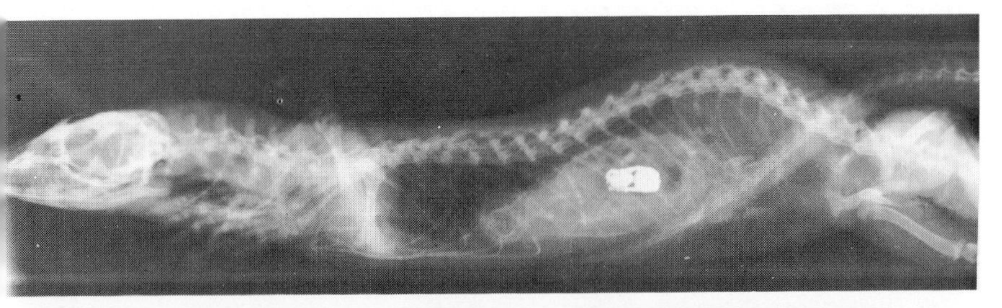

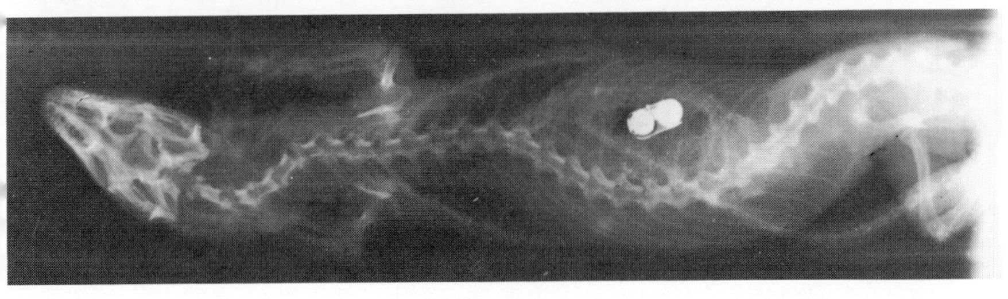

Fig. 5.12 A mouse containing a pressure transmitter was fed to a monitor lizard. In these radiographs the position of the transmitter, as well as the outline of the stomach contents, can be seen, for the mouse had been rubbed with a mush of barium. In this case a suitable transmitter geometry required placement of the battery beside the coil, into which the rest of the components were inserted.

plitude as temperature rose, and became irregular about four days after eating (which was still five or six days before passage of the transmitter and several days before eating again). The activity pattern was more regular in an herbivorous iguana. Thus in Fig. 5.13 are seen simultaneous temperature and pressure recordings telemetered from within *Ctenosaura pectinata*. Since the lizard is a vegetarian and eats rather uniformly, its activity pattern was less related to things other than temperature. Data such as this can be analyzed by counting contractions to give charts of the type shown in Fig. 5.14. The circadian rhythm of gastrointestinal activity was observed to follow temperature but ancillary tests, in which fixed temperatures were imposed, were necessary to indicate that there seemed to be a causal relationship. In the case of the observations recorded here, the animal was quite free to move at will in and out of a source of light and heat. It is interesting that the larger animal (the iguana) not only preferred a higher temperature, achieved by the way he regulated his pattern of activities, but that the small fluctuations of temperature during the day were at a higher frequency; that is, the loop gain of his temperature regulator was higher. Heating and cooling rates were not the same; they would not be expected to be in a nonlinear system, and this system was also not constant in configuration.

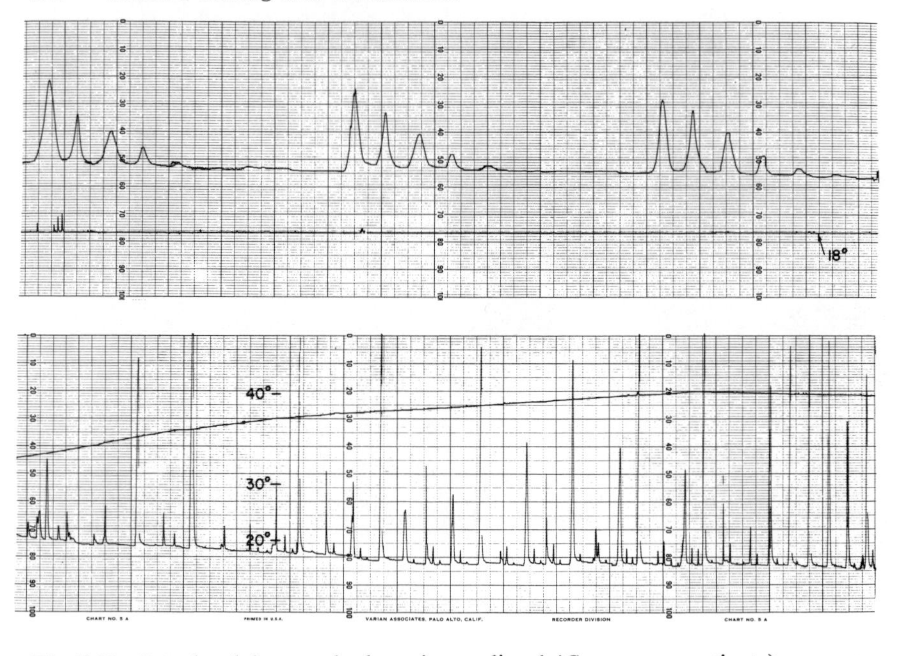

Fig. 5.13 Into the abdomen of a large iguana lizard (*Ctenosaura pectinata*) was surgically implanted a small temperature transmitter, and the animal was fed a pressure transmitter. On a recorder with overlapping pens the increase in contraction frequency and amplitude of contractions can be seen as temperature rises.

It is interesting to study these cold-blooded animals, because digestion does seem to be related to their temperature. The fact that their temperature does fluctuate much more widely than that of the so-called warm-blooded animals introduces some technical questions. These can become quite important in the transmission of absolute pressures. In the first chapter it was mentioned that some temperature sensitivity is purposely left in these transmitters so that a sip of cool water will cause a shift in baseline if the transmitter has not yet left the stomach, thus serving as an index of transmitter position. In the poikilotherm, large shifts in the pressure baseline are thus observed with gross changes in body temperature. Various components in the transmitter respond to temperature changes with somewhat different speeds, and thus complex transients are generated during sudden warming unless precautions are taken in transmitter construction. In Fig. 5.15 the pressure capsule starts to respond to a temperature change more rapidly than the temperature transmitter, there being first a rise and then a fall in the baseline. Where the pressure trace goes off the bottom of the scale, a switch on the recorder automatically introduces the bias (or else retunes the receiver by one step) to return the trace to the opposite side of the paper. This range-switching allows good sensitivity on a tracing of limited width. This tracing was specifically selected to depict and explain an apparent anomaly of performance that can appear on records if transmitters are used in applications for which they might not have been intended. The above record still demonstrates, however, that increasing activity goes with increasing temperature in the unrestrained iguana.

Psychological factors can affect the appearance of these traces. Whenever any human would come in sight of the iguana, all activity in his gut would cease. In the case of the monitor lizard, although he would challenge humans and freeze in place, the pattern of activity along his gastrointestinal tract would remain unchanged. Similarly, the stress of public speaking seems to leave some orators less affected than others. This was observed during lectures on telemetry in which the signal from an ingested capsule in the speaker was displayed for the audience by a projection oscilloscope (Mackay, 1963). Thus to study stress reactions it may not be necessary to go to astronauts, but merely to race-track drivers, or even bus drivers. In any case, the use of telemetry makes it possible to carry on experiments

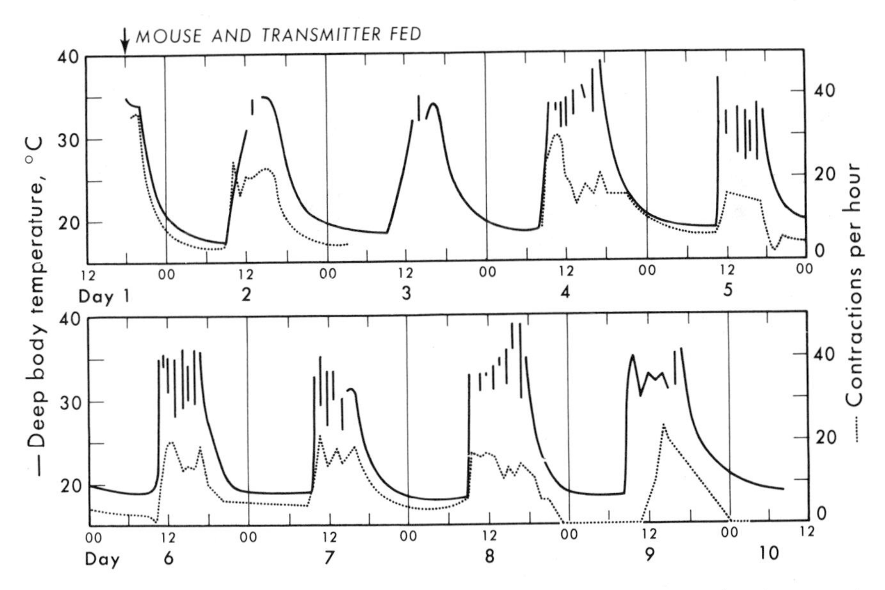

Fig. 5.14 Analysis of data from *Varanus flavescens* (a monitor lizard). The general pattern of contractions follows the circadian temperature cycle in the undisturbed animal over the 10 days required for the transmitter to pass, although contractions became weaker and rather irregular after the fourth day. On the third day of this run radio interference prevented recording the pressure signal with any accuracy.

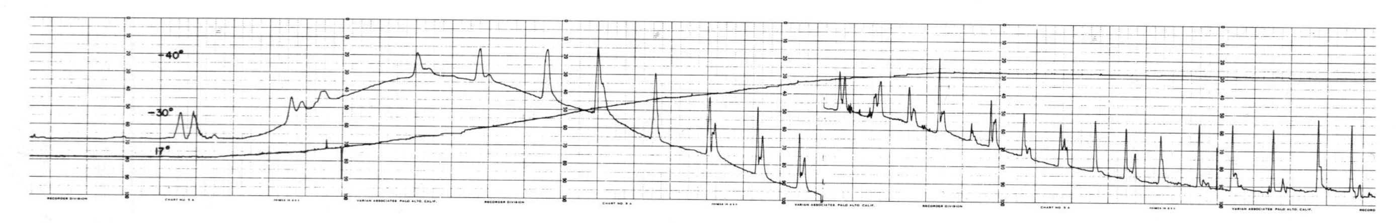

Fig. 5.15 Transients in a pressure transmitter which is not compensated against temperature changes can produce unusual base-line shifts with the large temperature excursions of some of the cold-blooded animals. In this case automatic shifting of the pressure trace was employed when the pen started to go off scale.

without interfering with the other activities of the subject. It might be mentioned that the feelings of stress sometimes noted by humans can be accompanied by a diminution of pressure changes rather than an increase. It is also true that some rather extreme cases of diarrhea are accompanied by few pressure fluctuations, the gastrointestinal tract apparently acting as a sort of open pipe. Subjects routinely showing the rapid passage of food may show unmodified patterns in the small intestine (Mackay, 1965).

Pressure fluctuations at other sites can also supply useful information. The first line in Fig. 1.10 illustrates the monitoring of the breathing pattern by an implanted transmitter. Activity in many areas of the body can be monitored by pressure fluctuation sensors, although in other cases with implanted transmitters it may be absolute pressure which is to be monitored. This is discussed in the last section of this chapter.

The problem at the start of the 1950's which led to the early experiments on passive transmitters was that of the pressure in the human bladder during micturition. Some researchers speculated that the pressure in the bladder might actually drop below atmospheric for a time. The early passive transmitters did not give a reliable enough signal at the receiving antenna to warrant this kind of experiment, but such observations have since proven quite simple using active transmitters incorporating batteries and transistors. It is considerably easier to insert an object into the bladder of the human female than the male, with sizes up to 24 F (8 mm) involving no strain, up to 30 F (10 mm) being relatively easy and probably not involving an anesthetic, and about 35 to 36 F (12 mm) requiring anesthetic. Using these methods it is not only possible to study voiding, but there is the potentiality for studying strictures or obstructions. Using the analog of Ohm's law for flow, we might say that the resistance is the difference in pressure from one end of the urethra to the other, divided by the flow velocity squared. (The flow velocity should be squared to take into account the fact that flow in the urethra is undoubtedly turbulent.) The velocity can be monitored by the rate of collection of fluid and thus, to compute the resistance, one need only monitor pressure in the bladder during voiding and pressure at the external end of the urethra. The latter pressure is not atmospheric, but is the impact pressure of the stream against a perpendicular surface. It can be recorded during voiding, and the internal pressure monitored as a function of time with the help of an inserted endoradiosonde.

Signals that change slower than the expected slowest ones of interest are considered as drift, whether this is an effect in the receiver, the transmitter, or an actual real pressure change either atmospheric or in the subject. These can automatically be eliminated in the record by having the receiver gradually retune itself to slow shifts in frequency. Such a very slow "automatic frequency control" is depicted in Fig. 11.8.

3. ACCELERATION EFFECTS

In the initial studies on pressure transducers with relatively heavy cores there was some worry about the effects of gravity as they would turn over. With lighter cores, this proved to present no problem, and with the arrangements of the type shown in Fig. 5.4, using only an aluminized diaphragm, the effects were negligible. In high acceleration situations it is still thought desirable to surround the diaphragm by a slightly projecting ring so that movement cannot throw the body of the capsule against an undigested lump in the stomach, thus perhaps concentrating the force from a larger area onto the diaphragm. These considerations, however, do not prove to be major, for reasons that will become apparent.

It was desired to study some aspect of motion sickness and the drugs to minimize such effects, and thus it was necessary to have a relatively reproducible means of producing this form of distress in human subjects. Accordingly, a swing was constructed. It was fabricated from an office chair to minimize the ability of the subject to move and influence the experiment. An 11-ft length of nylon rope was used for support. When the amplitude of swing was approximately 40 degrees on either side of the vertical, the change in level of the subject was approximately 2.3 ft. The period of this swing was such that there were approximately 17 cycles per minute. The back of the chair prevented the subject from leaning, and thus limited effective swinging movement. Motion was maintained by another person pushing smoothly at the end of each swing. So the subjects would not have a stationary reference to observe, they were blindfolded before a test. Of the ten young adults of both sexes tested, only one did not become sick within eight minutes. It might be added that this subject was not rendered sick by swinging for 30 min either before or after eating a meal. A typical experiment is shown in the photograph of Fig. 5.16.

A receiving antenna coil was looped around the subject, diagonally over one shoulder and under the other arm, in each case. From this 20 turn antenna, an 8-m length of coaxial cable ran up the swing rope and down into the receiver. The receiver input was not tuned, the capacity and inductance of the cable and antenna being too high even for a low-frequency transmitter. Pressure-sensing endoradiosondes were used both in and out of inflated covers, with similar conclusions.

As the subject swings, the trace moves up and down in synchronism with the motion. For the most part, this is *not* due to the effect of the acceleration upon the transmitter core and the diaphragm. The amplitude is considerably greater than when the transmitter is swung in a beaker of water. The extra forces are due to the weight of the column of tissue and body

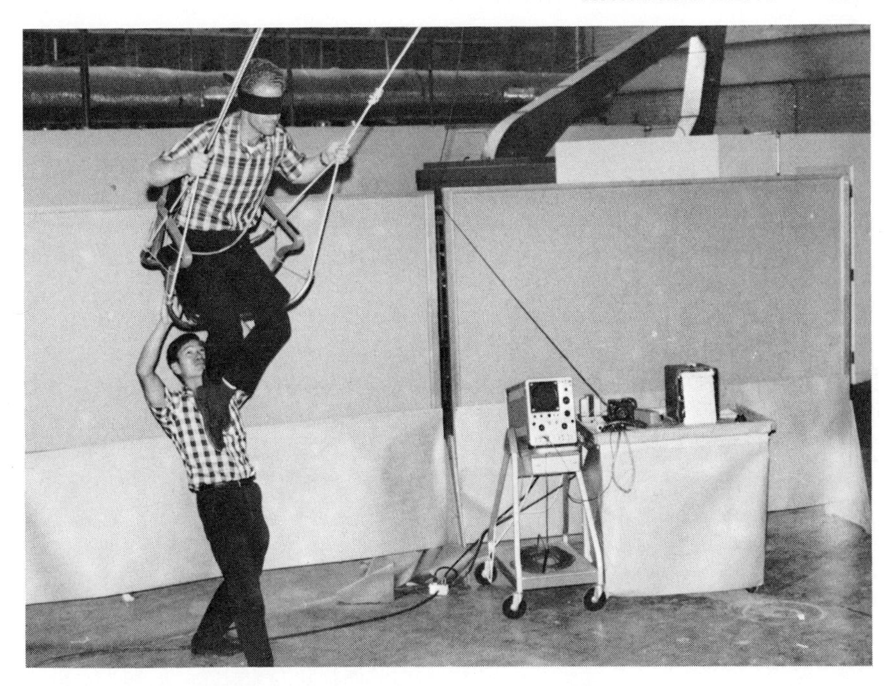

Fig. 5.16 Swinging while blindfolded was employed as a reproducible stimulus towards motion sickness. Notice the receiving loop antenna draped diagonally over the shoulder of the subject.

fluids above the transmitter. Representative tracings are shown in Fig. 5.17.

A tendency for the subject to breath in synchronism with the swinging tended to break up when a sick feeling arose. The various subjects were instructed to continue swinging until that stage of motion sickness was reached where they would feel warm and nauseated and start to pass gas by mouth. They were instructed, however, that they need not continue the experiment until vomiting took place. Of the four subjects tested in greatest detail the reactions were rather similar, but not identical. The three who became sick reported rapidly increasing nausea and "burpy" sensations and they became pale. Two mentioned a slight headache, warmth, dizziness upon stopping, and a sensation of dampness at the end of a run. Another emphasized extreme weakness during the swinging, extreme headache, coldness on stopping, and cold perspiration.

The number of tests run was sufficiently few so that the following observations must be considered as preliminary. The sequence of events is rather different when the pressure-sensing endoradiosonde is in the stomach than

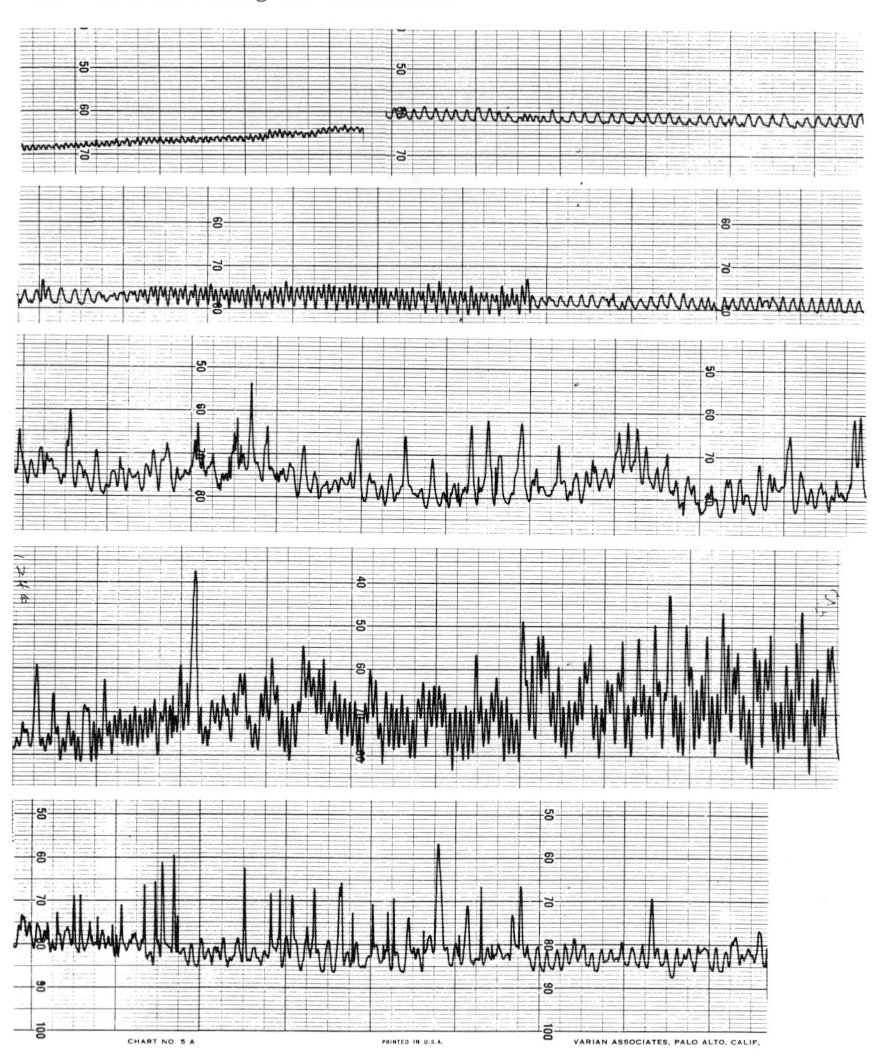

Fig. 5.17 Representative traces in the swing experiment. Line (1) transmitter swung in water and then in stomach before swinging. (2) Transmitter in the stomach, with part of trace before swinging, during swinging, and after. (3) Transmitter in the small bowel immediately before swinging. (4) During swinging. (5) Immediately after swinging. Lines (1) and (2) were made before lunch, from (3) on, after lunch.

when it is in the small intestine. It appears that if there is no activity in the stomach before swinging no activity will take place during or shortly after swinging. If there was moderate activity in the stomach before swinging then a slight decrease in activity during and immediately after swinging was noticed. The situation with the transmitter in the small intestine is a little less certain. If activity is present before swinging, then during and after there is perhaps a slight increase in activity in some subjects and a decrease in others. In two of the subjects in which there was very regular activity in the small intestine before swinging, there seemed to be about the same frequency of activity during and after swinging (approximately 8 contractions per minute for the large waves) with somewhat diminished amplitude of contraction. The subject who reported no sensations showed no significant change in pattern of the record.

Until the normal pattern of activity as modified by motion sickness is well known, it would be unsuitable to attempt to use these methods to judge the changes produced by drugs that are supposed to alter the course of this distress. Because all of these patterns are somewhat irregular, statistical conclusions on large numbers of subjects may be required in applying this approach at all.

There are certainly more elaborate and immediate methods for accelerating human and animal subjects into a state of motion sickness. If pressure sensors are to function during various vibrations and accelerations, their components should have low inertia. It is clear that real pressure changes, induced through the inertia of the components of the body of the subject, will often completely dominate artifacts in the measuring instruments. In some cases it may be more appropriate to monitor a related parameter such as electric activity within the gut, rather than pressure itself.

4. ABSOLUTE PRESSURES, INCLUDING BLOOD PRESSURE

In many cases it is necessary to continuously transmit the absolute value of pressure, rather than just to note the magnitude of changes therein. In general, by this is meant the absolute amount by which the pressure in question exceeds the momentary atmospheric pressure. Thus indications will generally have to be corrected for the momentary reading of the barometer, since an internal transmitter cannot connect with the atmosphere to automatically compensate itself. Another way of saying this is that pressure transmitters essentially are referenced to the pressure existing at the time they were sealed, and even evacuating them does not refer the pressure to the present ambient value. (We shall see in Chapter 12 that the effect of ambient pressure changes on an internal pressure transmitter actually allows calibration of the incremental sensitivity of such transmitters while still in the animal.) An important example of these considerations is the measurement of blood pressure. A person's indicated blood pressure is about the same if he is up in a balloon or down in the ocean in a diving suit. The absolute pressures are quite different in the two cases when referred to a fixed standard, but the tension in the arteries of the subject is about the same because the ambient conditions balance out. The term "absolute pressure" is thus used here to include what is often meant by "gauge pressure," which is pressure above ambient. The implication here is that the reference from which baseline readings are taken is not continuously drifting in an unknown fashion. (The same considerations apply to the section on temperature measurement where it will be seen that temperature changes can often be specified more accurately than can the actual temperature.)

One system for measuring absolute pressure consists of placing a compliant box within the region to be measured and noting changes. If the box is made of a material that does not slowly creep or spontaneously deform in a plastic fashion, and if the entire unit is relatively insensitive to temperature and other extraneous variables, the sensor calibration remains fixed, and both the transmitter and the receiver are stable, then absolute pressure will be indicated rather than just sudden changes therein. Metals, glass, and quartz have dimensions that are relatively stable with time and spring back to their original dimensions when small applied forces are removed. Thus these materials will generally supply the restoring force to a displacement transducer in systems for measuring absolute pressures. Since these materials are rather incompressible in the bulk sense, they are used in a hollow configuration with walls that bend. It is necessary that gases and vapors do not leak to or from the inside of the transducer, and it is desirable that changes in temperature not lead to large changes in pressure inside the transducer.

If the motion sensor is rather stable, results can be quite reliable. If the whole configuration is rather stiff, so that motion in response to pressure changes is small, the overall configuration will be able to respond to rapidly changing values; that is, the high-frequency response will be good. Such a unit can, for example, be placed in the brain to measure pressure there, or it can be placed in the bloodstream to measure pressure in that location. The possibility of positioning in the bloodstream has only come into existence since coatings for preventing the clotting of blood have become known (Chapter 4). Other sites also can be investigated, for example, in the esophagus lung pressure is approximated without surgery.

Various transducer types can be used, and an especially interesting one will be mentioned in Chapter 13. But the increasing commercial availability of small semiconductor strain gauges, both individually and in quartets or pairs for use in a bridge circuit, makes them increasingly appealing. They can be placed in various ways so that an increasing pressure on a diaphragm of bellows causes slight bending or stretching of the elements. Perhaps the most compact structure results when the elements are affixed directly to a diaphragm and a "can" sealed over the back to complete the structure. Such a unit is depicted in Fig. 5.18, where an increasing pressure stretches

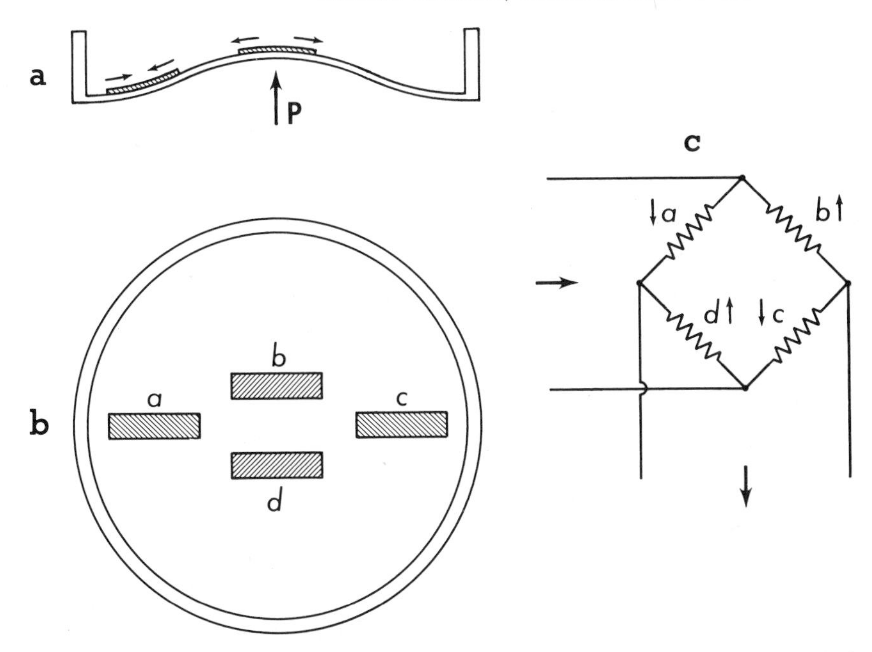

Fig. 5.18 a. Pressure on the outside of a diaphragm clamped at its edge will stretch centrally placed elements and compress those more peripheral. b. Placement of four strain-gauge elements so that pressure changes will increase the resistance of one pair and decrease that of the other. c. Bridge circuit which is unbalanced by pressure changes, but relatively little affected by temperature changes that cause all resistances to change in the same direction.

two elements and compresses the other two. The gauge material can be evaporated on to the diaphragm. Changes in temperature similarly affect the four elements and thus should not upset the indication. However, at maximum sensitivity, flow of fluid past the unit can cause a redistribution of heating that may produce a noticeable signal, and thus any such unit should be tested for flow sensitivity. Such units are rather delicate if very small, and the diaphragm may creep somewhat with time. The four wires must emerge from the unit, and some manufacturers have had problems with leakage at that point.

The trace in Fig. 14.10 made by VanCitters and Franklin was produced by inserting such a unit through a slit in a large artery, with the electrical leads left extending through the opening (VanCitters and Franklin, 1966). In that case a 500 Ω bridge was energized by a battery and the resulting dc output voltage was used to modulate a voltage-controlled oscillator serving as a subcarrier oscillator. This particular transducer unit was constructed in cooperation with a company, and is commercially available in a metal case 6.5 mm in diameter and 1 mm thick. The circuit of Fig. 3.3 is quite effective in connection with such units.

Placement of the transducer in the bloodstream opens up many possibilities, and for example, it may be possible to transmit directly from within an artificial heart valve. Placement in an artery not only somewhat restricts the area available for flow but occasionally will cause clotting problems. The best nonthrombogenic surface is still the lining of the blood vessels, and thus it is of interest to consider the possibility of blood-pressure measurement from outside the artery wall. Any such measurement should give absolute readings independent of changes in the elastic properties of the intervening vessel wall. There is a method for accomplishing this which allows working on vessels down to a few millimeters in diameter. The principle is that of the Mackay-Marg tonometer, which has been quite successful in measuring the pressure in the human eye in connection with glaucoma testing. On a larger scale Kleiber has used a related method to study bloat in cattle and Smyth to monitor the contractions of labor. In a related application, such units can be used to estimate pressure in a muscle bundle, in connection with clinical claudication questions, or to replace contractile force measurements in some cases. The development and theory has been given elsewhere (Mackay et al., 1960; Mackay, 1964A; and Schwartz et al., 1966).

The method can be understood from Fig. 5.19 in which is shown a body cavity pressed against a flat surface into which has been inserted a force transducer. The body cavity (eye, blood vessel, skin flap on head, and so on) is flattened to beyond the force-sensitive region out onto the insensitive

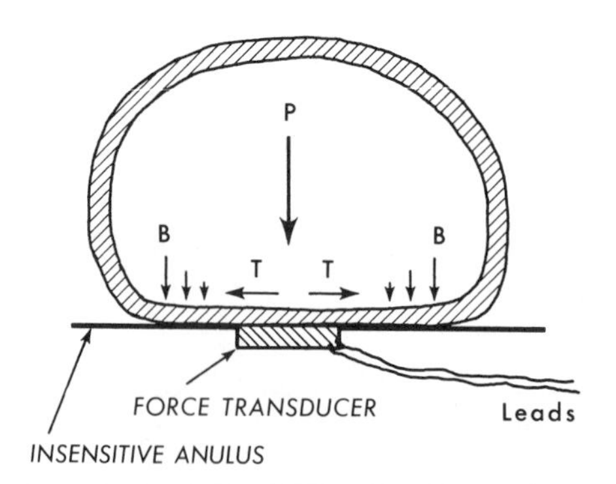

Fig. 5.19 If a body cavity, or a blood vessel, is pressed against a force transducer surrounded by a coplanar insensitive anulus, only pressure and not the various tissue forces and surface-tension forces will be detected.

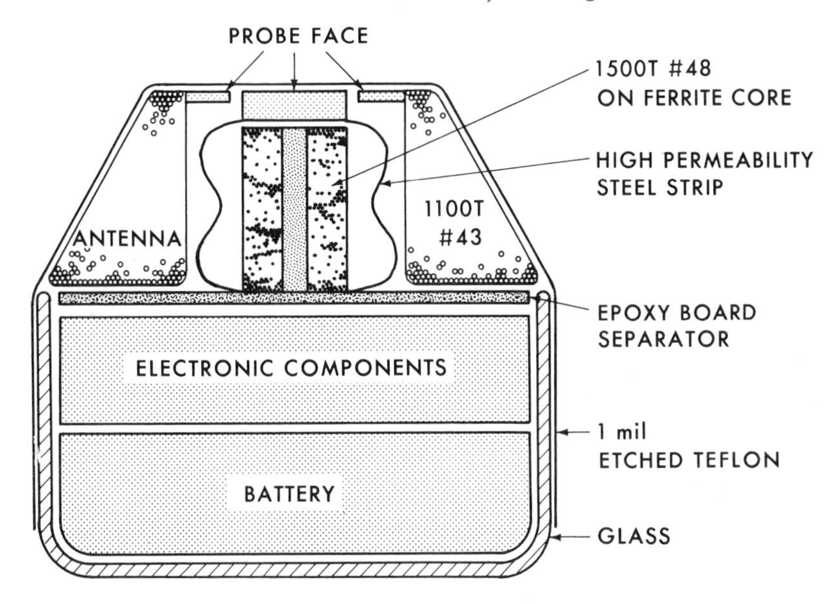

Fig. 5.20 A complete transmitter based on the principle of Fig. 5.19 and using the constant spring force of a piece of metal as a reference. Diameter is 1.45 cm; height is 1.35 cm.

surrounding annular region. It has been shown both experimentally and mathematically that the generated signal is essentially proportional to intracavity pressure alone. Qualitatively, it can be seen that the bending is done by the insensitive annulus, which thus largely takes up the bending forces, and the tissue tension is a radial force which can neither push nor pull on the transducer, thus leaving only pressure to generate an electrical signal. Since the forces associated with elasticity in the vessel are largely not sensed, changes in elasticity will not change the calibration of the unit, which can be considered absolute.

A specific arrangement is shown in Fig. 5.20 where a complete transmitter employing these principles is sketched. A metal strip supplies the spring force and is unchanging, so that creep in a plastic diaphragm does not introduce drift. This same strip is also an important element of the variable-inductance transducer. This unit functioned well, except that after a few weeks, diffusion through the Teflon film did somewhat alter the pressure within the cavity of the transmitter. A similar sized unit having no permeable components is shown in Fig. 5.21. A system of electrodes was used as the transmitting antenna rather than a coil. We were able to fabri-

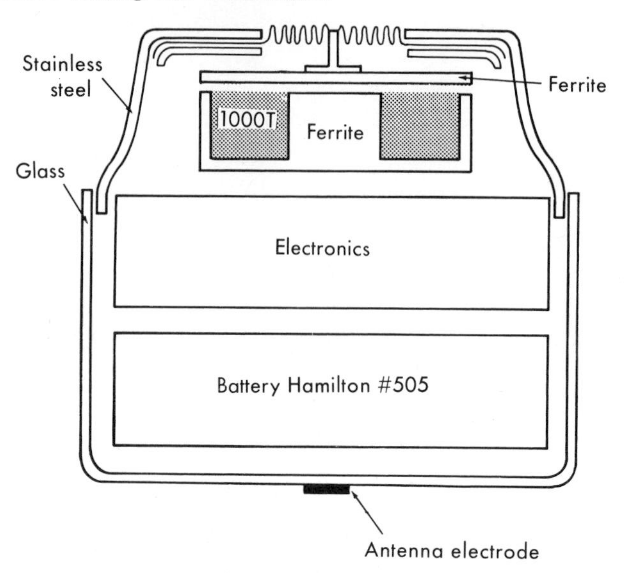

Fig. 5.21 Absolute-pressure transmitter in which no materials permeable to body fluids are exposed. The size of the device is approximately the same as that of Fig. 5.20; the corrugated diaphragm is approximately to scale. A pair of electrodes instead of a coil is used to carry the signal from the transmitter.

cate corrugated stainless steel diaphragms, with indentations as steep as shown for maximum compliance, with an active diameter of 2 mm. Such a diaphragm is seen in Fig. 5.22 as the second element from the right. The transducer in this case was associated with the rather complex output transformer shown in the center of Fig. 5.22, and around which the electronic components were placed in a circular pattern. In Fig. 5.23 is shown a similar unit having an elongated geometry in its transducer, and employing a variable-capacitor transducer in a plastic covering. Moisture affected this transducer adversely, although perhaps some silicone oil properly placed could have eliminated this problem. Other transmitters of this type were constructed in such a way that the transducer was remote from the transmitter, so that the transmitter could be positioned at a more convenient subcutaneous spot. Such units are shown in Fig. 5.24. In any construction of this type the major problem is to be certain that there is no leakage or breakage in the shielded wire connecting the two elements. The entire unit is covered with silicone rubber as a last step to prevent adverse tissue reactions.

This last transducer was actually cast into the silicone rubber sleeve, with the wraparound parts being formed over a Dacron mesh. When the artery was slipped into the notch in the casting and wrapped into place, suitable flattening was automatically provided, with a resulting decrease in cross-sectional area of the artery by under 20 percent. (Many agree that a reduction in lumen area to under half is needed before significant changes are observed in either flow rate or pressure drop, e.g., Walawender et al., 1970.) A radiograph of such a unit is shown in Fig. 1.9, and a representative tracing in Fig. 1.10. Such cuffs are sufficiently soft and flexible as to inflict minimum damage on an artery, and they do not present sharp corners. Also, attachment of the transducer to a telemetry transmitter seems to impose fewer mechanical strains on the associated vessel than does carrying the wires directly out through the surface of the skin. We have had such cuffs in place for over a year without damage to the enclosed artery. Tissue invasion under the cuffs was limited to the edges, and the flat spots remained.

In the description of this transducer system it was stated that the transducer was inserted flush into a flat surface, but the application of pressure must necessarily slightly deflect the transducer or else no signal can be

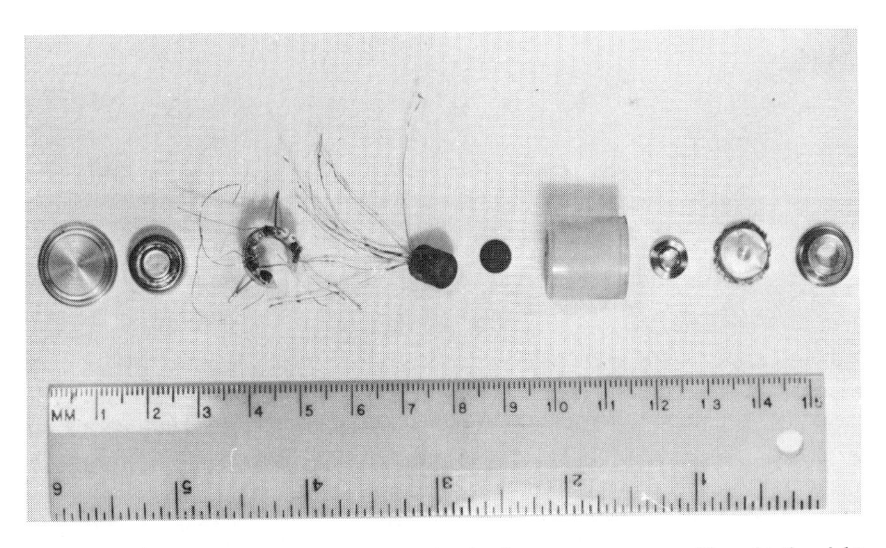

Fig. 5.22 Exploded view of components in absolute-pressure transmitter. At the right is the diaphragm with the two rings for holding it. To the left of the body of the case is the core, followed by the transducer coil and transformer, to the left of which is the electronics fashioned into an encircling ring. At the extreme left is the base of the case and the battery.

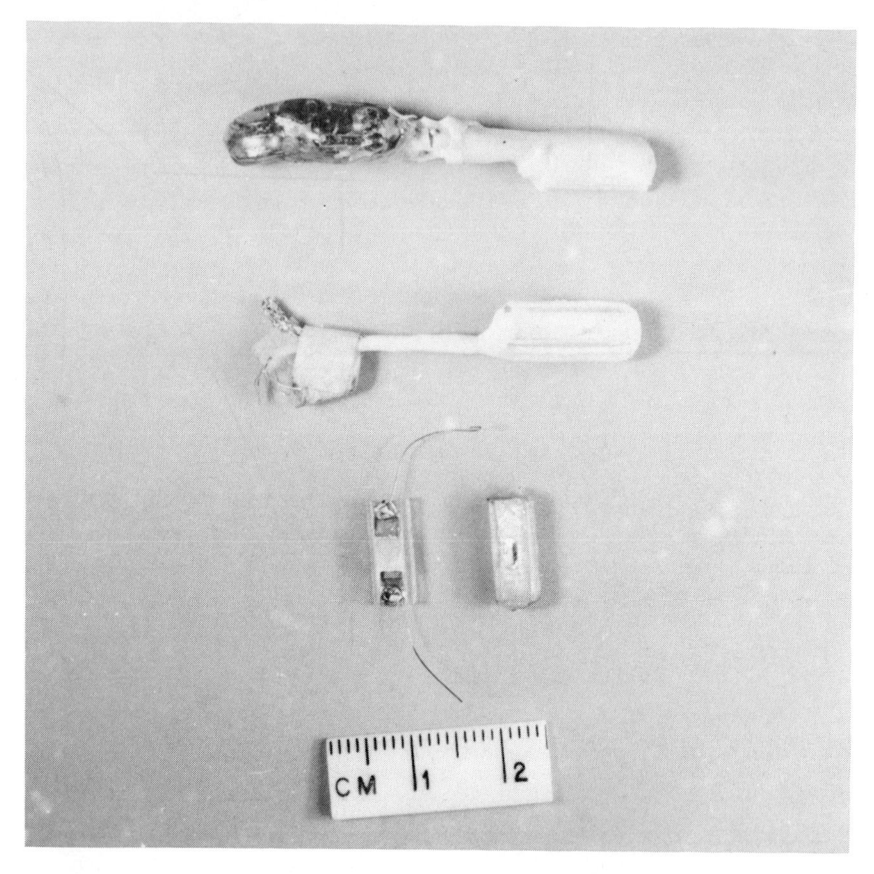

Fig. 5.23 Absolute-pressure sensors with elongated geometry based on a pressure-sensitive capacitor. The interleaved anodized aluminum structure of the transducer is shown at the bottom.

generated. As soon as there is some appreciable deflection or indentation of the transducer, the tension forces in the tissues can begin to subtract from the actual pressure. Thus it is necessary to know about the interaction of the transducer and the biological system being studied; that is, how much deflection can be allowed for the accuracy required in a given system. This amounts to determining the elastic properties of the vessel wall in question.

As in earlier studies on the eye this result was achieved by making, on actual artery walls, a series of measurements of transducer output as a function of actual pressure for a number of different initial settings of the degree of coplanarity; that is, observations were made with the transducer flush, or extended, or initially recessed. These measurements were made using a highly modified commercial probe (Biotronics, Redding, California). It was, in effect, infinitely stiff (total deflection of a few millionths of an inch) and thus did not move significantly from the initial setting in response to applied pressures. A strain-gauge unit was used to monitor momentary actual artery pressure communicated through a hypodermic needle, while

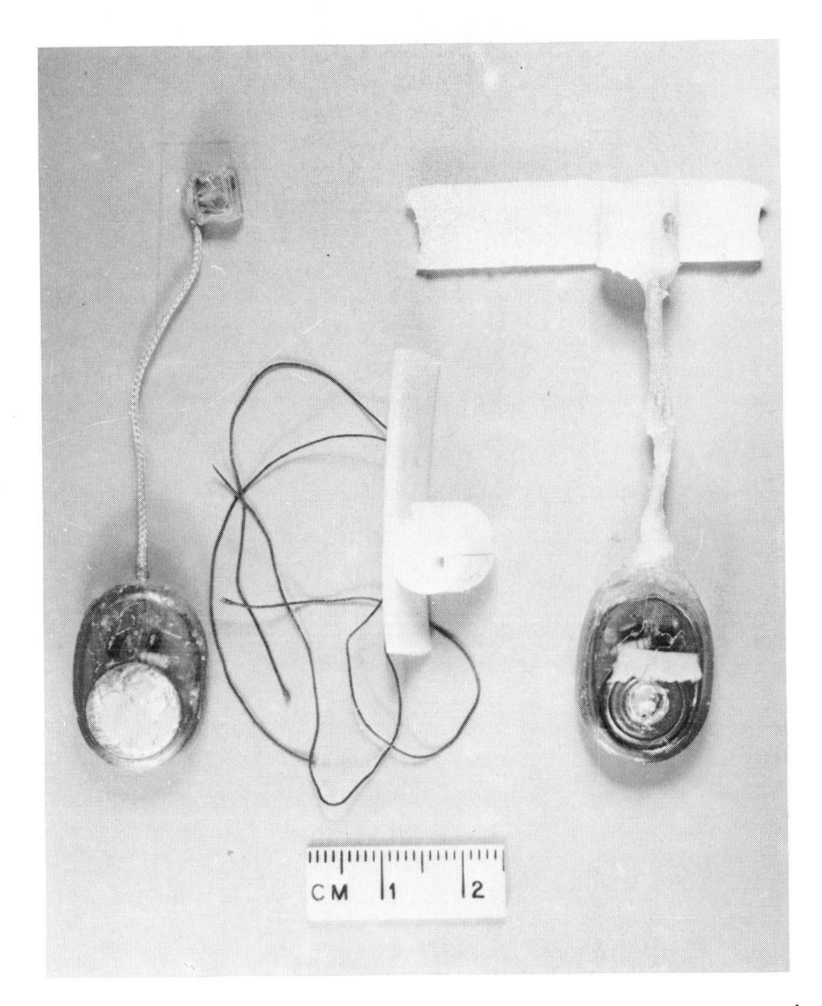

Fig. 5.24 A variable-inductance blood-pressure sensor with a separate transmitter for subcutaneous placement. The transducer at the upper left is fitted into the sleeve at the center to give the unit at the right. An artery is slid into the notch in the sleeve and the projecting sheets are wrapped around and tied into place.

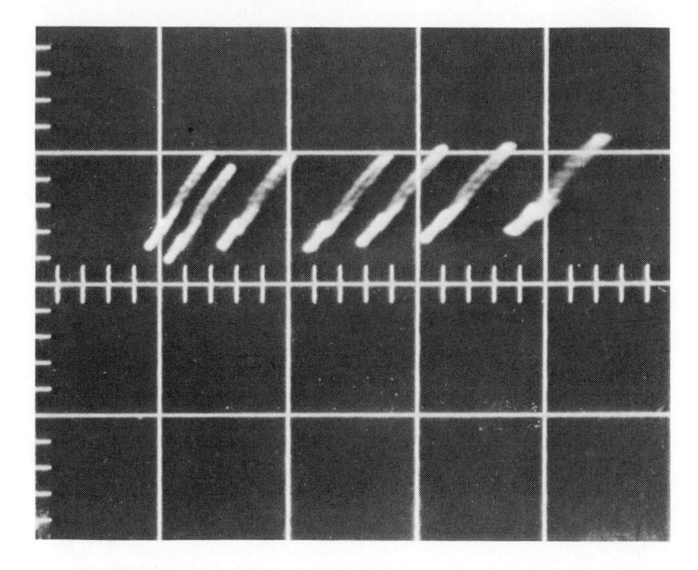

Fig. 5.25 An oscilloscope tracing of tonometer output as a function of actual pressure within an artery during the variable pressure of the heart beat. Various probe extensions, positive and negative, give the family of curves.

the signal from the transducer system under study was simultaneously recorded. In Fig. 5.25 are shown actual oscilloscope traces accompanying the beating of the heart of a dog on whose abdominal aorta the observations were being made; the strain-gauge and transducer signals were communicated to the vertical and horizontal axes of an oscilloscope to directly produce the resulting graph. The seven sloping lines represent different degrees of flatness. Since each line represents several dozen pressure cycles (heart beats), it can be seen that the short-term repeatability of an indication is good. Before these experiments were tried on a series of living animals, sections of excised artery (aorta and iliac artery) were similarly tested on a pump system which was able to generate a nonsinusoidal pressure sequence that imitated some aspects of the beating of a heart.

It was desired that the data taken in this way extend over a wider range of pressures than is normally associated with the beating of the heart. An increase in pressure above normal was produced by an injection of adrenalin and a decrease in pressure induced by the injection of histamine. The aorta seems not to be muscular and so the drug should have had only a small direct effect on the elasticity in these calibrations. The data to be summarized were taken on six mongrel dogs weighing 9 to 15 kg each.

In each case the artery was exposed and enough was freed of surrounding tissue so that the tonometer holder could be slipped around it. Positioning the tonometer was checked by tightening the holder strap and moving the tonometer up and down against the artery. A condition of no change in tonometer indication with additional flattening of the artery indicated proper position. Probe extension could be changed without moving the tonometer from the artery. Between each series of readings the zero was checked, and then the tonometer was positioned on the artery for readings to be taken. These were taken at each of the indicated extensions from an extension of 0.003 in. to a recession of 0.003 in. Several heartbeats were observed for each tonometer setting, and the results were superimposed on the photographs. Intraarterial pressure was indicated by a Statham pressure transducer with a polyethylene catheter inserted from the iliac artery to a point just distal to the tonometer. As hoped, tonometer position proved to matter little.

The results are summarized in Figs. 5.26 to 5.28. The response of an ideal tonometer (or transducer system) when plotted against actual pressure would be a straight line passing through the origin. This is observed to be the case when the initial extension of the transducer is zero, that is, when the unit is flat and undergoes no deflection. As expected (Fig. 5.26), if the sensitive area of the probe is initially extended, then readings will be too high since the tension in the artery walls will also serve to push inward on the sensitive region. The reverse is true if the sensitive region is initially recessed. The magnitude of this effect for the artery walls under study is indicated in Fig. 5.26. Actually, when a transducer is placed on a series of

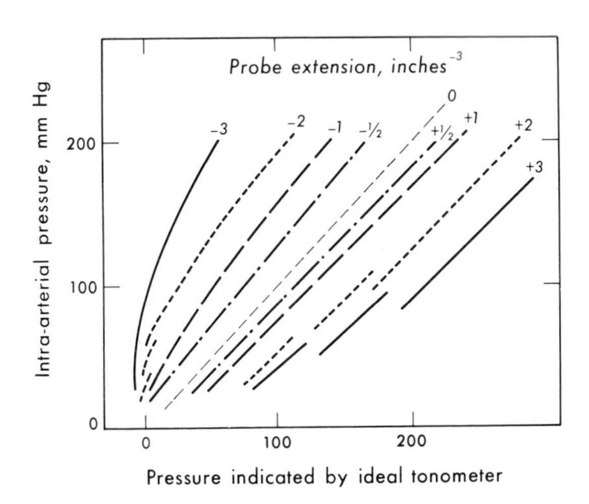

Fig. 5.26 A summary of averaged probe data.

arteries, there is some spread in indication for any particular probe adjustment. The maximum spread in probe readings observed is summarized in Fig. 5.27, where the band of response is indicated for each initial setting of the transducer coplanarity. Setting errors of degree of flatness can account for much of the spread seen. By referring to Fig. 5.26 it is possible to approximate what spread would be expected for an estimated setting error of ± 0.0002 in.

From the above data it is possible to indicate the type of response to be observed when an actual transducer of less than infinite stiffness is applied to a real artery. Under these conditions, with each beat of the heart the transducer will move slightly inward and outward. In that case, although all parts of the system are basically linear, the overall response will be nonlinear and can be predicted. The dotted lines of Fig. 5.28 indicate the response of a transducer that travels the distance indicated at the pressures indicated. In the left-hand section of the drawing are sets of curves from the experiments for probe settings, in thousandths of an inch, of -1, $-\frac{1}{2}$, 0, $+\frac{1}{2}$, and $+1$. In this drawing, if the transducer is at $+1$ for 15 mm of Hg and at -0.001 in. for 200 mm Hg, the response will be as shown by the dotted curve. The interaction with the tissue elasticity will give high readings at low pressures and low readings at high pressures relative to a linear system (a system in which the elastic forces are small). The other two parts of Fig. 5.28 illustrate the expectations from stiffer systems permitting less deflection.

The softer the suspension used in such a system, the milder are the requirements for stability and amplification, and indeed systems have been

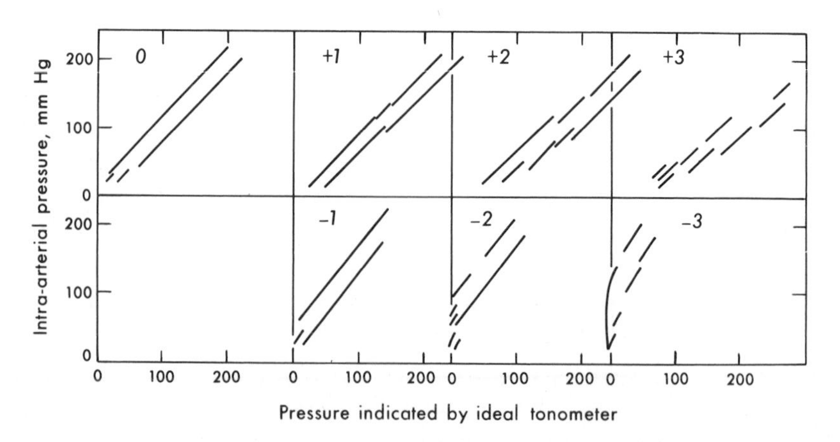

Fig. 5.27 The spread in data for any setting of the probe. Part of the spread is due to a small uncertainty in initial probe setting when working with an adjustable probe.

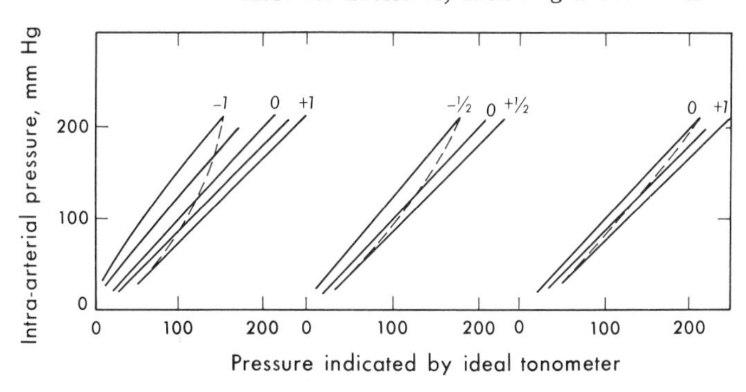

Fig. 5.28 The solid lines are the data for the abdominal aortas of medium-sized dogs. The dotted curves give the calculated response of a pressure-measuring device with less than infinite stiffness and undergoing excursions between the indicated limits in the measurement process.

made which grossly indent the artery walls in order to produce a large signal. Such systems also have a poor frequency response if they are allowed to move through a large easily sensed deflection. But from the above observations it is possible to say how stiff a system must be in order to yield blood-pressure indications to a specified degree of accuracy from the abdominal aorta of such dogs. A preliminary calibration is still required and can either consist of observations of spring stiffness and transducer area, or else indications received from observations on an inflated bag at known pressure.

A tonometer based on these principles can measure pressure in the vein of a human through the skin if the vessel is outward bulging due to being below the level of the heart. Thus far, less success has been achieved in transcutaneous pressure measurements on the deeper-lying arteries, due to positioning problems. These methods have potential application to absolute pressure measurement through the wall of any body cavity when one surface is or can be exposed.

6

Temperature Sensing and Transmission

It sometimes appears that temperature is measured in certain experiments for no better reason than that methods of measurement are known. However, in many experiments temperature is an extremely significant variable, with most life being able to exist over only a very limited range.

The comments on transducers at the start of Chapter 5 apply also to the measurement of temperature. One way of noting temperature changes is by the mechanical expansions that they produce. In the same way, changes in the dimensions of coils or capacitors can cause drift in otherwise precise circuits. Placing a volatile liquid in a pressure transducer gives enhanced sensitivity to temperature. An ordinary mercury thermometer can be set to modulate a transmitter either through optical effects or direct electrical effects. There remains a significant pressure sensitivity in general. Under a pressure of a few atmospheres, a thermistor may read high by 0.1°C, whereas a mercury thermometer may be off a full degree; a humidity observation based on a wet bulb can be considerably worse due to gas law problems with changing pressure. Thus mercury-in-glass thermometers can give somewhat misleading readings in high pressure physiology studies, for example, on divers.

As with many other variables that constitute energy sources, there are some self-generating transducers. A thermocouple can be used to monitor temperature in a biological system, and because of the low electrical impedance level there is little short-circuiting effect by body fluids and tissues. The very low voltages generated by these devices are perhaps best handled by converting them into alternating current with a "chopper," after which amplification and transmission take place. Such a system is a bit complex, but can be quite stable. In ordinary measurements with a thermocouple, the reference temperature may be the ambient temperature to which the rest of the circuit is exposed. In telemetry a reference junction can be placed outside of the animal and all readings are then referred to this possibly changing temperature. Significant amounts of heat can be carried along the wires of a thermocouple either to or from the site under investigation, thus changing its temperature from normal. In experiments with thermocouples applied to bats it is usually found that they tangle in the leads badly even when not flying, thus indicating the desirability of telemetry techniques. Similarly, mice are often observed to pull out their thermocouples by the leads. Because of the low level of the signals from simple thermocouples, it is generally true that they are not convenient sensors in the telemetry of temperature.

In some of the later sections we discuss the sensing of sounds and pressure changes by the piezoelectric effect in ceramics of barium titanate and related materials. It is sometimes found that a change in temperature of such a transmitter will cause a momentary large signal which may bias the input amplified into an inoperative saturated condition. Changes in temperature can generate a voltage in these substances, and there are perhaps useful applications for this pyroelectric effect.

There are also transducers of temperature in which changes affect some parameter of the circuit such as resistance, capacitance, inductance, or mutual inductance. In many applications it is especially convenient to use components that show a change in resistance for a change in temperature, as do some nonmetallic materials. Thermistors are temperature-sensitive resistors which decrease in resistance with an increase in temperature. They are available with resistances ranging from several ohms to many megohms and with various rates of change of resistance for a given change in temperature. In part, the convenience of these devices lies in the fact that relatively high temperature coefficients are available. It has been mentioned in Chapter 2 that thermistors also display a negative resistance because of their high negative temperature coefficient. This gives them certain amplifying characteristics but also places certain restrictions on their use and testing.

An early study of material having a negative temperature coefficient of resistance is Faraday's investigation of silver sulfide. Nernst later carried on experimentation with rare earths, but apparently did not obtain stable or reproducible units. Present raw materials were put to use at the Bell Telephone Laboratories in the early 1940's. The ceramic mix for modern thermistors usually combines oxides of manganese, nickel and cobalt. The materials are calcined, blended, formed into the desired shape, and fired into a hard ceramic form. Very tiny beads can be formed on fine wire to give a unit with a rapid response. The beads are often glass-coated or mounted in a glass probe. Larger thermistors are directly formed into discs, washers, rods, and the like.

Figure 6.1 depicts a representative squegging oscillator circuit whose pulse rate is controlled by a thermistor. In this particular circuit the battery is tapped down on the coil to give a stronger radio signal (see Appendix 1). This circuit clicks slowly enough so that timing of the rate can be done by ear.

Thermistors are only slightly affected by pressure, although higher resistance thermistors can give misleading information in the presence of moisture. With a thermistor it is possible to measure temperature *changes* down to about 0.001°C, in which case bridge circuits, as shown in Fig. 3.3, are convenient. Certain manufacturers, however, suggest that a thermistor may allow absolute readings over a longer period such as a year to perhaps an accuracy of 0.1°C. Repeated calibrations of a number of the blocking oscillator circuits suggests that the longer-lived ones do not drift by more than 0.1°C per month due to such factors as battery aging, and thus this might be taken as an approximate figure for expected accuracy, barring other problems. Such transmitters as these can be made rather small, as shown in Fig. 6.2. The complete temperature transmitter shown uses a

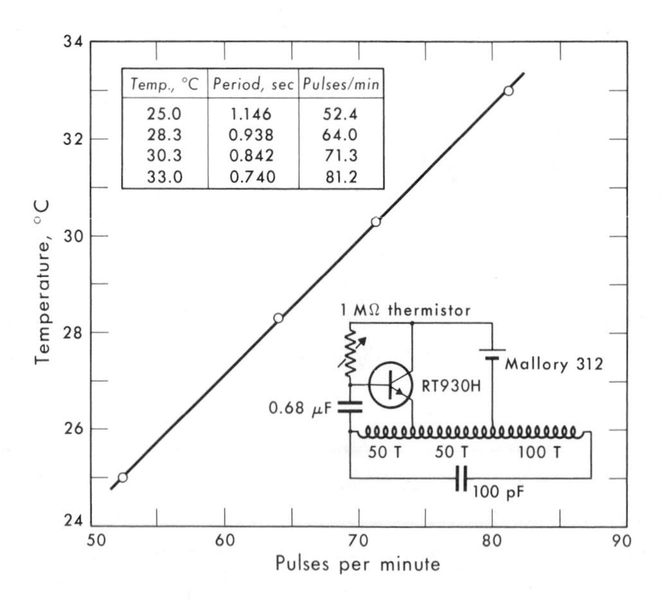

Temp., °C	Period, sec	Pulses/min
25.0	1.146	52.4
28.3	0.938	64.0
30.3	0.842	71.3
33.0	0.740	81.2

Fig. 6.1 Observed response of a blocking-oscillator temperature transmitter with a tapped-down coil for increased signal strength.

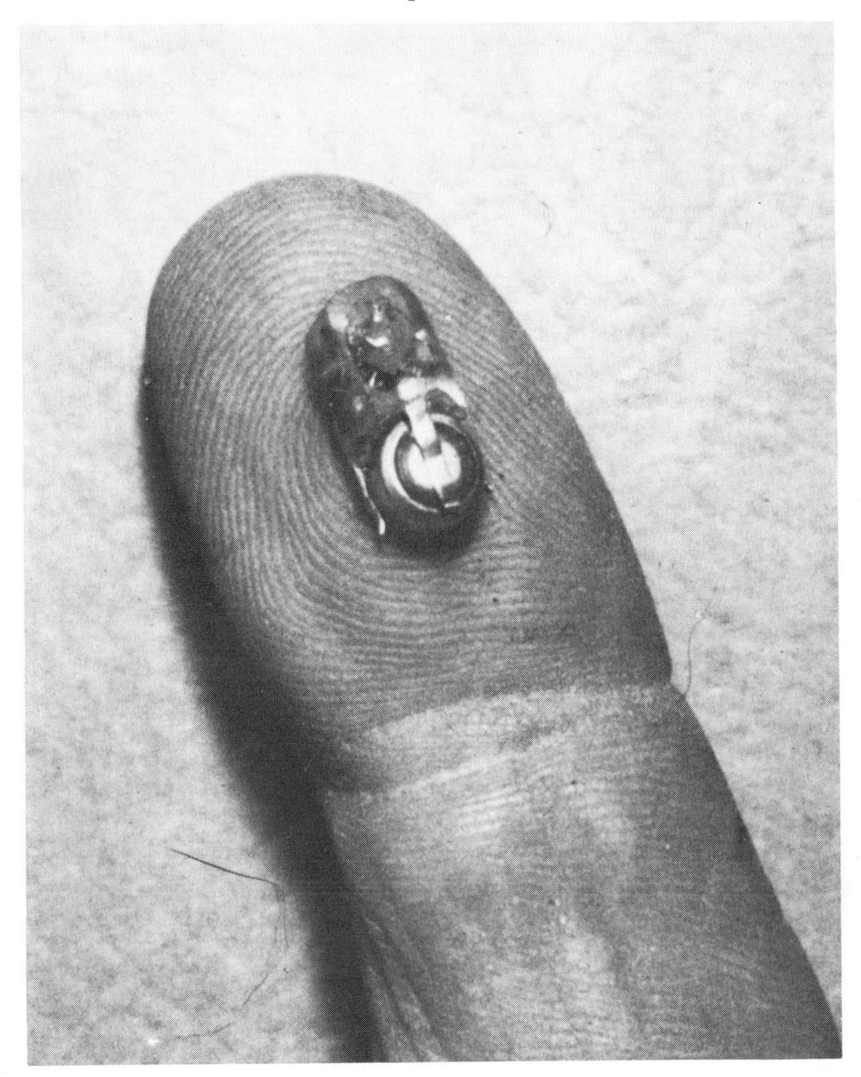

Fig. 6.2 A transmitter on a human index-fingertip. The volume is 0.3 cc, the weight 0.7 g, the range about ½ m, and the life 3 mos.

Mallory 212 cell in a spring clip for easy changing, and the entire unit is 12 by 4.5 by 6 mm, and weighs 0.7 g.

An example of the use of such a transmitter in an animal experiment is given in Fig. 1.8, with the data shown in Fig. 14.2. They are also applicable to human studies, sometimes in combination with surgical techniques. Thus in a recent study with Pegg and Baldwin it was desired to determine the temperature of the unanesthetized human brain. Figure 6.3 shows a pair of radiographs of the head of a man with the transmitter placed on the dura mater. There were approximately 2 cm of tissue over the transmitter. A somewhat smaller transmitter could have been placed in a sulcus.

The data from the experiment are summarized in Fig. 6.4. The data were

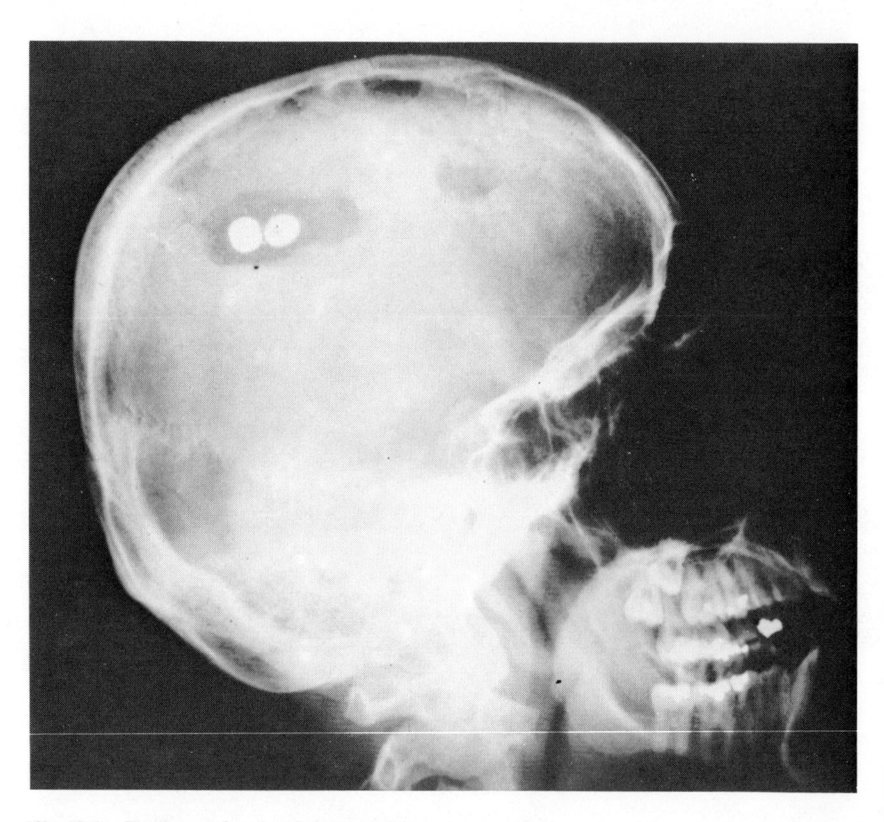

Fig. 6.3 Radiograph of a human with a temperature transmitter on the surface of his brain following a surgical procedure.

taken by nurses, without any special training, who were instructed to time 100 clicks while listening to a receiver. This should result in errors of under 0.2°C. The surgical procedure, during which the transmitter was both introduced and later removed, in both cases involved cooling of the body. The warm-up from surgery, diurnal temperature variations, and the cooling down to surgery are clear and correlated with rectal temperature. Somewhat less distinct is an elevation in temperature following epileptic seizures, with a hint of a temperature drop before seizures. This last might reflect a decrease in cerebral blood flow resulting from hyperventilation, which is known to induce seizures. It would be interesting if, among many species, large brains had the same average temperature.

The capacitance, inductance, and mutual inductance in any circuit can also be controlled by changes in temperature, as was the case with pressure

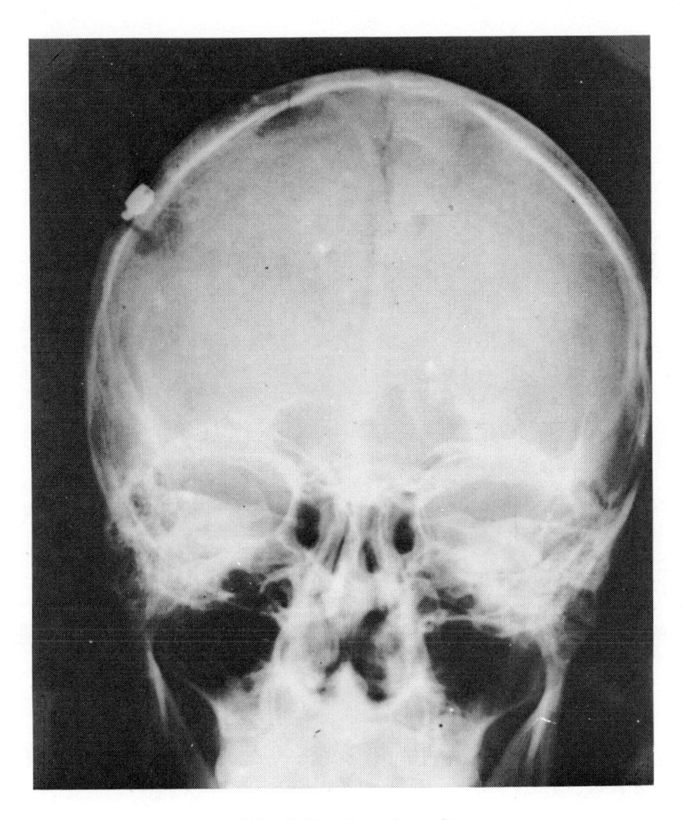

Fig. 6.3 (*continued*)

transducers. Most small ceramic capacitors change their capacitance with temperature changes. The materials that show the highest dielectric constant (giving the most capacitance in the smallest space) often show the highest temperature coefficient. A transmitter using such a transducer is shown in Fig. 13.12. Any capacitor selected for this application should, however, be checked for hysteresis; that is, when temperature is changed and returned to the initial value it should be checked to be certain that the capacitance

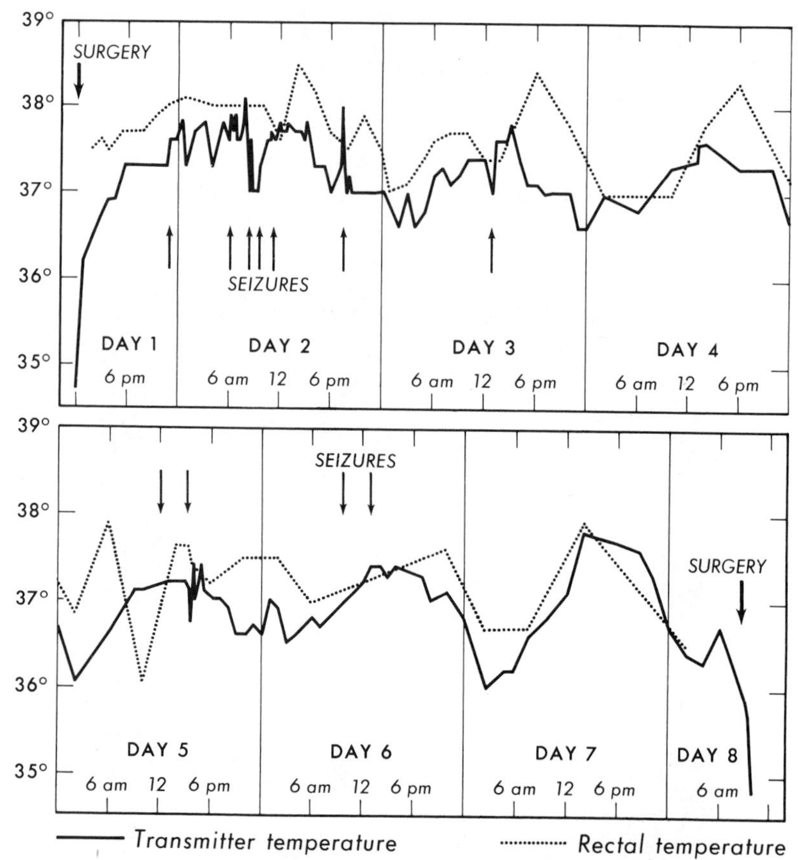

Fig. 6.4 Human-brain temperature following implant surgery, during which chilling was used. Some epileptic seizures may have gone unnoted.

also returns just to the initial value. Some capacitors are acceptable for use under small temperature excursions, but not when the expected temperature variations are large. The high dielectric constant shown by many of these materials disappears above a particular temperature, sometimes called the Curie temperature. Adjustment of the composition of the material allows setting this temperature in a desired range (deBretteville, 1953). Operation near the Curie temperature can result in extremely large capacitance changes for small temperature changes.

Similarly, the magnetic permeability of various ironlike materials, including ferrite, changes with temperature. A coil wound on such a material will have an inductance that changes with temperature. Wolff (1961) has fabricated units based upon the magnetic properties of metals specifically prepared to temperature compensate electric meters. Here also there is a temperature above which magnetic properties abruptly disappear, and it is referred to as the Curie point.

Oscillatory systems are affected by temperature in a way that can either be useful or a nuisance; for example, there are many ways in which an oscillator crystal can be cut from its parent. The frequency of a crystal may increase or decrease for an increase in temperature, depending on the orientation of the resonator plate with respect to the crystallographic axes. In Fig. 2.15 is shown the well-known effect of cutting crystals in sheets containing the X axis but at various angles to the Y axis. It is often desired that an oscillator be controlled by a crystal which changes its frequency very little with temperature changes, and it is seen that such crystals can be constructed. It is also possible to cut a crystal in such a way that it will be temperature sensitive. A pair of crystals can be bought rather inexpensively, one of which has little temperature coefficient, and the other having a rather large temperature coefficient but a similar frequency of oscillation at room temperature. Comparison of the difference in frequency of these two crystals allows estimation of temperature to something like 0.001°C. Such a pair of crystals can be incorporated into an oscillator with a beat frequency serving as the subcarrier, but in many cases it is more convenient to let one crystal control the transmitter frequency and incorporate the other crystal as a reference standard in the receiver.

Depending on the mounting of the crystal, the response of such an instrument may be relatively slow. Various schemes for speeding the response by feeding into the system an amount of heat necessary to always maintain a fixed temperature, and then using the heating or cooling as a measure of external temperature, have been considered for reducing the response time, but they require too much power for these applications. An alternative consideration is depicted in Fig. 6.5 in which a slow transducer with good long-term stability has its signal combined with a faster transducer with less reliable long-term stability. In this particular case we can alternatively view the operation as the incorporation of a slow precise reference standard for the occasional calibration of the thermistor. If the mounting of the crystal is such that it acts as a single time-constant thermal system, then the two signals can be combined directly to give an accurate indication at all times. There is a range over which the transients can cancel, although accuracy is apparently degraded somewhat because of imperfect matching. The commercial receiver indicated handles the two types of information simultaneously, but other receiver systems also are effective. The crystal signal controls the basic radio frequency; the thermistor signal controls the frequency at which the basic signal is amplitude modulated. In working with such systems one of the main problems is to find standards against which

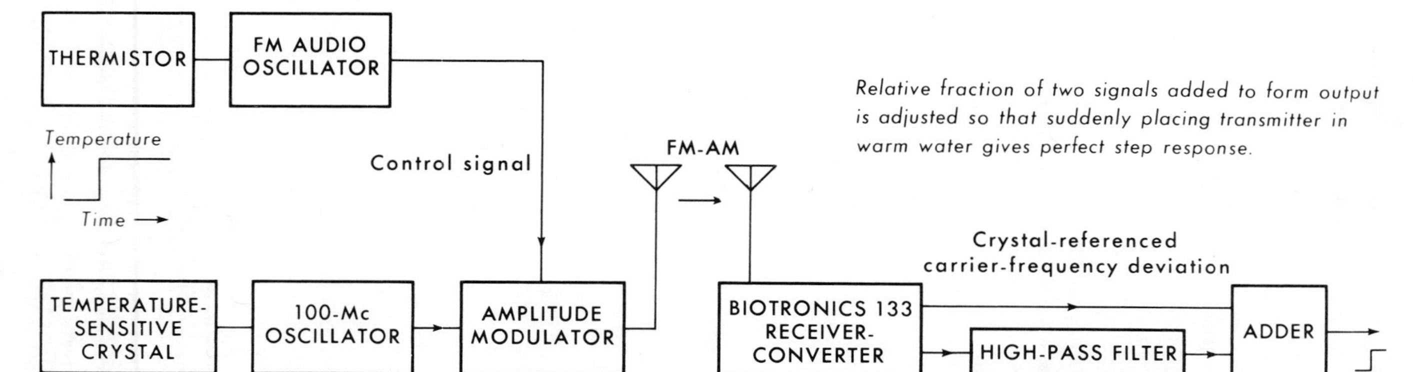

Fig. 6.5 The signal from a slow but stable transducer can be combined with that of a more rapid but less permanently stable transducer in order to give a more desirable response. The relative fraction of the two signals added to form the output is adjusted so that sudden placement of the transmitter in warm water gives a "perfect" step response.

calibration can take place with sufficient accuracy. Precision such as this apparently is not required for most presently proposed studies on animals. There are other situations in which this general scheme of combining a fast transducer with a slow stable one may prove valuable. For example, pressure measurement in a heart by catheterization can be made fast and accurate by combining two signals. The fluid column carries absolute pressure to a manometer (suitably adjusted in height) while a piezoelectric element in the tip senses fast changes.

Some general comments on calibration of various transmitter types are made in Chapter 12, and methods of estimating response time are also indicated. An added observation especially applicable to temperature transmitters is that of thermal capacity. If a tissue is to be monitored in temperature by a transmitter implanted therein, then it must supply or remove the heat necessary to change the temperature of the transmitter. An observation was made on one transmitter of known weight and initial temperature by dropping it into a thermally isolated beaker of water of known weight and different initial temperature. From the final temperature it was possible to calculate an average specific heat for the entire transmitter, and this turned out to be approximately 0.8. Since this is similar to, but slightly less than, the specific heat of water or tissue, in an experiment with this transmitter slightly less heat would have to be supplied than would have to be supplied to a similar mass of tissue in its place. In terms of the normalcy of the adjacent tissues, it should be noted that the thermal insulation of the transmitter also affects how rapidly heat can flow in and out, and this is also somewhat different than for an ordinary tissue. In some cases it may be desirable to place a thermistor at the end of a stalk or antenna away from the body of the capsule so that the tissues being observed do not have to slowly charge and discharge the thermal capacity of the entire transmitter in order to change a reading. Temperature in a biological system is often a slowly changing variable, and that is why a number of transmitters have been described which carry information by clicking slowly enough to count by ear. This allows the use of very simple receiving equipment and also allows transmission over a very small range of radio frequencies (limited bandwidth). In those cases in which significant changes can take place in fractions of a second rather than minutes, the above considerations are relevant and continuous demodulation and recording are required.

In principle, it is possible to tell temperature by the shift in the radio frequency generated by any oscillator. Temperature affects the transistor beta, bias, and current, and also its collector capacity. There is also an effect on the capacitance and inductance of the resonant circuit, and on the position of any diaphragm. However, with some transmitters, a slow creep in diaphragm position (e.g., because of the plastic properties of the diaphragm or humidity sensitivity of the diaphragm) can give a slow shift in baseline which may be mistaken for a temperature change. In other transmitters a change in the proximity of surroundings may cause similar shifts.

A diode changes its various properties in response to temperature changes, and this is especially true of germanium diodes. This can be used not only for measuring temperature, but in some cases it can be used to compensate an overall circuit against temperature changes. Thus a typical multivibrator slightly shifts its switching points and overall frequency in response to temperature changes. The circuit in Fig. 7.8 is compensated in this fashion, and the frequency of the sinusoidal oscillator in Fig. 3.2 is also compensated against temperature changes by a germanium diode.

If temperature difference alone, rather than the actual temperatures at two sites is of interest, then it can be best to transmit the difference from a single transmitter. A pair of thermocouple junctions produces a signal proportional to temperature difference, and a bridge arrangement of thermistors can give a similar result. Crystal oscillators with positive and negative temperature coefficients can be beat against each other to give a difference frequency depending on temperature difference. However, two temperature responsive blocking oscillators acting at different rates can have their signals combined (added) through a pair of resistors so that each turns on and off an output transmitter oscillator; such a subcarrier arrangement transmits the two temperatures individually with little extra complexity or power drain.

An extensive series of observations has been made of temperature measurements on human subjects by Fox, Goldsmith, and Wolff (1961).

"In this investigation, intestinal pill temperatures have been compared with simultaneous thermistor temperature measurements from stomach, esophagus, rectum, external auditory canal and sublingual regions, while exposing the subjects to a variety of thermal situations. These included sitting at rest in a comfortable environment, rapid heating and cooling by immersing the lower half of the body in a water-bath, and exposure of nude or lightly clad subjects, with or without physical exercise, to hot and cold conditions in climatic chambers."

Their conclusions were the following:

"There was no evidence that responsiveness to body temperature change or absolute temperatures recorded differed in different parts of the small intestine. Bulky intestinal contents, especially in the stomach or large intestine, reduced responsiveness. In general, pill temperatures in the small intestine accorded most closely with rectal temperature but were more responsive to temperature change. In cold conditions, with the subject peripherally vasoconstricted, temperatures in the mouth, ear, and esophagus all fell well below core temperatures as measured in the rectum and by the pill in the intestines. In hot climate conditions the differences in temperature between the sites were small, but the mouth and ear responded the most rapidly, the pill in the intestine less rapidly and the rectum the least rapidly to changes in environmental temperature."

As they indicate, there can be temperature gradients in the body. Thus gradients in the body are encouraged during anesthesia with cooling to minimize the chance of ventricular fibrillation of the heart. However, it is difficult to maintain two well perfused systems of a body at different temperatures. A human is almost a poikilotherm when deeply anesthetized.

Warm-blooded animals show a daily cycle of temperature like that suggested in Fig. 6.4 and also recorded in Fig. 10.5. A temperature pattern such as Fig. 10.5 is noticeably modified by the appearance or presence of a human. In addition, females show superimposed changes in temperature associated with ovulation. To increase or decrease fertility it can be useful to monitor these temperature changes using transmitters either inserted or implanted; this has been done in various species, including man.

Following death, the body cools. It might be effective in settling certain forensic medical problems of "time of death" to know something about representative temperatures at given times after death, in some sense of the word. If it were desired to do this with humans, it could perhaps only be done by feeding a small temperature transmitter to subjects obviously about to die.

A type of temperature cycle seen in a cold-blooded animal was shown in Fig. 1.13. The animal was free to move and adjust his temperature. Small oscillations during the day were recorded, as well as heating and cooling at the start and finish of the day. There is no reason to expect heating and cooling times to be the same unless the regulation system of the animal is linear; in this case the difference is largely due to changing surrounding conditions.

Many kinds of biological assay, drug testing, disease testing, pregnancy testing, and so on, depend on the injection of material into a test animal and the observation of a response. In some cases the response can be conveniently recorded or observed as a temperature rise, with the animal simply being fitted with a transmitter. In other cases where the freedom of an animal from disease is to be judged by temperature monitoring following a suitable injection, it should be possible to both give the injection and monitor the response by radio transmitting darts shot at the animal, thereby eliminating the trauma of an actual capture in cases where that is extreme.

The above comments are largely directed to measuring temperature relatively deep in the body, as a physician might do in evaluating health. Temperatures above the fatty layer are quite labile. There is a response by subcutaneous temperatures to ambient temperature changes, and they can also change in the time of a few heart beats in response to startling events, etc. A human blush or the red neck of a courting male ostrich can involve a degree temperature change. The more typical moderate alterations perhaps are associated with circulatory changes that could also contribute to psychogalvanic reflex.

Without discussing the problems of radiation measurement or the use of liquid crystals, it can be said that surface temperature measurements are difficult to make without upsetting the normal flow of heat. In some cases a more meaningful indication can be had by placing outside the sensor a "guard" region driven to have the same temperature as the sensor, for example, a thermistor beyond which is another thermistor and a heater. (If temperature were below ambient, a Peltier cooler rather than a heater would be needed; minimum interference with evaporation would be required.) Surface temperature does have biological interest. In humans, it indicates subsurface irregularities (thermography), and in animals generally, pigmentation probably affects temperature regulation.

The analogy of this last method with that in Fig. 5.19 should be noted. The application of several principles of measurement to rather different variables is indicated in this chapter and the previous, and it is hoped that the reader will appreciate the applicability to other parameters as well.

7

Bioelectric and Chemical Electrode Potentials

1. ELECTROCARDIOGRAM AND RELATED TRANSMITTERS

The functioning of many systems in a living organism, both plant and animal, is accompanied by electrical signals that can be monitored by suitable electrodes and circuits. From the observed pattern of bioelectric potentials, it is often possible to monitor physiological processes, both normal and pathological. A few examples might be mentioned. For instance a gold wire coated with polyethylene over most of its length can remain in the muscle of a cat for a year to continuously signal movements of various structures. Alternatively, an electrode suitably placed with respect to the round window of the inner ear can pick up a voltage having a frequency of any impressed sound signal and showing no time delay. Such cochlear microphonics, by indicating what sounds an animal might actually be receiving, could be valuable in studies on the echo-ranging sonar systems of bats by indicating the character of returning echoes, including the effects of external ear motion. Likewise, similar observations made on seals or sea lions subjected to changing pressures in a tank of water might indicate something about the rapidity of pressure equalization in the inner ear by detecting transient distortions of impressed sound signals.

Another example might make use of the relatively steady voltage developed across the retina of the eye (not to be confused with the fluctuating voltages observed in a retinogram). Because of this voltage, turning the eyes from side to side renders one side of the face positive with respect to the other by an amount depending on the angle of gaze. It is thus possible to follow movements of the eyes in a subject with closed lids, asleep, or sleepwalking. By placing a small television transmitter on a helmet on the head of a subject in order to telemeter the direction a subject is facing, and using these potentials of oculography to indicate the direction of a subject's eyes within his head, it is possible to tell where a subject is looking or focusing his attention. (Such studies have been carried on by Shackel in England, with human subjects.) These signals can all be compared with the simultaneous brain-wave pattern, an example of whose recording and processing is indicated in Fig. 1.10.

Perhaps the most widely familiar recording of this general type is that of the cycle of voltages associated with the beating of the heart, an example of which is also given in Fig. 1.10. This is generally regarded as a parameter of considerable interest, and several circuits suitable for its telemetry are used as examples in this section. With little or no modification, however, these same types of circuits can be used to follow the electrical aspect of the functioning of other body structures.

The voltages generated in these processes can cover a rather wide range. The potentials associated with the functioning of the brain, as monitored on the surface of the scalp, can be in the range of tens of microvolts, although as electrodes are placed into the brain the potentials can become at least an order of magnitude larger. The beating of the heart can generate signals at the millivolt level. Some electric fish generate signals ranging from a fraction of a volt up to approximately 600 V, but these are rather special cases. Certain plants, when disturbed, produce signals in the millivolt range. The signals associated with plants are generally somewhat slower than those from animals. To transmit an electrocardiogram from a human subject with little distortion requires the transmission of a range of frequencies extending from approximately 0.5 Hz up to about 100 Hz. The impulses associated with the functioning of certain muscles and nerves may require a range of frequencies extending an order of magnitude higher, or more. (In some cases a 20 Hz filter is used to reject muscle potentials when other slower voltages are being studied.) Workers have found potentials in the retinas of some fish that are steady when a light is on and disappear when it is off; to transmit these requires either a frequency response extending down to dc or else special arrangements (Fig. 12.5). In these various cases the input transducer is some system of electrodes whose potentials are supposed to mimic those otherwise existing at the placement site. These can vary from a piece of metal having a resistance in the range of a hundred ohms up to long thin glass capillary tubes (micropipettes) filled with fluid and having a resistance in the range of tens of megohms. Obviously, many specific situations will fall out of the range of these generalities, but this hopefully conveys an impression of what is involved.

In some cases the use of telemetry *per se* confers certain advantages. Consider trying to monitor the electrocardiogram of a subject walking about in a room and trailing behind him several wires leading to a fixed amplifier and recorder. Presumably the amplifier would have a balanced differential input between two electrodes in such a way that synchronous changes in potential of both electrodes would not be recorded, but only differences in the signal between the two electrodes. This would normally necessitate a third wire running to the subject in order either to somewhat fix or monitor his average potential. Problems will often be observed, however. It may be found that the entire body of the subject is raising and lowering in potential by ten volts in synchronism with the power wires in the walls of the room, or more if he touches a fixture. If it is a dry day, walking over a rug may cause changes in average potential of 10,000 V. Most amplifiers are incapable of rejecting these very large "common mode" signals while retaining sensitivity to the relatively small potential differences across the chest, which are of interest. In some of these cases it will be found that a small telemetry transmitter in or on the chest will, in its entirety, follow changes in average body potential, thereby leaving essentially only the differences in potential between the two input connections applied to the input stage of the amplifier. Being inside of the body confers something of an extra bit of shielding. Even in extremely difficult situations where a telemetry unit with a differential input is desired, it is usually possible for the input stage to have a sufficient voltage range of operation so that satisfactory results are achieved.

In some cases a differential input stage allows the suppression of unwanted neural signals arising away from a recording electrode.

Since the impedances of electrodes vary with time, depending on various surface phenomena, it is desirable that any electrode be coupled to an amplifier with an input impedance considerably higher than its own. The amplitude of the signal received at the input connection will then be less subject to significant variations. Thus in general for the applications of this section it is desirable that an amplifier have an input impedance above roughly 0.5 MΩ.

In any of these studies the electrodes generally constitute the most subtle and troublesome elements of the system. Books on electrochemistry or physical chemistry generally contain material relating to the theory of these matters. A few practical points are included here (see also Weinmer and Mahler, 1964; and Lykleme, 1964). Electrodes generally show a fixed difference in potential from the bulk of the solution into which they are immersed, in association with the formation of an electrical double layer. Slow changes in such electrode potentials can interfere with measurement of steady or slowly changing bioelectric potentials. Artifacts sometimes arise in connection with movements because of resistance changes at electrodes where currents are flowing because of electrode potential differences. In biological work, it would be desirable to use reversible electrodes, which can be of three types. There can be (a) a metal dipping into a solution containing its ions, (b) an unattackable metal in a solution containing ions in two valence states, or (c) a metal in contact with one of its insoluble salts immersed in a solution of a soluble salt of the same anion. An example of this last is silver chloride, whose preparation will be mentioned later. When current is passed through electrodes immersed in solutions of halogen acids or of halides (e.g., sodium chloride solution), the chlorine ions are discharged before the hydrogen and hydroxyl ions of the water; if the discharge potential of the chlorine were greater than that required for oxygen evolution, the electrode would not be reversible because of the layer of oxygen formed thereon. If they are to be used for recording slowly changing potentials, silver chloride electrodes should be stored in saline and interconnected by resistors equivalent to the input impedance of the amplifier. Because of questions of toxicity, it is often necessary to employ electrodes of platinum, gold, or stainless steel. Stainless steel is stainless in the presence of oxygen, and certain grades are quite inert in biological tissues. However, if a direct current path appears, then the positive electrode will rapidly decompose. A thin layer of chromic oxide normally present on this material restricts the low frequency response in many cases so that artifacts of drift are not noticeable.

Of course, a crude test of the functioning of a voltage transmitter is simply to grasp the input wires in the fingers of the two hands, thereby applying the electrocardiogram voltages to the input, without worry about the material of the wire. A more usual electrode for picking up voltages from the surface of the skin consists of a piece of silver metal which has been chlorided, and which is held in some sort of cup on the skin of the subject, with improvement in electrical conduction being effected by a paste made of concentrated Ringer's solution thickened with some inert material. In some special applications, it is found that results are better if the surface of the skin is lightly sandpapered (not to the point of bleeding) before application of the electrode with its conducting jelly. It was found by Patten, Ramme, and Roman that electrodes that worked well during exercise could be formed by spraying a conductive mixture over a thin wire and the skin. A mixture of silver flakes in Duco cement dries rapidly, leaving a thin flexible layer that holds the lead relatively firmly in contact with the skin, even while showering, and yet is easily removed with acetone.

Some circuits suitable for transmitting these kinds of signals are shown in the next few figures. In Fig. 7.1 are seen the components of the transmitter used for sending the brain-wave pattern of Fig. 1.10. The two input

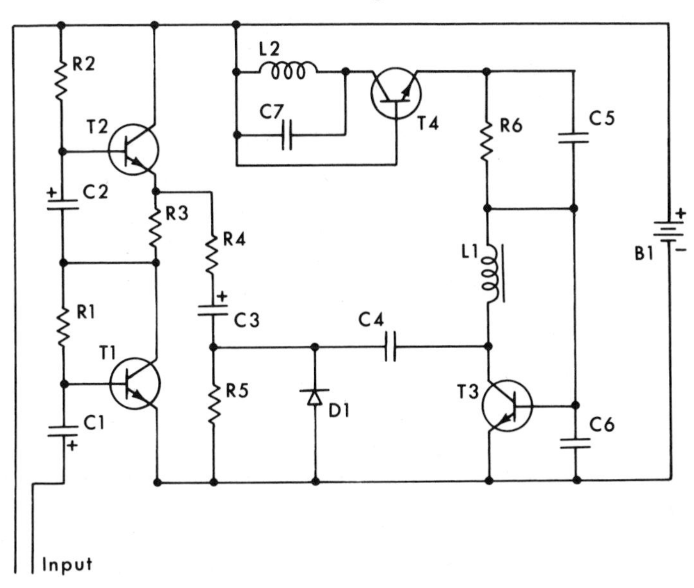

Fig. 7.1 Voltage transmitter with two input stages of amplification to drive the voltage sensitive capacitor *D*1. The frequency-modulated oscillator drives an output stage whose antenna *L*2 is tuned to resonance by *C*7. Representative components are: *B*1 is 1.5 V; *C*1, 2, and 3 are 2 μF.; *C*4, 5, and 6 are 430 pF; *R*1 and 2 are 10 MΩ; *R*3 and 6 are 200 kΩ; *R*4 is 500 kΩ; and *R*5 is 5 MΩ. Most small transistors are suitable.

transistors apply their signal through capacitor *C*3 to the variable capacitance *D*1 which frequency modulates the following oscillator stage. A separate output stage is shown to emit a signal in the frequency range of several hundred kHz. A lesser signal can be taken from the collector of the first transistor, in which case *T*2 can be omitted and *R*3 connected directly to the positive side of the battery. That circuit was used for the transmission of the electrocardiogram and for the transmission of the voltages generated by the accelerometer element in Fig. 1.10. Using the simplified circuit, we were able to obtain continuous transmission of electrocardiogram from within a dog for over a year. Presently outdated parts were used in constructing this circuit, but it is also quite effective when modern components are substituted.

Figure 7.2 shows a circuit suitable for use at a transmission frequency of 100 MHz. The high-resistance input stage is shunted with a 1 MΩ resistor, and delivers its signal to the second stage which serves as oscillator, amplifier, and frequency modulator. Tests on a number of transistor types were

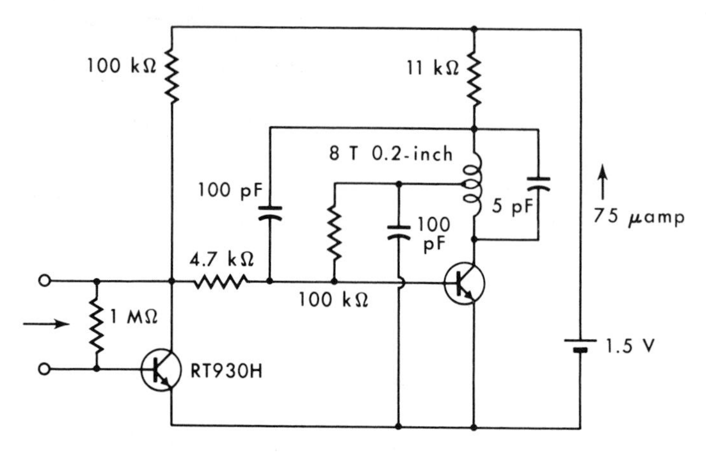

Fig. 7.2 Frequency-modulated voltage transmitter in which the oscillator also functions as amplifier and modulator for the signal from the input stage that presents a high input resistance. Several transistor types, including the D26G1, work at high frequencies.

made to determine which were capable of operating at this frequency with a low voltage and a low current. One of the best proved to be a transistor type 2n2857, but the smaller and cheaper D26G1 performs acceptably.

In Fig. 7.3 is shown a modified circuit containing some useful changes. For reasons that we shall discuss later, it is often true that decreasing the number of turns in the coil will increase the signal from it, especially if the area is also increased. A rectangular coil of two turns is shown. The entire circuit is built into this coil to save space. The battery will reduce the signal if placed in the coil, but causes no problem if placed beside the coil, outside of it.

We have found that the leads that bring in the low-frequency bioelectric potentials can also carry out some radio-frequency currents to beyond the extent of the transmitter case, thus somewhat increasing overall signal strength. Shifting the leads however, does slightly shift the radio frequency. In the present arrangement, a small single turn couples some energy from the basic oscillator coil, and this flows out into the tissues of the subject. By making the area of the output coil smaller than that of the main coil, it has the effect of being less than one turn in the transformer. Transmission range is approximately 5 m, with a life of about three weeks on a single cell.

In Fig. 7.4 is also depicted a 100 MHz frequency-modulated voltage transmitter using more customary connections. Once again the input electrodes are used to carry out some radio-frequency power. In this case modu-

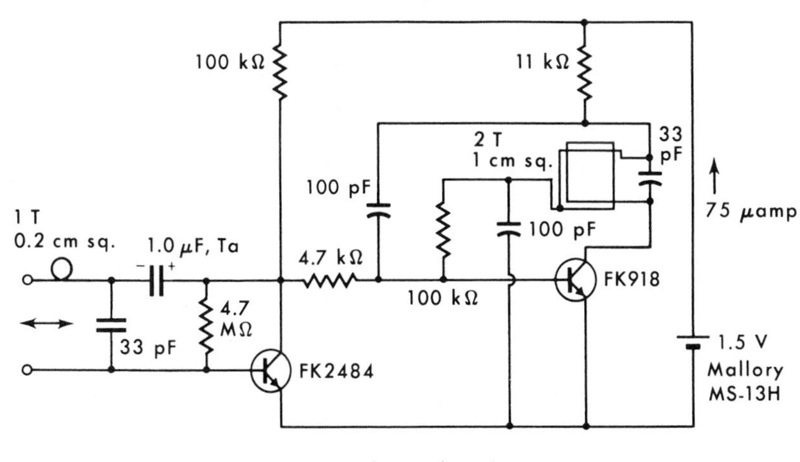

Fig. 7.3 A circuit similar to that in Fig. 7.2 employs a two-turn rectangular coil and couples some of its signal back out through the input leads.

lation is produced through the action of the voltage-controlled capacitor. This circuit was quite effective, yet small enough to be swallowed by a human subject. The steady current drain was 200 μA and the circuit was used in some experiments with a 60 mA hour battery, type RM-13GH.

Oscillators which are frequently modulated by a varactor (as opposed to using collector current modulation in response to base drive) have certain advantages. They can display a higher input impedance, there can be

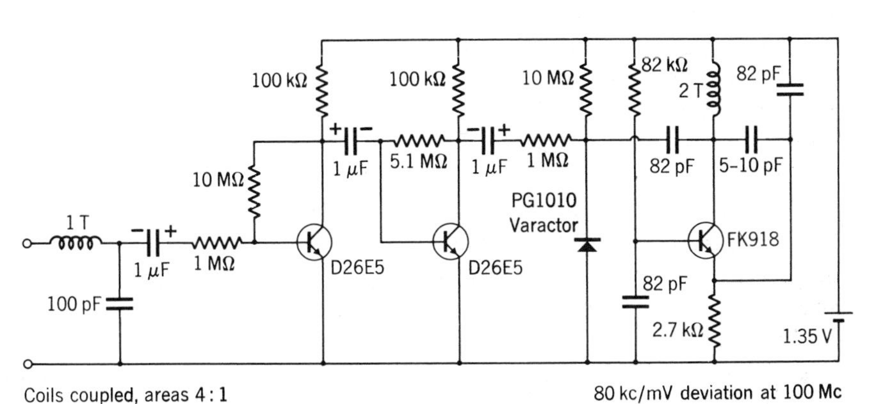

Fig. 7.4 A high-frequency circuit modulated by a voltage-sensitive capacitor and coupling some of its output signal back through the input leads.

smaller changes in oscillation amplitude with frequency changes, and they can display a higher sensitivity. But they are temperature sensitive, several extra parts are required, and some of these parts can be large. If the investigator wishes, a transistor can be used in place of the voltage-sensitive capacitor. Thus an FK3502 with the base attached to the emitter (using the collector junction) is small and quite sensitive, whereas a 2484 is not so good because of less sensitivity and a lower Q.

Generally, in a collector-modulated transistor oscillator, if there is low coupling in the feedback loop so that power output is low, modulation sensitivity will be high. Placing a voltage-sensitive capacitor from collector to ground in such an oscillator, which biases the varactor and yet effectively connects it across the tuned circuit, increases the frequency swing for a given applied voltage. In most oscillators the average voltage on the resonant circuit is the full battery voltage, and changes in base drive change the current and oscillation amplitude. Even though the average voltage is constant, replacing some of the fixed capacitance gives a different average capacitance with changes in amplitude of oscillation. At a given frequency a big inductance and a small capacitance tend to give large frequency shifts with minimum amplitude change. Amplitude modulation can affect the effectiveness of the limiter in the receiver, and also affect sensitivity to give extra frequency modulation there. At limiting range or with a poor receiver this can interfere with the data handling, and thus is undesirable. In brief, variable alternating voltages with a fixed average give frequency changes, and the more of the resonant capacitance that is nonlinear, the smaller will be the undesirable amplitude modulation accompanying a given frequency modulation.

The signal transmitted from the stomach of a canine subject after swallowing the transmitter type in Fig. 7.4 is shown in Fig. 7.5. From the time scale it can be seen that this activity is not associated with the beating of the heart, although such potentials sometimes appear and can be emphasized. Such gastroelectromyography could prove useful in studying motility of the gastrointestinal tract under conditions when pressure recordings were unsuitable. An interesting study would combine the simultaneous transmission of electrical signals and pressure signals from a single transmitter unit to give kinds of information which could only be approximated in the past (Daniel and Chapman, 1963). Notice that the absolute value of the transmitted voltage wave is recorded, and not just the relative pattern with time.

Along the gastrointestinal tract it might be possible to detect an injury potential at an ulcer. A different voltage might be expected at a malignancy due to a different growth rate from the surrounding tissue. An external magnetic field could be used to slowly spin a capsule and perhaps to make

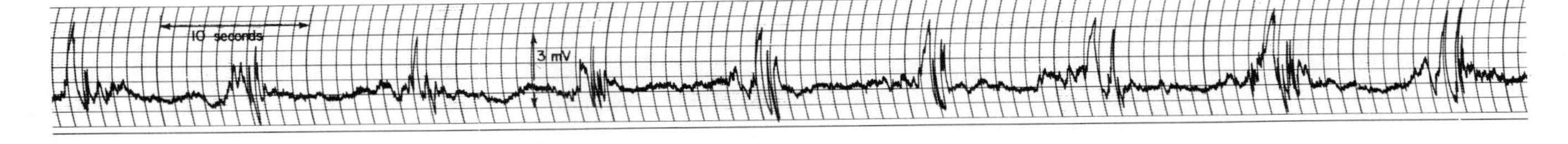

Fig. 7.5 Pattern of voltages transmitted from the gastrointestinal tract of a dog which had swallowed the transmitter in Fig. 7.4. This was during a time when heart-beat signals were not strong.

any potential differences into an ac or time-varying signal. Transmitters such as this, possible in conjunction with a transmitter of psychogalvanic reflex (skin-resistance changes) to indicate which thoughts are meaningful to the subject, might be useful during psychoanalysis. Telemetry of electromyographic patterns from various positions in some cases should elucidate the cause of lower back pain; undoubtedly the pattern of activity or spasm can change as the subject enters his home, a doctor's office, a church, and so on. In another area of investigation the source of sounds produced during dolphin vocal activity perhaps can best be clarified by the transmission of electromyographic potentials from several possible sites.

As discussed later in this chapter, there are some electrodes that are suitable for detecting various chemical species, and amplifiers such as have been here described are suitable in some cases. Acidity (pH), as determined by an antimony electrode, is not always properly interpreted, but relatively low resistance circuits (which have been described here) are suitable for use with such electrodes, if a dc response is provided. A silver chloride electrode, referenced to a similar electrode in a saturated solution, allows evaluation of chloride ion concentration. Here again the transmitter input resistance need not be extremely high. Such a transmitter might be fed to a dolphin to monitor ion concentration along the gut. This would be easier than monitoring sodium ion concentration in connection with water-balance studies in various aquatic animals. In these cases and others (e.g., the measurement of eye motion by oculography), it is generally necessary to follow very slow changes, and thus direct coupling must be used between amplifier stages. In the eye-motion case, however, if one merely wished to study such things as the nystagmus of a person with closed eyes who had partaken of too much alcohol or who had been spinning on ice skates, then ac coupling through tantalum capacitors can supply an adequate low-frequency response.

2. SURGICAL PROCEDURE EXAMPLE

The ability to surgically implant telemetry transmitters of various sorts depends upon the biological experience of the investigator. An example of the preparation and implantation of an electrocardiogram transmitter will be given here both to suggest a set of procedures for this particular case, and also to indicate to engineers the kind of treatment to which the apparatus will be subjected.

The transmitter was a 100 MHz unit like that described in Fig. 7.3. The subject was a rabbit and sterile procedures were followed throughout. Figure 7.6*a* to 7.6*j* depicts the sequence of events.

Across the top line is shown the preparation of the transmitter itself. The

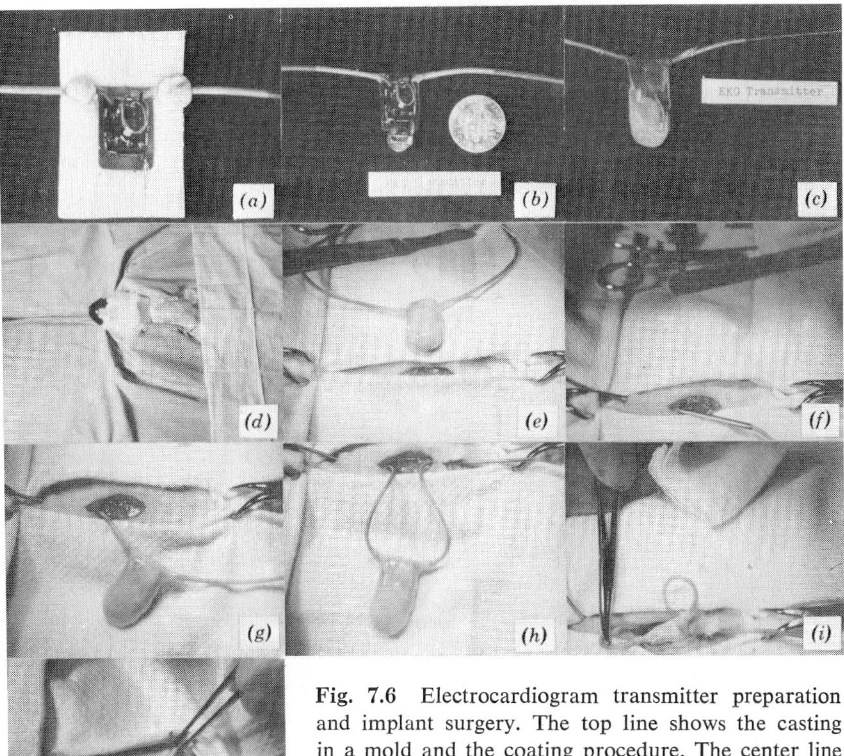

Fig. 7.6 Electrocardiogram transmitter preparation and implant surgery. The top line shows the casting in a mold and the coating procedure. The center line shows the preparation of the rabbit, and the line below shows the insertion of the leads and placement of the transmitter. The final picture illustrates the closure of the main incision.

leads were formed by tightly winding steel wire into a long narrow helix, using a piece of steel wire itself as the winding form. After the winding form was slipped out to leave the hollow "spring," this lead material was coated with silicone rubber. In Fig. 7.6*a* two such leads extend from the circuit, which is sitting in a hollow milled into a sheet of Teflon. Epoxy resin is then poured into this form to pot the transmitter into a solid block. In Fig. 7.6*b,* the transmitter has been removed and is shown with a Mallory 312 cell placed in the battery holder. For size comparison, a dime-sized object is also included. In Fig. 7.6*c* this transmitter is coated with wax as a moisture barrier and then with silicone rubber to reduce any possible tissue reaction. The leads in this case are shown extending to the sides, but in some cases insertion under the tissues is easier if they are both brought out straight from the top of the transmitter.

This unit is then thoroughly cleaned with soap and water to remove foreign matter. It then must be sterilized. This cannot be done by using heat, due to possible damage to the battery and other components. Sterilization with certain gases such as ethylene oxide is possible, but these gases can permeate plastics and, with later release, cause burns to surrounding tissues. We prefer an overnight soaking in Zephiran chloride diluted 1:1000; although not a perfect sterilizing agent, this has given us good results. Such an implant should be readily accepted by the host subject. If a lead should later break off, then an infection would be expected to result in that vicinity.

The ear of the rabbit is cleaned and a light injection of Nembutal is administered into a vein. This is then supplemented by a local Xylocaine anesthetic at the site of the procedure, such a combination of anesthetics being felt to be safer than a more massive general anesthetic. The animal is shaven and washed on the chest, after which he is laid on his back and draped as in Fig. 7.6*d*. In this procedure, the leads will be somewhat removed from muscles that generate interfering voltages by being attached near the top and bottom of the sternum (somewhat like Fig. 1.9).

A shallow incision is next made through the skin, as shown in Fig. 7.6*e*, in which the transmitter and the completion of draping also appear. The transmitter is inserted through this incision, and leads must extend from it under the skin to the right and left. The head of the animal in this figure is to the left.

A shallow incision of smaller size is made to the left of the initial one, and a hollow needle pushed through from one incision to the other. The left-hand lead is threaded through this needle (Fig. 7.6*f*). The needle is then slid away, leaving the lead in place, as shown in Fig. 7.6*g*. The process is repeated for the right-hand lead (Fig. 7.6*h*). The transmitter is then slid under the skin, as shown in Fig. 7.6*i*.

The lead wire, with rubber covering removed at the end, is straightened, threaded through the eye of a needle, and the needle pulled through the tissue just below the skin. The wire can then be knotted once and it will stay in place. In some cases we have instead attached a piece of platinum wire to the end of the basic lead, and knotted it around the silicone-rubber installation. This lead is then held in the correct general vicinity by a loop of suture material, but without sewing the metal of the electrode itself into the tissue. There seems to be somewhat less of a tendency for the leads to break in this case. The procedure is completed by sewing up the incisions, as shown in Fig. 7.6*j*. Generally healing has been more effective if most air is worked out from beneath the incision. The administration of antibiotics to minimize the possibility of infection is routine in most cases, and these final stitches are removed after a few days.

3. CHEMICAL SPECIES AND HIGH-IMPEDANCE CIRCUITS

The first endoradiosonde constructed for the purpose of a chemical determination followed acidity changes by the reversible mechanical expansion that these induced in a suitably synthesized copolymer (Jacobson and Mackay, 1957). The mechanical changes were sensed by a unit similar to the pressure-transmitting capsules. A similar principle might be used to monitor oxygen tension by the reversible mechanical expansion and contraction of certain chelates. Such systems, however, are inherently somewhat slow and their indications may be subject to alteration by competing reactions. Thus it was obviously desirable to develop circuits that would accept the signals from electrodes that would give electrical responses varying with the chemical compositions in question.

An electrode of antimony has long been known to yield potentials that depend upon changes in pH of a solution. In conjunction with a suitable reference electrode, the resulting monitoring systems display a low enough resistance to inject a signal directly into most simple oscillator circuits. Specific combinations were mentioned by von Ardenne (1960) and Noller (1960). The commercially available "Heidelberg Capsule" for acidity measurements apparently is based on Noller's work. In complex systems, it has been our experience that an antimony electrode may show changes in acidity, but that its indications do not always represent precisely pH; it is suggested that their indications might better be called "antimony numbers." As used by Noller, the transmitters are employed more to indicate the end point in a titration. An observation can consist of drinking a measured amount of sodium bicarbonate solution and noting the time required to return to an original reading. This allows an estimation to be made of acidity and secretion rate, without the disagreeable procedure of swallowing a stomach tube. Other kinds of information are obtained by observing whether secretion is influenced by stimulants such as caffeine. Some estimates can be made of the effectiveness of an antacid preparation in a given dose for a given subject, with a full meal usually appearing more effective for relief.

A chlorided silver electrode produces a voltage that depends on chloride ion concentration in the surrounding solution. This electrode can show a relatively low impedance, and can be used for monitoring chloride ion concentration if referenced to a similar electrode surrounded by a saturated solution in a permeable covering. Some workers have used a simple chlorided silver electrode as the reference electrode for a sensing electrode, thus incidentally introducing unwanted signals in any situation involving a possibly changing chloride ion concentration.

For studies of such things as the basic stool observed in cholera patients a more nearly absolute value of pH would be desired. It is generally accepted that the proper way for monitoring pH is to employ a glass electrode, referenced to a suitable indifferent electrode. An ion-exchange model is useful in explaining some aspects of the performance of such an electrode, and it is often assumed that there is an equilibrium established between ions in solution and in the hydrated glass at the interface between bulk solution and a swollen layer. In any case, the potentiometric behavior of such an electrode is rather well described by the well-known Nernst equation, with a change in voltage of approximately 60 mV/pH unit. The electrode must be fabricated from a suitable glass, a film of which is then filled with a standard solution. It is said that very tiny pH electrodes can be constructed for intracellular observations by using an open-ended micropipette pulled from the proper glass (Lavallee and Szabo, 1965). Such a measuring system can be considered as comprising a variable battery across which is interposed a thin glass partition, and thus very little current can be supplied. For these and other related measurements, it is thus necessary to have an amplifier system with a frequency response going down to zero, and an input resistance measured in hundreds of megohms.

Such resistances do not fall within the usual transistor technology. Several groups proposed to overcome this problem by having the electrode voltage control the capacitance of a voltage-sensitive capacitor, which would then modulate the frequency of an associated oscillator. Because of the high resistance of backward-biased silicon diode, or some grades of barium titanate, this approach works. Some aspects have been discussed by Wolff et al. (1962). If a diode is used as a variable capacitor, it is necessary that it does not go into conduction on any part of the radio-frequency cycle. This can be assured either by biasing the oscillator circuit on a separate conducting diode, or by using a pair of opposing diodes to whose common point the control signal is injected.

With the availability of field-effect transistors and related devices, superior possibilities exist. Figure 7.7 shows a transmitter circuit having an input resistance considerably above that of most glass electrodes. The signal goes to a field-effect transistor, which is combined in a feedback arrangement with an ordinary *PNP* transistor, as shown. The output from this combination, now at a lower impedance level, is used to control the base return voltage of an astable multivibrator, thus giving a variable frequency in response to input voltage changes. In the collector circuits of the multivibrator is placed a coil of approximately 500 turns to serve as the transmitting antenna. The capacitors (C) determine the general frequency range in which this transmitter works, the multivibrator directly generating radio-frequency currents.

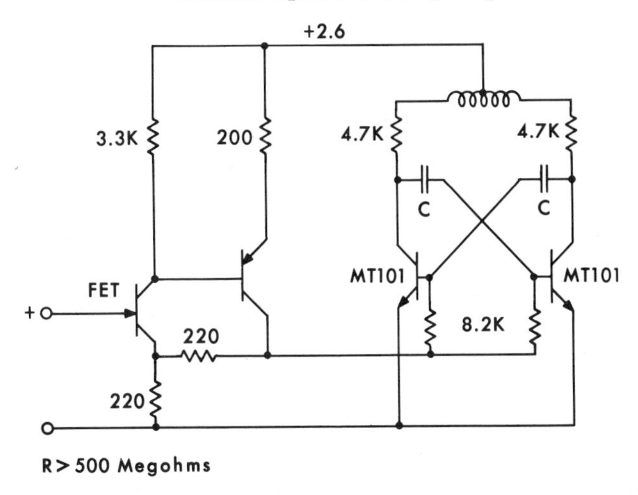

Fig. 7.7 A field-effect transistor can be combined with another to give a very high input-resistance stage whose output controls the frequency of an astable multivibrator. Voltage changes modulate the frequency of the signal emitted by the coil.

This circuit shifts its frequency somewhat in response to temperature changes. This is largely because of the effect of temperature on the multivibrator components, rather than on the input circuit. The entire circuit, however, can be compensated by the suitable placement of a germanium diode, as is shown in Fig. 7.8. This circuit also incorporates

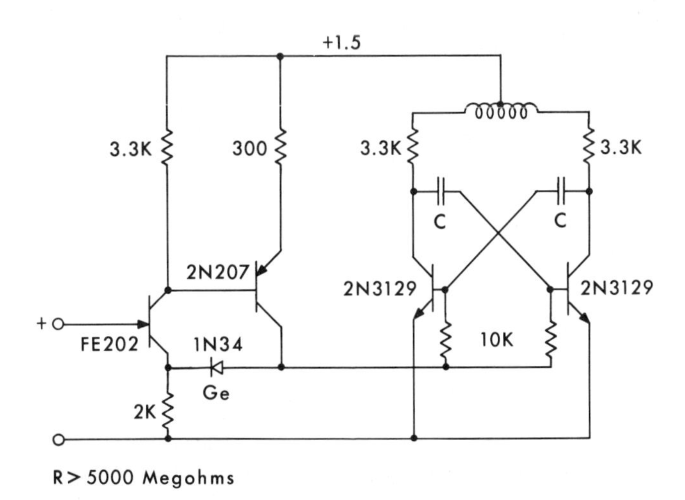

Fig. 7.8 A circuit similar to that in Fig. 7.7 but with slightly altered components, and the frequency of the multivibrator temperature compensated by the diode.

a somewhat improved field-effect device and slightly altered circuit constants to give an improved performance and a higher input resistance.

The same high-input resistance combination can also be used to control the frequency of a blocking oscillator, as shown in Fig. 3.7. The layout of an actual pH telemetering capsule based on a similar circuit is shown in Fig. 7.9. The sensitive components are closed off in a separate wax compartment. The reference electrode consists of a piece of chlorided silver wire in a recessed cavity containing saturated potassium chloride solution with a few extra crystals of potassium chloride. The cavity is

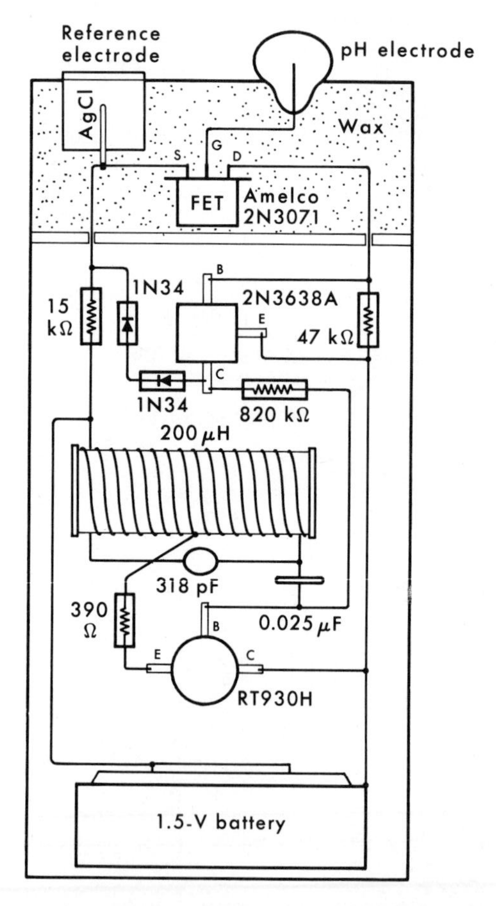

Fig. 7.9 Schematic of ingestible pH transmitter in which critical components are encased in wax. The saturated solution around the silver chloride electrode is easily changed between uses. The circuit is basically a variable-frequency squegging oscillator.

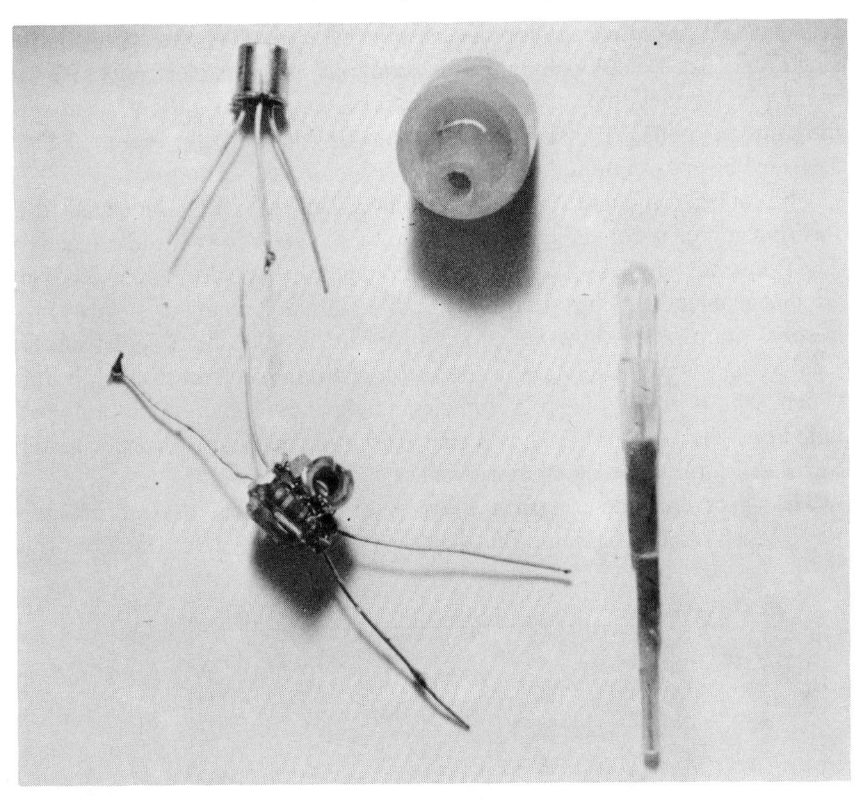

Fig. 7.10 Photograph of some of the components for use in a pH transmitter. At upper right is the case body of the capsule; to the left a field-effect transistor. At the lower right is an intact commercial glass electrode. The rest of the circuit components, without battery, are at the lower left.

covered over with a sheet of Parafilm into which a single small pinhole has been punched. This film is readily changed when needed. The case was cast from epoxy resin. Glass sensitive to pH changes is a bit difficult to work, but satisfactory electrodes can be fabricated by anyone with some glass-blowing experience. A commercial glass electrode, such as the Beckman 39042, can also be employed. With such an electrode, the components appear as in Fig. 7.10. The commercial electrode and the field-effect transistor occupy the majority of the space, while all the rest of the components, except the battery, are seen at the lower left. In the upper right of Fig. 7.10 is shown the cast case with the hole for the glass electrode, and the silver wire in the crescent-shaped well of the indifferent electrode. This unit was 1.19 cm in diameter and 3.7 cm long. The diameter could be reduced to 1.0 cm except for the battery, which is 1.15 cm diam for the single 1.5 V cell employed here. By cutting off the glass electrode and wax sealing in a buffer, the overall length can be reduced to 3.0 cm, even with the same field-effect transistor. With regard to this transistor, newer, better, and smaller units continue to appear.

The calibration curve for this unit is shown in Fig. 7.11. The signal from this transmitter is conveniently picked up by any nearby AM radio receiver. The frequency of clicks is a bit too high to count by ear, but simple decoding circuits can convert this frequency into a direct indication of pH. In a hospital atmosphere, however, it is convenient to take the output from the receiver and apply it directly to the recorder of an electrocardiograph unit, which will normally have a sufficient frequency range to record each individual click. From the known chart speed the number of hertz is readily calculated to give the momentary value of pH.

The silver chloride electrode is easily prepared from a piece of silver wire, which should be somewhat flattened to perhaps 0.025 in. This wire

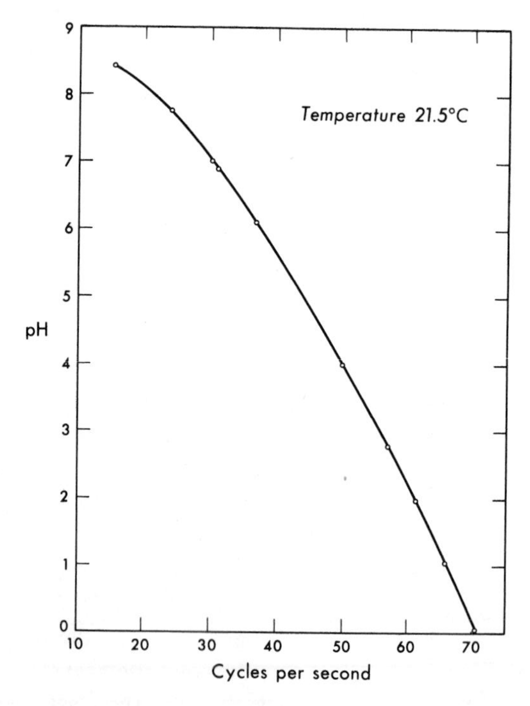

Fig. 7.11 Calibration curve of pH transmitter. The frequency is too high to time by ear, but it can be recorded in a hospital on a typical electrocardiogram recorder, or on a tape recorder and played back slowly if decoding by ear is desired.

is connected to the positive side of a 3 V battery through a 1500 Ω resistor with the negative side of the battery going to another silver or a copper wire. These are immersed in 0.1-*M* hydrochloric acid for approximately a half hour. The positive electrode will take on a flat black appearance after about five minutes. The silver chloride layer that builds up is somewhat soluble in the potassium chloride, which will be used as the surrounding solution, and so between uses it is desirable that the covering diaphragm be removed so that the electrode can be stored in distilled water, or even be allowed to dry out. The silver chloride can be cleaned off with ammonium hydroxide if necessary. Saturated potassium chloride surrounds the wire, since it is considerably superior to sodium chloride as the electrolyte in this application. Potassium chloride gives a smaller junction potential because the mobility of potassium and chloride ions are similar. It is tempting to consider using agar to eliminate physical currents and to stabilize the solution, but the acids and enzymes in the biological systems to be studied (e.g., the stomach) do not permit this. Experiments have demonstrated that a single hole of minimum size in the Parafilm cover on this electrode is quite sufficient. The hole is formed by pressing a sharpened needle into the film until perforation just extends through the back. This can be judged as complete when light pressure on the diaphragm causes a droplet of fluid to form on the external surface.

Glass electrodes should be soaked in water for several hours before use. They can be sterilized in Zephiran chloride. Such electrodes are somewhat sensitive to temperature changes. They are also slightly sensitive to sodium ion concentration, and for precise work, corrections can be applied for the presence of sodium and potassium ions.

The extent of the "alkali error" of pH glass electrodes is known to be a function of the composition of the glass. Advantage has been taken of this fact by various workers to produce glasses that provide practical sodium and potassium-sensitive glass electrodes. Commercial glasses are available which are approximately 100 times more sensitive to sodium ions than to other alkali metal ions, and other glasses show a moderate selectivity for potassium over sodium ions. More recently, glass electrodes sensitive to divalent cations such as calcium have been discussed (Garrels et al., 1962). Electrodes can be made sensitive to other ions if formed of solid state ionic conductors or of water-immiscible liquid ion exchangers held by a thin porous membrane. A commercial unit (Orion, Inc.) for calcium uses this principle by a filling of calcium chloride against an exchanger of a calcium salt of organophosphoric acid, for example. A recent summary (Rechnitz, 1969) mentions some new approaches, including immobilized liquid membrane electrodes, mixed-crystal membrane electrodes, enzyme electrodes, and antibiotic electrodes. In a number of cases, a simple ion released as a result of a specific reaction, for example a specific enzyme response, can be sensed by a standard type electrode to give overall detection. Such electrodes need not be implanted for all purposes, but may sense interesting species in sweat, saliva, or urine.

Various combinations of electrodes and semipermeable membranes, perhaps with intermediate reacting chemicals, allow for the analysis of diverse chemical species. In some cases it may prove practical to employ colorimetric methods by enclosing suitable "test papers" within various membranes while using the illumination of a lamp into which power has been induced from an external source.

Carbon dioxide can be measured with the help of a pH electrode. The pH of water equilibrated with carbon dioxide is a function of the partial pressure of carbon dioxide. Early use was made of this fact (Stow and Randall, 1954) for biological measurements by including within a rubber membrane some water and a pH and reference electrode pair. Carbon dioxide diffusing through the rubber membrane would combine with the water to form carbonic acid, which would ionize to hydrogen ion and bicarbonate ion. A number of the modern plastics, when used in thin sheets, make more suitable membranes in this application. It has been suggested (Severinghaus and Bradley, 1958) that a Teflon membrane is good, and that under it should be placed a layer of cellophane soaked in 0.01-*M* sodium bicarbonate and 0.1-*M* sodium chloride, for an increase in sensitivity and stability.

The input properties of circuits to be used with such electrodes are measured by first of all determining voltage sensitivity; that is, by applying known input voltages and observing the output. Then a high resistance (say, 500 MΩ) is placed in series with the input and a known voltage applied. The output will correspond to a changed input because of the drop in voltage across the resistor; from this the current and the input resistance can be calculated. The input resistance may not be constant for all applied voltages, but if it is always large with respect to the test electrode resistance, then changes in the test electrode resistance with temperature or with time will matter less.

Other kinds of observation require the measurement of very low currents. This is done by running the unknown current through a very high resistance, the resulting voltage then being measured by a suitable circuit. This circuit must have a resistance considerably larger than the current-indicating resistance. Thus these circuits are again relevant.

It is possible to continuously monitor the oxygen content of a solution or a gaseous mixture. If a platinum cathode is maintained at about 0.6 V with respect to a nonpolarizable anode, then the current that flows as oxygen is reduced is proportional to the oxygen content of the solution, under proper conditions. Clark et al. covered such an electrode with cellophane for use in whole blood (1953). Severinghaus and Bradley (1958) expressed a preference for polyethylene as a permeable membrane rather than Teflon or Mylar. A typical membrane thickness is 0.001 inch, and the contained solution is usually saturated potassium chloride, for use with a chlorided silver reference anode. Other electrode combinations require no external applied voltage, since the oxygen is spontaneously reduced at the cathode.

The smaller the platinum, the smaller will be the current that must be measured, but the response will be faster and there will be a smaller disturbance to the conditions otherwise existing at the site. Since oxygen is actually consumed (reduced) at the cathode, the observed current is limited by the rate at which oxygen can diffuse through the surrounding tissues. The effects of stirring and movement are reduced when a high consumption rate does not produce high oxygen gradients in the solution. In any case a tip a few microns across gives low sensitivity to movements. Temperature changes affect the rate of diffusion of oxygen through the tip-covering membrane, but also the percent saturation for a given number of parts per million of dissolved oxygen in a given liquid. At equilibrium the partial pressure of a gas above a liquid is the same as the partial pressure of the gas dissolved in the liquid. These oxygen sensors measure the partial pressure of oxygen, and this doubles if the total pressure on a given percentage composition gaseous mixture doubles. This pressure sensitivity is not troublesome, but can cause confusion, for example, in diving experiments.

Cardiac output and blood flow in tissues and vessels can be measured by dilution techniques in which something recognizable (heat, dyes, radioactivity, etc.) is injected at a known rate and the resulting concentration monitored after mixing; similar measurements are made from the time for desaturation. Hydrogen gas can be added by respiration or by infusion with hydrogen-saturated saline. As with oxygen, the result can be monitored polarographically by a platinum electrode polarized with a quarter volt, the resulting current measuring hydrogen concentration (Aukland, 1967). The indication can be telemetered as indicated above, care being taken to avoid an explosion.

The gas measurements are relevant not only to questions of respiration and flow but also, for example, to studies on the swim bladders of fish. Chapter 9 will mention measurements of flow and gas pressure that do not require circuits with special input properties.

As another example of the applicability of high input impedance circuits, the electrical activity of single cells as sensed with a microelectrode can be transmitted with one of these field effect transistor input stages activating any of the frequency modulated output stages that can provide a bandwidth up to 5–10 kHz. The input capacitance is low and the resistance high, while the noise level is relatively low. Overall system amplification is set so that signals ranging from 10 microvolts up to a few millivolts are reproduced. Across the cell membrane itself is a potential of approximately −90 mV with respect to the outside. The signal from a neuron can be sensed within roughly 200 microns but if the observed spikes have two amplitudes, then signals are probably coming from two cells. For such discrimination reasonable transmitter fidelity is required. At the present time the input circuits illustrated seem best for internal use, while external transmitters with higher voltage can alternatively well employ an MOS input transistor.

Single unit activity in the brain of an unanesthetized cat has been transmitted (Edge et al., 1969) and the position of the electrode advanced under radio control as the animal moves about (Findlay et al., 1969). For the latter, a small electric motor and worm gear drive was employed. Pain is not felt from brain tissue. Such studies on a given cell might be prolonged by the use of a neutrally buoyant electrode floating in the brain to further reduce mechanical disturbances to its tip with movement by the subject (Gualtierotti and Bailey, 1968), though recordings for several days have been reported with the use of a fine (60 μ) flexible wire alone (McElligott et al., 1969). Such an electrode may show a resistance of a megohm.

Other electrode combinations can be used for following other chemical and physiological processes, but the previous were selected as useful examples. Alternative experiments often require similar circuits in their transmitters.

8

Multichannel Transmission and Fetal Studies

As a direct extension of the matters discussed in the previous chapter, a particular series of examples will be taken from one area of research. A most valid area of application for these methods is to the study of life before and during birth. In some cases the observations employ the transmission of several variables from a single transmitter with a single antenna and battery. Such a unit can be surgically implanted in a fetus within the uterus of the mother animal, and the baby will later be born with the transmitter in place and monitoring already in progress.

Cardiovascular information has been of interest in studies with Ben Jackson, aided by George Pieseci and Marc Abel. One way of regarding the action of the heart is to assume that at any instant during its beat the summation of the distributed voltages can be approximated by a vector. This vector traces out a curve in space as the heart cycle progresses. The vector can be represented by its projections on three perpendicular directions which in the human can be taken as up-and-down, front-to-back, and side-to-side. Three electrodes plus a common electrode can be placed in these directions to pick up the pattern of activity. Transmission of this information requires the communication of the changes with time of three independent voltages. At a receiver the three voltages can be combined in pairs to give the three loops that constitute the projections of the spatial vectorcardiogram upon the sagittal, frontal and transverse planes. In some of the experiments to be described it was desired to do this for an unanesthetized fetal animal. Before birth, the lungs are collapsed and there are shunts for the blood. It was desired to assess not only normal heart rates, and the effect of various disturbances to the mother upon these, but also axis changes associated with birth, and in the period immediately after that, as well as the effect of major alterations in fetal dynamics, in particular axis changes in response to various stresses. As examples of the latter, a total aortic constriction proximal to the ductus arteriosus puts a pressure load on the left ventricle and a volume load on the right, while a total occlusion of the ductus arteriosus does the opposite.

One can study a relatively undisturbed fetus in an unanesthetized and unrestrained mother by these procedures which is a great advantage. In addition, in the present case it may be almost impossible to obtain similar information by other means. The difficulties in detecting the voltages generated by the beating of the fetal heart in the presence of the larger voltages generated by the beating of the maternal heart are well known when use is made of electrodes placed on the surface of the maternal abdomen. At times a human fetus is coated with a greasy somewhat insulating material known as vernix caseosa. In general, the voltage generated by the beating of the fetal heart does not spread uniformly through the fetus and mother but rather seems to emanate largely from the umbilicus and mouth (Kahn and Koller, 1966) and perhaps the rectum. This does not lead to the expectation of a good vectorcardiogram by externally placed electrodes.

A transmitter and electrodes placed within the fetus well communicates the beat of the fetal heart and *not* that of the mother. If the fetus dies, a trace of the maternal heartbeat can appear after a few hours.

Three separate transmitters could be packed together in one part of the fetus to transmit the three voltages, but they would then be unnecessarily bulky in comparison with a single three channel transmitter. In addition they would tend to interact with each other unless the transmitting antennas were mutually perpendicular, in which case there would be a tendency for one channel at a time to fade in any simple receiving system. One of the transmitters employed in these studies uses subcarrier oscillators (Chapter 3), and it exemplifies the simultaneous application of a number of the previously described circuits.

An analogy may be helpful to non-engineers in understanding the simultaneous transmission of several independent pieces of information through a single transmitter with the help of subcarrier modulation. Suppose one had a large auditorium and wished to send information about the temperature in the right-hand part, the center, and the left-hand part. One could place a man with a violin at the right, a man with a cello at the center, and a man with a bass violin at the left. Each would be instructed to play his high notes if he felt warm and lower notes if he felt cool. The signal from this primitive orchestra obviously could be received by a single microphone and transmitted from a single radio transmitter, the latter perhaps being an FM transmitter. A person in another state could receive this signal on a single standard FM radio receiver. By paying attention to the sound of the violin he could tell if it was warm or cold in the right part of this distant room. By listening to the sound of the cello, which is in a different range of audible frequencies, he could tell about the temperature at the center of the room, and an independent observation could similarly be made of the temperature at the left of the room by noting changes in the low frequency part of the signal. The three distinguishable musical instruments correspond to three frequency modulated subcarrier oscillators acting in three different frequency ranges. Their signals are added to frequency modulate a single transmitter, and thus this corresponds to FM-FM modulation. The separation of the three signals can be done automatically by suitable filters in the receiver after detection of the basic radio signal, and these three signals could then be separately decoded and recorded.

In the circuits shown in Fig. 8.1 a slight simplification is achieved by using two subcarrier oscillators for FM-FM transmission, while simply changing the frequency of the basic radio signal in response to the third voltage (FM modulation). Astable multivibrators (Chapter 2) are used as subcarrier oscillators in the top two channels, their frequencies being modulated in the vicinity of 20 and 12 kHz respectively by changes in voltage applied to the base resistors. The signals, sensed by stainless steel electrodes, are amplified before application to these oscillators. This circuit is to be used in warm-blooded animals, and thus the different channels are not stabilized against temperature changes. Any calibration should be done at body temperature. Simpler construction results if direct coupling is used between stages, but the blocking capacitors were employed to reject slow changes in voltage that might accompany such things as shifts in chemical gradients or movement artifacts. The input transistor of each stage is selected for lowest possible noise from the available stock.

Multivibrators constitute a simple voltage controlled oscillator (often designated VCO) but their output is almost a square wave containing many harmonics rather than a sine wave. The coupling circuits from these oscillators can be chosen to eliminate some harmonics by rounding the corners of the wave, or else bridge oscillators or phase shift oscillators can instead be used. The multivibrator in the center channel employs an extra transistor to return the voltage abruptly. The resulting extra symmetry in the wave goes with the elimination of second harmonic which can fall within the frequency range of the upper channel under certain conditions, thus producing "cross talk" that cannot be cancelled. The resistors R_1 are

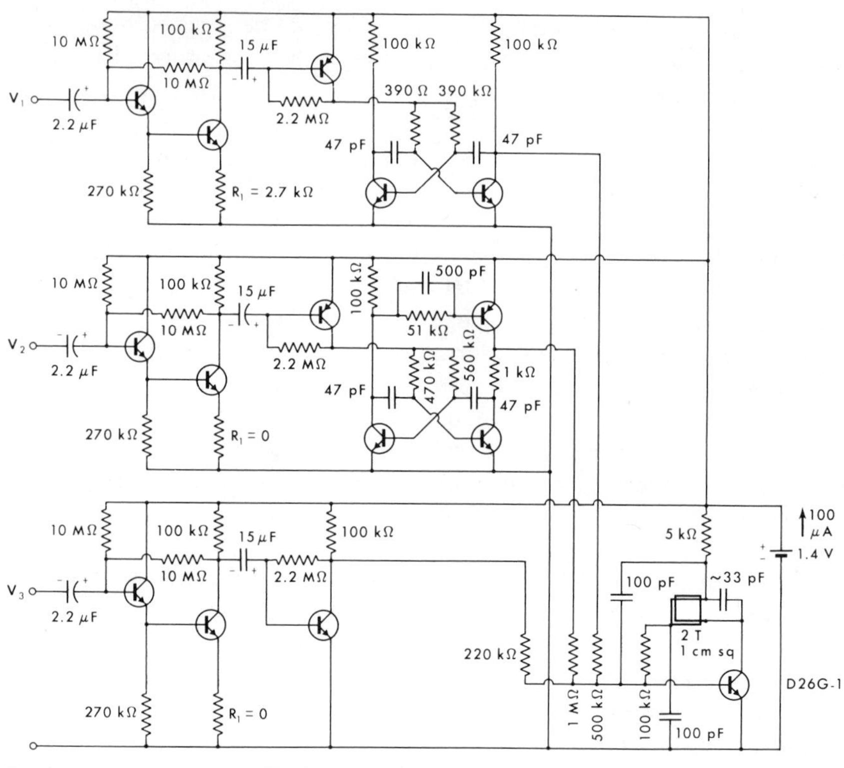

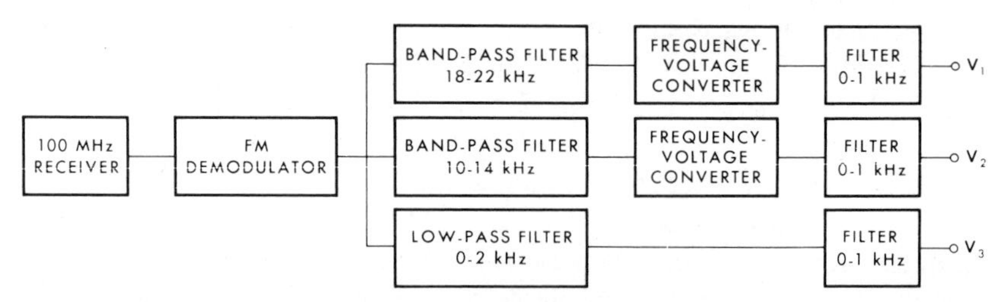

Fig. 8.1 Three-channel voltage transmitting circuit with receiving system shown below. The upper two channels employ FM-FM with nonsinusoidal subcarrier oscillators, while the bottom channel is straight FM.

adjusted to make the sensitivity through each channel approximately the same.

The signals from the three channels are combined (added) through three resistors into the base of the transistor at the lower right. It is the active element in the radio frequency oscillator which is frequency modulated by this combined signal. The general range of radio frequencies employed is in the 100 MHz region, which allows small components, inexpensive receivers, and reasonable tissue penetration.

We have used a number 400 mercury cell as the power source, but others could be employed. Teflon coated stainless steel wires are employed as leads from the stainless steel pieces that serve as electrodes. They are attached to the circuit with Certanium 34C flux-cored solder. The entire transmitter is potted in epoxy, and then coated with wax and then with silicone rubber. Using standard small commercially available components soldered together by hand, the complete transmitter can be quite small as is seen in Fig. 8.2. For maximum effectiveness, gas sterilization (ethylene oxide) is used, followed by a sufficient interval of outgassing.

Electrically, the transmitter has approximately the following characteristics. In the lower channel, 500 microvolts at the input causes a 25 kHz change in carrier (radio) frequency. The input impedance is approximately 1 megohm. The noise at 1 megohm is approximately 10 microvolts. The frequency response extends from approximately 0.2 Hz to over 1 kHz. In the center channel a 500 microvolt signal causes a 1 kHz change in the 12 kHz subcarrier frequency. The 12 kHz subcarrier deviates the carrier frequency ± 25 kHz. The properties of the top channel are similar, the subcarrier frequency being set at 20 kHz. With the largest expected signals simultaneously appearing on all three channels, the resulting deviation of the radio frequency carrier should not carry it out of the pass range of a standard FM entertainment receiver.

Various receiving antennas have been employed including tuned loops, a metal cage in which a metal shelf served as a second electrode, and a long length of "twinlead" laid out randomly on the floor in the vicinity of the subject. These matters will be taken up further in later chapters.

The receiver used was a good quality entertainment receiver made to cover the range of frequencies 88–108 MHz. In some cases stereo receivers are preferable in being able to respond to the relatively high frequency of the upper subcarrier. It is sufficient to take the detected radio signal directly through a pair of wires soldered across the discriminator (Chapter 11). This incidentally gives unlimited low frequency response in the bottom channel; it exists anyhow from the upper two because of the subcarrier frequency carrying the low frequency information. As indicated in the lower part of Fig. 8.1, the demodulated signal then goes through three filters to separate out the information in the three channels. The separated signals are then converted into three fluctuating voltages for recording by a penwriter or etc. If one does not wish to construct filters to separate the different frequency ranges, then standard filters can be purchased for this application if the so-called IRIG standard frequencies are instead employed.

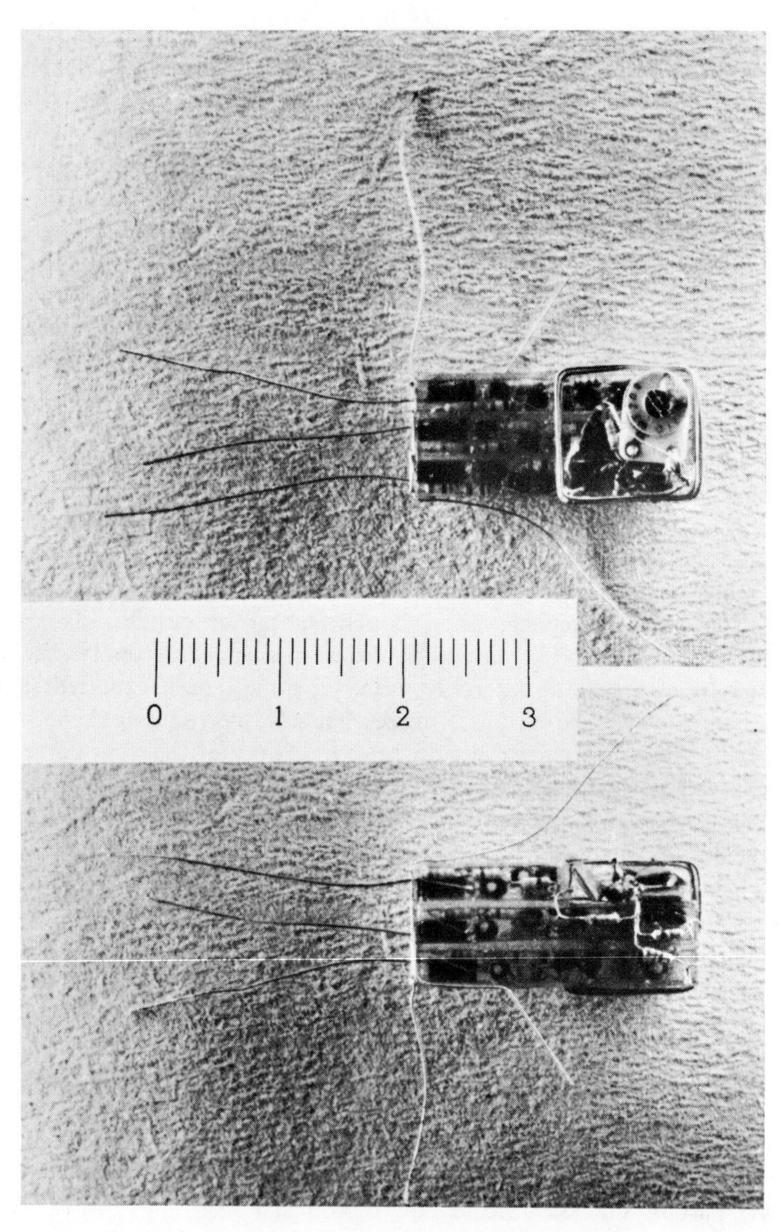

Fig. 8.2 Front and back views of a three-channel voltage transmitter constructed of standard components cast into plastic, compared with a centimeter scale. The three channels are in rows below, and above can be seen the radio frequency section with a variable capacitor inside the antenna loop. The thickness is approximately the width of one of the rows.

In this application relatively rapidly changing voltages are to be compared with each other instant by instant. The results will be meaningless if the total time delay of a signal through each of the channels is not the same. This applies to the overall system from an input at the transmitter to the recording on the output oscilloscope. Thus it can be necessary to cascade two filters in the lower receiver channel for this reason. A way of checking the equality of delay is to apply a changing voltage from an oscillator to two of the channels in parallel. The corresponding outputs from the overall receiver system are then applied to the horizontal and vertical axes of an oscilloscope. A diagonal line must appear with no trace of a loop when the input signal is changing at any rate up to the highest frequency of interest. Many oscilloscopes have an unequal time delay through their horizontal and vertical deflection systems, and this should similarly be checked in advance.

It should be noted that such a transmitter can be used for signals other than bioelectric potentials. Thus we are exploring the life of the oxygen tension measuring system indicated in Fig. 8.3, when in the moving liquid of an artery. Such a cell generates its own power with an output that depends upon oxygen tension, and being a low-impedance device, does not require the measurement of minute currents that can be disturbed by leakage currents (Smyth, 1967; Clifton and Parker, 1970). The traditional electrolyte in such a cell is potassium hydroxide but potassium bicarbonate or isotonic saline seem to have some advantages. The signal from this transducer can be applied directly through the base resistors of a subcarrier oscillator, or through a stage of direct coupled amplification if the electrodes are made very small. The resistor loading the cell is in the general range of 1000 ohms, depending upon electrode size. More channels can be added to the basic transmitter circuit that has been shown, but the three channel unit is capable of transmitting one electrocardiogram and two values of oxygen tension, or a single vector loop and one value of oxygen tension.

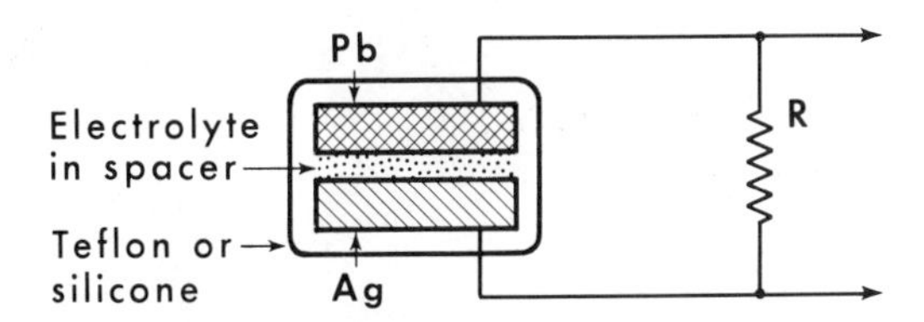

Fig. 8.3 Galvanic cell delivering a current depending upon the surrounding oxygen pressure. The consumption of appreciable oxygen results in sensitivity to both flow and content; response depending on the rate of oxygen arrival at the region. An alternative geometry employs a pancake spiral of fine silver wire against a cloth on an aluminum or lead sheet.

An alternative method of multichannel transmission is to rapidly cyclically sample each variable in turn for an instant (time-division multiplexing). Such circuits have been described (e.g., Fryer et al., 1969). The switching circuit that alternately connects the different inputs to the transmitter modulator can have its interstage advance pulse at one stage connected to the transmitter through a small capacitor so that the resulting transmitted impulse will mark or identify one channel in the cyclic sequence and thus prevent uncertainty in identification of the components in the received chain of signals. Alternatively, an extra channel in the cycle can instead periodically deliver a standard signal. This method could have been used in the above cases with comparable complexity. Care in filtering the final signals is required when display of the information is itself accomplished by sampling as in Fig. 8.9. Any one variable must be sampled at least twice in the period of the highest frequency to be observed on that channel (from that source). The corresponding frequency limitation in the subcarrier method of "frequency division multiplexing" is that changes in the signal can be followed that are slower than the corresponding sub-

Fig. 8.4 The fur of a fetus is seen through an incision in the exposed uterus of a pregnant beagle. Elevating the margins of the several layers minimizes loss of amniotic fluid.

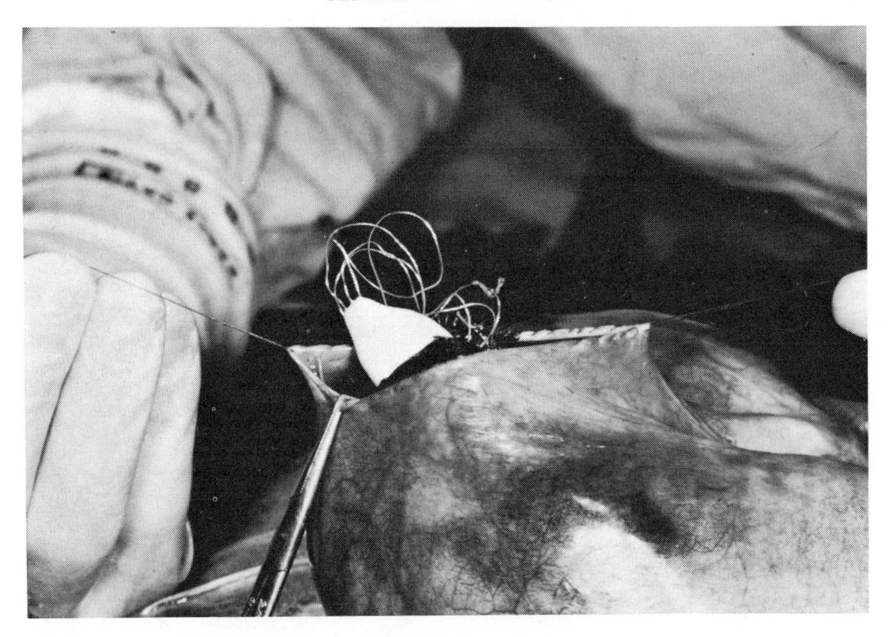

Fig. 8.5 Once the electrodes are in place, the transmitter can be slid beneath the skin of the fetus.

carrier frequency. Subcarrier multichannel transmitters have also been described by Fischler et al., Goodman, Robrock and Ko, Vreeland and Yeager, and others (see General Reading section). Other modulation combinations can be employed, and the early unit using FM and pulse-frequency modulation is mentioned in Chapters 1 and 11.

The surgical procedure used in the placement of such a transmitter is rather delicate. The generalities of such procedures have been given (Jackson and Egdahl, 1960). The uterus is exposed and, especially with dogs, a momentary spray of local anesthetic (5% Cyclaine) on the outer surface is helpful. This seems less significant in sheep or monkeys, where greater problems can be experienced with surgery close to term. Cutting through the several layers, emphasizing minimum disturbance to the placenta, exposes the fur of the fetus as in Fig. 8.4. The margins of the incision can be raised to stop the loss of amniotic fluid. An incision into the fetus allows the placement of the electrodes, and then the transmitter and leads are slid into place as in Fig. 8.5. In order to monitor the contractions of labor, and their effect upon the fetus, a transmitter of the type described in Chapter 5 is dropped into the amnion before closure (Fig. 8.6). This is a separate

Fig. 8.6 After the fetus is prepared, a separate pressure transmitter is dropped into the amnion before final closure.

transmitter acting upon a different frequency. After the birth of the fetus, the mother can again become pregnant.

In the received signal any one channel can be recorded alone to display a standard electrocardiogram for the fetus. The signal can be recorded on a magnetic tape recorder for later analysis, or on a penwriter (Fig. 8.7) or

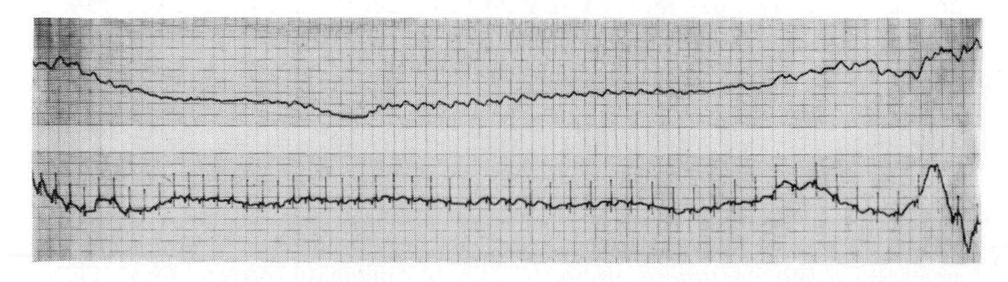

Fig. 8.7 Beagle fetal electrocardiogram and pressure recording from uterus during elimination by mother. Rapid pressure cycles are often observed during defecation. A pressure change of 20 mm Hg is here recorded with increasing pressure downward. Similar labor contractions are also thus recorded. The ECG was from the scapular-lumbar electrodes, and the chart speed was 25 mm/sec.

oscilloscope. In the latter case it is also possible to display a convenient indication of beat-to-beat changes in heart rate as in Fig. 8.8. In this case the recording was from the dorsal and ventral electrodes, and a capacitor connected in series with the oscilloscope further restricted the low frequency response so that gross activity would not tend to shift the trace in a vertical direction. The sweep of the oscilloscope was synchronized to start near the peak of an "R wave." The trace was slowly and continuously shifted upward by the application of the signal from a commercial triangular wave generator through an 80 K resistor to the input of the oscilloscope in parallel with the signal itself. The result rather clearly shows any instantaneous changes in the period of the fetal heart beat, and this method of display can be applied to many variables (Webb and Rogers, 1966). The pattern also emphasizes those cyclic aspects that are of significance relative to noise, though irregularities or isolated events are also quite noticeable. To save the cost of film, this pattern is normally recorded on a storage oscilloscope for observation, though in most cases a smaller spot size is achieved by using the same oscilloscope in a photographic mode of direct recording.

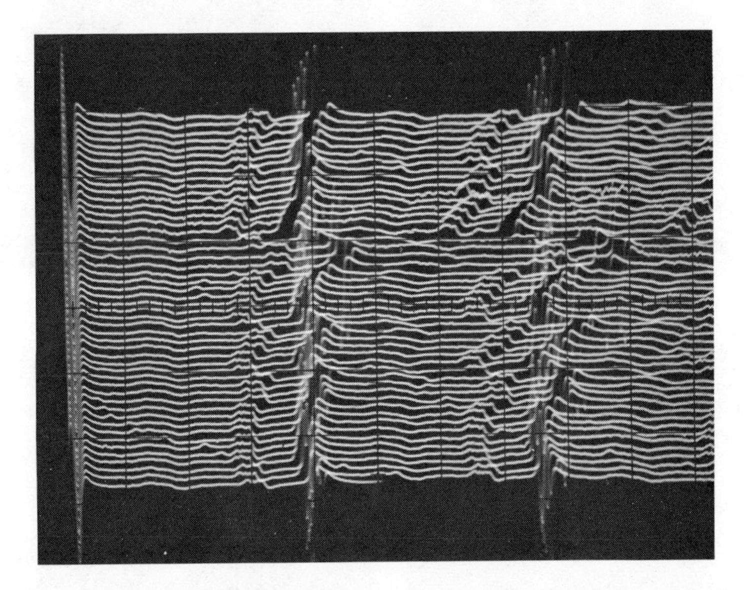

Fig. 8.8 An oscilloscope with sweep speed set to display two beats of the heart is synchronized to start at a particular part of the heart cycle. A slowly-increasing voltage is also applied to the vertical amplifier to spread the traces and allow easy judgment of the degree of change from beat to beat. The time deviations at the second beat appear approximately twice as great as at the first because the change over two cycles is added following the start of the sweep.

Each pair of voltages due to the beating of the heart can be applied to the vertical and horizontal deflection amplifiers of an oscilloscope to trace a loop that is the projection of the spatial vectorcardiogram on one plane. This has been done with one pair of voltages to produce Fig. 8.9. A 1 kHz square wave was applied to the "z axis" of the oscilloscope to cyclically change the intensity of the trace, thus allowing an estimation of the rate of progression of the signal at any moment. The coupling circuit was also arranged so that the actual signal had a somewhat triangular shape, resulting in dots that do not have constant width or intensity. It will be seen that each dot is in the form of an arrow that points in the direction in which the loop is being traced out, and in some cases this piece of information can be of crucial importance. If the origin of the vector loops is controlled, then the gradual shifting process of Fig. 8.8 can be applied to the family of loops to make changes in them noticeable.

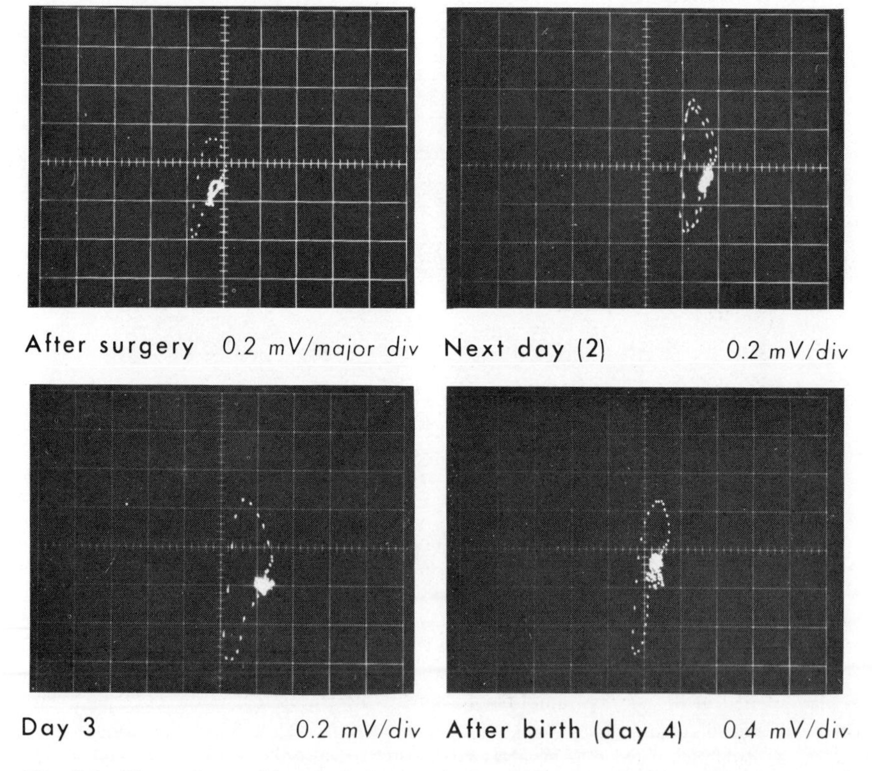

Fig. 8.9 Vector loop of beagle before and after birth. Successful monitoring experiments on this species have thus far been instituted from two to ten days before birth.

The overall system can be calibrated to indicate not just relative changes in voltage but the actual voltage levels as indicated. At about the time of birth we may see a relatively rapid rise in voltage as indicated, apparently associated with an increasing blood pressure. A diagonal loop (4:1 size ratio of axes) can change to a diagonal line in a few heart beats, and back. Overall length can change by a factor of two in a few beats, and rate by an even greater factor. This is with the mother sitting calmly. Radical loop changes need not be accompanied by significant rate changes. A sign of fetal distress has been an enlargement of the loop corresponding to the "T" wave.

By combining the three voltages in various pairs we can display the three orthogonal projections of the loop in space. Through the use of rapid switching, these loops can be made to simultaneously appear on the face of the oscilloscope as in Fig. 8.10. This display was produced with a Type 564 Tektronix oscilloscope with two Type 3A74 plug-in units. To allow both proper blanking between traces and also dotting of the loops, a 2.2 K resistor had to be soldered across the outer terminals of the switch at the back of the oscilloscope. To produce dots a 1 kHz signal of approximately 35 volt amplitude was inserted into the binding posts on the back of the oscilloscope. The figure is a photograph of such a display from a lamb three days before birth. Beat-to-beat changes can be monitored since the three loops are for a single beat of the heart.

Various electrocardiogram "lead" systems can be formed by resistors placed after the receiver rather than in the transmitter input. A recorder is not used here in place of a transmitter even when the application is to take data for later use (rather than monitoring momentary fetal well-being without disturbing the mother), because one of sufficient frequency response could not be found with this size and life. The received signal can be recorded for later study and analysis, with the data then being combined in various ways to weight the electrodes differently.

In brief, the observed patterns are consistent with a picture in which the right ventricle is as active as the left before birth. Normal changes in the pattern immediately after birth are relatively slow, though there can be rapid axis changes in response to various manipulations.

These same transmitter types can be used to transmit brain wave patterns with one more stage of amplification, and single cell responses if a field effect device is at the input. Using an EEG transmitter external to the fetus it should be possible to follow the development of the visual system *in utero* by flashing a light and noting the response (similar to the experiment in Fig. 1.10). The same would apply to sound or electrical stimulation. To obtain sufficient light intensity it would probably be necessary to induce electrical power into a small bulb rather than shining a light through the maternal abdominal wall. With precocious species it should be possible to

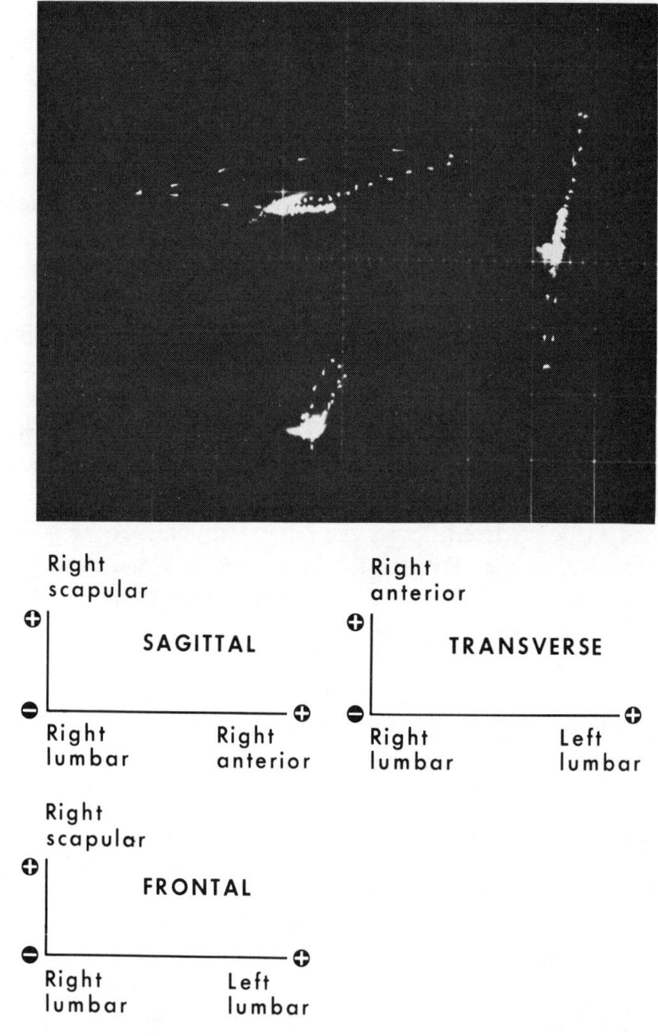

Fig. 8.10 Three perpendicular vector loops of a fetal lamb as simultaneously displayed on an oscilloscope from the electrode combinations indicated at the bottom. Each dot in all three loops is put in several times in a single beat of the heart. The sensitivity is 0.02 mV/major division. The loops have been chopped into dots at the rate of a thousand per second and pointed in the direction of progression.

follow the development of imprinting and pattern recognition in the first day or two after birth using an internal transmitter. Now that the legal definition of death is tending to involve brain activity, perhaps we will have to study the development of the EEG in the fetus in connection with the abortion question.

Exciting experiments of other sorts are also now possible. Thus the slowing of the heart rate of a dolphin upon prolonged breath holding and diving has been demonstrated, and it would be interesting to note if there are changes in the blood supply to the fetus and cardiovascular changes within the fetus as the mother dives. In a related observation, we see that when a mother sheep starts breathing pure nitrogen, the heart rate of the non-traumatized fetus starts to rise after approximately 20 seconds, probably due to the release of catecholamines by the adrenals, and then after approximately 2.5 minutes it falls considerably below the original rate. (Telemetry is helpful here since a distressed fetus will not show the initial rise.) The uterine circulation of dogs is no better protected against certain stresses (hemorrhagic shock) than is that of the kidney (Jackson and Egdahl, 1962); it will be interesting to see if the regulatory mechanism of aquatic mammals better protects the fetus during breath holding than in land animals, or if there are perhaps special protective mechanisms in the fetus of diving animals.

Other fetal parameters can be studied and other transmitter types employed, but these examples were cited to indicate the potential of the methods. In all the present studies, while our fetal subjects are monitored from the next room, the mother is alert and active, with visual observation being made over closed circuit television.

9

Sensors and Transmitters for Other Variables

1. SOUNDS

Any cyclic mechanical disturbance or pressure fluctuation can be considered as a sound. Thus all the comments of Chapter 5 are relevant here, as well as the forthcoming remarks on the measurement of activity with accelerometers. There are special advantages to monitoring frequencies normally detected by the ear, either of animals or humans. A microphone on an animal can pick up various environmental sounds, walking sounds, sleeping sounds, mating sounds, eating sounds, breathing sounds, heart sounds, the sounds of predation, laughing or crying, choking, gasping or coughing, wheezing or sneezing, and various forms of vocal activity, to name a few. An appealing project is to determine if adult cats and cheetahs purr at times when they are not in the presence of humans. The human ear covers a range of frequencies from approximately a few dozen hertz up to about 20,000 hertz. Over this range of frequencies, the ear is especially good at interpreting complex sound patterns. (If a higher frequency tone is played in through the jawbone, a fixed high frequency will be perceived which is known not to be analyzed in the usual way.) The usual analyses describe a sound on the basis of the time distribution of the various frequencies comprising the sound. Banks of tuned filters, and other methods of Fourier analysis, have been used by many workers to automate these procedures. Any such description of a sound source, however, leaves out the relative phases of the different frequency components. These affect the actual shape of a sound wave. It appears that this type of phase information could be useful to certain animals. Thus an emitted compression pulse from a bat could be returned as a compression from a solid object or as a rarefaction from an opening, the two echoes having the same frequency spectrum, but differing in a useful way. Similarly, the sonar system of a dolphin might distinguish a smooth stone from the swim bladder of a fish in murky water by using phase information. This is not to say that the usual analyses are not meaningful, but only that in certain cases other kinds of information might have significance to the subject, and should be considered in the analysis of records as well as in the design of transmitters.

With regard to the frequencies that may have meaning for an animal, this can cover a considerable range. Sharks perhaps are attracted by the low-frequency disturbances of an animal in distress. Perhaps the schooling of fish is controlled by pulsations propagated between neighbors. At the other end of the scale, the work of Scott Johnson shows that some porpoises certainly respond to frequencies above 125,000 Hz. In the air bats are known to generate frequencies almost this high. Piezoelectric transducers can be made to cover this range of frequencies in the water. Variable capacitance microphones are able to cover a similar range of frequencies in the air. Adequate sensitivity at the high-frequency end is a limitation in some research on bats, with the best microphones to date being made by cutting a shallow spiral groove in a metal base plate, and covering it with an aluminized sheet of plastic (Pye and Flinn, 1964). Perhaps the pressure sensitivity of certain electrical discharges would supply a better response, although the noise might be higher. When the question of importance is not what sounds exist, but rather what sounds the subject is probably sensing, then the transmission of cochlear microphonics by an electrode on the round window of the ear would seem ideal.

One point relating to internally placed microphones is perhaps deceptive. It is well known that a sound wave, in crossing an interface between air and water, suffers an attenuation of 30 dB. But the instantaneous pressure on the two sides of such an interface must be approximately the same. Thus we might ask whether the signal from a pressure transducer, which was initially responding to a sound wave in air, would give a similar indication when immersed below the surface of some water in the same vicinity, or would it be much less. A pressure transducer is a device which generates a signal in response to pressure changes without itself deforming much, and a piezoelectric cylinder is a fair approximation to this hypothetical infinitely stiff transducer. If such a transducer is attached to an oscilloscope, it will be seen that the response to a sound is about the same whether in or out of a basin of water, as indicated in Fig. 9.1. This may prove unexpected, and it is due to the fact that, although often used in air, these transducers

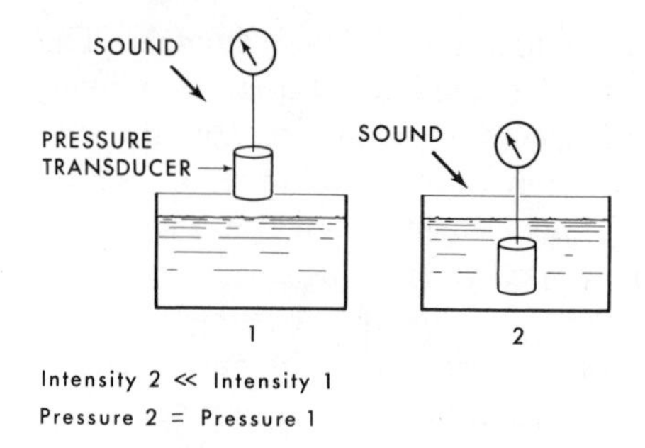

Fig. 9.1 The pressures on the two sides of a flat interface (water surface) are the same. Thus a stiff sound transducer will sense approximately the same signal after being immersed or implanted.

really are not especially well-matched to this medium. That is, part of the sound signal which would have been reflected at the surface of the transducer is now lost at the surface of the water, leaving the signal roughly the same. Actually, because of a reflection of sound waves at a less-to-more-dense interface, the signal can be twice as large when the bath of water is moved into place. If the extent of the liquid is small with respect to a wavelength, then the entire system can be considered as a small unit immersed in the sound field, and the pressure sensed will be just one times the pressure previously existing. This is presumably the case with the "bugged" olive placed in the martini by spies. This apparently started out as a joke, but raises serious questions about the ability of an immersed transducer to pick up nearby airborne sound waves. Similarly, in biological experiments employing certain transducer types, there should be no surprise if ambient sounds are picked up beneath the surface of the skin almost as well as before the transducer was implanted. Biological sounds can be quite noticeable to an internal detector, and so this effect may or may not be advantageous, depending on the application.

In connection with internally placed transmitters of sound, or any other variable, it is significant to note that the transmitter may be considerably smaller than an egg which might be laid by a female of the species, in many cases. Telemetry has especial significance in connection with sound monitoring from subjects displaying even small motion, in that wires trailing from an ordinary microphone can generate large interfering sounds by dragging over adjacent stationary surfaces (e.g., the skin, ground, covers, or etc.).

Hollow cylinders of polarized lead-titanate-zirconate ceramic, if silvered inside and outside, make useful sound transducers for many purposes. Deformation of the cylinder by a sound source generates voltages between these two electrodes, to which wires can be soldered. It is also possible to place much of a transmitting circuit within such a cylinder, as will be seen. The low-frequency response of such a transducer depends on how long it takes the electrical capacitance of the cylinder to discharge through the input resistor of the amplifier connected across the cylinder (i.e., the greatest period that can be detected is determined by the RC time-constant). The high-frequency response is determined more by the mechanical resonant frequency of the cylinder, although the properties of the electrical circuit can be made to influence the high frequencies also.

A transmitting antenna coil can be placed within a piece of metal if the

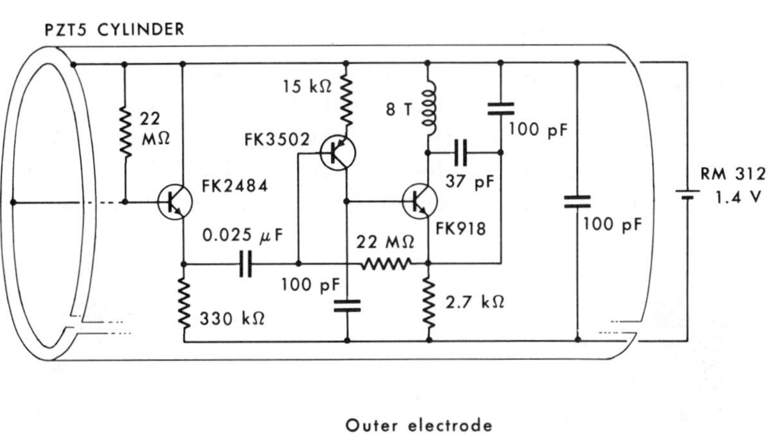

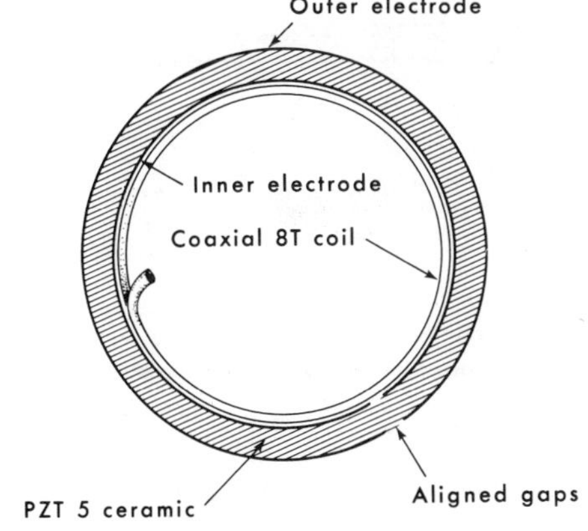

Fig. 9.2 The circuit of a sound-sensing transmitter, except for the battery, can be placed within a cylinder of piezoelectric ceramic. The transmitting antenna will be electrostatically shielded from disturbance, yet able to radiate if the electrodes on the cylinder have axial scratches that are aligned.

latter does not form a completed loop (a "shorted turn"). Thus a transmitting antenna can be placed within one of these transducers if an axial slit is scratched in both conducting surfaces. Because of capacitance effects through the high dielectric constant of the ceramic, these two gaps must be aligned, if they are to be effective. The transducer, while allowing the electromagnetic signal to emanate, serves as an electrostatic shield which prevents changes in the surroundings from affecting the basic oscillator frequency. Figure 9.2 shows a transmitter working near 100 MHz, with all components except the battery being inside the transducer tube. This transmitter is frequency modulated by sounds falling upon the ceramic tube. To give a low-frequency response extending down to about 2 Hz, a rather high input resistance is required at the first stage. Leakage into this transmitter must thus be avoided for proper functioning. The unit is slid into a plastic case having a clearance of about 0.001 in., which has been filled with molten wax. An alternative construction to the above places the coil and some of the components within a short transducer tube, whereas others of the components and the battery are beyond its end. Such a transmitter is shown in "exploded view" in the photograph of Fig. 9.3.

When this transmitter was swallowed by a human subject the signals

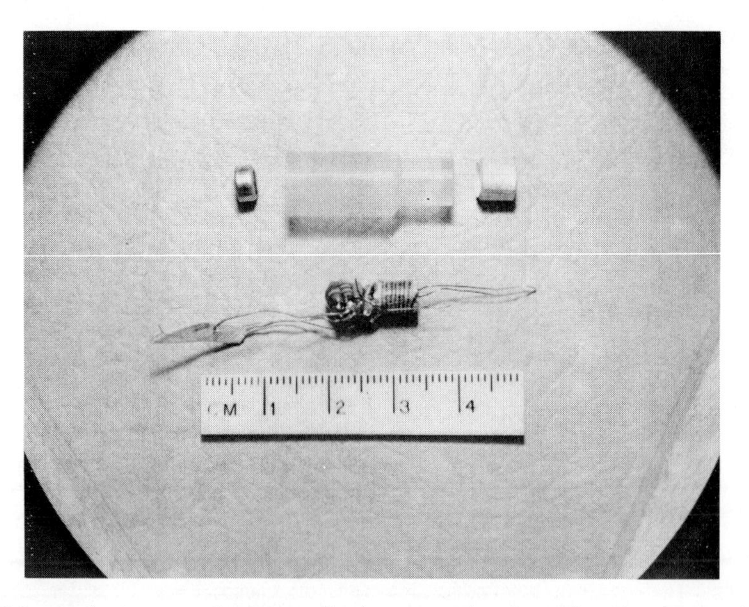

Fig. 9.3 At the top are a battery, a plastic case, and a short piezoelectric transducer cylinder. Below is the completed circuit; the coil and components on the right fit into the cylinder.

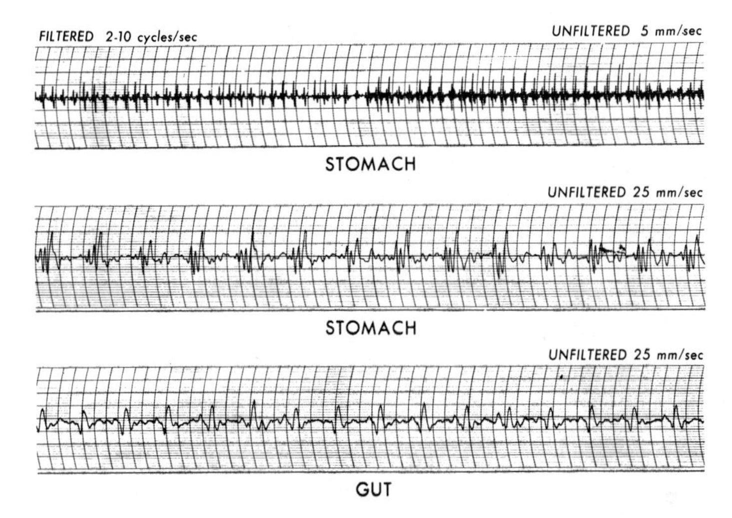

Fig. 9.4 Signal transmitted from a human gastrointestinal tract after the sound transmitter has been swallowed. Under conditions of small overall activity the heart rate is clearly indicated, as is the breathing rate.

were as seen in Fig. 9.4. It seemed like an interesting concept to feed a subject a pill and then be able to remotely monitor his heart rate. In the case of man the heart beat was clear except when the torso was moving vigorously; useful information was transmitted while walking but not while running. When listening to the demodulated transmitted signal, two of the heart sounds were heard. There were no breathing sounds, and neither eating nor a full stomach interfered with this observation. However, breathing in quiets the signal, especially the second sound, and the first sound is stronger at the end of an outward breath. The envelope of the pattern in Fig. 9.4 shows the breathing rate. When the capsule was down in the intestine, a single distinct sound was heard.

In the dog it appears more difficult to get heart rate from sounds in the stomach. Shivering and panting generate especially troublesome interference. However, the sound of breathing is generally clear. In the dolphin, for which a similar technique with a different transmitter type was used, the sounds of breathing were especially clear, but the heart beat was generally less distinct.

The absolute calibration of sound transducers (microphones or hydrophones) can become quite involved. Very often recourse is had to standard sound sources or relatively expensive calibrated standard microphones. None of this is required if all that is needed is a measure of sound fluctuations or relative intensities, and this is all that is transmitted in the usual entertainment broadcasting. In other cases, however, knowledge of absolute intensity levels is desirable, for example in studying the echo ranging systems of certain animals. A transducer calibration technique was developed which proved quite effective in many cases. The transducer under test was placed in a 10-cc glass hypodermic syringe with the leads being led out through the opening normally occupied by the needle. They were waxed into place. This chamber was then partially filled with water and the plunger was fitted into place, being certain that no air bubbles were present. A pointed weight was then balanced on the plunger. A handle with a hole was slid over a bolt screwed into the top of the weight. If this handle were suddenly pulled upward, it would abruptly remove the weight from the syringe upon coming into contact with the head of the bolt. Thus an extremely rapid decrease in pressure on the transducer could be generated while watching the electrical signal from the transducer on an oscilloscope. A representative trace appears in Fig. 9.5. The reproducibility of this test can be judged from the fact that this figure incorporates 10 separate traces. In the present case, from the known oscilloscope deflection sensitivity and plunger weight it can be said that this transducer had a sensitivity of approximately 4.0 μV/dyne/(cm^2). From the oscilloscope sweep speed it is

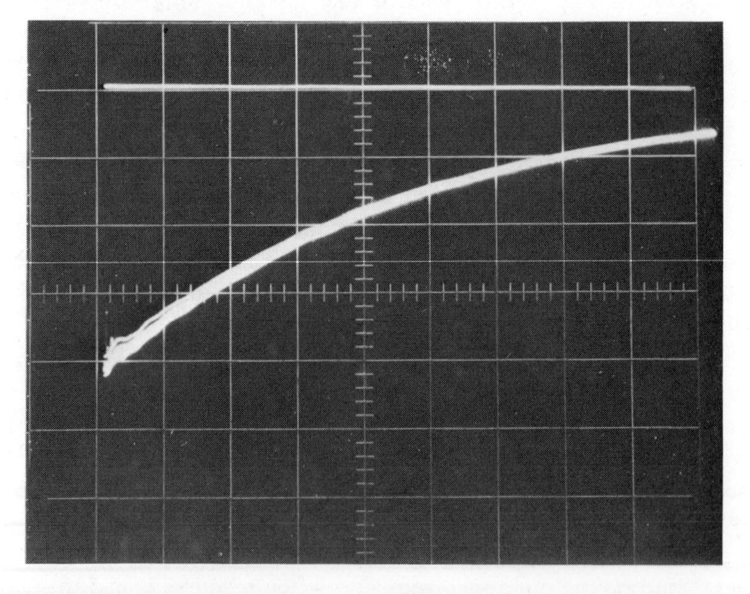

Fig. 9.5 The absolute calibration of a sound transducer and its associated electrical components. This is the response to the sudden removal of pressure within a hypodermic syringe in which the transducer is placed, and repeated 10 times to check reproducibility.

possible to estimate the time constant of the voltage decay, or the expected frequency response of the transducer and its associated circuit. The actual pressure change takes place in roughly 200 μsec which is considerably less than 50 msec response time of a typical 5000 pF transducer used with a 10 MΩ resistor. Friction did not interfere with the motion of the plunger when used with weights as small as 1 g.

Using this technique, it was also possible to evaluate various modes of mounting such a transducer. A cylinder coated with wax and having ends plugged with wax was vastly more effective than a totally exposed cylinder (used in an insulating fluid). An air bubble inside was more effective than solid wax. The transducer retained most of its effectiveness if a glass rod was waxed through the center of the cylinder while leaving a gap of 0.005 in. Thus it was felt that this thickness of wax was enough for isolation of a transducer from any rigid components which might be placed inside.

Such transducers and transmitters as these are sometimes more convenient than the pressure transmitters of Chapter 5 which have a low-frequency response extending essentially to zero hertz. It was noted that some of those transducers did pick up breathing and heart-beat signals. These same piezoelectric transducer types are convenient for generating and receiving higher frequency signals as well, for example, in ultrasonic telemetry under water (Chapter 15) or in acoustic exploration of living systems (the next section). In these cases the transducer will be vibrating at a very high frequency and will act as a very low impedance. In that case there is minimum worry about "short circuiting" the ceramic, and it is sometimes possible to implant such a transducer without any covering.

2. FLOW

The usual reason for wishing to measure a flow velocity is to monitor the passage of blood. Similar methods, however, might give further information about the water filtration and feeding of clams, or about gastrointestinal motility in higher animals. A urologist could perhaps study strictures in the urethra with a transcutaneous flowmeter. The velocity of a bird with respect to the air might be determined by comparing the resistance of a heated and an unheated thermistor, heat loss and resistance in the former going up with increasing speed. The same considerations apply to these and other possible examples, but the remaining discussion is especially directed to blood-flow measurements.

Some flow-measuring systems require contact with the blood stream. Observations of the pull on a projecting fiber, the passage time of an injected oxygen bubble, or the pressure in a venturi or Pitot tube are generally of this type. As already mentioned, the current from an oxygen electrode does depend on stirring or movement, and in some cases where it is not necessary to know the direction of flow, this deficiency can be emphasized to allow the estimation of flow rate.

Other methods are able to impart some distinctive property to a flowing stream without direct contact. Thus it has long been known that a comparison of temperature upstream and downstream from a heater on a vessel allows the estimation of flow rate. Some workers (Singer, 1960, A and B) have experimented with nuclear magnetic resonance methods in which either the rate of dissipation of oriented protons is followed to give a measure of volume flow, or the time of passage past two fixed points is noted for more accurate velocity measurement; these systems are rather complex for some telemetry purposes, but could be valuable in studying shock in humans. Mechanical vibrations transverse to the direction of flow invoke changes of momentum that allow estimation of mass flow rate. Thus Togawa and Suma (1965) vibrated a tube out of phase at two points and monitored any vibrations at the midpoint, but they could just as well have placed the driver at the center and monitored the two ends or perhaps have monitored electromechanical reactions on the driver itself. Various workers have employed systems in which injected sound signals are carried along by the flowing blood, thus arriving sooner or later than normal depending on the sense of flow. Very small time differences must typically be resolved (Franklin et al., 1962; Haugen et al., 1955; Noble et al., 1962; and Zarnstorff et al., 1962). Franklin (1970) has recently tested an FM sonar system in which the transmitted frequency is swept so that the outgoing and received signals beat to give a difference that is a measure of transit time, and where signals due to multiple reflections between transducers can be rejected by their very different frequency. Some presently mentioned sound systems use frequency shifts in back-scattered sound, though this is generally weak and requires extra overall power. At this time the two most relevant metering systems for the present purposes seem to be the electromagnetic flowmeter and the Doppler shift flowmeter.

The principle of the electromagnetic flowmeter goes back to Faraday, who realized that an electrical potential would be developed across a conductor (such as is blood) that is moving through a magnetic field. The magnitude of this transverse voltage is proportional to the magnetic field strength and to the velocity of the moving conductor. If the conductor is circular and the velocity profile axially symmetrical, the induced voltage is proportional to the average velocity. Thus, in many cases, the induced voltage is proportional to volume flow. Due to the work of Kolin and others such flowmeters have developed many forms (e.g., Kolin and Wisshaupt, 1963; and Wyatt, 1961).

If the magnetic field is supplied by a permanent magnet, then a steady output voltage results. This causes some problems with the electrodes placed on the vessel wall and also makes amplification of the signal difficult. If a sine wave is applied to an electromagnet, then the output signal will also be an alternating voltage. The alternating magnetic field, however, induces some signal into the pickup circuit by transformer action and by capacitive and conductive coupling from the magnet to the electrodes, even when there is no flow present. There are various methods for discriminating against these unwanted signals, but small changes in geometry of the probe unit can spoil the degree of rejection. Periodic reversal of a steady field (a square waveform) combines some of the advantages of both systems (Spencer and Dennison, 1959), and a somewhat more slowly changing wave form (trapezoidal) seems to minimize problems of induced transients. Such flowmeters can be made in small units, and their signals can be telemetered. There is a problem with gradual drifts in sensitivity and in the magnitude of signal corresponding to zero flow. The clamp for turning off the blood supply to a particular organ within an animal (to be mentioned in Chapter 16) was in part investigated because of a need to occasionally determine a signal corresponding to no flow in an animal.

A recently perfected ultrasonic technique (Franklin et al., 1966) seems to overcome some of these difficulties with very simple systems. These methods make use of the fact that sound returned from a moving object will be shifted in frequency. Such Doppler shift flowmeters have one important advantage in that if there is no velocity there will certainly be no frequency shift. The principle can be understood from Fig. 9.6 in which an oscillator supplies voltage to set a ceramic transducer into vibration, thus emitting high-frequency sound waves. Any particles in the fluid will scatter some sound back to the transducer. If the particle is in motion,

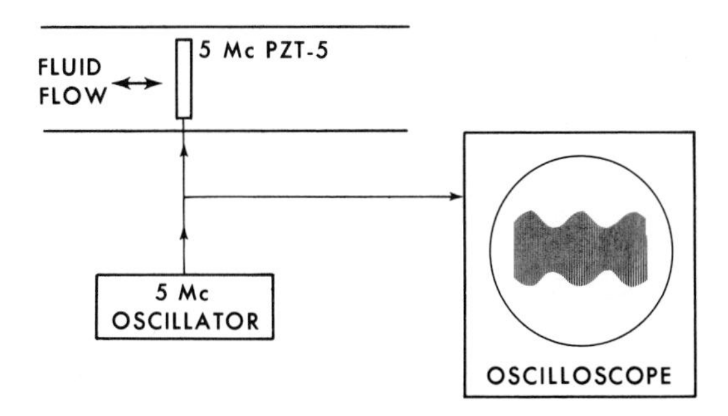

Fig. 9.6 Principle of a Doppler flowmeter. The piezoelectric transducer is excited at its mechanical resonant frequency by the oscillator, and the sounds returned by elements within the moving fluid will have a slightly different frequency. The combination of the two frequencies gives rise to an amplitude-modulation pattern sketched in exaggerated form on the oscilloscope. The modulation frequency is a measure of flow velocity, with ambiguity as to direction.

then the returning sound wave will have a slightly different frequency from the one emitted, because of the Doppler effect. The "crystal" will thus be vibrating under the influence of two different signals of slightly differing frequency and there will be small periodic changes in amplitude due to "beating" of these two frequencies. The voltage across the crystal will thus periodically change in amplitude, as is suggested in Fig. 9.6, and the frequency of this envelope of oscillation is a direct measure of velocity of fluid flow. Notice that the same difference frequency between the two signals would result whether the fluid was flowing towards or away from the transducer. A basic frequency in the range of 5 to 10 MHz is best for many purposes, and the crystals are then quite small.

An alternative explanation for what is observed can be seen by considering the case of very slow motion. In that case the reflected wave back at the transducer slowly goes in and out of phase with the direct signal; that is, very slow changes in amplitude result. The slowest motion which can be sensed depends upon the low-frequency response of the amplifiers in any overall system (and on slow drift or noise).

In Fig. 9.6 a scratch can be made across the metal coating on one side of the transducer to divide it electrically into two parts. The energizing energy can be inserted into one part, and the signal come from the other. This is observed to reduce the original signal component in the output by a factor of approximately 100.

In actual use there generally are two transducers, as shown in Fig. 9.7.

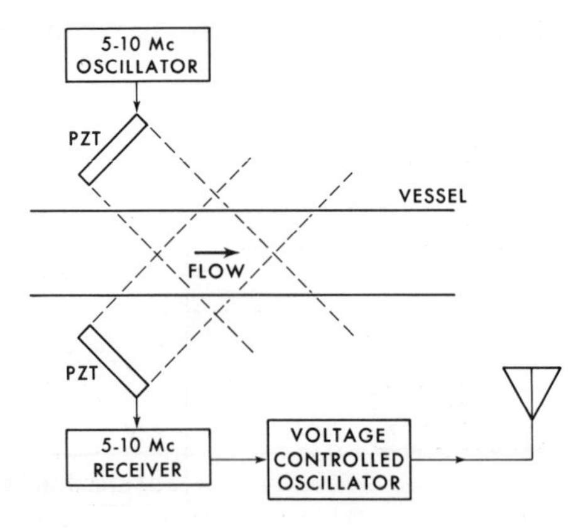

Fig. 9.7 Doppler flowmeter system for placement outside of a blood vessel or even outside of the body. The signal received at the second transducer can be sensed by an AM receiver whose output can be considered as a subcarrier oscillator signal. The receiver must have a wide range, and its low-frequency response limits the lowest velocities that can be measured.

They can also be placed along the vessel. These would be mounted in a plastic cuff which would be placed around the blood vessel. Power into the upper transducer would insonify the blood in the vessel of interest; red corpuscles would then scatter some sound out to the second transducer. This transducer also receives some direct signal, at the original frequency, from the first transducer. Thus a signal with cyclic changes in amplitude again results. Actually, the scattered signal is generally much less than the direct signal (e.g., 10 μV versus 100 mV) and thus the magnitude of the amplitude fluctuations are shown greatly exaggerated in Fig. 9.6. This signal appears like a typical amplitude-modulated radio signal (although it might more properly be termed a "single side band" signal) and it can be passed into an AM radio receiver which will then deliver a voltage that fluctuates at a frequency proportional to flow velocity, that is, proportional to the Doppler frequency. The frequency is zero for no flow. The receiver is tuned to the basic frequency and must have a wide dynamic range. Its signal can then be used as a subcarrier to modulate the transmitter oscillator, as shown in Fig. 9.7. With a basic frequency of 5 MHz, there will be a shift in frequency of about 70 Hz for each centimeter per second of velocity. Flow rates are generally under 200 cm/sec, and thus these frequencies typically fall in the audible range. If the transducer is not aimed directly along the direction of flow, the frequency shift will be reduced additionally by the cosine of the angle between the acoustical axis and the flow direction.

An example of a tracing from the studies of the above mentioned workers is given in Fig. 14.10. They report that the flowmeters do not work on an octopus, presumably because the blood does not contain any formed elements to scatter back sound.

In many cases relative velocities, or velocity changes, are of primary importance, and units of this type can perform this function in a simple and effective fashion. With pulsatile flow there may be some ambiguity, for a reversal of direction of flow does not cause the signal to move in the other direction. In many cases this does not matter, although the complication of an electrical reference signal in the demodulation circuit can be incorporated to give direction sense. For those interested in using the Doppler effect to measure flow (or movement or activity) there are more advanced relevant techniques described in the extensive radar and sonar literature. Sharp filters or phaselock methods can be made to separate frequencies higher than the basic frequency from those lower and thus distinguish forward from backward flow. There is also a technique convenient for telemetry that allows separation of these frequencies or side bands by some operations performed at the lower Doppler frequencies. It has been described by Kalmus (1955) but more details are given in the radio literature on "single sideband" reception (e.g., Norgaard, 1956). The basis is indicated in Fig. 9.8, where the demodulators can use the double MOS device of Fig. 15.5 or a transistor biased for nonlinearity. If a signal having a frequency, for example, 1 kHz different from that of the local oscillator is applied to the two demodulators, the output from each will be a heterodyne tone of 1 kHz and these two audio outputs will have a 90° phase relationship independent of the heterodyne frequency. If the signal has a frequency lower than the oscillator, then point A leads B in phase and if higher, B leads A. An analog frequency meter combined with a phase detector at points A-B provides the velocity signal with its sign if the local oscillator is also the driver of the sound source.

To the right in the figure is indicated one standard radio engineering scheme for processing the result by placing the signals back into phase either adding or subtracting. Proper performance here is for the range of velocities producing frequencies for which the difference in the second phase shifts remains nearly 90°. The total output comes from the upper or lower connection depending on flow sense, and thus the difference in these is the desired signal. The second phase shifters can be eliminated by multivibrators triggered from the two demodulators so that a signal appears in one channel if A precedes B and in the other if B precedes A, the difference in the two channels being the desired signal (McLeod, 1967). This method then tends to give instantaneous net flow without detailed flow structure information.

If there is a distribution in the velocities across the diameter of the vessel, a corresponding distribution in frequencies will emerge. It is generally

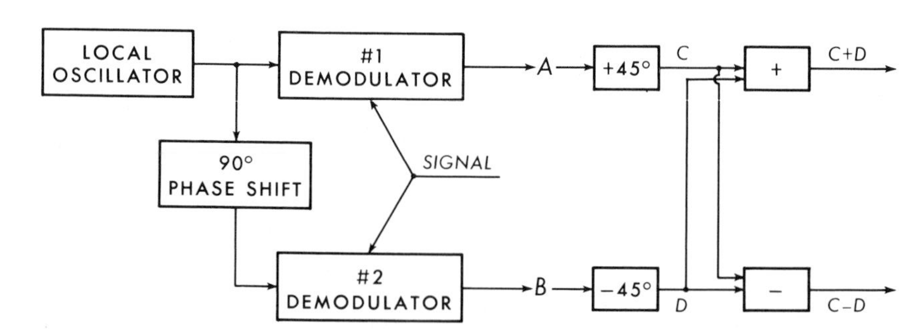

Fig. 9.8 Method for demodulating a signal, with an indication of whether its frequency is higher or lower than that of a reference oscillator. If the latter is also the power source in a Doppler flowmeter, then the electrical signal should dominate any unchanged frequency signal at the receiving crystal. A change in flow direction reverses the lead-lag relation at AB, which signal is converted to a voltage for recording over the range of Doppler frequencies for which the second phase shifters (shown as ±45°) maintain a difference of 90°. The flow signal then has magnitude *and* sense.

assumed that the signals from flowmeters of this type represent some kind of average of velocity over the vessel width that, in conjunction with a measure of cross-sectional area (which may be quite uncertain), gives an indication of volume per unit time. A more precise estimate of volume flow rates can be achieved with the help of a spectrum analyzer (a fast audio-frequency analyzer). However, with the geometry of many existing transducers a single particle (e.g., a bubble) moving at a given velocity will produce a different frequency depending on where it is both axially and transversely due to changing angles, and thus there is not always a direct relationship between flow distribution and frequency distribution. With turbulent flow some transverse velocity components may be sensed. In these units the movement of red cells is monitored because these scatter most sound. It is known that erythrocytes are not uniformly distributed during flow, tending to be concentrated along the axis where the velocity is higher. This gives an altered flow indication, though in some cases it is only the movement of the red cells that is actually of interest.

The sound transducers are driven at their mechanical resonant frequency, under which conditions they may act as an electrical impedance in the range of 50 to 200 Ω. Thus they are not readily shorted by fluids. Approximately two volts typically appear across one of these transducers. If too much power is applied to the transducer, then tissue burns can be produced, but it is not necessary to go to these levels. If the transducer holder is fitted around the blood vessel loosely, then damage to the vessel will not take place and yet tissue will quickly fill in any spaces to provide good acoustic coupling to the vessel.

The above are the most commonly used sensors of flow in cardiovascular studies, but different methods have proved useful in other investigations. Thus Birzis and Tachibana (1962) have measured changes in local cerebral blood flow by the impedance changes observed between a pair of suitably placed electrodes connected into an ac bridge circuit. In an animal it was not clear whether the impedance changes were due to a fundamental change in resistance with flow or to a changing geometry with each pulse, but the results appear to have been useful. In some cases volume changes rather than flow rates are of major importance. Estimates of the volume output of the heart can be had by measuring the resistance between electrodes set into the ventricle, and estimates of the size of a breath can be had from electrodes placed across the chest. So that no shock or other physiological effect will be experienced, the electrodes are placed in a bridge circuit that is energized with a relatively high frequency alternating voltage. In the case of breathing perhaps the most accurate flow rates are determined with a pneumotachograph. In this the subject breathes through a tube across which is placed a porous screen and the small pressure difference on the two sides of the screen is a measure of the rate of passage of the breath. Exotic systems based upon the frequency shift of Mossbauer radiation might be applicable in certain cases. Other special measuring systems can be adapted to special applications.

3. RADIOACTIVITY AND BLEEDING SITE

A clinical problem which might be aided by these techniques is that of localizing the site of bleeding along the gastrointestinal tract. It is estimated that in approximately 20 percent of those cases in which a patient shows a bloody stool it is impossible to tell where the bleeding is taking place, other than somewhere between the throat and rectum. It was thought that if an endoradiosonde could be made specifically responsive to blood, a radiograph taken in response to a signal would be helpful in establishing a general location. The injection of a tracer into the bloodstream suggests itself. It is known that ascorbate affects the current in an oxygen electrode, and many large molecules can be detected by the sensitivity of their fluorescence, but the original thinking centered upon the use of radioactivity techniques (summarized in Mackay, 1961). The assumption was that a sample of blood would be taken from the subject, the red corpuscles rendered radioactive, and the sample reinjected into the subject. The endoradiosonde would then be ingested, and it would have to be sensitive to the presence of a few radioactive erythrocytes.

It was with this objective in mind that the initial attempts at building a radioactivity transmitter were undertaken. It was observed that placing one of these transmitters near a piece of radium or in a moderate X-ray beam did not alter the output signal. However, removing the opaque covering from the transistor in one of the simple oscillator circuits resulted in a rather large sensitivity to illumination changes. Thus a form of scintillation detector was constructed in which a fluorescent material glowed in response to radiation, and the resulting visible light changed the signal from the circuit. A piece of X-ray screen is relatively sensitive to gamma radiation while the lighter scintillating plastics emphasize beta particles over gamma rays. If the resistor in the blocking oscillator circuit (previously described) is replaced by a cadmium sulfide cell of suitable resistance, then the blocking rate will change with changes either of visible light or X-ray intensity.

Circuits of this type are rather sensitive, but they do respond to the average flux rather than to individual radioactive events. Thus they are appropriate for dosimetry in a patient during therapy, but their use presently requires too high a patient dose in the above-mentioned application. Because of their small sensitive volume, it was felt that semiconductor detectors making use of effects in a junction region would also generally be somewhat inadequate. However, the steady improvement in solid-state radiation detectors suggests their periodic reconsideration.

More recently, Michael Grahmme has constructed a radiation-sensing endoradiosonde in connection with his master's degree project. The circuit is used with a very small commercial Geiger-counter tube, and is shown in Fig. 9.9. One form of oscillator converts the voltage from the battery pair at the left into an alternating voltage which is then stepped up in the transformer. The five diodes and capacitors constitute a circuit which will deliver a direct voltage of approximately five times the peak voltage of an applied sine wave. This is sufficient to put the counter tube in the plateau of its operating range. The diodes indicated showed a sufficient back-voltage rating, although they were a bit larger than necessary. The circuit constants were chosen so that the capacitors in the quintupler were not significantly discharged during each radioactive event, as this would unduly limit the maximum counting rate. The output transmitter was originally arranged to go on momentarily in response to each event, but it proved more convenient in adjusting the receiver to the transmitter signal to have the transmitter go off in response to each event. This unit was compressed into a package that could have been swallowed by a human subject, but the experiments with it were all performed on dogs.

In making observations such as these, it is preferable to use beta rays instead of gamma rays because gamma rays have a sufficient range to enter the gastrointestinal tract from the surrounding tissue. However, this simple concept is confused by the fact that the interaction of electrons with matter

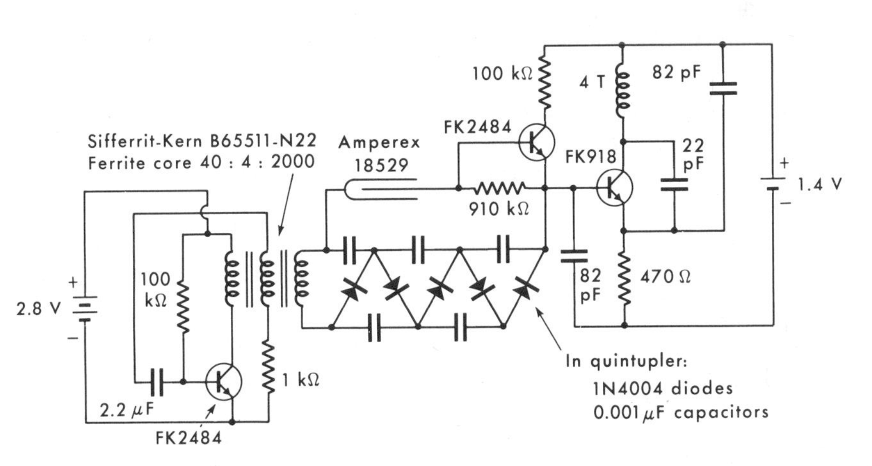

Fig. 9.9 A miniature Geiger-Mueller counter tube is supplied with an operating voltage by the arrangement shown and momentarily turns off the transmitter in response to each radioactive event.

gives rise to X-rays or gamma rays, and it can take a centimeter of lead to shield a counter tube against bremsstrahlung. In the present experiments the counter tube was coated with a layer of wax and thus the beta particles on the low-energy part of the distribution curve were unable to penetrate into the sensitive volume. Because of this, it was impossible to reliably sense cells labelled with phosphorus 32 when the total dose to the animal was moderate. Other useful biological observations were made, and it seems probable that this method will be successful in conjunction with a counter having an effectively thinner window.

Other methods for sensing the presence of blood can also be considered. The appearance of an unusual oxygen concentration does not seem promising both because of the confusion caused by the air that is normally swallowed by humans, and because bleeding could be venous with a minimum oxygen change. A wound potential perhaps could be detected, with background signals being reduced by atropine, but this is highly speculative. The use of a tracer, fluorescent or otherwise, which would not ordinarily diffuse into the gastrointestinal tract, would seem most desirable. Occasionally reverse peristalsis can cause the appearance of blood above the site of bleeding, and in these cases the repeated administration of a tracer holds the potential for localizing the site of most recent efflux.

Systems depending upon noting color changes within the body can be effected. Power can be inwardly induced to activate a small incandescent lamp, while the region of interest is monitored by a pair of phototransistors covered by different colored filters. Little application has yet been made of such a method in connection with blood detection.

Various chemical methods can be considered. A system for sensing the iron of hemoglobin, perhaps concentrated by certain chelates, could perhaps be devised. Pathologists and criminologists have long used certain reactions to detect blood. In general these are based on the peroxidase activity of hemoglobin decomposing hydrogen peroxide or sodium perborate. The decomposition products then produce a color change or luminescence in another reagent if blood was initially present. The primary reaction is also accompanied by the liberation of heat and thus can be monitored by a temperature transmitter. We have explored some of these possibilities, as has a group in Japan (Kimoto et al., 1964).

Because hydrogen peroxide is a fluid, it is natural to consider sodium perborate instead. This is a mild chemical which has been used in tooth powders. However, boron toxicity has been reported in the literature, with as little as 5 g of boric acid proving fatal to a human. Thus repeated observations might lead to cumulative effects of an undesirable nature. Calcium peroxide has proved capable of sensing blood in a similar fashion, although there is a more continuous liberation of heat with that material. These reactions are generally insensitive to the normal contents of the gastrointestinal tract, although such observations would normally be done while on a diet free of dark red meat (and soup therefrom), uncooked vegetables and aspirin, and after a day of fasting.

It has proven difficult to maintain an adequate supply of the reacting chemical without going to a series of successively dissolving miniature capsules. Recent successes in making water-insoluble enzymes (Katchalski, 1962) suggests the possibility of chemical binding of an organic peroxide to an insoluble polymer. For maximum reaction with fresh blood, it may be desirable to hemolize the red cells with a ring on the transmitter of saponin or any surface active agent. A survey of various organic and inorganic peroxides and persulphates may reveal a material that functions well and which can be incorporated in such a way as to retain its activity over an extended period.

Glucose-6-phosphatase is an enzyme in the red cells which is not normally found loose on the surface of the gastrointestinal tract. Thus one might coat a sensor with a substrate degraded by this material, and monitor the resulting electrical or mechanical changes. This could perhaps be useful above the stomach (where there would be little interference by proteolytic enzymes) to detect esophageal bleeding. For this region one might also make use of a specific immune globulin against the contents of the red cell, which would then form a precipitate. This part of the body is of considerable significance, and although accessible, such an examination might be more convenient than the use of an esophaguscope.

The detection of radioactivity or radiation has applicability beyond that of sensing bleeding. In some cases a radiation detector would be preferable during therapy to the estimation of dose using a phantom; this would especially be true when dealing with structures near the gastrointestinal tract which are themselves able to move a little. Sensitive transmitters might also be useful in such things as the diagnosis of gastrointestinal neoplasia following the concentration in cancerous tissue of intravenously administered phosphorus 32 (Nelson et al., 1963).

The movements of small animals can also be followed, even when buried or otherwise obscured, if they have been injected with a radioactive material; for example, a radioactive tantalum wire can be inserted in or attached to an animal for remote tracking. This aspect is discussed in Chapter 17.

4. OTHER SENSORS AND TRANSMITTERS

From the previous examples, it can be seen that essentially any variable for which a sensor can be conceived can have its value telemetered, if this is desirable. Because the number of possibilities is limitless, the remainder of this chapter is devoted to summarizing a few possibilities of somewhat general interest.

Ultrasonic energy appears to have a vast potential for monitoring the position and movements of various body structures. Some systems are rather like the sonar sets on boats. In one of these methods an electrical impulse is applied to a piezoelectric element which gives out a sharp click of sound energy. Some of this outgoing energy is reflected back successively from each interface between structures within the body, and these successively returning echoes generate electrical signals in the original transducer. This succession of echoes can be displayed on an oscilloscope to give a pattern that represents the cross section of the body. Moving of the transducer in a scan in synchronism with the oscilloscope trace allows the formation of an extended image which, in some cases, is considerably more detailed than a radiograph (especially in soft structures). An example of the kind of detail to be seen is the image of the densely packed structures of the human neck depicted in Fig. 9.10, supplied by Dr. Douglas Howry. Other examples and references are given in the several articles in the April 1967 issue of *Ultrasonics*. The sound beam can be fixed on a particular region, the motion of which will affect the time of return of echoes, and this signal can then be telemetered to follow movement.

Sound is especially well reflected at an interface between a gas and a solid or liquid and thus we were able to give an objective demonstration of the bubbles of decompression sickness. The velocity of sound in tissue is such that for a frequency of 15 MHz, the wavelength is 0.1 mm, and

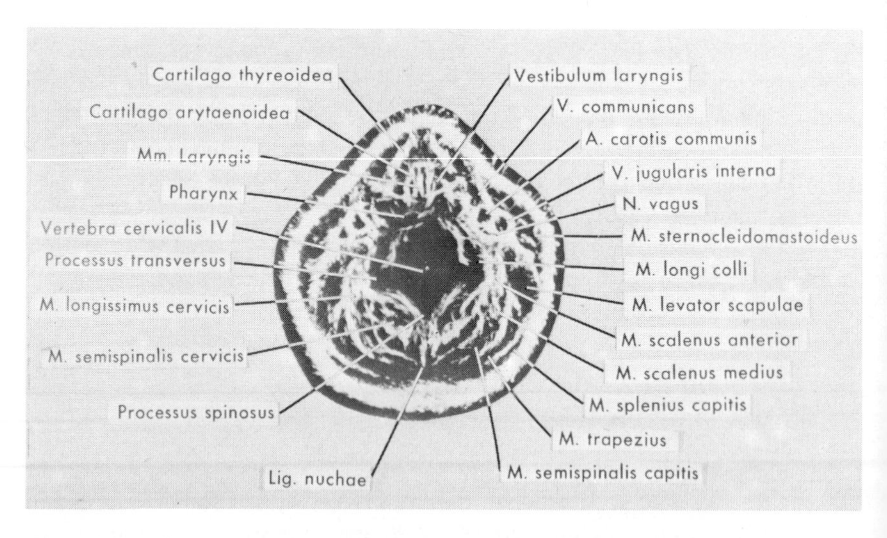

Fig. 9.10 Pulsed ultrasonic systems can show the details of even densely packed structures such as the human neck. Soft structures are especially well visualized and movements can be followed. (Photograph courtesy D. Howry.)

detail of this magnitude can be resolved. In connection with problems of decompression, it is possible to "pump up" a very tiny bubble to noticeable size with a more intense sound wave. Using these methods, for example, it may be possible to answer certain traditional questions about whether whales suffer the bends; under some conditions they might. From such images as Fig. 19.1, it should be possible to distinguish, for example, if it is pressure differences or the ratios that are of special importance in safe decompression, and possibly to more rapidly and reliably prepare decompression tables.

In some cases ultrasonic methods might substitute for radio telemetry. Thus the size of a small hollow bubble of glass (or a bellows) varies with pressure, and can be monitored ultrasonically. In some cases this could be used to transmit pressure information. In other cases local metabolism might be studied by the rate of change of the size of a bubble of a suitable gas introduced at the site. Comparison of two gases would allow correction for diffusion. Optical methods could be employed with bubbles in the eye or in some vessels, with the possibility existing of measuring pressure changes along the latter.

Perhaps it is possible to estimate the biological age of a subject by changes in the elastic properties (mechanical ultrasonic properties) of the lens of the eye. The presbyopia of human subjects suggests mechanical changes, and apparently the dry weight of an excised lens is sometimes used to estimate age in animal studies. Telemetry would not be involved here, but merely a different transducer possibility for a non-destructive *in vivo* method.

Pulsed Doppler sonar systems are known in naval work, and in biological or medical studies they will allow simultaneous measurement of position and momentary velocity. For example they would allow concentration on the velocity of a single moving structure or region among many. In principle, a structure could thus be mapped with the display indicating the velocity of different parts by differences in color. (We have already produced colored X-ray images in which different absorptions come out in different colors, and also colored electron micrographs from three wavelengths.)

A few general comments on cathode ray tube displays or images may be helpful. A variable intensity display is good for shape evalution. A color display of an image allows judging equal intensity regions or contours. Using the usual television horizontal scan raster, one can apply a small vertical deflection rather than an intensity change to obtain a different view in a monochromatic display. Small deviations or patterns may then be more noticeable. Roughness also has a different appearance or texture than the simple grayness of a variable intensity display. (Similarly, the ECG of Fig. 8.8 could consist of brightened traces to give a vertical wavy line showing time changes in the different complexes, but this would be less effective for some purposes.) An example of application is the finding by Rubissow in his doctoral research that a deflection display of ultrasonic images is very valuable. In many displays the information will be in more noticeable form after contrast and edge enhancement (Jacobson and Mackay, 1958).

General activity is a parameter of interest, and it can be monitored in a number of ways. If the subject is confined to a general region, then the easiest way can be to reflect a sound beam off of him, and monitor any return that is Doppler shifted in frequency due to movement. A subject carrying a sound transmitter of fixed frequency can also be monitored as to velocity by noting frequency shifts at one or two receivers. This may be especially valuable in judging the swimming speeds of certain of the cetaceans. The rate of change of velocity gives acceleration, which is a measure of work capability, or of deceleration which is a measure of drag. The time integral of velocity gives an approximate estimate of position relative to some starting point.

An accelerometer can be placed in or on an animal to monitor this parameter directly, or merely to give a general indication of activity. This is true even in the weightless state. If only the general degree of activity is to be monitored, for example, wakefulness, the accelerometer need not be especially precise. This was the case of the examples shown in Figs. 1.9 and 1.10. In Fig. 9.11 is shown a nocturnal animal of rather secretive nature

Fig. 9.11 A cockroach carrying a transmitting accelerometer on its back. In this case longitudinal accelerations are sensed and transmitted.

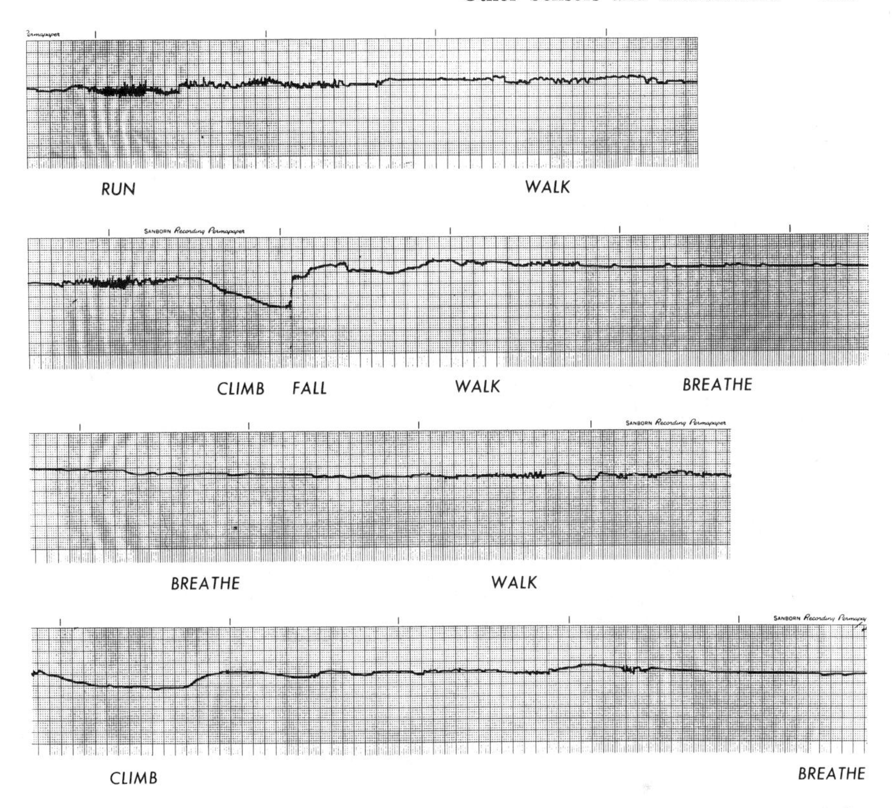

Fig. 9.12 Tracings from the accelerometer on the cockroach. Some aspects of the pattern of activity can be recognized without having to make direct observations on the animal.

carrying an accelerometer. The subject is an insect, specifically, a female cockroach *Brysotria fumigata.* A small gastrointestinal pressure transmitter was employed in this case with a weighted diaphragm. The indications are not especially precise but do allow interpretation of activity (Fig. 9.12) in the dark or from behind obstructions.

In the case of Fig. 1.9 no attempt was made to fabricate three-dimensional units which would indicate the three perpendicular components of acceleration. Instead, a piezoelement about a millimeter square and a centimeter long was weighted with a drop of solder at one end and clamped at the other. The voltages generated in response to motion were telemetered by a transmitter like that described in Chapter 7. For sensing accelerations in any direction, it is possible to employ a commercially available hollow sphere of piezoelectric barium titanate, into which (or around which) is placed a mass. A hollow cylinder can be used similarly for transverse accelerations. A tendency for a solid mass to roll over the inner surface is reduced if it is irregular, or there can be a complete filling of mercury.

The motion of a particular structure can be followed in a number of ways. The use of an accelerometer or seismograph is one. Attached to the tail of a dog, such a unit might allow the evaluation of a "happiness function," and it could be most interesting to study the activity of a Tahitian dancer using a suitably placed accelerometer. More serious projects might include monitoring the wing motion of a bird or the ear motion of certain bats. Other systems for monitoring motion employ the transducers discussed in Chapter 5 on pressure sensing, but with less of the springy restoring force being supplied to the moving element. Other methods of movement sensing employ transducers discussed in the section on flow measurement. Also, useful measurements can be made by attaching a small permanent magnet to the moving structure and monitoring the voltages induced in a nearby coil.

As indicated in Fig. 1.13, the pattern of signal-strength changes from any transmitter can be used as an index of motion. The actual pattern of small intensity changes can be recorded, or the frequency of "drop-outs" that appear in some other record can be noted. Although this does not give a precise indication of the six-dimensional character of all possible translations and rotations, it can often give all the information about activity that is actually wanted. Very simple observations can sometimes be useful. A previously steady signal can indicate moment of waking, for example.

Social interactions might be studied by monitoring the separation of animals rather than their positions. A unit can be designed which will give out a sound click each time it "hears" a click. If one of these is placed on each of a pair of animals, then the resulting click rate decreases with separation. General proximity could be sensed by a magnet and transmitting magnetometer, or related field strength measurements.

Special projects often call for a special combination of transmitters. Thus we may be interested in the hatching of eggs in the nest of a bird and not wish to disturb the normal behavior of the parents. In that case a blocking oscillator of the type previously described can be used to monitor temperature. A similar circuit can be used to monitor humidity, if a suitable variable-resistance humidity sensor is used in place of the thermistor in the blocking circuit. Humidity sensors can consist of a pair of electrodes in a gypsum block, or an anodized aluminum unit related to the device of Fig. 5.23, or a gold-grid lithium chloride element. Variations in intensity of the signals can be used to obtain an indication of egg-turning activities. Actually, in some situations nesting apparently is a time of high thermal stress for the parent birds, and so telemetering from them as well could supply an interesting overall study. This is an example of a case in which extremely short-range telemetering can be valuable, since the receiving antenna coil can be placed directly under the nest.

Measurements based on a change in resistance are often valuable, and a number of circuits for transmitting them with varying degrees of precision have been given in the preceding chapters. In the case of breathing it is common to measure the motion of the chest of an immobilized human subject by a variable-resistance band which is stretched with each breath. However, largely because of the change in geometry, the electrical resistance of the chest itself changes with expansion of the lungs (e.g., for a summary see Baker et al., 1966). Thus a suitably placed pair of electrodes can be used to monitor the frequency or the pattern of breathing, and to give some estimate of the volume of air involved. Actually, in many cases the pattern of breathing (e.g., fast in and slow out, coughing, or choking) is almost as useful as a more precise volume measurement. To prevent sensation to the subject this impedance pneumography is usually done with alternating currents in the range of 20 to 100 kHz. Four electrodes can be used instead of two with some advantage. In that case power is supplied to one pair of electrodes by an oscillator which passes the alternating currents through the chest. The second set of electrodes merely senses the resulting voltage drop; these electrodes have none of the major currents flowing in them. Thus contact resistances exercise much less effect and many motion artifacts are eliminated. A summary of some of these matters has also been given by Pacela (1966). For a chronic preparation, stainless-steel wires can be attached to a rib or elsewhere within the chest in a way that has been described in connection with electrocardiogram transmission; movement artifacts are then less than with surface electrodes. Either the unbalance of a bridge circuit can be telemetered or a constant alternating current can be passed through the electrodes, and the changes in the rectified voltage across these electrodes can then be used to modulate any of the transmitting circuits.

Galvanic skin response or psychogalvanic reflex is usually measured as a small change in resistance of electrodes on the hands to either small dc or ac inputs. Though there are a variety of interpretations, such measurements have been used as indices of alertness, emotions, or generally how meaningful the thoughts of a subject are to him. Extended measurements might become practical if an automatic compensating circuit, related to those to be discussed in the chapter on receivers, were incorporated at the sensor to balance out slow drift.

Similar considerations apply to the measurements of various pulsations in tissue, which is sometimes referred to as plethysmography. Thus, external electrodes on the head have been used by various workers to estimate flow and artery obstruction, and this is called rheoencephalography. In some cases even simpler sensors can be used to derive information about the cardiovascular system. Thus a flashlight bulb can shine through the tip of the human finger to give intensity changes at a light sensor which are related to cyclic changes in pressure. This constitutes one way of obtaining heart rate. If a subject is very sick or has gone into shock, then flow may be so reduced as to make the usual physician's method of blood-pressure measurement impossible due to the absence of any significant Korotkoff sound, although the use of a Doppler flowmeter could be effective in this case, or in any noisy environment. In the recovery room of a hospital a check on heart beat can be obtained by taping a small accelerometer transmitter to the chest. Simpson has described a variable capacitance unit in which one plate is a thin sheet of aluminum which is mechanically resonant, and thus eliminates the necessity for some of the electrical filtering of undesired frequencies at the receiver.

The velocity with which a pulse wave travels outward along the cardiovascular system depends on the elasticity of the vessels and the pressure. By a comparison of the electrocardiogram and a pressure tracing towards an extremity, an estimate can be made of propagation rate, although a more precise measurement is possible by putting a sharp impulse into a vessel at one point and noting its arrival further along. From observations of this kind in conjunction with a pressure measurement, changes in vessel tone could be made. This could be important in the prolonged weightless state or in the testing of certain conditions associated with aging. In such cases the pressure sensor might also be used to inject the test impulse. It is not necessarily true that all frequencies will travel along the vessel with the same velocity, and thus the form of this impulse can be important. From regular changes in the blood-pressure pattern it is also generally possible to estimate breathing pattern.

Under vigorous conditions perhaps the best heart signals can be obtained from the esophagus, whether the sensor is electrical or otherwise. In man, such procedures have proven relatively undisturbing when telemetry was employed. A useful index of myocardial contractility in intact animals is the isometric time-tension relationship which is readily computed (Smith and Schwede, 1969) from the electrocardiogram and left ventricular pressure.

Various gases can be analyzed in a number of ways. Those involved in respiration have special significance and thus are often emphasized. Methods for measuring oxygen and carbon dioxide have been mentioned in Chapter 7, but there are others. A mouthpiece can be fitted with a warm nichrome wire on one side and a transparent chamber on the other to absorb the infrared radiation. A mechanical chopper can be interposed also. Carbon dioxide placed in the chamber will absorb energy at characteristic radiation frequencies and expand, causing a rise in pressure (cyclic if chopped) which can be telemetered. Carbon dioxide in the breath will subtract out energy at these same radiation frequencies and be monitored. Other gases will interfere only to the extent that their absorption spectra overlap. Woldring has worked with such a nondispersive infrared analyzing system, and noted that the pressure transmitter can be at the end of a tube remote from the absorption chamber. Other gases and vapors can similarly be analyzed by simply placing a sample of the material in the chamber. If a gas can be stimulated to emit the wavelengths it absorbs, then this light source plus a detector can sense changes in this specific type gas interposed in the beam, for the same reasons.

Oxygen is not well analyzed in this way, but it is known that it can be sensed by several means, in addition to those mentioned in the section on oxygen electrodes. It is magnetic, and the cooling of a heated wire by the "oxygen wind" in a nonuniform magnetic field is a measure of amount. Dewhurst and Kirsner (1969) have noted oxygen measurement by changes in inductance of an oscillator coil through which the gas passes, with stability being obtained by alternating the stream with air ten times per second as a reference. The voltage from an oxygen-depolarized battery is a measure of oxygen concentration, as is the voltage from a fuel cell having a hydrogen supply in a small reservoir. Certain chelates expand and contract with oxygen changes. These methods have been discussed in various places by different workers, and occasionally they show an advantage over the polarographic techniques previously discussed. A lead-silver cell is rather good for oxygen measurements (MacKereth, 1964) if the temperature coefficient is neutralized.

Various gases can be analyzed with one of the several types of mass spectrograph, though size limitations presently restrict the application in connection with telemetry. Gases monitored can include oxygen, carbon dioxide, nitrogen, neon, argon, and krypton. Gas pressures in blood or tissue can be recorded through permeable membranes of rubber stretched over polyvinylidine (Saran) catheters that are directly connected to the analyzer (Woldring et al., 1966).

Various groups are now working on photodiodes arranged in a matrix to provide very compact television transmitters. Such a unit affixed to the head of an animal might show the scene that it sees. As we noted before, the potentials of oculography then allow one to tell on which part of the scene the subject is concentrating attention. Such a method could perhaps be useful in studying the homing activities of certain animals which might navigate, for example, by watching the sun. A continuing question is whether the animal can see the sun, or, indeed, if visual cues are especially important.

In some of these studies the orientation of an animal can be significant.

In some cases a combination of the orientation of the animal and its momentary velocity, derived perhaps from one of the flowmeters, could be especially instructive. A telemetering magnetic compass can be constructed. A Hall-effect device is not necessarily the most convenient sensor because of its temperature stability and the nature of its signal. A fluxgate unit, some of which are commercially available, is often quite convenient. These consist of two high-permeability segments of wire, each wound along its length with an excitation winding through which is passed an alternating current to cause partial saturation of the wires. The two wires, placed side by side, are together wound with a secondary winding into which is normally induced no net voltage. A magnetic field along the axis of the element causes one of the wires to saturate before the other, with a resultant voltage being generated in the secondary of twice the frequency of the excitation. This output voltage is directly proportional to the axial component of the magnetic field, and inverts in phase when the field is reversed in direction. A pair of units can measure two components of the earth's magnetic field, and telemeter these. A variable inductance device can also be constructed for measuring magnetic fields. In the so-called "thin-film detector" a permeable metal film is deposited in a field, and has a hard and an easy magnetizing axis. Driving a current through the "hard" axis results in a changing inductance in response to fields along the easy axis. Such a device gives a very low noise performance because of uniform switching of domains over the sensitive area. When any of these devices are used as a compass in a telemetry arrangement, it must be remembered that several battery types have magnetic cases or components, and these can give distorted indications.

Attachment of a magnet to an animal is sometimes a convenient way of following his movements, and in this application, such magnetometer sensors are also useful. In that case there are no telemetry attachments. This application of these components is mentioned in Chapter 17.

In concluding this section, it should be noted that in making measurements the use of indirect correlations may be useful, once their range of validity has been ascertained. Thus for some purposes it is probably sufficient to monitor sodium ion concentration in the cervix rather than trying to determine estrogen level. Interference by mucous with the electrode would there have to be considered for extended telemetry. Similarly, it may be possible to estimate heart stroke volume from time intervals picked up by a low frequency microphone. Agress et al. indicate that stroke volume is rather proportional to the ratio of ejection time to isovolumetric contraction time.

Having covered a few examples of sensors and transmitter systems, it is now appropriate to go on to other aspects of the overall telemetry system.

10

Frequency and Antenna Selection

1. NEAR AND FAR FIELD

Light or radio signals can be considered as electromagnetic waves composed of alternating electric and magnetic fields. In a vacuum, these waves travel with the velocity of light which is close to 3.00×10^{10} cm/sec. In air the velocity is almost as high, and in other media the velocity is reduced by the square root of the so-called dielectric constant or relative permittivity. Some approximate values of the relative permittivity of polyethylene, ice, glass, and water are 2.3, 3, 5, and 80, respectively.

For a traveling wave, velocity equals frequency times wavelength. For those not used to thinking in terms of wavelengths, a few values of wavelength in air corresponding to useful frequencies have been tabulated in Table 10.1. The interaction of a wave with various structures depends upon the positions of the points of constructive and destructive addition, that is, on the size of an object compared with a wavelength, no matter what the frequency is. Thus it is often useful to speak of the size of an antenna or its height above the ground in units of wavelengths, this distance being calculated for the actual frequency of operation.

As mentioned, radio waves consist of an associated alternating electric and magnetic field. The well-known equations describing the strength of these consist essentially of two terms which are added. These can be considered as representing two superimposed fields; a field that starts off strong but which falls off rapidly with distance, and a weaker field that changes more slowly. Close to the source the so-called "near field" only will be

Table 10.1

Frequency (Cycles per second or hertz)	Wavelength (in meters in air)
50 k	6000
300 k	1000
3 M	100
30 M	10
100 M	3.0
300 M	1.0

sensed, while farther away the so-called "far field" will largely affect a receiver. The two terms become equal at distances from the transmitter of approximately a sixth of a wavelength. At distances a few times this the far field is dominant, and at a fraction of this distance, the near field is dominant. The properties of these two field regions are quite different in several respects, and some biological experiments are done in one, and some in the other.

Of course, when working with some low power transmitters, the signal can become useless before even reaching the transition range. In some cases, it is most useful to discuss field strength, and in others, the energy in a wave. The latter is proportional to the square of the former.

In the near field, the signal strength is quite sensitive to source position, that is, there is appreciable range information. Thus an array of detectors can well localize a source or emphasize the signal from one region while rejecting interfering signals from another. There can be a sensitivity null over a given volume rather than along a given direction, as in the usual radio cases. This has been applied in electrocardiography to sense the voltage from a single part of the heart (Fischmann and Barber, 1963), and similar methods can be used to discriminate against sources of noise in the aquatic experiments of Chapter 15.3, or to determine the position of an internal transmitter.

In this chapter, a number of differences in the two field regions will be mentioned. How some of the various possibilities differ is indicated in the advance summary of Table 10.2, which should be considered as a convenient guide rather than an absolute series of rules. Some of these considerations are especially applicable to the field configurations produced by the small antennas in many of the present experiments, which patterns are described in the next section on "dipoles."

Table 10.2

	Near Field	Far Field
Usual use	In laboratory or at low frequency	Outside or at high frequency
Extent	Within a tenth wavelength of transmitter	Beyond a few wavelengths from transmitter
Omnidirectional possibilities	Field exists everywhere; full omnidirectional reception possible	Field generally has nulls into which a receiver may fall, with signal loss
Locating transmitter	Direction of field at receiver depends on transmitter orientation; good range information included; look for strong signal region	Standard direction finding; direction of maximum signal is toward source (perpendicular to field lines)
Antenna combinations	Field mostly in original form (magnetic or electric); receiving antenna usually resembles transmitting array	Magnetic and electric energy equal, so various combinations effective
Optimum receiver antenna size	Loop of diameter equal to separation, or near a long wire	Resonant length
Field strength fall-off	Cube of distance	First power of distance (energy drops with square law)

We sometimes also speak of the near and far fields of a radar system. The transition distance depends on the diameter (D) of the reflector and the wavelength (λ), being approximately D^2/λ. Within that range the beam is collimated and the energy density is rather uniform. For a representative "K band" radar having a wavelength of 1.5 cm the beam does not show spherical spreading up to about 60 m.

2. LOOP AND OTHER DIPOLE ANTENNAS

Before continuing, the idea of a "dipole field" should be mentioned. If a positive electric charge is placed near a negative charge, lines of force will connect the two in a characteristic pattern. Likewise, current flows from a positive electrode to a negative one in a solution over similar paths. Most people are familiar with this steady dipole pattern from having observed iron filings sprinkled in the vicinity of a bar magnet. The magnetic field set up by a steady current in a loop of wire is also somewhat similar to this in character. These fields have certain properties that are relevant here. They fall off in strength with the cube of the distance from the pair creating the field. At a given distance the field strength along the axis is just twice that to the side (e.g., the total magnetic field of the earth at the North Pole is twice that at the equator). There is some field at all points near a steady dipole, with the direction being along the lines of force. Thus at a given spot in space the direction of maximum influence changes as the axis of a nearby dipole is rotated. A number of the transmitting antennas used in or on animals effectively are dipoles, and their near-field components can be visualized as having this form. The far-field components can be considered as associated with ac changes in the dipole. The lines of force associated with the far field do not end on the source but instead can be visualized as closing on themselves in a sort of kidney-shaped pattern. This radiated field falls off more slowly with distance and vanishes along the antenna axis.

From a consideration of equations such as those given by Stratton (1941, especially pp. 431–438) one can describe a dipole in free space, and gain considerable insight into the situation. All distances of observation must be large with respect to the size of the dipole, that is, ten times the length of an electric dipole or ten times the circumference of a magnetic dipole. (It is because this is true for most transmitters that one is often dealing with a dipole field.) An electric dipole is formed by a current element of magnitude I and length l, and its effectiveness is proportional to their product. The near field corresponds to the static or induction field of the electric-current dipole mentioned above, with a magnetic component given by the familiar Biot-Savart Law. In the far field the field has no radial component, both magnetic and electrical components being transverse; there is a radiated spherical wave. The magnetic dipole consists of a current loop, and its effectiveness is proportional to the area, the number of turns and the current. The near field is the same as would be obtained with a very slowly varying or steady current in the loop. In the far field the magnetic lines "break away" from the dipole and propagate, there being no radiation along the dipole axis (in contrast to the magnetic near field which does not vanish). The electric and magnetic dipole situations are alike if the electric and magnetic fields are interchanged in going from one to the other.

In Fig. 10.1 is seen a sketch of the electrical field of a small oscillating vertical dipole at an instant of time. In the center at right is seen the relatively tiny pattern associated with the near field. The transition field, to a somewhat different scale, is at the left. A few extra loops are sketched in the far field curves to show the directions of the field at intermediate positions between those of local maximum field strength. The magnetic lines of force in this case are simply circles perpendicular to the page and centered on the dipole axis. Most of the energy is radiated near the

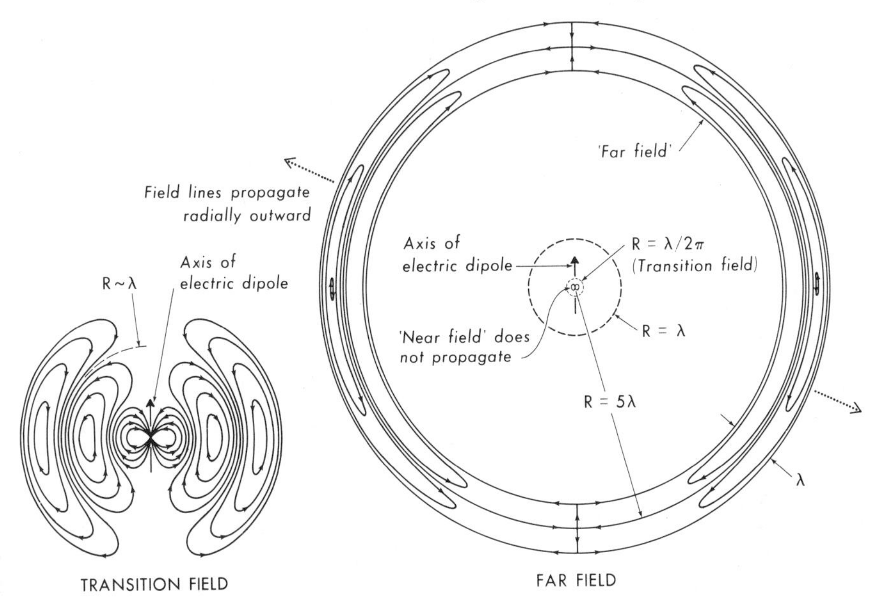

Fig. 10.1 Approximate shape of radio fields around a dipole. The strength of the far field approaches zero along the axis. The electric field of an electric dipole is shown, but the magnetic field of a small coil is similar.

equatorial plane and none along the axis where the amplitude becomes zero. A small coil source with axis vertical gives a similar pattern except the electric and magnetic fields are interchanged.

For purposes of comparison, we can compute the ratio of electric to magnetic energy density in the various cases. In the near field of the magnetic dipole, most of the energy is in the magnetic field, while in the near field of the electric dipole, most of the energy is in the electric field. From this we can conclude that, in the near field, an antenna sensitive to magnetic field is appropriate and should be used to receive from a magnetic dipole, and an antenna sensitive to electric field should be used with an electric dipole. In either far-field case the electric and magnetic field densities are equal, and so either antenna type is suitable (although at high frequencies a large loop antenna may display too much inductance to perform well).

With regard to its far field, a single-turn loop can be considered in another way. The effect of the current in the front of the loop will not cancel the effect of the current at the back of the loop because of physical separation of the wires. It is then clear why the signal comes out mostly at the edge of the loop and none along the axis. (As noted, in the near field or low-frequency case there is a field along the axis.) When acting as a far-field receiver, the loop will similarly have most sensitivity to signals in its plane. All current around a loop must be in phase and thus the circumference length must be under a tenth of a wavelength; it is a relatively low-frequency device for this reason. A current (ac) flowing through the inductance of a horizontal wire (one side of a loop) generates a horizontal voltage difference; the wave from a horizontal loop is termed "horizontally polarized," the associated magnetic field being set up in the vertical direction (both field components being perpendicular to the direction of travel in the far field).

Most biological telemetering experiments done at frequencies less than approximately 25 MHz take place in the near field, although some long-range studies do take place at lower frequencies. In the near-field cases the ideas of transformer design are more relevant than the usual considerations of radio transmission. Also, direction finding in the usual sense will not be effective, although completely omnidirectional performance is possible. At higher frequencies, it is more likely that the receiver will be in the far field, to which the more usual ideas of radio apply.

To generate a maximum signal near a magnetic dipole, we require a maximum area and number of turns with a given current, independent of frequency. This suggests a maximum number of turns and greatest possible area, while still giving a usable inductance value. In many transmitters, however, there is a voltage limitation imposed by the battery supply. In that case we obtain the most magnetic flux per volt with the fewest possible turns. Spreading the flux over a greater area helps, and thus it is found that a single large turn is often the best antenna. Tapping down on a multiturn antenna coil is helpful in raising the output power, but adding more turns beyond an existing coil does not help. This is discussed in Appendix 1.

The far field of a magnetic dipole is well received by an electric dipole oriented into the plane of the original loop. Thus the "whip" antenna on the top of a portable FM receiver, if placed horizontally, will receive the signal from a small horizontal loop energized at a high frequency. The field strength at a given distance varies with the square of the frequency, so the higher the frequency the better. The near field is independent of frequency, with a fixed dipole strength or moment; changes largely affect the extent of the region in space over which the near-field approximation exists. The radiation (far) field falls off with $1/r$, whereas the induction (near) field falls off with $1/r^3$. In spite of this, more power can be transferred through a near field at small distances than through the radiation field at large distances; for example, at 100 MHz an electric field at 3 cm can be 0.03 V, whereas in the far field at 15 m it is 4 μV; similar considerations apply to magnetic fields.

In most useful cases there is very loose coupling between the transmitting and the receiving antennas. The two antennas do not interact in general, and thus can be analyzed separately. (This is not necessarily true in those cases where appreciable power is to be induced into the body of an animal to energize a piece of apparatus.) On the basis of any of several mathematical arguments, it can be shown that there is an optimum diameter receiving loop for the signal emitted by the tiny loop in a typical endoradiosonde. The radius of the receiving loop should be approximately equal to the separation of the plane of that loop from the transmitting coil (Mackay, 1960A; Schuder and Stoeckle, 1962; and Collins, 1966). This is a practical but not critical near field consideration.

A configuration rather like a loop for transmission in the far field is obtained by placing a conducting cylinder just inside a conducting loop to give what is called a strip transmission-line ring antenna (for example, see Bassen, 1967). These can be extremely rugged mechanically and are very predictable, even when a variety of materials, including conductors, is within the antenna. Proximity effects are small and there are no significant effects of movement when placed around tissue.

In the case of loop antennas in particular a considerable improvement results for a specified size antenna by using ferrite rod cores. Typical flux concentrations by a factor of 10 result over free space concentrations. Other factors considered in later sections include the avoidance of loss of coupling between transmitter and receiver with changes in antenna orientation due to animal movements, direction finding, and the effect of placing a loop in a higher dielectric medium such as fresh water or any partly conducting medium such as ocean water.

3. DIRECTIONAL ANTENNAS

It is often desirable to know the location of a transmitter from a remote position. Methods in which an animal walking over the surface of the earth changes its relationship to a matrix of receiving antennas on the ground will be discussed in Chapter 14. Other methods are mentioned in Chapter 15, in which the difference in arrival time of an emitted signal at several stations allows the investigator to calculate where the transmitter is. But the most usual procedures employ some form of direction finding to indicate the location of a transmitter. The situation is not quite the same when performing this kind of operation in the near field of a transmitter as when working in the far field.

In almost all human studies in which an endoradiosonde is swallowed the signal is picked up with some sort of a small loop antenna, either with or without a ferrite core, and this receiving antenna is in the near field of the transmitter. In that case the apparent direction to the transmitter depends on the orientation of the transmitter, as well as its actual location. If we orient a receiving loop in such a way as to receive a maximum signal, then the axis of the loop will point along the lines of force. By analogy, consider a compass placed near a bar magnet. If the axis of the magnet passes through the compass, then the compass will point directly at the magnet. If the axis of the magnet is then turned through 90 degrees without otherwise moving either magnet or compass, then the compass needle will turn through 90 degrees and point in a direction parallel to the magnet. Thus approximate location of the position of an endoradiosonde in a human subject is not by direction finding, but rather by moving the antenna until a position of strongest signal is obtained, whereupon the transmitter is almost below the antenna. By moving the antenna in several directions, a rather good estimate of position can be obtained. Jacobson has arranged a two-dimensional servosystem to approximately follow the motion of an endoradiosonde through the gastrointestinal tract of a human subject (Jacobson and Lindberg, 1963) in order to do the above operation automatically. Characteristic patterns of activity in various parts of the gastrointestinal tract have been noted using the method (Bárány and Jacobson, 1964). When using this type of apparatus, tilting of the capsule will appear as an apparent displacement on the trace, which is otherwise a two-dimensional projection of the actual motion of the capsule. In this near field there will be no orientation of the transmitter for which the receiving antenna in any specified place will be unable to pick up some signal, if the receiving antenna can be angled to point in a variety of directions.

Direction-finding techniques are also used to locate an animal moving freely in the field. This is a common requirement in "home range" studies or to follow the path of migration of herds of animals, for example. In this case longer ranges of transmission are generally employed, and operation usually is in the far field, in which the more ordinary concepts of radio transmission are applicable. The apparent position of a transmitter does not depend on the orientation of the transmitter, although for some orientations the signal at a receiver can disappear. In these procedures some form of directional receiving antenna is rotated to give either a maximum or a minimum signal, thus indicating the direction of the transmitter in or on the animal. From two of these stations with known positions, a simultaneous observation allows triangulation to pinpoint the transmitter location on a map of the area. The alternative is for the investigator to use the directional properties of some portable receiving equipment to allow him to approach arbitrarily close to the subject at will. Since the range becomes quite short at the end of the location process, the angular accuracy required in this last method is considerably reduced. To some extent the angular accuracy with which the direction to a transmitter can be specified depends upon the size of the receiving antenna (in terms of wavelength) for reasons similar to those that indicate that large-diameter telescopes have more resolution than small telescopes. It is of interest that one can often locate a transmitter that is out of sight beyond large trees or low hills, though directional accuracy and signal strength usually suffer.

A common form of direction-finding antenna is a loop with its plane vertical (its axis horizontal). In many cases this is used to pick up the signal transmitted from a similar loop placed around the neck of the experimental animal in the form of a collar (e.g., Fig. 1.16). As we noted previously, a loop antenna is most sensitive edge on, and has no sensitivity along its axis, when only the far field is considered. Thus it makes a very satisfactory direction finder. On the other hand, a good transmitting antenna is a good receiving antenna, and vice versa, and the directional properties are the same in each case. Thus, as has been mentioned, no signal will travel out along the axis of the coil from a collar oriented in a vertical plane, and thus for some orientation of the subject no signal will be received no matter how the receiving loop is turned. This lack of the possibility of reception under all conditions goes with working in the far field.

A vertical whip antenna as a transmitter would allow a suitable signal to be directed outward along all directions on the surface of the earth, although for use on a bird, we should then have the problem of no signal going upwards or downwards.

To continue this aspect another step, consider a loop transmitting antenna wrapped around the neck of a giraffe. A signal would be radiated in all horizontal directions, but it could not be picked up by a loop antenna with its plane vertical. To have coupling in the far field between two such loop antennas the receiver antenna must also have its plane horizontal, but because of the symmetry of the receiving loop, just as with a vertically oriented whip antenna, there can be no horizontal direction sensing; that is, a signal is equally well received or transmitted in all horizontal directions. It is possible, however, to add the signals from two horizontally oriented loop antennas which are spaced in a horizontal plane by somewhat under one wavelength, and the loop pair can be swung in a horizontal plane for a maximum signal in order to find the direction to the transmitter. (A pair of vertical whip antennas can similarly be used on a vertically polarized signal.) A horizontal electrical dipole or monopole such as a whip antenna can be used to do direction finding on a horizontal loop transmitter, and there are many more sophisticated arrangements. As another alternative, the phase of the signal at two spaced receiver antennas can be compared in suitable circuits to give a quick indication of the direction to a transmitter.

Repeated in brief, a vertical whip and a horizontal loop both radiate in all horizontal directions, but the former is picked up with a vertical whip and the latter with a horizontal loop or a horizontal whip or dipole; the last combination allows direction finding.

By contrast, it should be remembered that two loops, when closely spaced (when working in the near field), have strong coupling if their planes are parallel when they are coaxial. For this reason, when one approaches very close to a subject while direction finding with a vertical loop at a range of approximately one wavelength, all directional sense tends to disappear, and then, closer in, the properties tend to reverse with a minimum in the loop's plane and a maximum perpendicular to this.

A direction-finding loop need not be circular, but it should be symmetrical. Large-area loops approach the performance of a quarter wavelength-long whip antenna with regard to range, but smaller loops generally give sharper null patterns. A single-turn loop should be carefully tuned to resonance at the operating frequency of the transmitter, and moisture should be kept away from this element so that the loop retains its properties as a high Q tuned circuit. The loop should be under a tenth wavelength in circumference, and resonated with a capacitor between its ends, from across which is taken the signal. Peculiar directional properties may be seen if the loop is not shielded. This can be done by forming the loop from a length of shielded cable (for example, RG-11/U) with a break in the outer conductor opposite the capacitor so that the shield does not form a shorted turn. The capacitance of the two sides to ground should be made the same. In a later section will be mentioned the matching of the characteristics of a loop to that of a receiver input for improved signal transfer.

In some cases it is best to look for a maximum signal while rotating the loop antenna, and in other cases it is most effective to look for the minimum signal or null. These two directions are approximately perpendicular to each other. In a high-noise situation the maximum may be most useful, while close to a strong transmitter the null will undoubtedly be most precise. Usually the null is most precise. In using a loop antenna there is a front-to-back ambiguity; that is, we can say that the animal is on a certain line, but it is not possible to say whether he is in front or behind. On occasion, this has led to amusing interactions with wild animals. It is known that the nondirectional signal from a nearby vertical whip antenna can be combined, after a suitable phase shift, with the signal from a loop antenna to cancel the reception by the loop from one direction. This does remove a front-back ambiguity, but generally slightly reduces the range of effective operation.

Much has been written about the design of radio antennas, in many cases with special emphasis on their direction properties in order to reject interfering signals. Since a highly directional receiving antenna is equally directional when used for transmission, these same antenna types are often used to beam signals over long distances to selected receiving sites. A directional antenna familiar to many workers because it is often used with home television receivers is the so-called Yagi array. A diagram of one example of such a unit is given in Fig. 10.2. At the center of this configuration appears the element which is actually connected to the receiver.

This is a horizontal wire into whose middle a connection has been made, and whose ends have been short-circuited by another parallel strand of wire, thus forming what is known as a "folded dipole." The total length of this object is approximately a half wavelength at the operating frequency, although the actual length is made slightly less because it is found that an electrical half wavelength is about five percent less than the physical half wavelength for resonance. This folded dipole would normally display an impedance of 300 Ω, and nicely match into a cable having this property, but placing other metallic objects in the vicinity somewhat alters this. Consequently, a coil and capacitor are also incorporated, as shown, to allow the signal to pass smoothly from the antenna down to the receiver with a minimum of wasteful reflections. This extra tuned circuit also helps to reject unwanted interference on nearby frequencies. Behind the dipole is placed a metal rod, for instance a piece of hollow aluminum tubing, which is about five percent longer than the dipole. This functions as a reflector, and the combination not only has rather good directional properties, but there is little ambiguity between front and back directions. The performance is further sharpened by placing in front of the dipole a so-called "director" which is also a metal rod, but approximately 5 percent shorter than the dipole. This three-element array is fairly effective in direction finding,

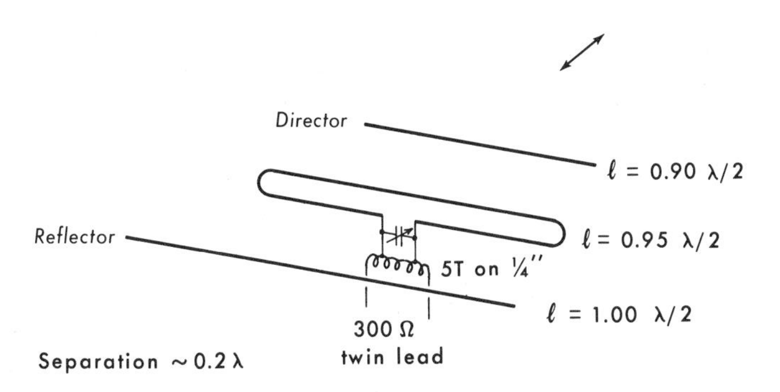

Fig. 10.2 A folded dipole antenna with a slightly shorter metal rod in front and a slightly longer metal rod behind to sharpen the directional pattern. More directors can be added. The matching network shown somewhat increases the efficiency and further emphasizes the rejection of unwanted signals.

and a photograph of one is shown in Chapter 18, as it was used in various zoo exhibits. To heighten the directivity of the array, more directors can be added, each having the same size and spacing as the one shown. This will sharpen the direction of main sensitivity, and increase the sensitivity of the antenna, while also introducing several directions of intermediate sensitivity (so-called "side lobes") which can prove confusing in certain applications.

Direction finding is most effective when the subject is at least several wavelengths distant, in order to be in the far field. Many antenna types which might be placed on a tower become quite ineffective at direction finding if the animal moves near the base of the tower so that the signal approaches the antenna from below. In any electrical engineering text, especially relating to electromagnetic theory, other antenna types are described. The present samples should serve not only as examples, but they also represent two antenna types that have proven valuable to biologists.

4. SPINNING FIELDS AND CIRCULAR POLARIZATION

There are various relative orientations of a transmitting and a receiving antenna for which the signal vanishes; that is, all coupling between the antennas disappears during some reorientation of a transmitting antenna in an animal or during some movement of an animal with respect to a stationary receiver. For example, suppose that a receiving coil has its center in the plane of a transmitting coil, and is oriented so that its own plane is perpendicular to the plane of the other coil. No signal will then couple from one to the other. To maintain a continuous flow of information, it is desirable that this decoupling be relatively unlikely. With a strong transmitter, the angle over which no signal is transmitted may be sufficiently small so as to be quite unlikely, and no further consideration need be given. In systems in which there is not a lot of excess power, however, one might prefer greater reliability at the limiting range. A way of improving performance somewhat in certain cases is to set the outgoing radio wave to circling.

If two coils are placed with their axes perpendicular to each other and are fed with alternating currents 90 degrees out of phase, then a rotating magnetic field, spinning at the supplied frequency, will result. (This is the principle on which many induction motors are built.) A similar procedure can be used to set up an electromagnetic wave in which the field vectors rotate. By facing in a number of directions they are more likely to couple into a receiving antenna of fixed orientation. By using the ideas of the induction of power from one coil to another it can be seen that the region of no coupling between two coils of fixed orientation is a spherical ring segment, that is, a band of constant width at a given distance. If instead the transmitter contains coils arranged perpendicular to one another and their currents are 90 degrees out of phase at the operating frequency, the only points at which the signal will vanish are the intersections of the two perpendicular rings associated with the two perpendicular coils; that is, with this circular polarization, only two spherical segments yield no signal.

There are a number of ways for generating a circularly polarized signal, or for feeding power to two coils with a phase difference of 90°. A resistor and capacitor can be used to take a basic signal and shift it in phase, after which the two signals can be used as the inputs to a pair of transistors driving the coils. A method that proves more effective is suggested by the fact that the currents in "critically coupled" circuits are out of phase by 90 degrees. In Fig. 10.3 is shown a transmitter operating in a pulsed mode near 500 kHz which was constructed to test out these ideas. Several different transmitter configurations were designed and tested to produce the proper current division and phase shift in the two coils. A relatively simple solution eventually proved to be the most satisfactory. A standard transmitter type was constructed in the usual way, and a second coil was arranged approximately perpendicular to the first. The second coil was connected only to a resonating capacitor. By arranging the two coils slightly off of perpendicular, some energy was coupled from the first to the second. If the tuning capacitor was properly adjusted, then the currents in the two coils would in fact become 90 degrees out of phase, and the energy in the two coils would be approximately the same. At any time, the advantage to be gained from the second coil was assessed by simply disconnecting its capacitor.

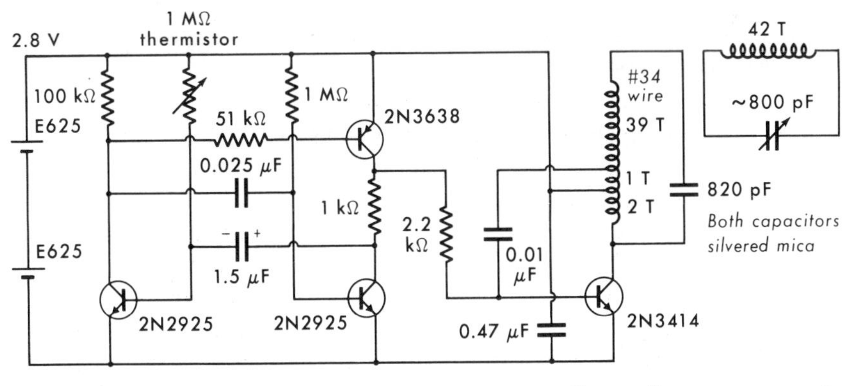

Fig. 10.3 A transmitter that sets up a spinning or circularly polarized field to minimize the likelihood of loss of signal at a receiving antenna due to changes in transmitter orientation. The frequency of turning on the transmitter is controlled by the thermistor in an astable multivibrator.

A number of experiments were performed to judge the advantage to be obtained by this extra complexity, and the results were as follows. If a receiver is fixed so that its antenna axis passes near the transmitter, then we can discuss the percentage of time that a randomly oriented transmitter might be expected to give no signal, all of this being specified in terms of different ranges. If the ranges are 10, 20, 30, 40, and 50, the corresponding percentages with a single antenna coil are 2.4, 12, 29, 43, and 63. With the two coils active, the corresponding percentages are 0.03, 0.75, 4.4, 9.5, and 24. This represents an improvement in the expected continuity of signal by a factor of 80, 16, 6.6, 4.6, and 2.6, respectively. It might be noted that the ratio decreases inversely as the square of the distance. At half the usable maximum range, approximately a tenfold decrease in "drop outs" is to be expected with the circularly polarized transmitter type.

The circuit of Fig. 10.3 has a number of generally useful aspects. At the left is a multivibrator which periodically turns on the transmitter for a brief interval. In this case the interval between pulses is controlled by a thermistor. The problem of generating short pulses separated by long intervals has been mentioned. In the present case, during the short interval of the pulse, the upper right transistor in the multivibrator is used to help rapidly restore the proper initial conditions for the timing of a long interval. The output transmitter, operating in the near field, is a Hartley oscillator, and falls within the present legal restrictions on frequency and power. Silvered mica capacitors were used because of their low loss characteristics at radio frequencies, and because of their relative stability against temperature changes. The two perpendicular coils are shown, with the second being connected only to a capacitor which is left fixed after initial adjustment. Large inexpensive components were used throughout, with the entire circuit, except for the batteries, being placed within the crossed coils. It is necessary to be certain that the battery leads, or other connections, do not form a closed low-resistance loop (shorted turn), thus accidentally reducing the outgoing signal. The battery can be placed close to the outside of the coils.

The fabrication of this transmitter is shown in Fig. 10.4. The coils are wound by hand on paper forms and fit over the rest of the circuit before the entire unit is cast into a block of plastic. Access is maintained for final adjustment of the tuning capacitor. Several of these transmitters were constructed for use in the zoo displays discussed in Chapter 18, in which application they were swallowed by various sea lions and caimans. In some experiments three batteries rather than two were employed, in which case reception at a central portable receiver was continuous for over a week from a pool and island area measuring roughly 50 ft by 100 ft.

In choosing these coils, the following considerations were involved. It was quickly found that a three-turn coil of ¾ sq in. area, which was reso-

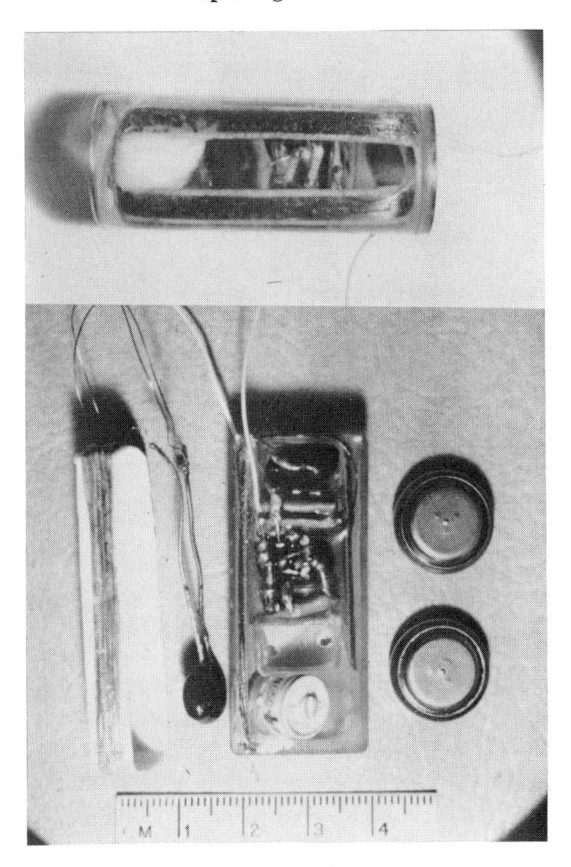

Fig. 10.4 Construction of the circuit in Fig. 10.3. Top: The crossed coils imbedded in a solid block of plastic, contain the rest of the circuit. Bottom: At the right is a pair of batteries which would be placed external to the coil. At the center is the circuit encircled by one coil, with a tuning capacitor at the bottom. To the left is the other resonating capacitor, and the second coil, which is slipped over the main assembly with the approximate orientation shown.

nated with a 0.2 μF capacitor, could give a range of approximately 30 ft at 500 kHz. This transmitter was to be pulsed and so the automatic frequency control circuit of a receiver would not help to maintain the continuity of a signal if the transmitter frequency tended to drift with a change in temperature. But a low temperature coefficient capacitor (silvered mica, or polystyrene or low temperature coefficient ceramic) in this capacitance range is rather large. Accordingly, turns were added to the top of the coil, thus allowing a corresponding decrease in capacitance at the same frequency. Adding turns to an existing coil does not increase the range as does tapping down the drive on an existing coil (see Appendix 1). It is important that there be close coupling with the first three turns of the tapped coil.

The general situation after the addition of the turns can be summarized by the following type of argument. Suppose that a turns ratio of 10 exists between the whole coil and the part driven by the transistor. The capacitor can then be 10^2 less for a given frequency since the inductance is 10^2 more. The capacitance reflected into the transistor looks about the same. The voltage of the resonant circuit is increased by a factor of 10 while the circulating current is decreased by a factor of 10. The number of ampere turns is the same, thus leaving the range approximately the same. The wire will be approximately 10 times as long and can have approximately one tenth of the cross-sectional area (one third of the diameter for a given loss), thus leaving the mass of copper fixed in the transmitter. If a capacitor had been selected which just matched the required voltage rating in both cases, then with a given dielectric material, we should find that the physical size of the capacitor or its energy-storage capability would remain unchanged, noting that energy is proportional to the first power of capacitance and to the second power of voltage. In the present case the addition of the extra turns allowed the use of a smaller available capacitor since stable low-voltage high-capacitance types did not exist; the alternative was a high-capacitance unit which was large because of an unnecessary high-voltage rating. The indicated capacitors also display low loss, which gives increased field strength or range for a given power delivered by the oscillator. After making these changes, the frequency and range are approximately the same, and the cyclic impulse from the transistor to the tuned circuit is the same.

Some details of the initial adjustment of the double-coil system should be given. Geometrically, the coils are identical, with one slightly deformed to slip through the other. With the help of a Q meter they are closely tuned to the same frequency. Then the angle between the coils must be set slightly off of 90 degrees to give the correct degree of coupling, and also the tuning of the second coil must be finely adjusted. If either of these adjustments is wrong, the field will be irregular as the axis passing through the two intersections of the two coils is circled, the probing being done with a small loop attached to an oscilloscope.

If the two coils are too closely coupled because of their axes being too far off 90 degrees, then there will be an abrupt shift in operating conditions between two distant states for small capacitance changes. Exact 90 degrees orientation will, of course, result in no power being induced into the second coil. The currents of "critically coupled" coils at resonance are just 90 degrees out of phase, which is the condition for a rotating field or circular polarization. When all adjustments are proper, the signal from the directly driven coil will be slightly greater than that from the second coil, thus giving an elliptical signal-amplitude distribution pattern about the pair. The minor axis should be within about 0.8 of the major one.

To check alignment, the antenna coils can be set approximately perpendicular with the axes of both coils horizontal. Then the small search coil is placed beside them with its axis passing through the center of the other coils. The signal induced into this probe can be monitored for amplitude on an oscilloscope or on a suitable radio frequency voltmeter. This probe is moved around on a cylinder with vertical axis to check the final symmetry of the overall field, but at this stage it is placed at 45 degrees to the axis of the two transmitter coils. Then the coupling between coils is increased in small steps, and after each increment the second coil is retuned, care being taken not to touch the coil. (If a screwdriver adjustment is involved, stray capacitance effects are reduced by using a sharpened length of wood rather than a metallic implement.) When there is too much coupling, a sudden jump in output is observed (actually a switch of signal to the other 45 degrees position). The transmitter coils are then backed off a little towards perpendicular. The tuning capacitor is then adjusted so that there are only two minima in the circumferential pattern. This is easily judged by tuning until the major axis of the ellipse is along the driving coil axis. Then there is also a good signal along the minor axis. The loose coil is glued into place and the trimmer capacitor is given a final check. All transmitter components other than the batteries can be placed within the coil pair at this point, and potted into a solid mass.

The foregoing results in a system that works in the induction field to induce a signal into a receiving loop of fixed orientation with a rather small chance of accidental signal disappearance. It might be thought that a third coil could be arranged perpendicular to the other pair to eliminate even this small possibility. Various phase-shifting techniques do not work, but a slightly different frequency can be applied to this third coil to wobble the previous pattern in such a way as to always couple a signal. Similarly, the three perpendicular coils can be fed by three slightly different frequencies to give a resulting field which cycles through all directions.

The signal from a single transmitting loop can be more effective in a suitably arranged receiving system. Two sine waves of the same frequency and 90 degrees out of phase cannot cancel no matter how their relative amplitudes are changed. Thus, if two perpendicular receiving loops are employed, and the signal from one is shifted in phase by 90 degrees before the two signals are added, then a near-field transmitter can have its axis rotated in the plane containing the axes of the other two coils, and the signal will not vanish. This can be useful when working with an animal that will spend most of its time walking or sleeping on the ground without often changing its orientation to the vertical. Techniques involving three perpendicular receiving antennas are mentioned in the next section.

The problem being discussed in this section is the polarization of the receiving with respect to the transmitting antenna, with circular polarization by one or the other being suggested. In some radar systems for locating the position of an object, the same considerations apply. Thus a resonant length of wire is quite noticeable in the return on a radar screen, but if plane polarized radio waves are used then there will be an orientation perpendicular to the beam for which there will be no return. Various antennas are known which provide for circular polarization on both transmission and reception. To provide for circular polarization on transmission alone can require twice as much power as plane polarization. There can be a special advantage to these methods when using unmodified standard receivers, particularly in public displays.

5. OMNIDIRECTIONAL SYSTEMS

In long-range experiments in which the investigator is working in the far field, it is quite possible for a receiving antenna to momentarily fall within the null of the radiation pattern of the transmitting antenna, resulting in a loss of signal no matter which way the receiving antenna is oriented. In that case absolute continuity or omnidirectionality is impossible unless there are several receiving antennas spaced over the countryside in such a way that the signal cannot simultaneously vanish in all of them. (Not all antennas radiate in a dipole-like pattern, for example, a large square loop with eighth wavelength sides radiates both in its plane and perpendicular to it.) Confusion can result from radio books that refer to an omnidirectional antenna as one giving a signal in all horizontal directions. However these generally do not give a signal upward from an animal to a passing airplane; they must carefully be placed on an animal during the preparation rather than randomly oriented, and the animal must then not roll over if the signal is not to vanish.

In near-field experiments, which constitute a large percentage of those presently being done, it is possible to surround the subject with three antennas whose axes are perpendicular, and the signal can never simultaneously vanish in all three. Specific arrangements include ferrite rod antennas placed along three edges of a cage and meeting at a corner, or loops placed against three perpendicular walls of a cage. Another arrangement involving three perpendicular loops wound on a ferrite mass is shown in Fig. 1.12.

The question remains as to what to do with these three signals. Clearly, they could be recorded on three channels of a penwriter and, in reviewing the data, the investigator could simply move his eye from channel to channel where the data was being recorded. However, such a system is obviously cumbersome and wasteful. Instead, we can arrange a switching circuit which, by means of diodes that are either biased into or out of conduction, allows any one of the three antennas to be connected to the input of the radio receiver. The circuit can be set to cycle between three antennas at a relatively high rate (which introduces some noise) so that there will never be a significant period during which data is being ignored, or switching to the antenna with the strongest signal can be arranged. There is an alternative which simplifies the equipment.

The cycling of the switching circuit can be interrupted whenever a signal is coming from the radio receiver. If reorientation of the transmitter within the subject causes loss of signal, or reorientation of an implanted transmitter causes a loss of signal due to reorientation of the subject during various activities, then the circuit that blocks the cycling action will be "disabled," and the input to the receiver will be connected to the next antenna. If a signal is found there then cycling will stop; if not, the third antenna will be connected to the receiver input, where a signal will surely be picked up, thus preventing switching back to the first antenna. The scheme is indicated in Fig. 10.5. Many receivers have an automatic gain control circuit from which the controlling signal is conveniently derived.

An example of a tracing from such a unit is also shown in Fig. 10.5. In connection with fertility studies at the Oregon Regional Primate Center with Lusted and Charters, it was desired to transmit the abdominal temperature of female monkeys. A blocking oscillator temperature transmitter was constructed having a relatively low repetition rate. When the monkey would reorient herself during play, there would be an appreciable interval before the next signal would be acquired, and during this time the recorder would fall back towards zero. Thus Fig. 10.5 shows the pattern of activity during a typical night and day, in the same way that was seen in Fig. 1.13. This representative pattern tended to repeat itself each day, with maximum activity being during the day when temperature was somewhat higher. Here again, the use of telemetry tended to produce a minimum interference with the normal pattern of activity and temperature.

There is an alternative method for combining the three signals. Each is passed through a full-wave rectifier. Then for any one signal the positive and negative halves of each radio-frequency cycle will become indistinguishable and will not cancel if combined with another similar signal (Mackay, 1960A). This frequency-doubling process can be considered as squaring each of the three perpendicular components of the magnetic field since $\cos 2\omega t = 2 \cos^2\omega t\text{-}1$. Taking the sum of the squares of the three orthogonal components of the transmitter field in space gives the square

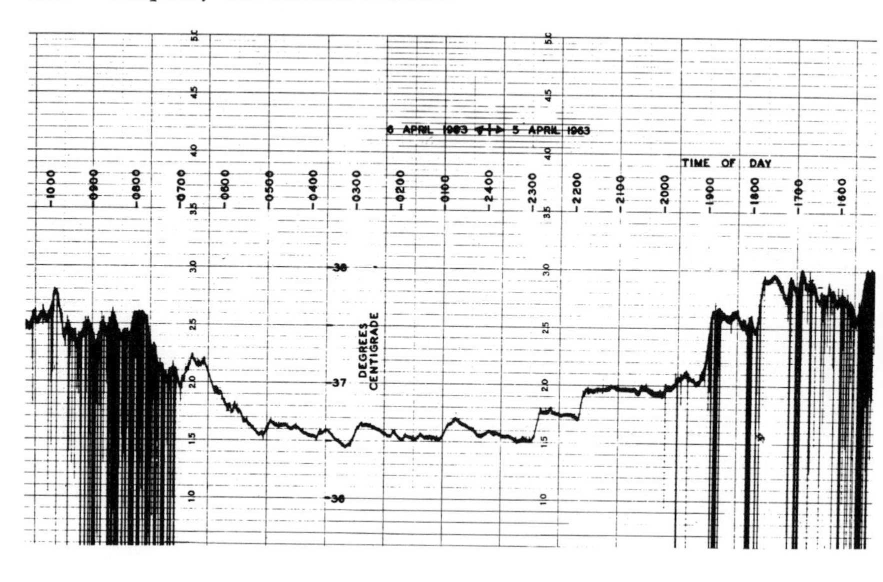

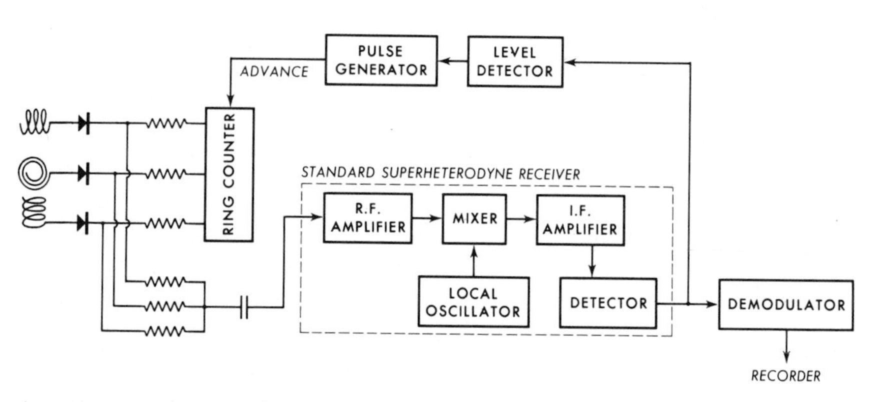

Fig. 10.5 Top: Pattern of temperature changes in the abdomen of a female *Macaca mulatta* monkey by night and day. This pattern repeats itself each day, with slight shifts associated with periods of fertility. Momentary loss of signal in the omnidirectional receiving system leads to the vertical lines, whose frequency is a measure of activity. This last aspect can be graphed automatically as a second line or viewed directly as shown. Below: A circuit type that uses a standard receiver switched between three (or more) antennas until a useful signal is found. Switching is made fast compared with changes in the variable transmitted to prevent information loss. The circuit is equivalent to a series of switches activated sequentially by a cam and motor which start when the receiver signal vanishes.

of the magnitude, independent of direction. We might note that similar considerations apply to the use of four electrodes on the body, if the conductivity of human tissue is used to bring out a signal that is propagated from a pair of electrodes attached to the ends of a transmitter that uses this different type of antenna.

Various arrangements can be used to give an output that is proportional to the square of an applied signal. One can employ a tunnel diode biased to the trough of its characteristic curve. Another method uses diodes in the arrangement sketched in Fig. 10.6. Each of the antenna signals is slightly amplified before going to the rectifying action of the diodes, since they are unable to act in a nonlinear fashion when the applied voltage is extremely small. The final receiver is thus tuned to twice the frequency upon which the transmitter is radiating. Information in frequency-modulated or other form is not lost. Such a system can simultaneously work over a range of frequencies, and the six tracings of Fig. 1.10 were actually recorded through such an omnidirectional system.

It can be demonstrated that similar considerations apply to any even harmonic of the original signal. The signals from the three coils can each

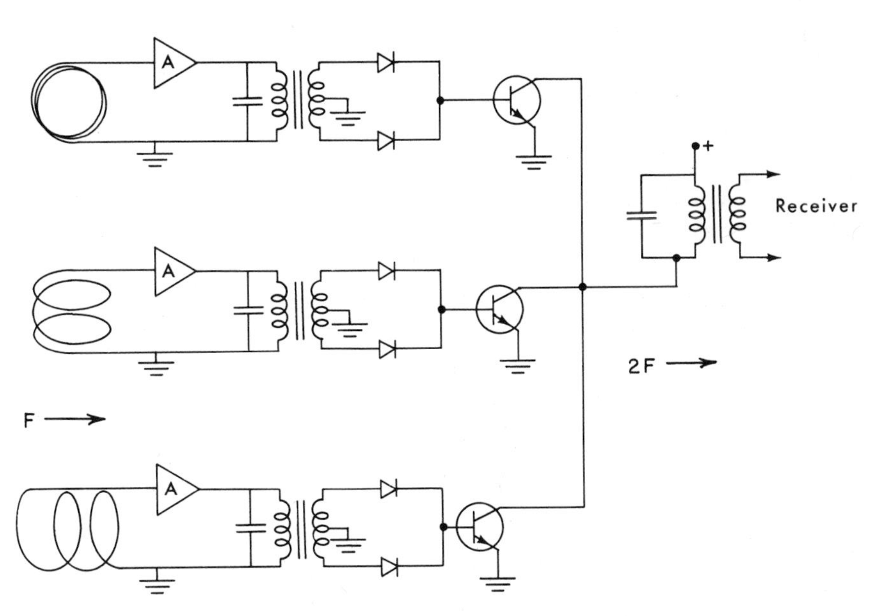

Fig. 10.6 The signal from three perpendicular antennas can be added after passing through frequency doublers to give a signal having approximately constant intensity in spite of transmitter orientation changes.

be passed through a half-wave rectifier (single diode) and then be added or combined, with a suitable filter to give out a steady voltage whenever the transmitter is turned on. This is a convenient system for counting clicks from a blocking-oscillator type of transmitter, but any subsequent amplification cannot use tuned filters to reject interfering radio noise. Such a method, however, can be used in an internal device which is to pick up power from an external field, no matter how the orientation of the device changes with respect to the external field. It might be thought that the impulses from a blocking oscillator would always be distinguished, no matter what the orientation, if we simply took the signals from the three antennas and applied them directly to the input of a receiver. It is possible that the transmitter could orient itself so that no signal was being received by one antenna while equal signals were being received by the other two but exactly out of phase and giving no output. Thus there is a need for a nonlinear element which can make it impossible for a positive half cycle in one antenna to cancel a negative radio-frequency half cycle from the other antenna (or antennas).

It is possible to build omnidirectional systems for use with the passive transmitters to be discussed in Chapter 13. By using spinning external fields, or three internal pickup coils, it is possible to induce power into a capsule for all orientations. The reradiated signal can then be picked up as if the transmitter were a typical active transmitter. In general, it is simpler to place the three coils around the transmitter and to cycle between them a connection to a unit in which a complete transmission and receiving function is taking place.

These methods are often especially important in a short experiment where a brief interruption in the continuity of information flow could seriously jeopardize the total result. However, these methods can sometimes be totally ignored since many experiments remain completely valid even if there are momentary interruptions or "drop-outs." Also, if the transmitter signal is strong, then the range of angles over which no signal is picked up at the receiving antenna can be so small that a fortuitous loss of coupling is extremely unlikely and is practically never noticed. On the other hand, for those systems that are working near their limiting range, the signal may be sufficiently weak that any small reorientation or reduction in coupling can cause total loss, and then these methods are needed. We might note that these methods also allow the combination of signals from several antennas which are oriented other than in a perpendicular sense, for example, to cover a larger area with some overlap between individual domains.

6. OTHER ANTENNA CONSIDERATIONS

A loop antenna can be tuned to resonance for a given radio frequency with a capacitor, but many well-known radio antennas are resonant because of their geometry. This requires that they be an appreciable fraction of a wavelength in size. Thus these resonant antennas can be quite large and are not normally used in connection with implant transmitters. Because of the high permittivity characteristics of water, the velocity of radio waves is significantly slowed, thus reducing the wavelength for a given frequency and the necessary size of an internally placed resonant antenna. This is a rather complicated matter, with related effects being touched upon in Chapter 15. It is found that a pair of electrodes setting up currents in the body of an animal can be a rather effective transmitting antenna. It should be mentioned that, although we have discussed dipoles at some length, in cases where antennas are located near dielectric or conductive discontinuities, images will be formed which will give rise to quadrupole and higher order fields; these are beyond the scope of this book.

The size of a resonant antenna can be reduced by combining suitably placed lumped elements such as coils with the basic structure. Thus a so-called base-loading coil placed in a break in a whip antenna allows considerable shortening whether the antenna is used for transmission or reception. This coil must be carefully tuned with the rest of the structure at the transmitter frequency if satisfactory results are to be achieved. The combination constitutes an extremely high Q resonant system and proper tuning, which is extremely sharp, is essential. If proper tuning cannot be maintained under changes in ambient conditions, for example if the temperature changes, then there is one other possibility that is sometimes useful. If such a system is being used as a transmitting antenna, then the transmitter frequency can be cyclically shifted through a small range, perhaps in synchronism with some variable to be transmitted. During each cycle of this frequency shift, the antenna will go into resonance and emit a burst of power, thus periodically assuring proper radiation. Although it sometimes means undue complication, various combinations of inductors and capacitors can be combined with various dipoles of reduced size to give effective operation at frequencies lower than those in which the structures measure a large fraction of a wavelength.

It has been mentioned that an antenna that is a good transmitter will also be equally effective as a receiver, and with the same geometrical distribution of sensitivity. This idea has significant implications in connection with the fields close to a straight wire. If a current is passed through a straight wire, a magnetic field will be set up which encircles the wire in a sense indicated by the well-known "right-hand rule." The field is strong close to the wire and weak away from it. Similarly, a wire is sensitive to changing magnetic fields set up nearby. An extremely effective receiving antenna system (to be discussed in more detail in Chapter 14), for use with low-power transmitters placed in animals crawling over the ground, consists of a series of long "hairpin" loops. Although covering an extremely large area, the transmitter is never very far from a wire (unlike a single large loop covering a similar area) and thus a strong signal is always induced. Being a near-field effect, generally all dimensions are limited to less than a wavelength. A small transmitting loop with axis vertical will induce no signal into a single strand of wire when directly over it, but will induce a strong signal when displaced slightly to either side. If the axis of the transmitting loop is instead horizontal and the wire is in the plane of the loop (the loop is straight above the wire) on the other hand, a strong signal will be induced. This last can be seen by considering the effectiveness of the reverse process in which an alternating current passed through the wire on the ground would induce a signal into a small loop nearby, since induction that is effective in one direction will also be effective in the other direction.

Similar considerations apply to the placement of a battery in the vicinity of the transmitting loop antenna in an endoradiosonde. If the battery is to be placed within the transmitting loop, then the loop should be several times the diameter of the battery case and the battery should be centered if the "shorted turn" effect is not to reduce the radiated signal too much. If the battery or other continuous masses of metal such as transistor cans are enclosed in a ferrite tube, then the magnetic field will bypass the conductors, and antenna or transducer operation will be satisfactory (or actually improved by the presence of the ferrite). A battery can be placed outside the loop without a ferrite cover, and it has been found that even very close placement does not much degrade coil performance. The same considerations apply to an observation in Chapter 4: that a thin layer of "Q-dope" over a wire before potting in plastic greatly reduces losses. This is true even though this separating layer is very thin and the bulk of the plastic subject to appreciable power loss at high radio frequencies.

In working with these small transmitters, it is sometimes true in certain geometries that the interaction between adjacent turns in a transmitting loop is small. This need not cause trouble unless a specific transformer action is required. Unwinding several loose turns to a bigger area leaves the frequency of transmission about the same (in some cases). This process results in a stronger signal, for reasons that have been indicated. The various comments, however, about the utility of reducing the number of turns to strengthen the signal hold when the final single turn has considerably more inductance than resistance, that is, still acts predominantly as a coil.

In general it is found that a small helix makes a better transmitter than a pancake spiral, and these are generally improved by the presence of a piece of ferrite.

Many of the small transmitters which have been described radiate their signals from a small coil. In many cases the range of transmission can be greatly increased by attaching a trailing length of wire to one side of the coil. Thus in Fig. 3.2 the wire would be attached at the junction of the coil and transistor. This can be quite effective with some types of animals, and the results are best determined by experiment.

If an animal is in a conducting cage while under study, an antenna placed near one of the walls may receive very little signal. If the cage is large and a high frequency of transmission is being used, it can prove sufficient to coat the inside of the cage with a material that absorbs electromagnetic radiation so that reflections will not set up a standing wave pattern into the null of which a receiving antenna can fall. In a number of cases we have had extremely satisfactory results studying animals in metal cages by using the cage itself as the receiving antenna. In these cases, we have used the metal shelf upon which the animal was supported as a separate electrode by placing plastic washers between the shelf support and the rest of the cage. The two wires to the receiver were then run from the shelf and from the rest of the cage. In several cases the apparent impedance of this unusual combination was approximately 50 ohms, which is quite convenient for attachment to a standard receiver.

With a group of adjacent animals all transmitting on one frequency, isolation of the signals from different cages to a receiver being cycled between subjects is uncertain when using most antenna configurations. The problem is simplified by using two receivers and two frequencies, and alternating the transmitters and cage connections so no two adjacent cages are on the same frequency.

A loop of wire can be tuned to resonance with a capacitor to detect distant signals, and a pair of wires or a shielded cable can extend from this capacitor to the input of a standard radio receiver. This works but is often not as effective as a simple alternative. A tuned loop acts as a very high impedance which is unable to deliver much power to the considerably lower input impedance of many receivers. Various schemes can be used to match the two components, but in some cases a real improvement can be achieved by simply placing an extra capacitor in series with the loop as indicated in Fig. 10.7. In this case loops of 12 cm diameter were receiving a signal at a relatively high frequency. The resonant com-

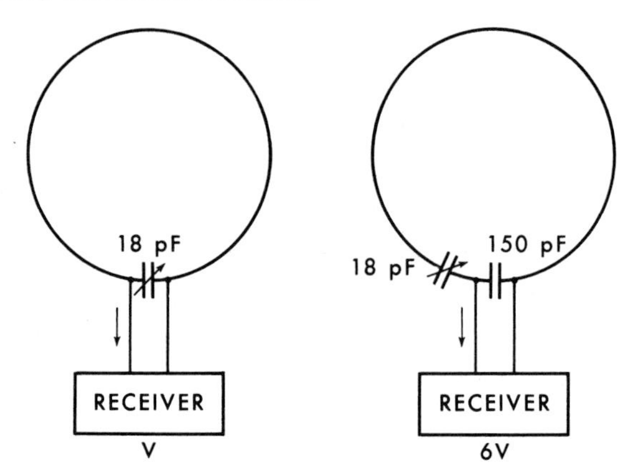

Fig. 10.7 A loop resonated with a capacitor has too high an impedance to effectively transfer energy into all receivers. In the case shown, six times the signal voltage is transferred by the addition of a component that improves the match between the two.

bination at the left gave a useful signal when fed into a standard simple receiver, but approximately 40 times as much power entered the receiver under the conditions at the right. Seemingly one is dividing the available voltage with a capacitor voltage divider, but actually approximately 6 times as much voltage appears when the impedances are matched. Properly tapping into the capacitors of a Colpitts oscillator similarly results in a good transfer of energy from collector to base on the associated transistor. In some transmitters a similar antenna configuration allows a stronger signal through radiation of increased power (Fig. 17.3). In that example, the antenna is tuned slightly off resonance so as to be reactive, and in combination with the capacitor appears as 5 KΩ rather than 40 KΩ. The signal from a receiving loop can alternatively be taken from across a suitable length of the loop at a place opposite the capacitor; Cochran (1966) has given some practical details of construction with a matching "stub" at this position. In practice, the use of the tuned loop of Section 3 and a field-effect transistor at the receiver input is most effective. The question of impedance matching is discussed in various radio books, including Goodman, 1966.

Mechanical movements of antennas can be considered. Thus sensitivity to polarization of different orientations at a receiver can be achieved by continuously spinning the antenna. This is generally relatively inconvenient in practice, but is instructive in principle. Some small external transmitters can be floated in a plastic sphere to always maintain an approximately vertical orientation, and this may prove useful in practice.

If the investigator is transporting an instrumented animal, it is important that it be placed in a shielded cage if it is to be flown. It is a simple mistake to overlook a radio transmitter inside of an animal. Breathing holes can be shielded with fine mesh wire screen, but it is important that electrical continuity exist over the whole cage, including the seam where the door closes. Some small transmitters will run down their batteries faster than usual if a conducting object is placed very close to the transmitting antenna, thereby increasing the power output. Thus there is some advantage to having the shielded region a bit larger than necessary, and the antenna or animal centered within.

The rejection of radio noise interference is an extremely complicated subject, but must receive brief mention somewhere. Interference by commercial radio stations, which can be especially troublesome at night when signals bounce in from more remote localities, is best avoided either by performing experiments within a shielded room or else by choosing frequencies outside of the usual bands. A more troublesome problem is the irregular noise often found in modern buildings. Thus elevator motors can cause periodic disturbances in their vicinity. Fluorescent lights are always suspect, and should be switched off in many cases. If the receiving apparatus consists of several parts, then all the equipment should be placed upon a single metal sheet and grounded at one point which gives minimum noise. Some noise comes in over the power wires, in which case a few turns of heavy wire in series with the line cord and a 1.0 μF capacitor across it will cut out some of the problem. Many forms of radio interference are airborne as well as conducted by the power source. In the laboratory it is well to take a portable receiver and search for the sources of interference to see if they can be switched off. One common receiving antenna consists of turns of wire in a loop around the cage of an animal. In some cases a layer of tin foil over the loop can be grounded to a water pipe to reduce interference. The foil must not be joined at the ends, the gap preventing formation of a continuous loop over the receiving loop. Such an antenna often feeds into a receiver through a transformer winding which is not connected to the rest of the receiver. Sometimes balancing the receiver loop as regards capacitances to ground can be an extremely effective way of significantly reducing noise. Such a situation is depicted in Fig. 10.8, where the extra capacitors to ground are individually adjusted until the noise that is being picked up has been reduced to a minimum. In some cases interchanging the two connections to the receiver can make the difference between success and failure, since this can greatly change the overall distribution of accidental capacitances to ground at the antenna and within the receiver itself. Battery-operated equipment is sometimes less prone to noise pickup than equipment that is plugged into a wall, and occasionally

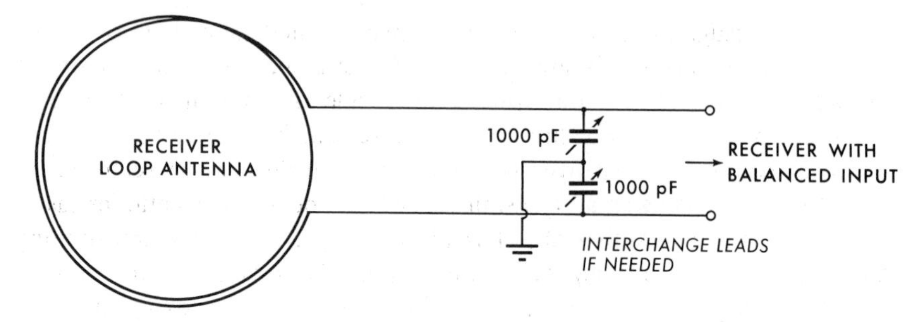

Fig. 10.8 Suitable balancing of a receiving loop with respect to ground can minimize radio noise and interference. Part of any unbalance can exist within the receiver input itself. The shunting capacitors should be as small as possible and adjusted for minimum noise.

noise is somewhat reduced by reversing a wall plug. All these measures involve considerable trial and error, although in bad cases a sensitive voltmeter can be used to pick up small differences in voltage between the chassis of the various components of the receiver system, and thereby suggest more appropriate grounding arrangements.

7. FREQUENCY FACTORS

A number of factors influence the choice of operating frequency; some are technical and some are practical. The early experiments using active transmitters were all performed at frequencies of a few hundred kilohertz because transistors effective at much higher frequencies were not then available. Although transistors do have problems when required to operate at high frequencies with low voltages and low currents, this performance limitation has generally been eliminated for most of the present purposes. Although they have other limitations, tunnel diodes are able to function at even higher frequencies and over broader temperature ranges. Some materials that display a bulk negative resistance are presently under development, and as semiconductor technology advances, any remaining limitations should disappear.

One should also choose a frequency where mutual interference with commercial radio stations is unlikely. In general, the present transmitters are of such a low power that their signals can be completely blocked by a typical commercial station. This can become a troublesome situation at night when signals from more remote transmitters can be picked up at any particular location. Frequencies below about 500 kHz or above about 1.5 MHz avoid the most crowded frequency region. It is sometimes tempting to work in the FM band frequency range extending from 88 to 108 MHz, if for no other reason than that extremely inexpensive receivers are readily available to cover this band. However, these receivers can be modified to accept signals in other frequency ranges.

It is well known that the higher frequency signals are not well reflected back to earth, thus giving transmission that extends essentially only to the horizon. Such "line of sight" transmission places a limitation on some extremely long-range experiments, but in other cases it means that there will be interference from fewer remote transmitters.

A receiver may be constructed having very low noise, but as soon as an antenna is attached, extra noise appears. Atmospheric noise is the principal limitation of radio service on the lower frequencies. At frequencies above about 30 MHz, the noise falls to levels generally lower than receiver noise (IT&T Co., 1963). Such noise depends upon the time of day and weather, and generally decreases with increasing latitude.

Of the simple transmitter types, a relatively low-frequency unit will tend to be more stable against such factors as changes in nearby stray capacitances because the capacitances already in the circuit are relatively large. Thus we have found it somewhat more difficult to build a stable pressure-sensing endoradiosonde to work in the 100 MHz region, than to build circuits working at 300 kHz. However, since both can be made to function satisfactorily, this is not a major consideration.

The components involved in the generation of low frequencies are larger than the corresponding components used to generate high-frequency oscillations. Thus transmitters can often be smaller if they are to work at the higher frequencies, although in some cases a change to a different circuit type can overcome this restriction. Also, at high frequencies certain variable-impedance transducers become more effective.

It has been mentioned that in the far field the signal strength increases with frequency. It is also well known that a given strength of an alternating magnetic field will induce a larger voltage in a sensing loop if the frequency is raised. There are also factors involving the required size of a resonant antenna. Whether a wave falls in step or out of step with currents in an antenna depends on the size of an antenna with respect to a wavelength, and thus smaller antennas can be effective at higher frequencies. If crystal control is required, small units become relatively inefficient above a few hundred megahertz, and so the frequency would not be increased even though a few inches of wire could then become an efficient resonant antenna. The higher frequencies sometimes reduce a tendency for a wave reflected from the ground to cancel the direct signal from an antenna that is on a small animal and therefore close to the ground, but radiation at low angles is then more attenuated by the ground.

These last considerations suggest going to a high frequency, but there are opposing factors which suggest a lower frequency. A surface that is large with respect to a wavelength can be capable of acting as a mirror for reflecting radio signals. Thus at very high frequencies small rocks and trees cast sharp shadows within which a receiver may be unable to pick up the signal from a moving transmitter. Similarly, signals can be reflected from objects and thus appear to be arriving at a receiver from a different direction, thus confusing direction-finding procedures.

Even more important is the role of losses that increase with increasing frequency. Thus a signal passing through woods, or through or over lossy ground, will be more strongly attenuated at high frequencies than at low. An especially important case is that of an animal which has ingested a transmitter that must propagate its signal outward through the partial shielding action of the body tissues, which are somewhat conducting. The same considerations apply to the passage of a signal through ocean water, which is typically somewhat more conductive than animal tissues, and also to natural pools of fresh water. These pools can display an electrical conductivity ranging from extremely low up to values approximating physiological saline when the mineral content is high.

A field passing through a conducting medium is attenuated in an exponential fashion. The signal is reduced by approximately ⅔ in a distance which is referred to as a "skin depth," the name deriving from the fact that most of the field is in this surface layer. The attenuation factor depends on the square roots of the conductivity and frequency, increasing with both. Placing the appropriate constants in the equation indicates that, for typical ocean water, the skin depth in meters is given approximately by 250 divided by the square root of the frequency in hertz. As shown in Chapter 15, this implies an attenuation of approximately 55 dB/wavelength. The lower the frequency the more effective is the penetration of a signal through a partially conducting medium. This would suggest using the lowest possible frequency.

In some cases antenna or propagation effectiveness requires a frequency as high as possible, whereas attenuation effects require use of the lowest possible frequency. This suggests that there is an optimum compromise frequency in those cases in which these considerations hold. An expression for such a frequency is derived in Appendix 2. In some near field cases, with certain detectors of electric or magnetic fields, an increase in frequency does not increase the "antenna" signal. Then the lowest possible frequency is preferable, at least to the degree that the skin depth is several times the desired range.

In the case of aquatic animals in ocean water the attenuation factor proves to be an extreme limitation and the lowest frequencies must be considered.

As discussed later, an electrode pair can be used as the transmitting antenna with the same shielding equations applying. But with frequencies below about 20 kHz it was observed that a sensation was produced on the human tongue with each electrical impulse, and thus to avoid the possibility of including distracting sensations within the subject, frequencies several times this are employed.

A signal arriving at a discontinuity will be partially transmitted and partially reflected. Different tissue types have different electrical properties, and there is certainly an extreme discontinuity between a tissue and air. It is for this reason, for example, that a conducting film considerably less than one skin-depth thick can act as an extremely effective shield. Thus the strength of a signal emerging from a particular region cannot be calculated simply by computing attenuation alone. When considering a signal being generated within a watery medium and emerging, it is necessary to take into account not only the initial attenuation over the lossy path but also the reflection at the surface (which can be somewhat frequency sensitive), the refraction along the surface, and also the altered effectiveness of a given sized radiator when immersed in a different medium.

In summary, some factors suggest a high frequency and other factors suggest a low frequency. In general, we have worked in a frequency range extending from approximately 50 kHz to a few hundred MHz. The lower frequencies have been used for dolphin experiments, and the higher frequencies in experiments where the transmitter would be external to a land animal. Some workers have made extensive use of model airplane control equipment which operates at 27 MHz. Frequencies in the vicinity of 90 MHz have been found to be an extremely good compromise for many studies, although many of our gastrointestinal studies are still done at a few hundred kilohertz.

In addition to the above factors, legal restrictions by the Federal Communications Commission limit the use of various frequencies for the protection of everyone. Excerpts from a few of the relevant rules applicable at the time of this writing are included in Appendix 3, but a complete summary of legal factors is beyond the scope of this treatment. Laws do change, and it is presumed that various frequencies will be set aside specifically for bio-medical studies and for intensive-care monitoring in hospitals. Generally, a shift in operating frequency by 20% will not significantly affect any of the above factors other than the legal ones.

8. POWER LIMITATIONS AND RANGE

In the near-field case the field strength tends to fall off with the cube of the distance; that is, a small range change is equivalent to a big power change. Thus much more power is required to double the range of transmission from a particular transmitter. Some extra attenuation (such as might be due to intervening tissues after swallowing a transmitter) can appear not to change the useful transmission range much because it can be compensated for by moving the receiver only a little closer.

Signals are relatively readily transmitted through pure water or ice and through many kinds of dirt. The properties of dirt and sand are highly variable; in some cases their major electrical difference from free space is not in their dielectric properties, but in their magnetic properties (an increased permeability). To indicate the possible variability a case might be noted in which unusually high radio attenuation was observed, apparently due to a quartz sand being piezoelectric and jumping around in response to the electric field, thus dissipating an unusually large amount of energy. It is generally possible to telemeter a useful signal through several meters of sand or soil in studying burrowing or hibernating animals.

The many theoretical and experimental studies carried out on ordinary radio transmission apply to the longer range biological observations carried out in the far field. The literature is voluminous as it applies to each range of frequencies. It is well known that at a low frequency, especially at night, the transmission of information over extreme ranges can be achieved, whereas higher frequency signals seldom go beyond the line of sight (to the horizon).

Procedures for computing the power required to achieve a necessary range are helpful, even if they do not always work precisely. General expectations can be illustrated by observations from a few specific experiments. In one case a small transmitter having a class C output stage was powered by a 3-volt battery supply which delivered 2 mA. The frequency of transmission was 100 MHz, and the antenna was a piece of vertical wire 20 cm long. No loading coil was used with the antenna. When used approximately 2 m above open ground, extremely reliable transmission was always achieved at ranges of 80 m. In another case, 100 μW gave a range of 30 m at 100 MHz. Another series of observations illustrates a common problem. Again a small whip antenna was used and, although it is difficult to maintain in resonance, it was base loaded in this case and gave an overall appearance of 50 Ω. Driven with 0.1 W, the range of transmission from the animal on the ground was about 200 m. But when the animal climbed 3 m up into a tree, the transmission range become more nearly 4 km. It is generally felt that in relatively open country, one can transmit to the horizon with almost any antenna type if the power is 150 mW, although in biological work there can be disturbances to the antenna which will reduce the transmission range even down to zero.

The crystal-controlled temperature transmitter described in Chapter 14 draws a little under 1 mA and its small loop antenna sets up a field of approximately 50 μV/meter at 50 ft. (These mixed units are customarily used in several countries in expressing field strength at different ranges.) Thus we might expect a signal of one microvolt at a half mile, which would be a useful signal, but in fact with short animals, the range was generally observed to be more nearly 500 ft.

In the literature both experimental and theoretical treatments of this problem of expected transmission range may be found, from which fairly precise computations can be made. Some of the factors are the transmitter power consumption and its efficiency, the gain of the transmitting and of the receiving antennas, the noise power at the receiver input (receiver noise temperature), the band width of the receiver, the required signal to noise ratio, the frequency of transmission, and the expected signal loss due to various causes over the path. Notice that some details of the form of the antennas must be included. The range equation for the far field is sometimes stated as necessary receiver sensitivity in dBm (i.e., relative to 1 mW) equals transmitted power in dBm $-$ 37 $-$ 20 log frequency megahertz -20 log range in miles $+$ transmitter-antenna gain $+$ receiver-antenna gain. In these terms a typical receiver might be -30, and reception would be good above this. A receiving antenna such as the 5-element Yagi arrays often used with television sets gives a gain of about 16 dB. Such a computation is a highly simplified approximation.

More complete books on the theory of radio transmission indicate how such computations are carried out more precisely. Cochran and Lord (1963) have worked out several examples of particular interest to biologists, and Pienkowski (1965) has prepared a series of curves from which a simplified calculation allows an estimation of range.

However, it should not be assumed that a precise calculation can be made in all cases. Thus an interesting region for which an exact calculation is extremely difficult is propagation in a jungle or rain forest. Some observations have recently been made by Burrows (1966) on this situation. He finds that the predominant characteristic of propagation in what is called the "very high frequency range" is the presence of a reflected wave that tends to cancel the direct wave and results in the received field being proportional to the product of antenna heights and inversely proportional to the square of the distance, with the radio gain being independent of the frequency. He indicates no exponential attenuation of ultrashort waves in a jungle. Our experience suggests that in some dense areas, range may be reduced after a rain.

Legal requirements (Appendix 3) are sometimes specified as limitations to the power input to the final radio stage, and sometimes given as the maximum field strength produced at a given distance from an antenna. With a proper receiver antenna, field strength can be measured. We can calculate the voltage appearing at a receiver antenna terminal when placed in a 50 μV/m field intensity from the following considerations. An antenna in the field of a linearly polarized electromagnetic wave has a received power available equal to the effective area times the power per unit area carried by the wave. Combining the relationship between power density, rms value E of the electric field, the effective area of a half-wave dipole, and knowing that the antenna impedance is 50 Ω, the voltage appearing at the antenna terminal can be calculated to be 0.132 $E\lambda$. Thus, using a half-wavelength long dipole at 100 MHz, and measuring the voltage output with equipment having a 50-Ω input impedance, we should observe 19.8 μV if the field has the critical value of 50 μV/m. An important aspect is to have enough frequency selectivity in the measuring instrument to reject the signals from nearby commercial stations, which may generate much larger signals. Similar calculations can conveniently be made for specified sized loops.

One other aspect of propagation might be mentioned here. In some cases the predictability of the rate of propagation of a radio signal affects the accuracy of localization of a transmitter. For example, if the arrival time of periodic impulses is observed at three points, then from transit time differences it is possible to say where the transmitter is (just as the position of an earthquake can be inferred from seismograph records). Possible accuracies are suggested from well-known observations on some ocean navigating systems. In the 3-to-5-MHz frequency range signals carry beyond the line of sight. Signal attenuation over the sea is low, but the sky wave is attenuated. The ground wave without compensation for the immediate atmospheric properties seems predictable to one part in 20,000, that is, a limiting accuracy of 5 yd in 50 miles.

11

Receivers and Demodulators

Having mentioned the generation and propagation of the information carrying signals, it is now appropriate to consider their reception and conversion into a form for study. In many cases cheap entertainment receivers are employed with satisfactory results. The range of a given transmitter, however, can often be increased several-fold by using special low noise receivers. By passing a very narrow band of frequencies in the receiver for any given setting of the tuning dial it is possible to reject interfering noise on adjacent frequencies. In some cases it is useful to expend considerable effort on receiver selection and improvement. The narrowest band receivers can be used with unmodulated tracking transmitters. However, the biologist will often find routine equipment quite adequate. Once the radio-frequency signal has been amplified, its modulation must be decoded to reveal the contained intelligence. Comments on this aspect are also contained in this chapter.

1. CLICK COUNTING

Some comments in this section have general applicability and others are specifically directed towards simple apparatus used with a restricted class of physiological variables. In *some* experiments parameters such as temperature, humidity, illumination, and pH may change slowly enough so that averages over a minute or two are adequate. Then it is sufficient to construct a transmitter that slowly clicks at a rate depending on the variable, and to time these clicks while listening through any receiver. Thus a number of transmitter types have been described which put out a short pulse every second or so for convenient counting. These transmitters also can make minimum demands on battery power and radio bandwidth. Recording of such signals can be made automatic with circuits to be described, but the use of the ear does have certain advantages.

Any standard receiver is suitable in these applications, but some are more sensitive, selective or display lower noise than others. Especially convenient are some of the cheap little pocket transistor radios. They are generally rugged, well engineered, and reliable and use readily available batteries. Some cover the AM band (530–1600 kHz) and others also cover the FM band (88–108 MHz). For timing the clicks, a stop watch is convenient, simple, rugged, reliable, and relatively insensitive to temperature changes. Such a combination of components not only allows telemetry at minimum cost but has the advantage of reliability; this can be especially important in field work where parts are hard to replace and experiments difficult to repeat. Such a method is inconvenient however, in the sense that someone must be present to take each observation day or night.

Receivers should be selected by entering a radio store with a small transmitter. Different receivers should be turned on, and it will be obvious that many are either rather insensitive or not very selective in the sharpness of their tuning. These receivers should be immediately rejected and the remaining few taken outdoors with the transmitter. A typical building will provide some shielding from nearby radio stations, but when the remaining receivers are taken outdoors several will usually be found to be swamped by a multitude of stations, if the observation is being carried on in a typical large city. From those that are best able to reject the overlap of adjacent stations, a few are finally compared for the maximum range at which they can distinctly pick up the portable transmitter, and from this observation a final selection should be made. It will usually be found that there is little correlation between size, price, and quality.

The previous is a simple procedure but, of course, precise measurements upon receivers can be made and their "noise figures" compared quantitatively. Such methods are described in many books and have been summarized (IT&T Co., 1963). Occasionally it will be found that an improvement in receiver quality beyond a certain level is not justified because of noise and interference when a real antenna is attached. Operation on higher frequencies may be advantageous in this case.

Receivers can be modified to work in a different frequency range from normal. Small changes in coils or capacitors are not difficult for an engineer to introduce. It is also possible to construct frequency converters which cause large shifts in an incoming frequency for acceptance by a given receiver. In these the incoming signal is beat or heterodyned with a local oscillator. In a nonlinear element there will appear some different frequencies which can be made to fall in the pass band of the receiver by tuning the oscillator. (A nonlinear element is required because in these, the presence of one signal modifies the effect of the other as far as contribution to the output is concerned; a linear element leads to a cyclic amplitude change or beat note without generation of other frequencies such as those corresponding to the envelope.) An example of such a circuit is given in Fig. 15.5. Actually, if several turns of wire are wrapped around a portable receiver and driven by an oscillator of adjustable frequency, then varying the oscillator will usually cause the receiver to tune from station to station without being adjusted; nonlinearities in the ferrite antenna and following circuit are sufficient for this action.

The transmission range of a particular combination of receiver and pulsed transmitter can often be somewhat increased by placing within or near the receiver a small steady oscillator of suitable frequency. This can significantly increase the noticeability of each pulse. If a steady transmitter is turned on and an ordinary receiver tuned to its frequency, nothing will be heard. (When a radio singer becomes quiet nothing is heard, though the radio station transmitter is still on.) Only modulation normally leads to a sound at the receiver. If the transmitter is switched off or on, a click will be heard at the time of switching. If a local radio-frequency oscillator is set so that its signal combines with the transmitter signal to give an audible note (difference frequency), then the presence of the transmitter signal is noticeable whenever it is on. This type of beat-frequency oscillator (BFO) is used in receivers for Morse code. The oscillator can either be placed in the receiver and connected to it, or it can be a small separate unit placed near the receiver, in which case its signal is induced from its oscillator coil. Most receivers radiate some signal of their own, and thus this effect can be observed by placing a suitably tuned second receiver beside the first one. The length of the transmitter pulse must be several audio cycles to work in this way. Very short pulses are not passed by typical receiver intermediate-frequency amplifiers anyhow. In several cases we have found that pulses begin to lose their effectiveness if reduced below 20 msec.

Some details of the adjustment of the frequency of the BFO might be mentioned. Any little oscillator can be constructed and placed near a receiver and retuned until a signal from a transmitter is heard in a fashion that is judged to be as good as possible. But there are some details worth noting. A typical receiver is a so-called superheterodyne (see Fig. 10.5) in which an incoming signal is beat with the signal from a local oscillator to give an intermediate frequency. This frequency is amplified and goes to a filter that passes a fixed narrow range of frequencies. Retuning the receiver consists of changing the local oscillator frequency so that different station frequencies will heterodyne to the value passed by the intermediate-frequency filter, and will be passed for demodulation and listening. For a typical AM receiver, the intermediate frequency is 455 kHz, and many circuit components are available to work in this range. A beat-frequency oscillator can be adjusted without a transmitter being present in the following way. The receiver is turned on but away from a station, and the oscillator is tuned until the receiver quiets (or in some cases, suddenly becomes more noisy). Then the radio frequency from the BFO may be entering the antenna and passing through the intermediate-frequency amplifier to cause quieting. Tuning the receiver will then move it from the BFO frequency and quieting will be lost. The BFO can then be retuned to a new frequency in which quieting is observed. If actually set at the intermediate frequency (IF), there will also be quieting, and it will remain as the receiver is retuned. Adjusting the BFO a little off the IF frequency will then cause an output tone for *any* received signal frequency. The tone will change in pitch as the receiver tuning dial is shifted. On the other hand, with a fixed transmitter frequency, the first alternative is sometimes preferable. If the BFO feeds into the receiver antenna at a frequency near the transmitter radio frequency, then changes in receiver setting, or drift, do not affect the beat frequency that is heard. This is because the intermediate frequency generated from an actual transmitter is not constant as the receiver is retuned, because the IF pass band is the fixed frequency. These considerations will also prove important in the next section where the monitoring of frequency shifts in a transmitter is discussed.

A similar effect of increased pulse noticeability can also be achieved by mounting an audio oscillator in the receiver and connecting it as a modulator or chopper. An audible tone will then come out whenever a transmitter is on. A multivibrator (Fig. 11.1) can be constructed from any small pair of transistors and will run over a wide range of voltages supplied by the

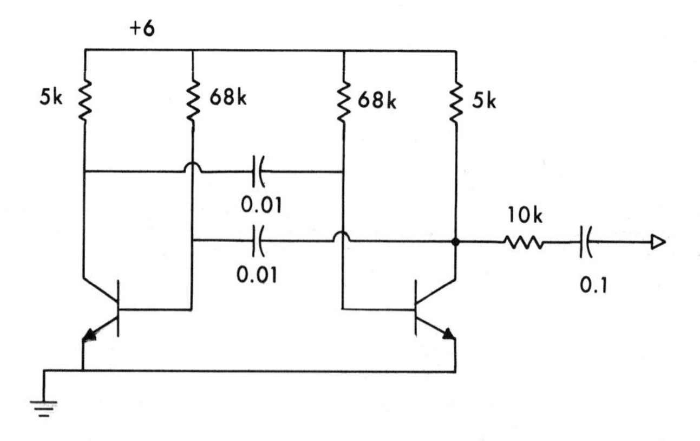

Fig. 11.1 An audible frequency multivibrator that can be attached to the back of a pocket radio to modulate a steady incoming signal into a noticeable whine. Any pair of small transistors is suitable.

receiver battery. Familiarity with circuits is not necessary to attach such an oscillator properly. The voltage-input leads are connected across the battery or batteries of the receiver. The receiver is tuned to a radio station and the oscillator output touched to various points on the back of the receiver (Fig. 11.2). The wire is then soldered in place when the desired result of "screaming" in the presence of a signal is observed. Figure 11.2 shows the appearance of the back of a pair of receivers, one of which is modified by the addition of such a circuit over a piece of tape for insulation from the components below. For the more sophisticated a similar result can often be achieved by twisting together the input and output wires of the intermediate frequency amplifier, thus causing it to oscillate.

In counting the clicks with a stop watch, a certain procedure should be used for maximum accuracy and minimum time. The investigator should listen for a moment to sense the rhythm, and it can be of significant value to wave an arm in time with the signal. Pushing the start button on the watch, the investigator should say 0, 1, 2, 3, . . . (not 1, 2, 3, 4, . . .) in judging the number of counts. Then he should time a given number of counts rather than count how many pulses come in during a specified interval.

The reason for this last suggestion incorporates a factor of rather general importance. Consider an example of taking a temperature to an accuracy of approximately 0.1° near 20°. Depending on the transmitter sensitivity calibration, this may mean an accuracy of something like one part in 200. If clicks are coming out approximately once per sec, then this degree of precision is roughly approximated by counting the number of clicks that fall in an interval of 200 sec. Notice that the resulting number will be in the vicinity of 200, but will be uncertain by plus or minus one count (somewhat depending on whether the watch was started in synchronism with the first count). This is because a final count could have come in just before the termination of the interval or else it could have been about to arrive just after the interval terminated. Because it is unknown which of these alternatives exists, such an uncertainty will usually be inevitable (see also Chapter 12). The alternative procedure is to start the watch on an impulse and stop the watch in synchronism with the arrival of an impulse at an approximately suitable time later, that is, after a certain number of counts. A typical stop watch ticks five times per second (i.e., indicates to the nearest fifth of a second) and a person's response time is generally such that he can push the button in synchronism to the clicks with this degree of accuracy. An interval of only 40 sec need be timed to the nearest fifth of a second to give an accuracy of one part in 200. Thus this same observation could be made in one fifth of the time by accurately timing some specified number of counts in the vicinity of 40. The latter measurement corresponds

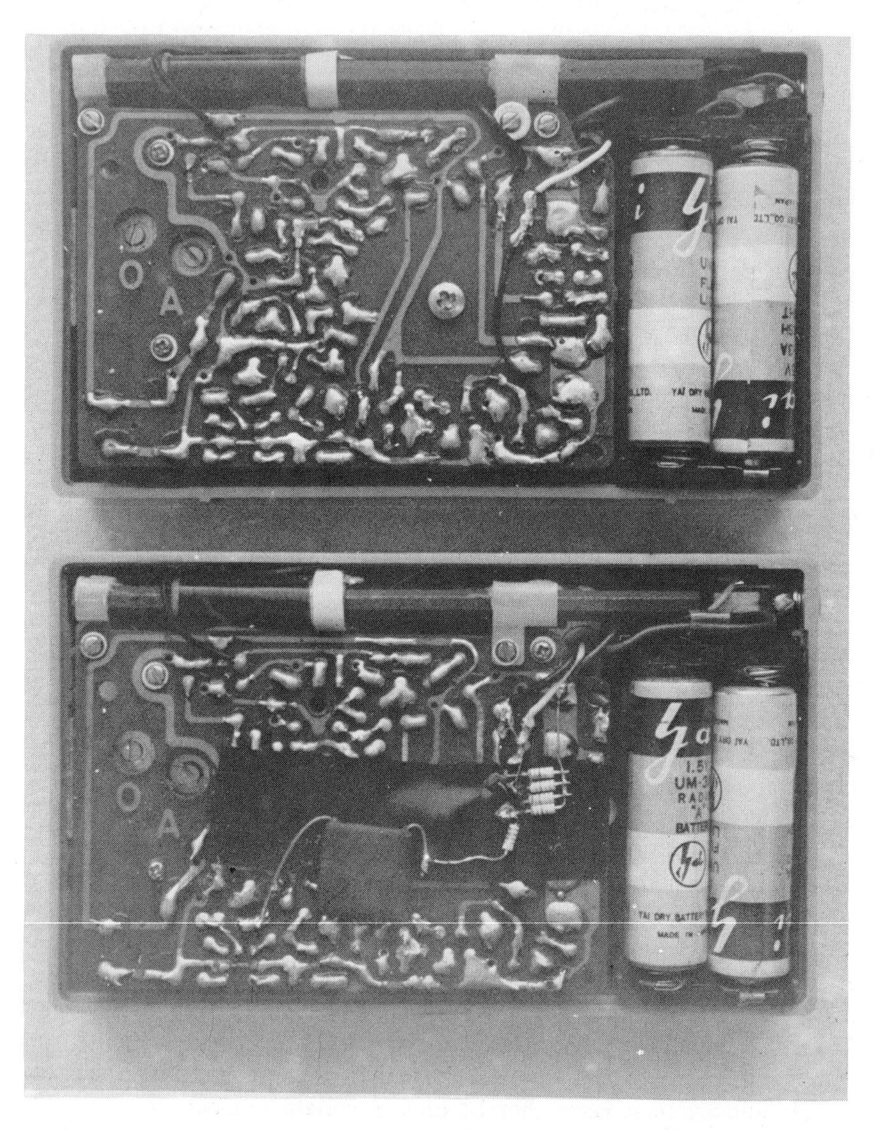

Fig. 11.2 Top: The appearance of the back of a typical pocket-radio receiver. Bottom: The connection of a multivibrator modulator is indicated. The output from the large square capacitor is applied to different points on the back of the circuit until the desired result is achieved. It is then soldered into place without actually determining the receiver construction.

to a period measurement (the average interval between pulses), while the former measurement was a frequency measurement (the number of cycles in a given interval), although either can be stated in the final result as frequency or as period, that is, as cycles per second or seconds per cycle. These same considerations often apply when using high-speed electric counters which can either directly determine the interval between two impulses or else count a number of rapidly occurring impulses in a predetermined interval. In these counters there is a basic high "clock frequency" to which all measurements are referred, and this is generally the vibrations of a quartz crystal, which corresponds to the ticking of a balance wheel in a watch. In general, it can be stated that it is preferable to count cycles in a given interval if time increments occur less often than information cycles, and if the opposite is true, then it is preferable to measure the time between cycles.

By listening we can count up to approximately ten clicks per second, although lower rates are considerably easier. The use of listening to a receiver has some very great advantages aside from simplicity and cheapness. The human ear is extremely good at picking sounds out of noise, and will often be much better at picking up a weak signal than relatively complex electronic circuits. The ear is also able to concentrate on one or the other of two simultaneous transmissions which might be difficult to separate with simple electronic circuits. This is related to what is called the "cocktail-party effect," in which a person is able to stand among a number of simultaneously speaking individuals and concentrate on and understand any one of them, while switching attention to another at will. In a train of clicks, in which the noise increases and decreases, the human ear and brain are actually able to fill in several totally missing clicks if this does not happen too often.

It is interesting to speculate what kind of system might give a comparably good performance. The performance to be considered is a separation of weak signals, such as periodic impulses, from noise. By listening, we seem able to count pulses in noise that could not be noticed on an oscilloscope. The use of an oscilloscope is generally more effective than the setting of a trigger level of a counting circuit which is to count pulses and yet not be triggered by noise. However, one can build a correlator circuit (e.g., Schwartz and Spindel, 1967) that averages over a number of pulses to accomplish a similar result on weak signals. We can visualize a possible process by again considering the periodically pulsing temperature transmitters. Suppose that we approach the transmitter in order to start with a strong signal and compare the received pulses in a fast cross-correlator with those from an adjustable-frequency pulse generator. The frequency of this generator can be changed, the adjustment being associated with a time constant of roughly 100 cycles. The average signal from the correlator controls the rate of the local pulse generator to make it match the external signal, which must not change much in 100 cycles. As we move away from the transmitter, the received signal degenerates to a poor one, but the local pulse generator should still follow pulse frequency changes in the remote transmitter; a scanning process could replace the original strong signal in finding the proper initial frequency.

Such a method senses when the next pulse should arrive, and it can fill in several successive weak or missing ones. The analogy between the rhythmic arm waving and the local pulse generator is strong. The system comprises sort of a phaselock loop for pulses, and a similar effect could, in principle, be achieved with a set of adjustable tuned circuits. The frequency of these circuits would have to be adjustable and match the harmonics (spaced by the repetition rate) of the pulses. The output of all circuits would have to be added in proper phase to give the signal to control the tuned circuit frequencies. The design of such filters to match a given pulse would be a prohibitive task in most cases.

2. HETERODYNE TO RATE CIRCUIT

With frequency-modulated signals it is necessary to observe or record small changes in frequency, and unless some subcarrier oscillator is used this frequency will be the radio carrier frequency itself. This was the case, for example, with the various transmitters that were to monitor gastrointestinal pressure fluctuations. Rather small percentage changes in frequency are usually involved. (In a legal sense transmitters are not allowed to swing through wide deviations in frequency, and in some cases subcarrier oscillators have the advantage of a given frequency shift being a large percentage change of the information-carrying frequency, yet displacing the overall radio frequency by only a small percentage.) Consider a specific example. As a result of a change in pressure, a gastrointestinal endoradiosonde may shift its frequency from 400 up to 401 kHz. This change of one part in 400 is rather difficult to record directly. But we can heterodyne or beat this signal with a local oscillator (BFO) of a frequency of 399 kHz. In that case the resulting difference frequency shifts from 1 to 2 kHz. The resulting doubling is readily noticed, and this heterodyning process can be considered as subtracting out a particular amount of the unchanging part of the frequency. A given absolute change is thus converted into a larger percentage change. It should be noted that small changes in the local oscillator frequency are similarly magnified, and thus that oscillator should be made as stable as possible.

Changes in the rate of this lower frequency signal are readily indicated or recorded by a number of types of circuits. These circuits usually consist of two parts, one to eliminate the effect of changes in signal amplitude, and the other to convert the variable frequency into a proportional direct voltage. A simple example of such a circuit is given in Fig. 11.3. In this case a pair of wires from the receiver (for example from the loudspeaker connection) applies a signal of several volts amplitude to the first transistor. Notice that the input transistor is switched from fully off to fully on by a change of base voltage of a small fraction of a volt, which results in the output signal from the collector swinging from approximately 0 to -6 V. This 6 V swing is applied to the rest of the circuit, and changes in amplitude of the input signal have relatively little effect as long as the input signal amplitude remains large. Any transistor type can be used, although those employed in chopper applications are more likely to have their collector potentials go very close to the reproducible condition of zero volts when the transistor is turned on; ordinary transistors more closely approximate this condition if the collector and emitter connections are interchanged.

The signal that is adjusted to an approximately constant amplitude by the first transistor is then applied to the rest of the circuit. With each cycle the charge on the first capacitor is fed to the output circuit through the action of the two diodes, and the higher the frequency the larger is the average current. A circuit of this type is sometimes referred to as a "bucket-dipper" circuit because with each cycle some charge is transferred from the little capacitor to the big capacitor, which is discharged continuously by the parallel resistor. Alternatively, the circuit can be regarded as passing periodic bursts of current from the input transistor through the resistor, with the output capacitor merely filtering the voltage drop to a steady average value. The circuit can also be regarded as a capacitor-coupled rectifier configuration which takes the input square wave and converts it

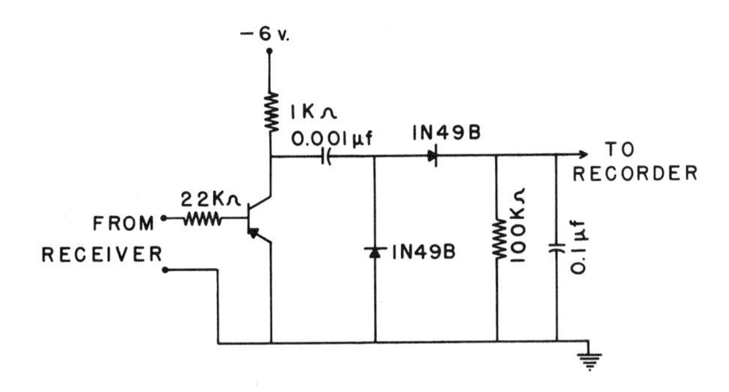

Fig. 11.3 The signal from the speaker connection on a receiver can be fed into the circuit to give out a voltage approximately proportional to frequency and relatively independent of amplitude.

to direct output voltage. In any case, as long as the output voltage remains small with respect to the six volts powering the circuit, the output response will not be too far from linear. The first diode can be replaced by a similarly oriented transistor whose base voltage is supplied by the lead to the recorder and then "bootstrap" operation takes place in which there is linearity of voltage response over a much wider excursion.

This circuit is useful for a wide range of frequencies, although generally in the present applications these will lie in or below the audible range for humans. With a so-called continuous wave (CW) receiver of the type used for code reception, which has a built-in beat-frequency oscillator, the circuit can be used to record small shifts in carrier frequency such as are involved with the gastrointestinal-pressure endoradiosondes. With a simple amplitude-modulation receiver, these circuits can record the pulse rate from one of the blocking oscillators or other periodically pulsed circuits. If the output capacitor is made very large so that the circuit responds very slowly, then we can directly record the rate of arrival of pulses which otherwise would be timed by the ear. If this circuit is used with an FM receiver, then it can record the subcarrier frequency of a transmitter using FM-FM modulation.

If the signal should become quite weak, then the input transistor might not be switched from fully on to fully off in a small fraction of the total cycle, and indeed might not be fully switched at all. Then the signal passed to the frequency determining section would not be of standard size and the resulting output would be modified. If we were transmitting an electrocardiogram, for example, then the amplitude of the pattern could appear altered, although heart rate could still be inferred from this marginal signal. In many cases, however, it is desirable, if the circuit is not working perfectly, that it not give any indication at all. Thus we might prefer no temperature indication to a faulty one. In this case it is desirable that the input circuit be of a regenerative form which switches in a precise way if it switches at all, thus suggesting a monostable multivibrator (Chapter 2). Monostable multivibrators are sometimes not reliably triggered by slowly changing signals, and thus it can be desirable to precede the monostable circuit by a bistable circuit that will always switch at a particular input-voltage level. Some bistable voltage-sensing circuits are referred to as Schmitt trigger circuits, and they are essentially emitter-coupled bistable multivibrators. Although a monostable multivibrator can often be used alone, it is sometimes preferable to precede it by a Schmitt trigger circuit in order to precisely fix the triggering level at which the fixed impulse from the monostable circuit will be generated, thus also allowing the setting of a threshold where noise will not be effective, but the useful signal will be. Similarly, a bistable circuit alone can precede the frequency-determining section, but sometimes a nonsinusoidal signal from a speaker input can cause multiple switching or erratic switching within a cycle, and thus it is often desirable to have both the bistable and the monostable multivibrator preceding the frequency-determining element, as shown in Fig. 11.4. The input-voltage level at which switching takes place is determined by the input potentiometer, and a high capacitor is shown for blocking out any steady voltage. In some cases the capacitor can be omitted entirely for direct coupling. If a meter indication of changes in frequency is desired, it is often sufficient to place a microammeter in place of the collector resistor of the right-hand transistor in the monostable multivibrator. In the circuit shown, the monostable multivibrator is very similar to the bistable input circuit, with each deriving some feedback from the common emitter resistor, rather than exclusively by a connection from the output collector back to the input base. In both cases the resistor carrying the signal from the first collector to the second base is shown bypassed by a 250 pF capacitor which is used to overcome small capacitances associated with the second transistor, thus assuring a fast transition.

With regard to the frequency-sensing circuit, its indication is proportional to the amount of voltage that is switched by the previous circuit, and thus the supply voltage should be stable. In some cases it is convenient to supply this from a separate battery. Because the action of a multivibrator is generally somewhat temperature sensitive, such a decoder of frequency is also somewhat temperature sensitive. Without precautions, a 1.0°C temperature change in the surroundings of this circuit can cause an apparent shift in a temperature indication of 0.1°C. The multivibrator and other components of this circuit can be compensated, or the circuit can simply be placed in

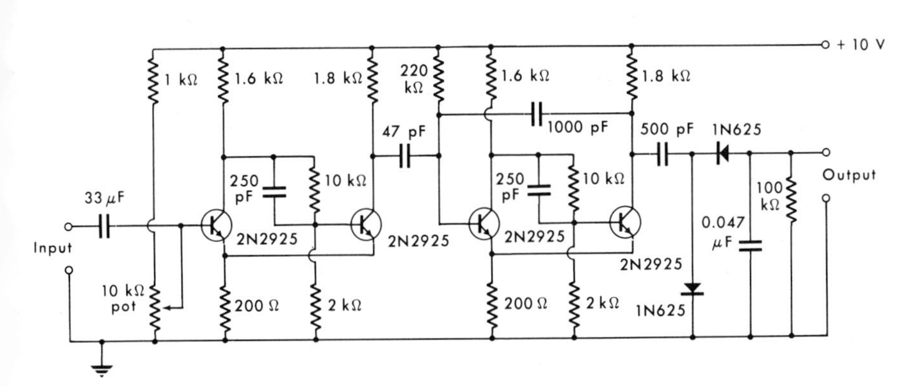

Fig. 11.4 A frequency-to-voltage converter containing switching circuits that give a relatively precise result if activated at all. At the input a bias level can be set to reject noise but accept the useful signal.

an approximately constant temperature surrounding. Thus the components are protected from large excursions by placement in an insulated box or in the investigator's armpit in field work.

Slightly more complicated feedback arrangements can be constructed which will switch a precise impulse of current for each cycle. With such an arrangement, a very high degree of precision and linearity is achieved. *A typical electric meter or recorder is neither as accurate nor as linear as many of these circuits.* The output voltage from one of these circuits can be read more accurately, although slowly, with a potentiometer technique in which a sensitive null meter indicates when an adjustable fraction of a known voltage equals the unknown; this fraction is then read.

Circuits of this type record an average frequency over a number of cycles. Their response is thus necessarily somewhat slow. However, it must be made fast with respect to any expected changes in the actual variable under study. In a noisy situation the loss of a few cycles, or the introduction of a few spurious ones, will cause a transient imperfect reading when the bias level cannot be set to reject this interference. Since this transient can take some time to die out, it is sometimes preferable to record the information cycle by cycle. If the period between cycles is observed, and the reciprocal displayed, then there is a linear indication of frequency which will only be disturbed for one cycle by a spurious impulse. Simple methods for accomplishing such a computation will be mentioned in Chapter 12.

A few further facts about receivers are relevant. Depending on the frequency range in which they work, and their adjustment, typical receivers will pass a range of frequencies of a few kilohertz for any given dial setting. A representative 500 kHz receiver may pass a range of 3 kHz or more, depending on its intended application. In the present mode of operation, once the dial is set, changes in transmitter frequency due either to drift or large deviation signals must not carry the signal out of the range of the receiver. If this is not the case, unusual artifacts can result; for example, Fig. 11.5 shows a tracing in which a human gastrointestinal tract appears to be generating double pulses of pressure. The actual situation is that an increase in pressure causes an increasing deviation in frequency until the edge of the passband is reached. Then the signal disappears, with a tendency for the recorder to return towards zero. Decreasing pressure and frequency eventually bring the signal back into the range that is passed, with an abrupt return of the recorder to its deflected position, after which the recorder will reproduce the further decreases in pressure back to the baseline. Slow drifts in either the transmitter or receiver can totally remove the signal from the band that is passed. (In the next section of this chapter, circuits will be indicated which can totally eliminate this problem.)

It might be noted that the local oscillator frequency (BFO) can be placed

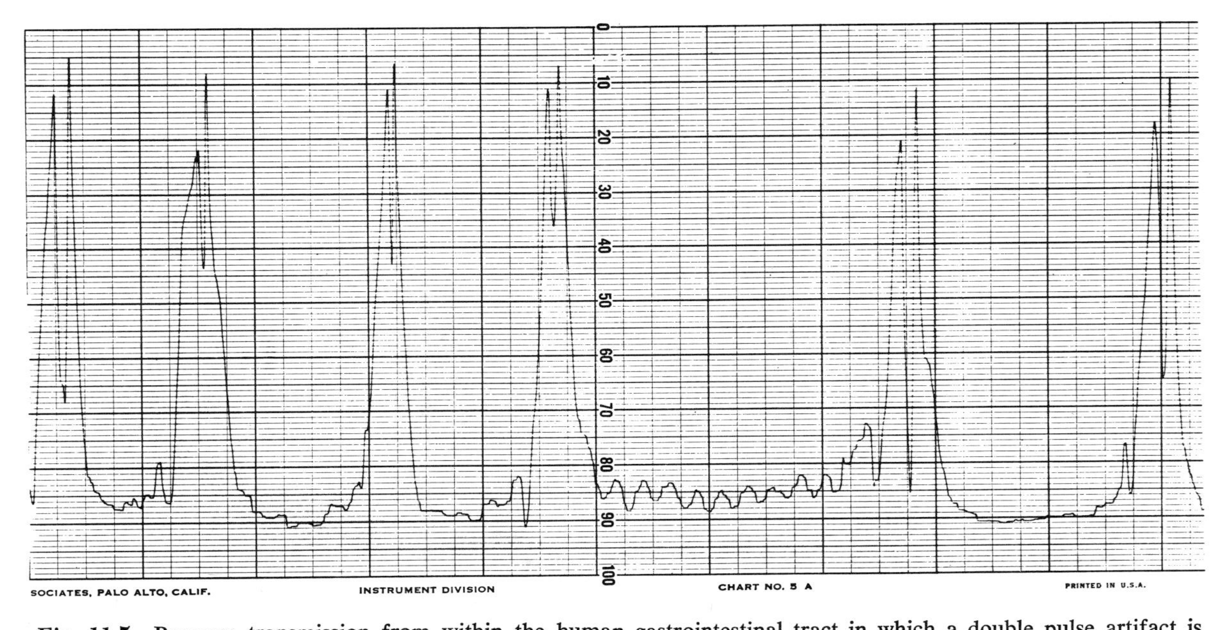

Fig. 11.5 Pressure transmission from within the human gastrointestinal tract in which a double pulse artifact is generated by a wide frequency excursion that momentarily removes the signal from the passband of the receiver.

either slightly below or slightly above the transmitter frequency in order to give enhanced sensitivity to small changes. In one case, however, an increase in frequency will produce a decrease in the beat frequency, and in the other case an increase in the beat frequency will result. This need not cause any confusion if we notice to which side of "zero beat" the original setting is made, and indeed this can be a useful possibility for reversing the direction of recorder deflection in some cases. It is generally undesirable for an incoming signal to pass from a lower frequency through the local oscillator frequency to a higher value, because then the beat note goes from a high frequency down to zero and then back up to a high frequency for a change in transmitter frequency all in one direction. The same is true if a signal passes in the other direction through zero beat.

For a number of years we have carried on our gastrointestinal studies by using war-surplus arc-5 receivers that contain built-in local oscillators for generating beat-frequency notes. To avoid interference to these low-power experiments, we used the receiver covering the frequency range of 200 to 500 kHz. This receiver is a superheterodyne whose local oscillator frequency is controlled by the tuning dial on the front. This signal beats with an incoming signal to give a difference frequency which is passed if it is in a small range of frequencies centered at the intermediate frequency of 85 kHz. The beat-frequency oscillator has its frequency set by a screwdriver adjustment, and it is set near 85 kHz. This gives an audible beat tone that changes as either the dial setting or the transmitter frequency is changed (as we discussed in the preceding section). This is an adequate mode of operation, although the frequency range of acceptance is limited. If the frequency of the BFO is made exactly 85 kHz, and then if the dial is turned for a zero-beat tone, the transmitter signal is set for the middle of the passband of the receiver. It then is unlikely either to drift or to deviate with a large signal out of the range of the receiver. Operation cannot take place in the vicinity of zero beat, in general, and a number of the above frequency-indicating circuits work well in the vicinity of 1000 Hz. Thus the BFO is shifted by 1000 Hz so that a 1000 Hz tone is generated in response to the average transmitter signal. Small frequency changes produce big changes in the audio tone at the loudspeaker connection.

We have fitted these receivers with a push button which connects a parallel capacitor into the BFO (not the local oscillator involved in the superheterodyne action) to momentarily deviate its frequency. The BFO is normally set off from 85 kHz by 1000 Hz; this deviation is removed by pressing the button. The button is pressed and the tuning dial turned for zero beat when setting up a transmitter. This adjusts the transmitter well within the receiver passband, and removal of the finger from the button shifts the BFO frequency so that a convenient 1000 Hz basic signal is delivered to the demodulator circuit. If all expected incoming signals are in one direction (never a voltage reversal, a negative pressure, or the like) then we can set things so a transmitted signal runs the intermediate frequency from one side of the passband to the other, rather than in either direction from the middle. Drift can then quickly eliminate the signal entirely, however, if it takes place in the wrong direction.

We can obtain large percentage deviations in a subcarrier oscillator by beating the signal from a sensor-oscillator combination against a fixed-frequency oscillator incorporated into the transmitter. The resulting radio-frequency carrier will undergo only small excursions in frequency, and the overall process can be considered as equivalent to moving the beat-frequency oscillator from the receiver to the transmitter.

3. DISCRIMINATORS AND AFC

Every FM receiver incorporates a device for converting small changes in frequency into corresponding output voltage changes. Discriminators for doing this are based upon the properties of properly tuned resonant circuits. They are often used to serve a further function beyond the conversion of a frequency deviation into an audible response. Useful signals for listening are those that take place at an audible rate, that is, not less than a few dozen hertz. Any slower changes constitute drift, probably in the receiver itself. Thus any slow variations in the output from a discriminator can be fed back to automatically retune the receiver in such a direction as to eliminate the discriminator signal, thus locking the receiver to the transmitter frequency as far as slow changes are concerned. Note that any slow but real change in transmitter frequency is rejected as drift, and not recorded in the signal output. Such an automatic frequency control (AFC) is found in most commercial FM receivers. In response to a shift in frequency all rapid components are taken out as representing useful information, and slow changes are rejected by gradual retuning through internal processes. It is because of this that an FM station, once captured, will not be lost if small changes are then made in the tuning dial of a good receiver.

Some discriminators employ a pair of tuned circuits which are slightly shifted from each other with regard to their resonant frequency. The resonant response from one is subtracted from that of the other, and at one frequency the two signals will cancel to give zero output. Small shifts in frequency in one direction will cause an increasing output from one circuit and a decreasing output from the other, to give a voltage in one sense, while a frequency deviation in the other direction will give a voltage deviation in the other direction. There are a number of related configurations for accomplishing a similar result, one of the most common being the so-called

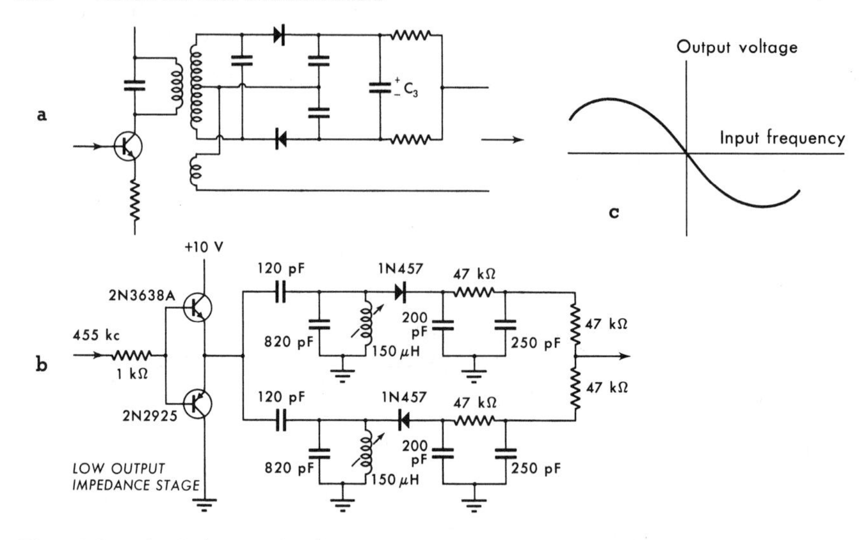

Fig. 11.6 Discriminator circuits, based on resonant combinations, are often used to convert a frequency-modulated signal into a variable voltage. At *a,* a ratio detector and at *b,* a more rapidly responding combination; *c* is the form of the output in general.

"ratio detector" depicted in Fig. 11.6*a*. The form of the response of this circuit, and that of many other discriminators, is shown in Fig. 11.6*c*. This circuit is widely used but its speed of response is necessarily limited by the presence of the output capacitor C_3.

For an application to be mentioned presently, a fast discriminator is required and thus the unit in Fig. 11.6*b* was designed. It was made to operate at 455 kHz, which is the intermediate frequency for many standard AM receivers. The incoming signal must drive a low output impedance stage to effectively activate the discriminator. An emitter follower is appropriate, but can show a different output impedance for increasing and decreasing voltages, and thus the double unit shown is employed. The discriminator itself uses a pair of resonant circuits slightly detuned from each other, but each having its resonant frequency in this general range of frequencies. The signal from each is rectified and filtered for subtraction at the output. The overall response is again of the form shown in Fig. 11.6*c*. This discriminator can prove useful in many applications.

Discriminators are useful in decoding some bio-medical telemetry signals, but they do have certain limitations that should be clearly understood. Proper functioning depends on the incoming signal saturating the circuitry which then drives the discriminator; otherwise the output amplitude depends on the signal amplitude. If the signal strength is not great, limiting action may be inadequate, and the entire response curve changes (becomes smaller in all directions); that is, the response of a discriminator depends on amplitude as well as frequency, and detuning or a weak signal may change the output when a discriminator is used to demodulate a frequency-modulated signal.

The center part of the response curve of a discriminator appears to be a straight line, and this is the region in which operation should take place. The precise straightness of this line depends on the relative properties of the tuned circuits. The ordinary alignment procedure on a discriminator need not give a precise straight line at all, and listening to music transmission, for example, would never uncover small irregularities. A linearity of better than 1 percent is very difficult to achieve. Temperature can continuously alter the frequency determining elements. In the previously mentioned circuits where a given current is switched with each cycle, if the circuit is working at all the response is linear and reliable, while in the present case small changes in both response and linearity can take place continuously. In some cases this limitation will in no way be prohibitive.

If a receiver is set to a transmitter and the transmitter frequency is slowly increased and decreased, a changing signal will appear in the automatic frequency-control circuit. This signal might, for example, be applied to voltage-sensitive capacitors in order to bring about a retuning of the receiver. As the receiver is retuned, this output control voltage cannot vanish completely or the capacitors would return to their original value and detune the receiver. Thus this output voltage remains and can be used as an output signal of degree of frequency deviation. The reliability of this voltage is limited to a few percent by the previously mentioned properties of the discriminator.

Suppose that we are working with a transmitter that has extremely wide deviations in frequency, actually extending beyond the band of frequencies the receiver could normally pass. In some cases this can provide a signal less affected by noise, but the problem remains in regard to the receiver that will accept it. Assume that we have a receiver containing circuits that will indicate the momentary deviation of transmitter frequency from the frequency set on the receiver tuning dial. Then, if an automatic frequency-control circuit is turned fully on, the receiver will partly follow the transmitter, allowing its signal always to pass, and its frequency-indicating circuit will show some fraction of the actual frequency deviation. The actual accuracy of the frequency-indicating circuitry, however, is now somewhat degraded by the performance of the discriminator which is doing the compensation. An example, which makes use of a commercial receiver, is given in Fig. 11.7. This receiver has 0.1 percent circuitry when used in the normal fashion, that is, with the automatic frequency control turned off. If the AFC

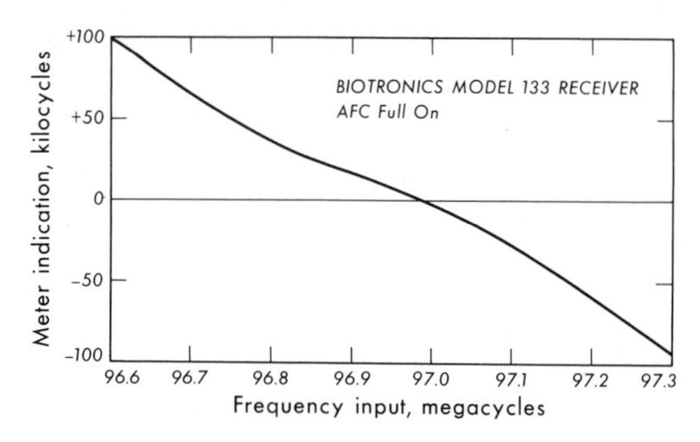

Fig. 11.7 An unusually wide frequency-deviation signal can be compressed to be accepted by a normal receiver by turning up the automatic frequency control. The linearity of the discriminator, however, which is generally relatively poor, then affects the overall demodulation process and can limit the stability and linearity of response. This commercial receiver would normally pass a range of frequencies of 200 kHz for a given setting with a high degree of linearity in its demodulation.

is turned on, the response of the receiver is as indicated in Fig. 11.7. The limitations of a relatively good discriminator are clearly seen. However, this may be quite an adequate response. We can see that, for a given receiver setting, a much higher than normal frequency deviation is accepted; that is, a much wider range of frequencies will pass through the circuit.

A number of these concepts can be illustrated by considering the design of a receiver suitable for simultaneously decoding the frequency information and the burst repetition-rate information in the signal from one of the blocking oscillators. As has been noted, it is usual, when considering pressure information, to bias the transmitter into continuous oscillation and, when considering such things as temperature information, to use a fixed average radio frequency. But we can achieve simultaneous transmission of these two variables, with some saving in transmitter power, if we are willing to go to a more complicated receiver system. The following circuits were made to work with the transmitter depicted in Fig. 3.4.

One circuit for this purpose was formed as in the block diagram at the top of Fig. 11.8. Pulses are sensed in the amplitude-modulation receiver (which is simply a modified pocket receiver). These are counted to yield the blocking rate, and each turns on a frequency-discriminator circuit, consisting of a pair of tuned inductance-capacitance circuits, in order to monitor the radio frequency during the pulse. The output from the tuned inductance-capacitance circuits is held over from each pulse until the next in order to yield a continuous radio-frequency indication. The fast discriminator in Fig. 11.6*b* was the one used to demodulate the radio-frequency information.

The Schmitt trigger circuit switches each time a fixed voltage level near zero is reached (a threshold detector). Amplitude changes shift the phase at which switching takes place, and this phase modulation appears as frequency modulation. Specifically, a 1-degree phase shift in the 100 μsec pulse rise time due to amplitude modulation caused a 1 percent error in the output. Thus cascaded limiters are preferred. In the second version of the circuit at the bottom of Fig. 11.8 a fast narrow band limiter was used instead. Some design problems were involved but the performance was good once the limiter was built to have a phase shift relatively independent of amplitude, and the same impedance for increasing and decreasing signals.

It was felt that more sensitivity (use of weaker signals) could be achieved by sensing the onset of a pulse in a different way. It is a common observation that FM receivers sound noisy when not tuned to a station but become quiet when tuned in. In the lower half of Fig. 11.8 the disappearance of noise (in the frequency range higher than the information-carrying frequency) was used to signal the circuit that a pulse had started; that is, quieting indicates the start of oscillation. Actually, with the standard receiver being used, because of its narrow band there was not much noise to disappear and thus reliable switching was slow. Although this system of sensing the presence of the pulse is good, in the final unit the limiters of the second system were used with the amplitude sensing of the pulse as in the first system.

The use of the fast discriminator to determine the radio frequency has been mentioned. It might be noted that similar circuitry can measure the frequency of a pulse radiated by a remote tuned circuit after being shock excited by an induced impulse. The operation, however, is much simpler then because the returned pulse will arrive at the receiver "on command" instead of being sensed at a random time. Thus these same circuits can be much simplified to be used for that form of passive telemetry (Chapter 13).

In most of these blocking transmitters a change in radio frequency is accompanied by a small change in the blocking frequency, and vice versa. The two received channels can be combined in a pair of resistors at whose common point an indication of pressure alone or temperature alone can be taken. This assumes that the interaction is reasonably linear, and the resistors are adjusted to give the desired compensation. This can be considered as the solving of two simultaneous equations involving radio frequency and blocking frequency to give the two unknowns of pressure and temperature. Two resistor pairs are required and it will be seen that the members of a pair are rather unequal since the pressure indication depends mostly on radio frequency while temperature is determined mostly by blocking frequency.

In both circuits there was incorporated a very slowly responding automatic frequency-control circuit that caused the receiver to automatically

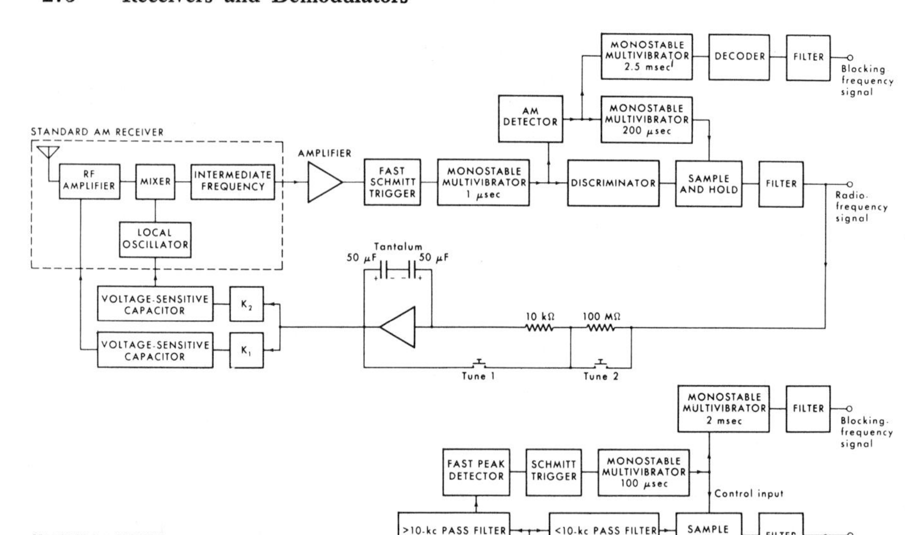

MONOSTABLE MULTIVIBRATOR 2 msec
FILTER
Blocking-frequency signal
FAST PEAK DETECTOR
SCHMITT TRIGGER
MONOSTABLE MULTIVIBRATOR 100 μsec
Control input
>10-kc PASS FILTER
<10-kc PASS FILTER
SAMPLE AND HOLD
FILTER
Radio-frequency signal
STANDARD AM RECEIVER
RF AMPLIFIER
MIXER
INTERMEDIATE FREQUENCY
CASCADED NARROW-BAND LIMITER
DISCRIMINATOR
LOCAL OSCILLATOR
Tantalum
50 μF
50 μF
VOLTAGE-SENSITIVE CAPACITOR
K2
VOLTAGE-SENSITIVE CAPACITOR
K1
10 kΩ
100 MΩ
Tune 1
Tune 2

Fig. 11.8 Modification of a standard receiver to simultaneously record the radio frequency and blocking rate of one of the squegging-oscillator transmitters. Very slow changes in the radio frequency are purposely rejected as probably being drift. The receiver was a standard pocket superheterodyne intended for entertainment purposes and was modified as indicated for electrical retuning.

track or follow, by retuning, any changes in transmitter radio frequency. Thus all slow changes, including drift in transmitter or receiver, creep in the pressure-sensing diaphragm, and the like are canceled out, and a continuous received signal is ensured. By making the response of this system slow (about an hour) actual contractions which cause frequency changes extending over several minutes are recorded instead of being retuned out. The retuning is accomplished with the action of the two voltage-controlled capacitors, which are adjusted to track each other so that all of the receiver is retuned together. To lock on to a station, the "Tune 1" button is pressed and the receiver tuned, in the usual way, to the station. That button is released and the "Tune 2" button pressed, whereupon the feedback system quickly locks on to the station. Release of the button then allows following slow shifts in carrier frequency, while recording more rapid ones. A similar system can, of course, prove valuable in other unattended monitoring situations in which pulsed operation is not being used and in which slow drift in either the transmitter or the receiver must be overcome.

4. PHASELOCK LOOPS

A phaselock loop system can be used to pick out very weak signals from the output of a receiver. In such a system an oscillator in the receiver generates a "clean" sine wave. A feedback arrangement (Fig. 11.9) compares the phase of this local signal with that of the incoming signal and forces the local signal into step with the incoming signal from the transmitter. A filter that will pass only slow changes (low pass filter) prevents any rapid changes in the local sine wave in response to momentary disturbances in the received signal. (The filter, however, must be set to pass changes as rapid as any real ones associated with the information.) A clean signal comes out with an indication of frequency and phase changes due to modulation. The inability to shift suddenly is equivalent to rejecting noise by a very narrow band filter adjusted around the transmitter carrier frequency. If fading suddenly removes several cycles entirely, they will be inserted at their proper place by the local oscillator, and a few extra impulses added by noise will not appear in the output. A meaningless signal comes out if the real signal is totally lost for an extended period since the local oscillator continues at its own rate.

In a representative case a superheterodyne receiver would feed a limiter and phase detector which would activate the voltage-controlled oscillator

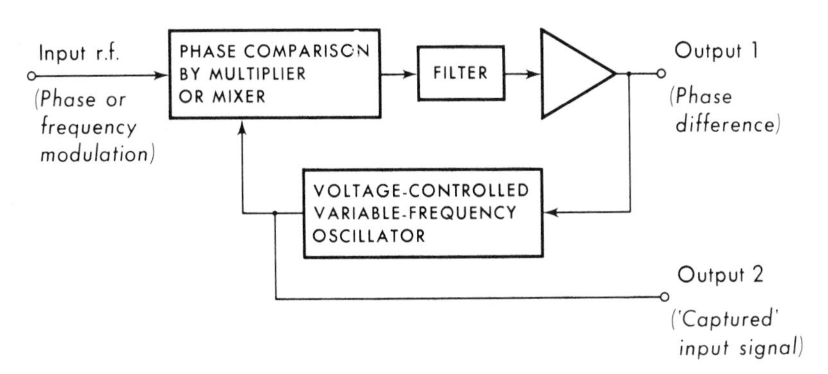

Fig. 11.9 A phaselock loop system is effective at rejecting interference and effectively provides a very narrow passband that is able to follow slow shifts in the transmitter frequency. Some signal comes out if the input is totally lost.

via the low-frequency filter, it being the intermediate frequency being followed. The actual phase difference between the local and incoming signals is generally less than 1 degree and depends on the amplification in the feedback loop. The overall effect is as if the receiver band width had been reduced to 1 or 2 Hz, which is tremendously effective in rejecting the noise that limits the overall receiver sensitivity. Commercial receivers note sensitivities of the order of −155 dBm. Not only does such a system give the effect of a difficult-to-construct narrow band filter, but unlike this fixed filter these devices can shift to follow frequency changes due to movement of either the transmitter or the receiver (Doppler shifts). Thus noise rejection can take place even though the transmitter is on the back of a moving animal, or the receiver on a pitching ship.

Baldwin and Ingle have used a pair of such receivers placed beside each other in an interesting way. The difference in phase between the two receivers gives an indication of the direction to the transmitter, which is being carried on an animal under observation. The direction of any frequency shift then tells whether the animal is moving toward or away from the observer, and generally tells of the activities of an animal through the continuous measure of motion. They have been able to transmit approximately 40 miles with 1 W of transmitted power at a frequency of 8 MHz. Higher frequencies are generally advantageous in retaining the full effectiveness of a low noise receiver if other factors do not limit the choice.

A number of situations can be thought of that can be considered as involving phaselock. Thus the demodulation of FM by observing the AFC signal might be considered as a low-frequency example. Relatively good sensitivity and low noise can result. This same general method can be used in the frequency-to-voltage conversion of Section 2 of this chapter. The previously discussed circuit types can be combined in an effective way that also makes the operation clear. A bistable multivibrator is driven to the left state by the incoming signal which is suitably squared for effective switching. It is driven to the right state by an astable multivibrator which serves as the local oscillator, and thus the amount of time spent in the right-hand state depends on the difference in instants of arrival of the peaks from the two sources. The frequency of the astable multivibrator is controlled by the filtered voltage coming from the bistable unit. This voltage depends on the difference in phase of the two signals, and also is the frequency demodulated output.

The double MOS device of Fig. 15.5 is valuable as a relatively pure multiplier for phase comparison in these receiver systems. Gardner (1966) has given an extensive treatment of phaselock techniques, and many practical and theoretical details have been covered by Tausworthe (1966). See also Appendix 5.

5. OTHER SYSTEMS

An estimate of the frequency of a transmitter can be had by using a "crystal set." As can be seen in Fig. 11.10 this can consist simply of a tuned circuit, a diode rectifier, and some device such as a microammeter for indicating signal strength. A small transmitter placed within the coil will cause an indication on the meter which will be a maximum when the capacitor is adjusted for resonance. Although this does avoid some uncertainties of more complex receiving systems, the receiving coil should not be coupled too tightly to the transmitting antenna since a pair of closely coupled tuned circuits does have a combined response that can prove confusing.

The technology of digital circuits has progressed to the point where it is possible to construct circuits based on cascaded bistable multivibrators that will count up to approximately 100 million equally spaced events per second. Some comments on the possibility of the elimination of the "inevitable" uncertainty of plus or minus one count in digital timing systems have been given (Mackay, 1960B). Heterodyne procedures can be used to achieve a digital indication of events occurring at rates greater than 100 MHz. Thus it is possible to use counter techniques to monitor any of the frequencies that have been discussed thus far.

In the calibration of endoradiosondes, we routinely take the signal from a closely spaced loop antenna and feed it directly into a commercial counter. With a weak signal we could cascade a series of tuned amplifier stages (a "tuned-radio-frequency receiver") in front of the counter to amplify the signal and reject noise or other radio stations. Instead, we can receive the signal on a standard superheterodyne receiver which beats the signal down to a frequency that passes through the intermediate-frequency filters. We can then take what comes from these filters and recombine it

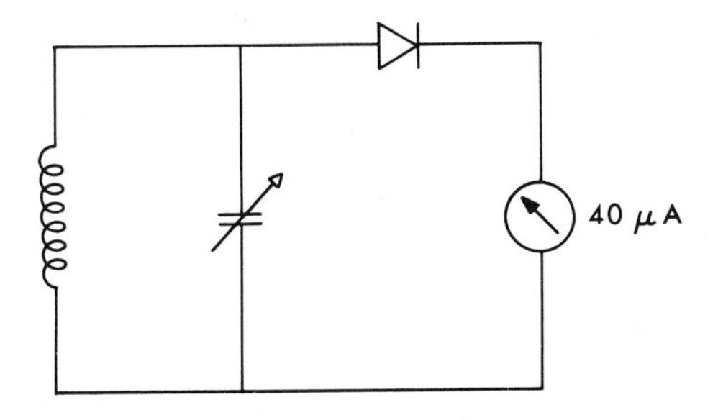

Fig. 11.10 A crystal set consists of a tuned circuit and a diode feeding some form of indicator. When tuned to a nearby transmitter frequency, the meter indication is a maximum.

with the local oscillator to raise the frequency back up to exactly the original frequency for application to a counter. Note that drift in the receiver does not matter, since the signal is returned to its original value.

Counter techniques generally do not show the presence of side-band frequencies due to modulation, while tuned circuits do. A single tuned circuit can be swept up and down cyclically in an interesting frequency region to note at what frequency a maximum response is elicited. The same frequency analysis can be made more rapidly by applying the signal simultaneously to a large group of filters in parallel, each being tuned to a slightly different small range, and noting from which a maximum response occurs. In the scanning case, the response of the circuit can be applied to the vertical axis of an oscilloscope and a horizontal voltage made proportional to the momentary frequency. Then a mask can be placed over the oscilloscope face, as in Fig. 11.11, to produce a moving spot of light which will trace out a graph of frequency as a function of time on a continuously moving sheet of film. This method was used in recording from some of the early passive transmitters to be mentioned later. Some of the more complicated frequency analyzers, such as produced in Fig. 3.9, can be considered as a receiver plus demodulator, with motion of the display indicating values of a transmitted variable. For absolute accuracy it is desirable to periodically apply a signal of standard frequency.

The frequency of pulses coming from one of the blocking oscillators can be judged by displaying the pattern on an oscilloscope. The sweep of the oscilloscope is synchronized to start with an incoming pulse. Motion

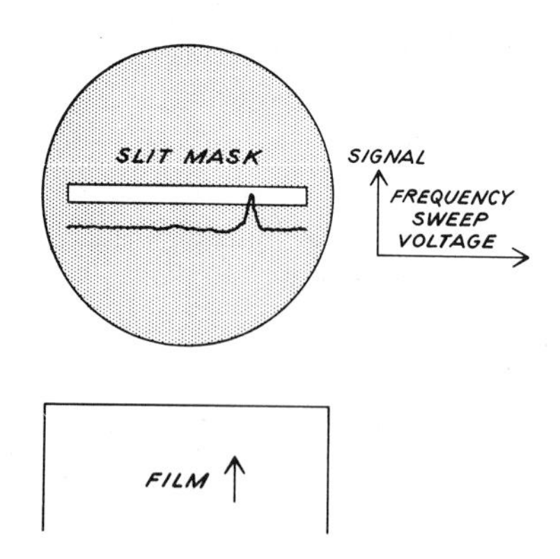

Fig. 11.11 The response of a receiver as a function of momentary frequency of tuning is displayed on an oscilloscope to indicate momentary transmitter frequency as a horizontally moving spot. Continuous motion of the film produces a direct graph of frequency versus time. The frequency of the receiver can be made to track the transmitter or simply be cyclically scanned through a range of values.

of the next pulse can be observed to signify frequency changes. If the sweep speed is slowed so that 10 pulses appear on the screen rather than one, motion of the final pulse will be considerably more noticeable, as will any "jitter" or irregularity in the overall process.

Receivers will generally not track changes in frequency of oscillators that are periodically turned off, and thus, in order to make use of the AFC, it is sometimes preferable to cause a periodic diminution in amplitude without complete signal disappearance. The circuit of Fig. 3.2, which was to be used with inexpensive portable receivers, was designed with this in mind.

The presence of a steady signal properly tuned can be sensed in an ordinary receiver not only by the appearance of a new sound, but by quieting of the normally present noise coming from a radio that is not tuned to any station. In part this is due to a decrease in overall gain caused by automatic volume control (AVC) circuits which are intended to reduce changes in loudness in receivers that are moving about. In fact, the AVC voltage is a crude measure of signal strength of a transmitter and can be so used in some measurements. Comparison between signals in such applications as testing different antennas for relative effectiveness is best made by continuously adjusting an attenuator at the transmitter to bring this signal back to a fixed value. In circuits that have an automatic volume control, pulse signals are sometimes best noticed as a periodic loss of noise if there is not present some sort of a beat-frequency oscillator.

In Chapter 10 some precautions to be taken in setting up a receiver to minimize interference were discussed. Metal cages containing animals can be grounded, or aluminum foil can be placed around wooden cages. Metal table tops should be grounded or aluminum foil should be placed under the receiver and the subject. If several table tops are involved, they should be connected together electrically. Antennas should be balanced with respect to ground, even though the receiver has a balanced input connection; this is done with variable capacitors that have a minimum value in order to avoid excessive shunting of the incoming signal. Nearby flourescent lights or large motors should be approached with a portable receiver to see if they are generating interference.

If information is arriving erratically, monitoring the period rather than the average frequency is best because then data points are occasionally achieved distinctly, undistorted by signal dropouts. The conversion of the time of the cycle to a frequency indication will be mentioned in the next chapter. It is sometimes useful to note that the average duration of a number of individually timed cycles comes out to be just the time of arrival of the final cycle minus the time of arrival of the initial cycle divided by the number of cycles that took place in the total interval.

After decoding, data must be displayed for a maximum ease of assimilation. One convenient technique for the compression of masses of physiological data into a compact form for visual review has been mentioned by Webb and Rogers (1966). Each experiment has its own individual requirements, and the generalities of the problems of display are extremely involved. In many cases the best procedure is to automatically process the data so that only the aspect of interest is saved in maximally reduced form. At times, this will result in the loss of a piece of unexpected new information, but it will reduce not only the number of aimless experiments, but also the vast amounts of data which are taken annually and never actually used. In many cases it is desirable that the signals derived from a subject be preprocessed so that only the data of interest is transmitted in the first place, thus also conserving radio bandwidth and battery power, with corresponding savings at the receiver end.

12

Calibration and Response Control

1. CALIBRATION

One can measure the response of each element in an overall telemetry system (that is, the output for a variety of applied inputs), and from this calculate the overall response of the system. In designing some systems, this methodical approach is essential so that the various components remain compatible. However, with some systems that receive careful attention in use from the investigator, and which are not destined for mass production, a simpler approach is often employed. This consists of simply applying to the input sensor a series of values of the variable in question, and at each step noting the indicated output from the final meter or recorder. If inputs covering the entire expected range are provided, then the overall system will be calibrated, including the effects of any nonlinearities or other characteristics of the intermediate components. The effects of such things as temperature changes in various parts of the system which cause drifts, for example, will have to be considered, but the procedure is in principle quite uncomplicated.

If a temperature transmitter is involved then a series of increasing temperatures is applied, in each case waiting until equilibrium is achieved, and the outputs observed. If pH is to be studied, then the transmitter should be immersed in a series of buffers. With some transducers of some variables, it is desirable to run through an increasing series of conditions and then a decreasing series to be certain that the indication at a given value does not depend on the previous history of the system; that is, that some part of the system does not show hysteresis. An example of a calibration curve is shown in Fig. 12.1. In this case a gastrointestinal-pressure transmitter was placed in a flask in which the pressure could be successively raised through a series of steps. It is extremely convenient to use the rubber bulb from a blood-pressure measuring cuff to pump up pressures in a controlled fashion. In this case successive steps correspond to increases in pressure of 10 mm of Hg, as judged by a manometer on the flask. The transmitted signal is picked up by a loop antenna, fed through a receiver and demodulator, and applied to a recorder set to advance at slow speed. It is seen that each step is not of the same height, with the sensitivity falling off slightly at the higher pressures. Although overall output plotted against an applied input is not a straight line (is a nonlinear response), this piece of apparatus is perfectly useful since for each output there is a corresponding specified input. In a later section we shall see how such a response could be converted to a linear one, or to a more nonlinear one, if either of these conditions were desired.

The degree of straightness of an overall response can depend upon the degree of cancellation of oppositely curving characteristics of two intermediate subassemblies. In the pressure sensor example, most of the nonlinearity of response happened to take place in the input sensor where a given change in pressure did not produce a given diaphragm motion at all levels of pressure, and a given change in diaphragm position did not produce the same change in inductance for all diaphragm positions. In many cases

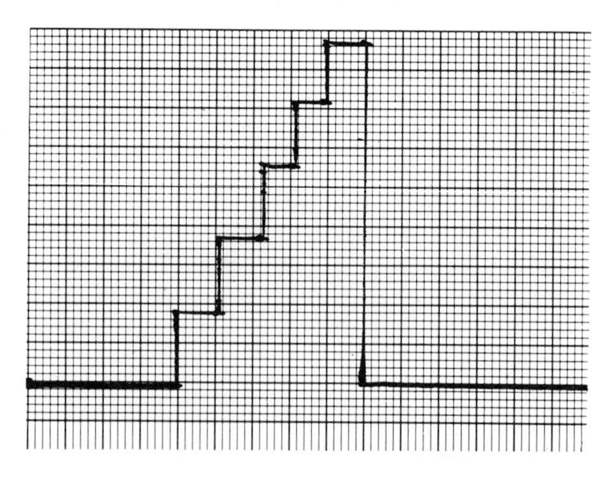

Fig. 12.1 The recorded signal as pressure on the transmitter is increased in a series of steps. The response of the overall system from sensor to recorder in this case is slightly nonlinear. Equal inputs, decreasing or increasing, gave the same outputs, thus making the system useful.

it is important to consider the conditions under which operation will actually take place, so that the range of applied variables will truly represent the range over which the system must work. To take an extreme example, consider the use of the transmitter of gastrointestinal pressures within a high-pressure chamber during diving experiments. One might expect a specified change in output for a given change in pressure, as indicated by the calibration curve made at normal atmospheric pressure. But at high ambient pressures, the sensitivity might not merely have dropped off a little, but the core could have been forced in all the way to the plastic at the back of the coil and be unable to move further, thus resulting in no change in output for any further change in pressure. This total saturation is an example of an extreme nonlinearity or discontinuity in response. It is, of course, obvious that in the particular case mentioned it would be advisable to simply momentarily equilibrate the air behind the core with the ambient air pressure, before starting the experiment.

2. *IN SITU* CALIBRATION

Even after a transmitter is in place within the body of an animal, its responsiveness can be checked, and in some cases actual *in situ* calibration is possible. For example, if a person swallows a pressure transmitter, then changes in ambient pressure around the person will give corresponding changes in the pressure indication from the transmitter within the body of the subject (Mackay and Jacobson, 1957). In experiments such as are depicted in Fig. 1.11 the animal can simply be given a gentle squeeze to be sure that the transmitter is responding at all. Then every few days it can be placed within a box or bag that is airtight and the pressure increased in a series of steps. The resulting recording indicates if there has been any drift in sensitivity. Note that this procedure calibrates the incremental sensitivity of the device without giving an indication in any drift in the "zero setting." In many cases this is no objection since absolute pressures are referred to the momentary barometric pressure, and may be of no interest if only contractions or changes are being monitored.

However, absolute pressure calibration can also be achieved by occasionally performing an absolute pressure measurement. This may mean inserting a hypodermic needle into the structure of interest and performing a direct measurement. In the case of the eye pressure measurements of Figs. 1.14 and 13.6 where one surface of the body cavity being investigated is exposed, it is sufficient to make a daily absolute pressure measurement without penetration with the help of a relatively accurate tonometer, such as one based on the principle indicated in Fig. 5.19.

Other transmitters than those for pressure can also be checked for responsiveness and calibration. Thus a pH electrode in a well perfused tissue will give a reproducible change in indication if the subject takes a breath of air which is enriched in carbon dioxide. In some cases a drift in a baseline reading is more likely or troublesome than changes in sensitivity to a small change in input. Some electromagnetic flowmeters display an approximately constant change in indication for a given change in flow rate, but the indication corresponding to no flow may drift. One way to check this is to periodically stop flow in the vessel being monitored. Thus the device in Fig. 16.3 was originally conceived for turning off the blood flow in an artery without disturbing the subject in order to check the indication of a flowmeter, though clearly it has more general applications.

3. RESPONSE TIME

One especially important parameter characterizing a system is the rapidity with which it can respond to changing conditions. In most of the cases relevant to the present discussion, this is limited by the input transducer, although occasionally a recorder will cause a limitation at the other end of the system. It is important to have an estimate of this quantity in order to know whether certain kinds of data will be correctly displayed, or somewhat distorted, or not recorded at all. Response times are sometimes specified as a frequency response; that is, the maximum number of cycles in a given time which can be accepted before the output reproduction is significantly reduced in amplitude from the value observed when the cycles are extremely slowly introduced. Alternatively, the observed response to a sudden change in input conditions can be specified. There are mathematical methods that in many cases allow conversion from one specification to the other.

In some cases observation of a response gradually approaching the final value has been observed to form an estimate of an "equilibrium time" that represents a guess at the length of time required to arrive at essentially the final value. Such an estimate separates things that are happening in minutes from those happening in days and weeks, but otherwise it is relatively useless in communicating precise information about an experiment to other workers. If the response to a sudden input is an irregular change, which perhaps starts and finishes slowly, then it is sometimes useful to specify the length of time required to go from 10 percent of the final value up to 90 percent of the final value, or between other specified limits.

Some processes are described by an exponential equation. On semilog paper such a process is described by a straight line. These processes are characterized by the same fraction of the remaining change taking place in each successive interval. Thus such events are well described by a characteristic time. In the case of radioactivity, a "half-life" is specified, which is the time required for half the remaining atoms to undergo a transformation. In the case of a capacitor charging or discharging through a resistor, a "time constant" is often specified, which is the length of time required for the capacitor voltage to change to within $1/e$ (where e is 2.718 . . .) of the remaining interval towards the final value. A half life is shorter than a time constant for a given process by the factor 0.693. In principle, such a process is never fully completed, but specification of one of these times is a rather precise description of what is going on and what is to be expected after any given interval.

From the slope of the line on a semilog plot, one can readily see how long it takes for the dependent variable to drop to half of a previous value, and thus determine the half life of one of these processes. A sketch can instead be made on ordinary graph paper to determine how long it takes for the process to complete half of its expected excursion. In some cases another graphical construction is extemely convenient for determining the time constant of such a process, that is, the time required to complete approximately ⅔ of a total change. Figure 12.2 shows a hypothetical graph of the output from a telemetry system as a function of time, with a sudden change in input conditions taking place at the start. It is not necessary to know the actual output values, just so they are equally spaced. The final value, which is approached after a very long time, is estimated and sketched in as shown. A straight line is then drawn tangent to the experimental curve

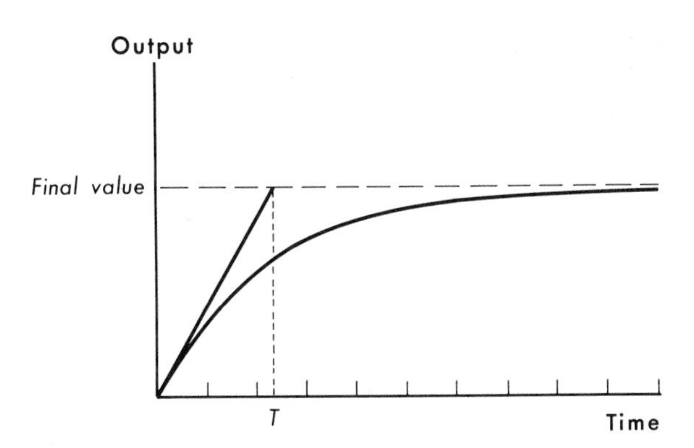

Fig. 12.2 A line tangent to the initial slope of the curve of an exponential process cuts the final value at a time that is one "time constant." This is a useful parameter for specifying the response speed of many systems.

at its start, as shown. This straight line will intersect the final value level of output at a particular time which is just one time constant.

Many observations of response time are made in this fashion. Thus the speed of response of a temperature transmitter can be estimated by suddenly dropping the transmitter into a glass of warm water, and recording the resulting warming curve. It is not necessary to know the initial temperature of the water, just so it is different from the initial temperature of the transmitter, which also need not be known. In that case the response time will generally be different if the water is stirred than if it is stagnant. This same experiment can be done by dropping the transmitter into a glass of cool water. For a linear system rise times and fall times should be equal, but this is not necessarily true of a nonlinear system. (A practical example is the case of an emitter-follower circuit in which an increasing voltage may be followed at a much different rate from a decreasing voltage, because in one case the transistor may cut off and become temporarily inactive. Similarly, animals often do not warm and cool at the same rate in response to changes in temperature of their surroundings.) The same kind of a description can be made of the response of a pH transmitter by suddenly shifting it from one buffer to a different one; stirring may have less of an effect here, the limiting speed perhaps being controlled by processes in the electrode rather than beyond it. Sound transducers are often conveniently described by their frequency response, but, as was indicated in Fig. 9.5, these units can also be tested by recording their response to a change that is rapid with respect to their ability to respond.

In certain cases related parameters may be of interest. Thus the disturbance to a biological system provided by the placement of a temperature transmitter depends upon the amount of heat that the system must supply in order to effect a change in transmitter temperature. This will affect the rate at which the biological system under study can actually change, and it thus enters the frequency response question in quite another fashion. In temperature monitoring it is thus sometimes appropriate to measure the specific heat of a transmitter and its thermal conductivity to its surroundings. The specific heat can be measured by dropping a transmitter of known mass and temperature into a small pool of water of known mass and different known temperature. From the final equilibrium temperature, if heat loss to the surroundings is either small or compensated for, the amount of heat necessary to change the temperature a given amount can be calculated. In a number of cases, this has come out to be somewhat less than for the same mass of water. The conductivity to the surroundings depends upon the surface treatment and geometry of the transmitter. The matter is somewhat complicated, but it often appears that the effect of a transmitter is not too different from that of a corresponding mass of tissue similarly situated.

4. LINEARITY

An overall response in which output as a function of input is a straight line is sometimes quite convenient. For example, then at all levels a particular change in output corresponds to the same change in input. On the other hand, in some cases in which a particular range of variables is of interest, a nonlinear response is desirable which spreads the indications in a particularly useful range.

The overall response of a particular system can be modified in an essentially arbitrary fashion to meet desired conditions. Suppose that the output from a demodulator was linear with respect to changes in input. Then by the scheme outlined in Fig. 12.3, this response can be made to curve either in the concave downward direction, or in the concave upward direction. Similarly, an overall response that was initially concave downward could be modified to a straight line or actually made to curve upward, and vice versa. The method uses a series of diodes which are successively biased to different voltages. In the top part of Fig. 12.3, the second diode is biased to a more positive voltage than is the first. When small voltages are applied to the input of this circuit, both diodes are biased in the backward direction and act as open circuits, thus passing any increase in voltage to the output unchanged. When the input voltage increases to the value V_1, the first diode

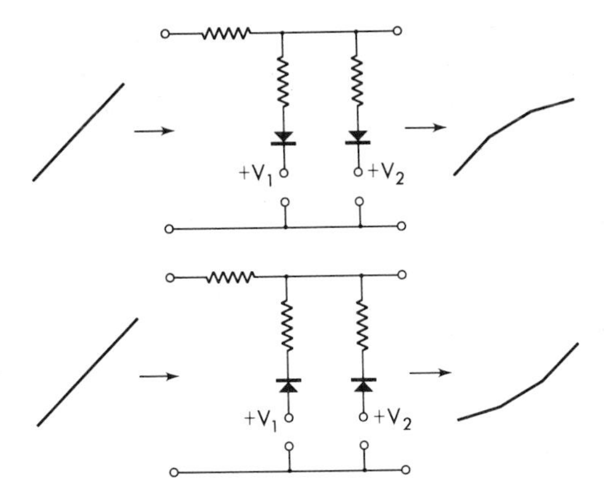

Fig. 12.3 The overall response characteristic of a system can be made more or less linear in either direction by incorporating, somewhere in the system, a network consisting of several diodes biased to different intermediate voltages. There a linear response is converted to a curved one, but a curved response could as well have been converted to one that is approximately linear. The more diodes, the smoother the corrected response.

suddenly begins to conduct, thus connecting its resistor in parallel with the output terminals. The series resistor and this resistor thus act as a voltage divider to attenuate further increases in voltage. This accounts for the first "break" in the shape of the output curve. When the input voltage eventually reaches V_2, a second resistor is connected in parallel with the output, thus further dividing down any more changes in input. A series of steadily increasing voltages applied to the input thus results in an output curved as shown. The "break points" are determined by the bias voltage settings and the corresponding changes of slope are determined by the sizes of the resistors. The output response curve is actually a series of straight lines, but almost any given curve can be arbitrarily closely approximated by increasing the number of diodes.

In the bottom part of Fig. 12.3, the diodes are shown as reversed, and a similar situation exists in which any input characteristic is rendered more upwardly curved. Various combinations of biased diodes can be used to give quite complicated overall responses (although it is impossible, by such a method, to generate a negative resistance). An arrangement of this second type applied between the demodulator and the recorder of Fig. 12.1 could be used to make the overall response of the system linear, if that were desired. In that case, adequate perfection could probably be achieved by the use of only one or two diodes, although for high percentage precision a dozen might be employed. The various bias voltages could simply be taken from a voltage divider across a battery, there being no need for several batteries or power sources.

It has been mentioned that circuits that record the time interval between cycles in a telemetry system are more effective in noisy situations than those that count cycles occurring over a longer interval, because a proper reading is available any time two successive cycles take place without corruption by noise. However, an indication proportional to the interval between cycles may not have the desired linearity if the indications are labeled as frequency (that is, cycles per second). This is because frequency is the reciprocal of period, and if one is linear then the other cannot be linear. There is a way in which the reciprocal of period can automatically be calculated so as to give out an indication directly proportional to frequency from each successive interval measurement. It might be noted that any such system allows an extremely rapid response (a response after one cycle) and might be referred to as an extremely rapid frequency demodulator. If an initially charged capacitor is started to discharge with the arrival of one cycle, and it is stopped with the arrival of the second cycle, then the final capacitor voltage will be linearly proportional to the frequency of the cycles if the capacitor is discharged through a device in which current is proportional to the square of the voltage (Mackay, 1962). Such a nonlinear current-voltage characteristic for the capacitor-discharging element can be achieved using materials called thyrite, but for less temperature sensitivity it is best to use a suitably adjusted biased-diode network of the type described above.

As discussed in Chapter 11, Sec. 1, for maximum rapidity it is sometimes best to take data by timing a certain number of cycles, and in other cases it is best to count the number of cycles arriving in a prespecified interval. In one case we are essentially observing frequency and in the others we are observing period, but either type of observation can be portrayed as a frequency measurement or as a period measurement. *With a given set of data,* one of these portrayals will generally appear more linear than the other. This can perhaps best be understood by considering a specific hypothetical example. In counting clicks with a stop watch, it was noted that the most accuracy was achieved in a limited time by timing a prespecified number of counts (Chapter 11). Some representative calibration data for three temperatures might then be as tabulated in Fig. 12.4. Any given temperature is then specified by the ratio of a number of counts to a time interval, but this ratio can be taken in two different ways. From the given data we can calculate cycles per second or seconds per cycle, that is, frequency or period. In Fig. 12.4, these two graphs of the same data are seen, both using similar linear scales. It is seen that while one graph is a perfect straight line, the other graph is not. In fact, if one graph is a straight line, then the other *must* be a hyperbola. (Similarly, if times are "normally" distributed, frequencies can not be.) In the case shown, for example, the midpoint

CALIBRATION DATA

TEMPERATURE, °C	COUNT, CYCLES	TIME, SECONDS
30.0	35	35.0
36.0	45	30.0
42.0	70	35.0

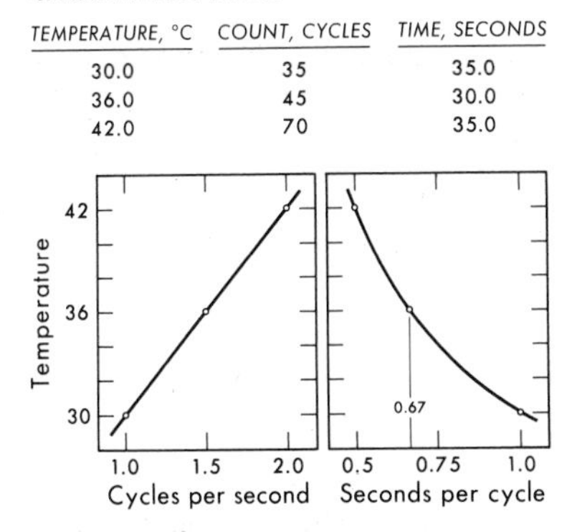

Fig. 12.4 A given set of representative data can appear linear on one graph and curved on another, although the coordinates are equally spaced in both cases. The data above are linear on a frequency graph, but the half-height appears at 0.67 rather than 0.75 on the plot of temperature versus period of the transmitter. A straight-line assumption in the case of the second drawing of the same data would result in considerable error.

temperature of 36°C falls at the center of the frequency axis, but at 0.67 rather than at 0.75 (the mid-point) on the period plot. On the second drawing, an assumption of linearity could cause significant errors. This point is perhaps obvious, but it has been known to cause confusion. If an actual calibration is slightly curved, then one of these methods of handling the given data will generally result in a closer approximation to a straight line than the other.

5. EXTENDING THE FREQUENCY RESPONSE

In the first two sections of this chapter the measurement of overall calibration and response time of a system were mentioned. In the previous section, the modification of calibration by various methods was discussed. In this section we shall discuss the possibility of altering the response time of a particular system, or altering its frequency response.

There are a number of aspects to this topic. A given transducer is characterized by a maximum rate at which it can change. But we can alter the rapidity with which measurements can be made by the suitable use of feedback and other techniques. Suppose that one has a rather large thermistor which can only respond sluggishly, and one wishes to measure rapidly changing wind velocities with it. The thermistor can be placed in a bridge circuit whose energizing voltage then raises the thermistor temperature. Any change in unbalance signal can be fed back to change the energizing power in such a direction as to maintain the temperature of the thermistor approximately constant. Changes in passing wind velocity then cool the thermistor to a different degree, but the feedback mechanism maintains the temperature approximately constant. The power applied to the resistance bridge circuit is then a measure of passing wind velocity, and since the thermistor essentially does not have to actually heat or cool at all, having been maintained at an approximately constant temperature, it can respond extremely rapidly. The response is in much less than one normal time constant, with the overall response speed depending upon the gain of the feedback system. (In an actual case, there are some well known problems in stabilizing the feedback system.) If a system is unable to rapidly respond to an injected signal, then a more faithful reproduction of the applied signal can be had by mixing with the output some of its own rate of change (derivative). This compensation for a restricted high-frequency response can be used in applications ranging from the achievement of sharper boundaries in photographs to the increased frequency response of wide-band amplifiers (so-called "shunt peaking" being a kind of example of the latter).

If it is desired, electrical signals can have their high frequencies emphasized. A simple filter is shown in Fig. 12.5*a*. Very high frequencies are passed almost without attenuation by the capacitor, and very low frequencies are more strongly blocked by the series capacitor. If the capacitor and resistor are interchanged, then the lower frequencies can be emphasized. More complicated filters can be built to emphasize high frequencies or low frequencies or bands of frequencies to various degrees.

In some systems, it is the low-frequency response, or the ability to pass very slowly changing variables, that is a fundamental limitation. Thus a changing pressure applied to a piezoelectric element generates a corresponding voltage, but the charge on the element soon leaks away through the resistance of the piezoelectric element itself and through the input resistor of the following amplifier. Thus it is generally impossible to record steady or very slowly changing pressures with such a transducer. Similarly, many amplifiers have a low-frequency response that is restricted, in some cases rendering them inadequate. In between stages of a multistage amplifier, there are often coupling elements that are essentially capacitors and resistors (as in Fig. 12.5*a*). A restricted low-frequency response can prevent the transmission of slowly changing information. A common example is a limited low-frequency response causing the distortion of an electrocardiogram pattern. In some cases the frequency response of a magnetic tape recorder is inadequate, and other examples could be cited. Such a restricted response can be compensated for by a suitable set of components, although this might at first seem to disobey certain fundamental laws of information theory or thermodynamics.

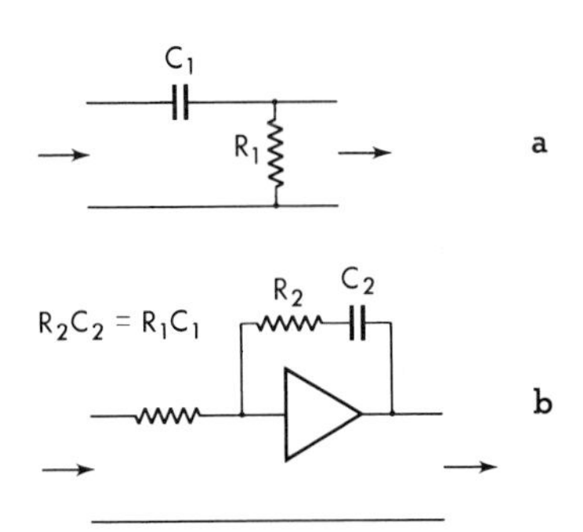

Fig. 12.5 *a:* Passage of a signal through this network limits the low-frequency components and emphasizes the higher. Such a combination is often placed between stages of an amplifier and also represents the equivalent circuit of some transducers with limited steady-state response. *b:* The effect of the preceding combination can be corrected with a suitable network that incorporates an amplifier. Playing a previously recorded signal through such a network can restore information not thought to be present at all. Some cases require cascading several correction networks to correspond to the number of degrading elements in the original overall system.

A frequency-restricting time constant (Fig. 12.5*a*) can be canceled by a similar time constant placed in a feedback loop (Fig. 12.5*b*). This can be considered as mixing with an output signal some of its own integral, thus partially undoing the imperfect differentiation of the other limiting response (Collins and Mackay, 1960). The integrator neutralizes the effects of the partial differentiation, but since the differentiation is not perfect, some of the unmodified signal must be made to appear in the output. Notice that the imperfect differentiation is not counteracted by a process of imperfect integration, but rather by combining the transient information of the signal with some of the steady information of its integral. Alternatively we can regard this process as restoring the original response by placing a model of the original structure in a feedback path, which is a procedure that is sometimes used in servomechanism work. In a multistage amplifier there are cascaded a number of short time-constant coupling circuits (such as in Fig. 12.5*a*), and to undo the distortion produced by them at all frequencies, a similar number of circuits (Fig. 12.5*b*) must be cascaded somewhere in the output. (Compensation for more complicated situations is possible, but in practice, this is usually sufficient.)

By this means it is actually possible to take a magnetic tape recording of an electrocardiographic signal previously recorded through an inadequate amplier, and at some later time to replay the recording through processing circuits which can restore the perfect form of the signal. It might thus appear that we were obtaining low-frequency information that was not even recorded in the first place, but the information is, in fact, there in terms of changes in signals, rather than their actual level. Perfect correction of known distortions can only take place in the absence of noise. Although there are limitations of the extent to which these processes can be carried in terms of noise and drift, they are quite useful in some actual situations. Thus, for example, the apparently shifted *S-T* interval in an electrocardiogram recorded through an inadequate amplifier is properly restored when the signal is later (or simultaneously) run through a suitable set of cascaded compensating circuits. Similarly, the use of these circuits allows very slow pressure changes to be monitored with piezoelectric elements, as long as the compensations are maintained. (It is possible to feed the signal from a piezoelectric element into the high resistance of a field-effect transistor, or perhaps to build one into the other, in order to achieve a good low-frequency response, but there would still be limits.) Similarly, the position of a so-called "velocity transducer" such as a coil moving in the field of a permanent magnet, can be monitored by passing the output signal generated in response to each movement through one of the circuits before display. The signal from a piezo element passing through a capacitor-coupled amplifier with one interstage coupling would require two cascaded compensating circuits before the display device for complete compensation.

Thus it can be seen that both the calibration and the apparent response speed of either a transducer or an overall system can be modified at will, within certain limitations, in order to provide the most useful possible display of the information being sensed.

13

Passive Transmission

A passive transmitter is considered here to be one that does not contain its own power source, but rather accepts power from a remote location and then reradiates it in usefully modulated form. But even among transmitters that do not contain a battery, there are several different types. Some of these transmitters are constructed containing only truly passive elements such as inductors, resistors, and capacitors. Others have no built-in source of power, but do contain active elements such as transistors or tunnel diodes. In these, power is induced into the circuit at one frequency or time, whereupon it is converted into a steady voltage to power an oscillator for reradiation on a different frequency or at a different time. The distinction becomes somewhat further confused with those transmitters which contain a storage battery, which is periodically reenergized by an oscillator and coil placed outside of the body of the subject. And then there are a number of methods that employ radar sets, and which obviously fall in this category. Ordinary transmitters, which are switched on by an impulse from outside of the body, or various active transponders, which reply to an incoming impulse, are not of this circuit type and are taken up in Chapter 17.

In the early 1950's, when serious consideration was given to the possibility of placing physiological monitoring transmitters within an animal, the junction transistor was not yet available. Some circuits were built up using the original point-contact transistors, but they were too difficult to power in the proposed applications. Thus passive methods were conceived in which there would be neither active elements nor internal power sources. One observation of interest was the monitoring of pressure from within the human bladder during micturition, and thus a tuned circuit consisting of a capacitor and coil was studied in which pressure changes would change the resonant frequency through motion of a core of powdered iron.

To measure the momentary frequency of the tuned circuit without actually contacting it, it was proposed to use what is called a "grid-dip" meter. These circuits have long been known to electrical engineers, and are traditionally used for measuring resonant frequencies of circuits. In this type of device, a coil is driven by an oscillator whose frequency can be adjusted. The coil is placed near the circuit under test so that some energy is coupled into the remote circuit by transformer action. The frequency of the oscillator is slowly changed, all the while observing a meter which indicates the current being drawn by the grid circuit of the oscillator tube. When the frequency to which the remote circuit is tuned is reached, the current in that circuit will rise, thus increasing the losses in the resistance of that circuit. This extra power must be supplied by the oscillator, thus leading to a shift in operating conditions which is accompanied by a dip in grid current. At this point the frequency of the oscillator is noted, since it is the same as that of the remote circuit. It was expected, by cyclically scanning the oscillator frequency up and down, that such a method could follow changes in the tuning of a remote circuit and thus provide a system of passive telemetry.

With this objective in mind, a number of circuits were designed, some incorporating the rather little-known properties of capacitors of barium titanate which can change their capacitance with changes in applied voltage. These circuits were constructed by Bob V. Markevitch as an undergraduate research project and later studied by Franklin Battat. Some figures from the 1954 undergraduate thesis of Markevitch are combined in Fig. 13.1. Figure 13.1*a* shows a thyratron circuit causing a cyclic change of voltage to be applied to a voltage-sensitive capacitor, which then causes the frequency of a radio-frequency oscillator to change. The grid-current indication of the oscillator is applied to the vertical deflection plates of an oscilloscope, to whose horizontal deflection plates is applied the voltage that controls the frequency of oscillation, thus giving an automatic plot of grid current as a function of instantaneous frequency. In the vicinity of the oscillator coil is placed the resonant circuit which serves as the transmitter, a movable core providing frequency modulation. Figure 13.1*b* shows a somewhat more sophisticated arrangement of nonlinear capacitors which was successful when tested. A number of these barium titanate capacitors (six in the case shown) are placed in series as far as the oscillator tuned circuit is concerned. Thus no one capacitor has very much radio-frequency voltage applied, and each looks rather linear (constant capacitance) to the oscillator; this provides a desirable sinusoidal oscillation. If, however, a single high-voltage capacitor were used in a low-voltage oscillator to accomplish this result, then extremely high voltage swings would be required of the slowly changing voltage that causes the sweep in frequency. In the arrangement shown the several capacitors are effectively connected in parallel with regard

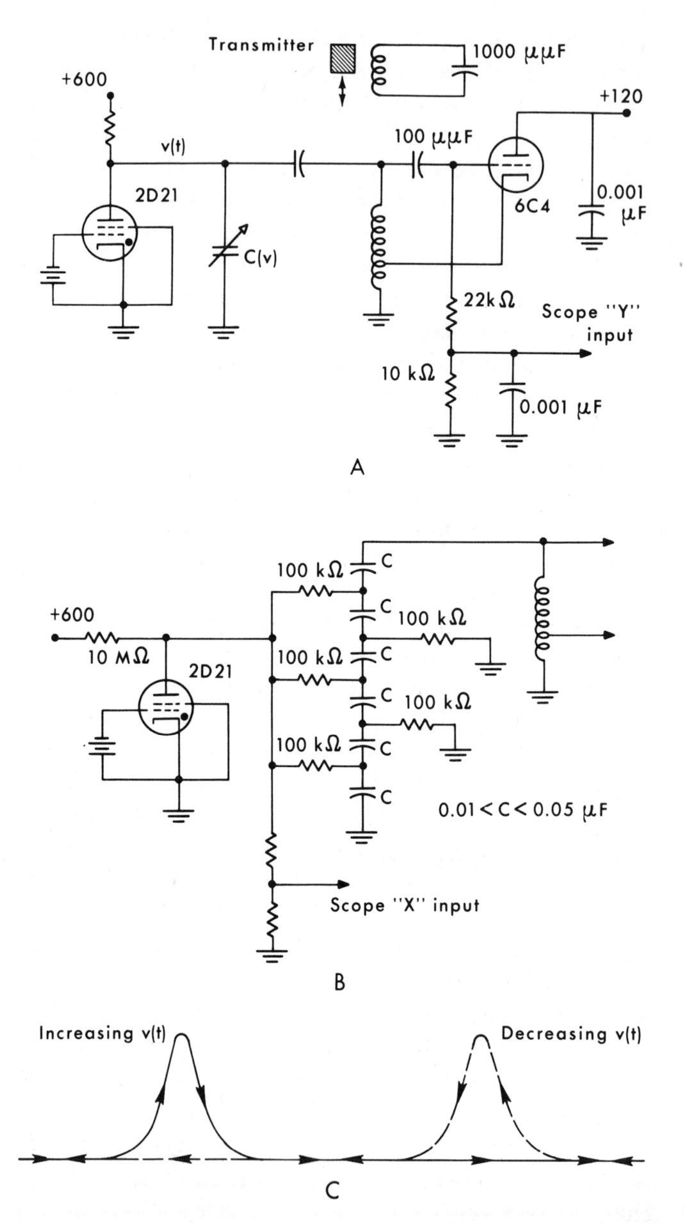

Fig. 13.1 An early passive system in which the transmitter was a resonant circuit alone. The momentary frequency of tuning was sensed by a cyclically scanning grid-dip meter. The scan in frequency of the meter was produced by a barium titanate voltage-sensitive capacitor or a combination of several in series. Motion of a peak on an oscilloscope trace indicated momentary transmitter-core position.

to the control voltage, and thus are swept with a relatively low voltage. The control signals are carried in through the resistors shown, and they are sufficiently high to prevent a gross reduction in Q of the tuned circuit. Figure 13.1*c* shows a representative trace from the face of the oscilloscope. As voltage increases across the sweeping capacitor there appears a peak somewhere along the baseline; the position of this peak shifts from side to side with modulation of the remote transmitter due to core motion. Decreasing voltage yields a somewhat displaced peak, which also moves in response to modulation. The presence of the two separate peaks is due not only to a difference in time delay through the vertical and horizontal oscilloscope amplifiers, but also to certain fundamental properties of highly nonlinear resonant circuits. The simplest use of this display was made by simply blanking out the return trace. Recordings from such a display are readily made by the method of Fig. 11.11.

These circuits were tested and functioned satisfactorily. However, as will be seen, the signal from a passive transmitter falls off extremely rapidly with distance. Thus the signal tended to be weak, and to disappear if the transmitter reoriented itself or made a sudden excursion away from the external coil. A number of tests on the transmitter were made from within the mouth and it seemed likely that continuous recordings could not be obtained from the bladder even with several differently oriented receiving coils placed near the abdomen. Thus, although the method was demonstrated, this application was postponed until the development of small active transmitters. As we shall see, for certain applications this type of system remains the most convenient of all.

In connection with grid-dip meter systems it should be noted that the cyclic scan frequency can be brought about by mechanically moving capacitors (e.g., a motor-driven tuning capacitor from a radio receiver) rather than by an electrically swept voltage variable capacitor. By the use of suitable feedback arrangements it is also possible to make a grid-dip meter circuit track changes in a modulated resonant circuit. Transistorized equivalents of a grid-dip circuit can be constructed. Thus the tuned circuit of a Colpitts oscillator can be placed near the unknown resonant combination, and the oscillator tuned. At the resonant frequency of the remote circuit some energy will be taken from the oscillator tuned circuit, across which an emitter follower is connected. A meter connected across the emitter resistor dips at the proper frequency. Such circuits, however, have little advantage and generally seem slightly less reliable than a grid-dip meter using a single vacuum tube.

In 1957 Magondeaux was awarded a patent on a signalling system for industrial uses. An impulse entering an antenna would set a quartz crystal into vibration, and the energy thus stored would slightly later be reradiated from the same antenna if a switch had not been opened in series with the antenna or closed in parallel with the crystal. Thus discontinuous or binary information could be transmitted. In 1956 Marchal and Marchal extended this idea and proposed its application to biological problems in a French patent which was granted in 1959 (Marchal and Marchal, 1959). One of their suggestions (Marchal and Marchal, 1958) was to study the process of digestion by placing between the springy contacts of a switch a suitable solid (a piece of meat) which would then dissolve and allow switch closure. Again the time of this event was to be determined from a passive transmitter consisting of a vibrating quartz crystal across which would be connected a coil, the return of an oscillation following an inwardly directed impulse indicating whether the switch had closed yet or not.

It is interesting to note that if such a vibrating crystal were mounted with a support at its center in a suitable fashion this transmitter type could also be activated by sound impulses, and could return either sound impulses or radio impulses, with incoming radio pulses and outgoing sound pulses also being a possibility. The support would have to be arranged so as not to seriously damp the crystal and yet allow it to be set into vibration by inertial or other forces at the resonant frequency. Such a system might also be capable of transmitting continuous variables such as temperature or pressure which are known to be able to modify the vibration frequency of some crystals. In such an arrangement, the crystal will generally show less than its normal high Q (will vibrate for fewer cycles) because of losses from its associated coil to the adjacent tissues.

Haynes and Witchey (1960) described a passive transmitter in which changes in pressure would move a core to retune a resonant circuit. In this case an impulse from outside of the body would set the resonant circuit to "ringing," and the frequency of the reradiated signal would then be observed as a measure of pressure. Farrar et al. (1960) used this system to record pressures from within the human gastrointestinal tract. In such methods a somewhat lossy receiving antenna must be used which will not superimpose its own characteristic frequency upon the returning signal.

The possibility of distinguishing individuals in a group of animals by their carrying differently tuned resonant circuits is mentioned in the next chapter. If this is done, it can be convenient to time a fixed number of cycles rather than attempting to count cycles in a given time for a frequency measurement. The number of cycles readily available from the different circuits will be about the same, being approximately the Q of the circuit.

There are other methods of transmission using purely passive elements. For example, Fig. 13.2 depicts two different ways in which a pair of coils can be placed so that energy introduced into one will not couple into the second. In Fig. 13.2*a* the two coils are perpendicular, whereas in Fig. 13.2*b*

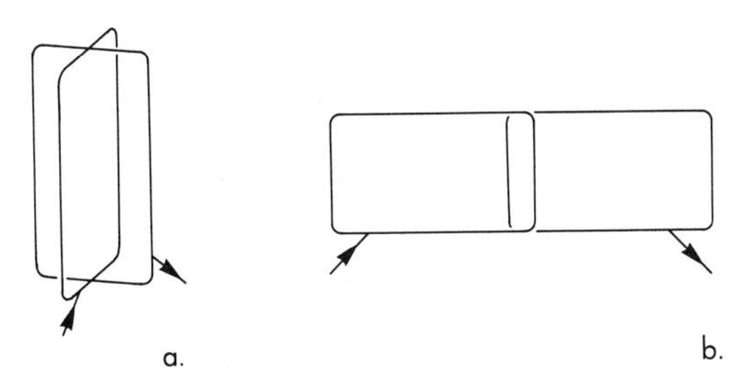

Fig. 13.2 Pairs of coils displaying little direct coupling. At *a* the coils are perpendicular; at *b* they overlap and are approximately in the same plane.

the pair of coils lie essentially in the same plane and overlap by a carefully adjusted amount. Intermediate relative orientations between the conditions of *a* and *b* are possible. If a tuned circuit is placed in the vicinity of such a pair, then it will be able to couple energy to the output coil. Frequency-modulated telemetry is again provided by sweeping the input frequency and noting at what frequency a maximum output signal is fed through. Action is effective even when the transmitter is about a coil diameter outside the decoupled coils. In these systems there are directional problems since orientation changes that cause a loss of coupling with either of the coils of a pair causes loss of signal. Such methods were studied by Arthur Chen in his master's degree research at Berkeley. Chen was one of the several students who considered the problem of the continuous monitoring of pressure within the eye, in connection with the glaucoma problem. From his considerations it was clear that it would be possible to build a passive transmitter that would work from within the eye of an experimental animal, but that the receiving antenna would have to be close to the head of the animal in order to achieve a sufficient signal.

A modification of the above method holds out several useful potentialities. The phase of the received signal, rather than the frequency, can be fed back to control the frequency of the driving oscillator, thus assuring that the driving frequency will always be that to which the passive circuit is momentarily tuned. This action has been demonstrated by placing the coil pair in the input and output of a standard amplifier circuit, and the effect is to convert the entire system into an oscillator controlled by the momentary frequency of the tiny passive element. This not only ensures good driving efficiency but radiates a strong signal in all directions. Thus this system also serves to carry information to a remote data-collection point; that is, it functions as a "booster transmitter" such as will be described in Chapter 17.

All these transmitters have a somewhat limited range, beyond which their signal can become weak and unreliable with small changes in orientation.

Thus, although for some operations they are irreplaceable, in other applications they are inappropriate. The general reason for the somewhat limited range is readily seen. A signal from an energizing loop, acting as a dipole, can fall off with the cube of the distance. The remote resonant circuit (passive transmitter) takes some of this energy and reradiates it in all directions. That part returned to the original loop falls off with the cube of the distance to the passive circuit. Thus the net effect is that the signal falls off with the sixth power of the separation between the two coils. If the separation between coils is not large with respect to coil sizes, then the dipole approximation is not valid and more detailed transformer theory must be used. In that case, it is often true that the fall-off in signal strength is approximately exponential, which can be even more rapid. One can arrange the energizing coil to produce a uniform field over a large region (e.g., by the use of Helmholtz coils), and then the variation need not be as drastic.

In spite of these limitations on range, there are some applications for which passive transmitters are extremely appropriate. For a given range the antenna is generally slightly larger than if a battery were also present, but it is true that at present, the smallest and lightest transmitters are the passive ones. They can perhaps be made small enough and light enough to allow the monitoring of the activities of honey bees within a hive, for example. Similarly, it has been possible to measure temperature changes with small ceramic capacitors which were temperature sensitive, and whose leads were connected into a small springy loop; absolute stability was not high, but temporary flattening of the loop did allow injection of the transmitter through a medium hypodermic needle.

Another application can be envisioned in which the overall package must be as inert as possible. Thus we might wish to monitor certain aspects of the long-term recovery of a hip operation or tissue transplant rejection, or tumor temperature using an implanted transmitter. Perhaps a more immediate application would be to monitor the pressure within the head. This would be relevant in cases of hydrocephalus and also after brain surgery when a rising intracranial pressure might give an early warning of impending danger and possibly death. The control of artificial limbs by electromyographic potentials is being studied by workers in several parts of the world. Perhaps the sensing of the control voltages can best be done by an implant. At this time probably only a passive transmitter could be designed which could be left in place permanently for the life of the individual.

A tiny passive voltage-sensing transmitter was constructed by connecting two diodes in series opposition (not a transistor) across a coil, and applying the monitored voltage from the common diode point to the coil. However, the signal was also sensitive to energizing voltage strength, as is necessarily true when nonlinear elements are present. It seems simpler to stabilize the energizing signal at the passive unit after conversion to dc, and thus some form of Fig. 13.12 is to be preferred when there is possible relative motion between transmitter and receiver.

If the signal strength can be made adequate in a particular experiment, then the overall reliability of one of these passive transmitters may actually be higher than with an active transmitter. This is because small batteries can fail unexpectedly, and in an active transmitter they can be inaccessible to change. With a passive system the energizing components are out in the open, and larger, more reliable batteries can be employed to start with. It should be pointed out, however, that passive transmission does not necessarily give a longer experiment in all cases. For some variables we have had continuous transmission for over a year, and for many experiments, this is long enough. After that length of time it takes extreme care to be certain that changing interactions between tissues and sensors are not rendering the data somewhat inexact, in any case.

A good example of the effectiveness of these methods is the recent studies by Carter Collins. In connection with his doctoral research on pressure in the eye of the rabbit, he constructed passive transmitters, some being as small as 2 mm diameter by 1 mm thick. These were implanted in the eyes of animals for continuous pressure monitoring. A photograph of the experiment is given in Fig. 1.14, and he has found that all sensory modalities can induce a transient rise in pressure within the eye of the rabbit. The following four figures (Figs. 13.3 to 13.6) are all from Collins' Ph.D. thesis. A transmitter is depicted in Fig. 13.3, where a tuned circuit is seen enclosed in a small plastic box. The tuned circuit itself is formed from two pancake spirals with their planes parallel to each other and closely spaced. Even if totally disconnected, this type of system shows a characteristic resonant frequency

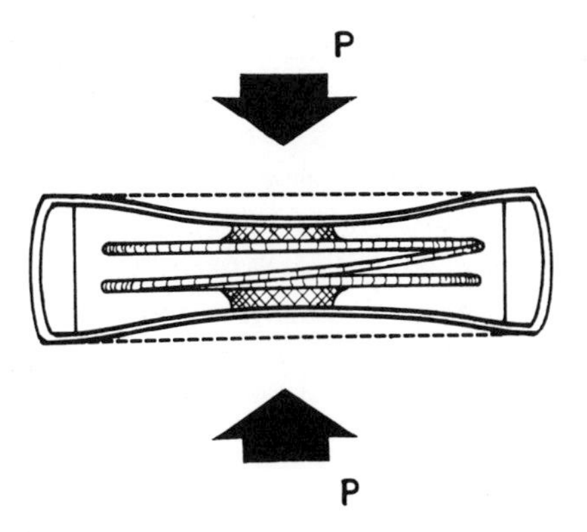

Fig. 13.3 A passive pressure transmitter consisting of a pair of pancake coils connected only at their outer edge and enclosed in a plastic bubble. The resonant frequency of this distributed system decreases with increasing pressure. (Figure by C. Collins.)

because of the mutual inductance of the coils and the capacitance between them. It is customary to connect the two outer ends to give a lower frequency, while leaving the inner ends totally disconnected. We can envision a number of other resonant combinations of loops and plates, but this geometry does give considerable ruggedness with a relatively low frequency in a small space. The frequency of operation is generally in the range of 50 to 100 MHz. The plastic (a laminate of polyethylene and Mylar, called Scotchpak) is permeable to body fluids, and such a unit does drift for about 100 hr after placement. After that further drifting is quite slow. The absolute pressure within the animal's eye can be measured occasionally with a Mackay-Marg tonometer for calibration purposes. The incremental sensitivity of the transmitter is monitored at any time by placing the animal in a box and slightly raising the pressure, which change must be faithfully reproduced by the transmitter within the eye of the animal (as mentioned in Chapter 5). We constructed a similar unit by gluing sections of microscope cover slips to the ends of a short length of glass tubing with wax and epoxy to explore the possibilities of reducing permeability, but found that adequate sensitivity was achieved only if the diameter was about 1 cm. Collins has now made all glass units in which the diaphragms are sealed into place by melting a lower temperature glass frit having a suitable thermal coefficient of expansion, and they have some information on the acceptance of such units by the brain (Olsen et al., 1968).

Figure 13.4 shows the circuit arrangement Collins used in monitoring and recording the changes in resonant frequency. Once again, it is the sweeping grid-dip meter circuit, this time scanned in frequency by a war-surplus vibrating mechanical capacitor. The characteristic response curve is electrically differentiated to give a curve that passes through zero at the point of interest, rather than trying to record the peak itself. Actually, at different ranges slightly different curves result, but they all pass near a particular point, and thus motion of that point is recorded to get frequency indications independent of range, as in Fig. 13.5. In Fig. 1.14 is a photograph of an actual experiment in which two coils were monitoring two transmitters in the two eyes of an alert and seeing rabbit. In that case, the rabbit was restrained by having his head project from a box. Above the ears of the rabbit can be seen the vibrating capacitor. In the sample trace of Fig. 13.6 can be seen the pressure fluctuations in the eye associated with beating of the heart, with breathing, and with the response to a stimulus. In this case the stimulus was a sudden sound.

Operation of a grid-dip meter circuit at these high frequencies is somewhat different than at lower frequencies, since the baseline is not generally perfectly flat. This is usually because of minor partial resonances at various frequencies within the circuit itself. It should be mentioned that such a

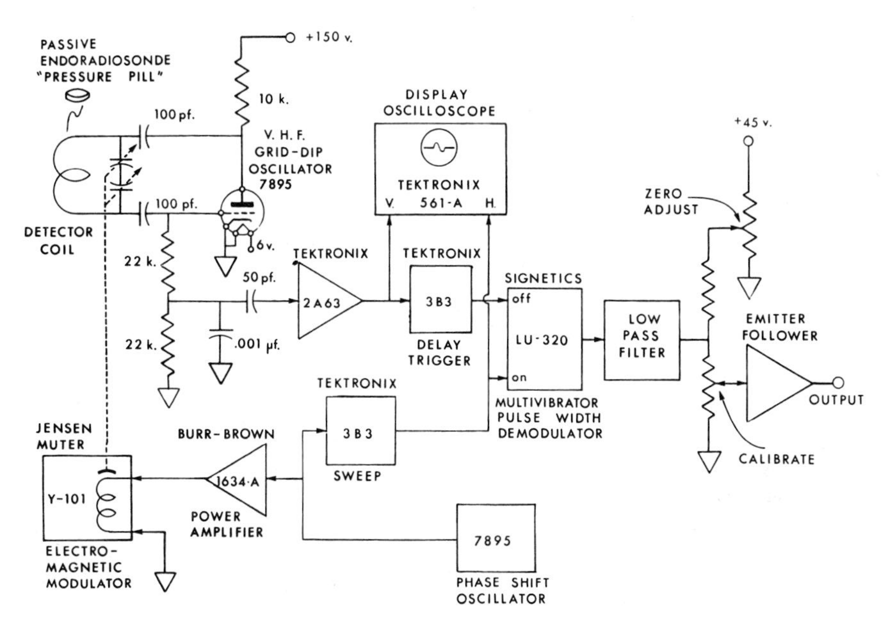

Fig. 13.4 Components of grid-dip meter circuit for use with the transmitter in Fig. 13.3. The circuit is scanned in frequency by the action of a vibrating mechanical capacitor. (Figure by C. Collins.)

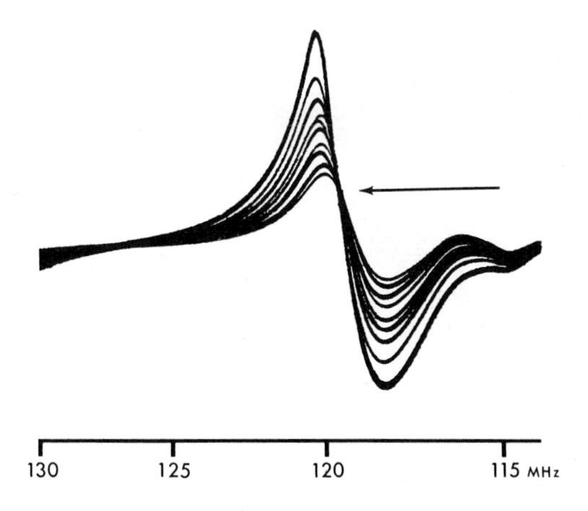

Fig. 13.5 The response peak from the grid-dip-meter circuit is differentiated electrically to give the curve shown. Changes in distance affect the appearance, but all curves pass through one point whose motion is recorded to give frequency indications independent of range. (Figure by C. Collins.)

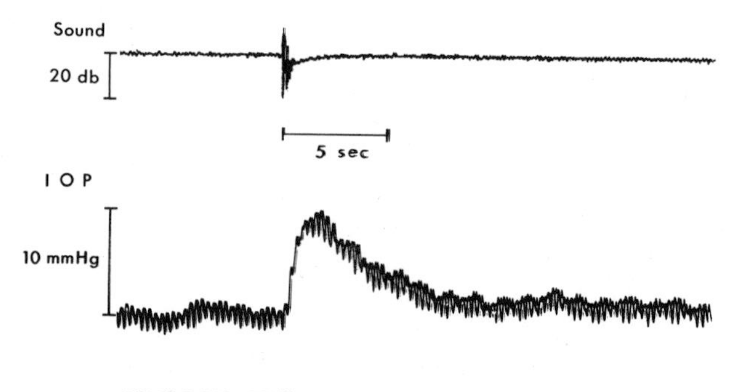

Fig. 13.6 Sample tracing of instantaneous intraocular pressure of a rabbit, with transient increase in response to a sound. The small pressure fluctuations are associated with the beating of the heart, whereas the slower alternations are associated with breathing. (Figure by C. Collins.)

method is much more appropriate at high frequencies than is a system in which a pulse is given out and a return ringing frequency monitored slightly later. A tuned circuit surrounded by a capsule and placed within a mass of tissue may show an electrical Q of perhaps 20. This means that the circuit, once excited, will continue to oscillate for approximately 20 cycles. At a frequency of 100 MHz, the entire observation would then have to be completed in approximately 0.2 μsec. Times would have to be resolved to a small fraction of this interval for accuracy, and thus the measuring process

Fig. 13.7 Henry Cabot Lodge at the United Nations pointing to the position of a passive sound transmitter secreted in a wall hanging.

can become difficult. Either the crossed-coil method of Fig. 13.2 or the sweeping grid-dip meter methods are still quite effective, however. At these higher frequencies, some of the voltage-sensitive capacitors display significant electrical losses, thus rendering sharpness of tuning less precise. It is for this reason that a mechanically modulated capacitor is often to be preferred. For greatest precision, a digital counter circuit can then be used to indicate the precise oscillator frequency at the significant time in the overall frequency-modulation cycle.

A larger passive transmitter is capable of giving greater range. In 1960, H. C. Lodge reported to the United Nations that a passive sound transmitter had been planted in a wall-hanging at one of the American Embassies abroad (Fig. 13.7). It is instructive to consider how this transmitter was constructed. This is shown in Fig. 13.8, where the unit is seen to be a

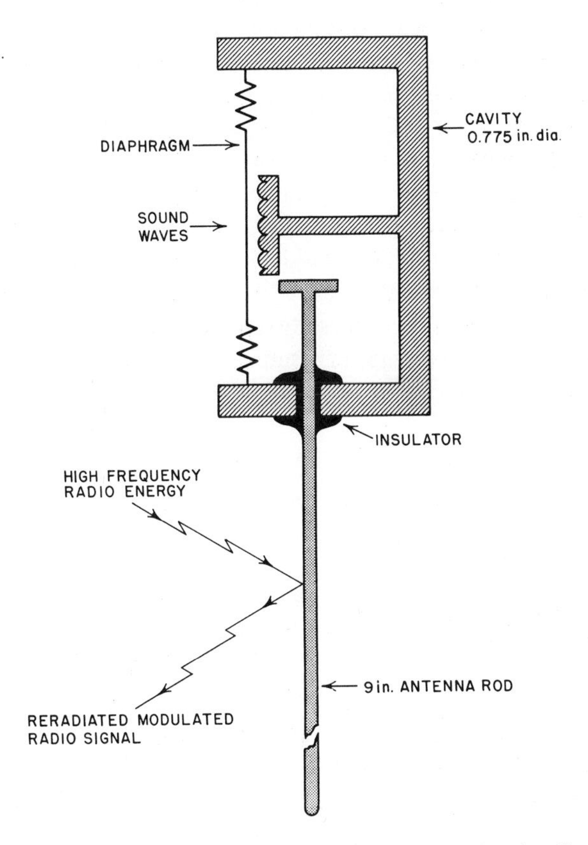

Fig. 13.8 Construction of a passive transmitter of sounds. When irradiated by a radio field, the amplitude of the reradiated signal depends on diaphragm position.

resonant cavity which is frequency modulated by diaphragm motion, and which is coupled to a small projecting antenna. The sensitivity of such a transducer (that is, change in resonant frequency with change in cavity size) is readily calculated, and such a system is useful for the transmission of sounds. If such a unit is placed in a steady radio field, then periodic detuning by diaphragm motion will affect the amount of the reradiated signal, resulting in amplitude modulation in the returned signal. If the incident radio frequency were scanned for maximum response, then it would be possible to instead infer the actual position of the diaphragm at any instant independently of signal amplitude changes, that is, it would be possible to transmit such things as absolute pressure rather than merely frequency of pressure fluctuation.

The previous discussion has involved transmitters incorporating only linear elements. If nonlinear elements are permitted, then other possibilities arise. Thus radiation by two different frequencies from outside can be accompanied by continuous transmission on their difference frequency. Either a highly nonlinear system or a system in which not all parts are fixed with respect to each other can sustain major circulating currents at a frequency different from the energizing frequency. This opens up the possibility of the continuous induction of energy in at one frequency and its simultaneous reradiation at a different frequency. A tunnel diode connected across a coil also provides several possibilities. If this combination is placed in a magnetic field that suddenly changes, the diode will switch to a high-current state, and then later abruptly return to a lower-current state; the temporal separation of these two pulses is a measure of inductance.

Another possibility for the use of a tunnel diode is indicated in Fig. 13.9. In this case a parallel-resonant circuit is connected across a tunnel diode

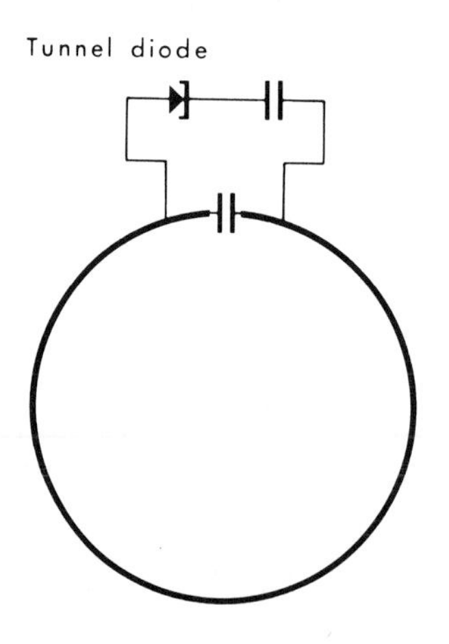

Fig. 13.9 Radio energy into the tuned circuit is rectified by the diode to charge the storage capacitor at the upper right. When the external source is turned off, the capacitor serves as a power source to set up oscillations in the resonant circuit. The signal can be reradiated at a slightly different frequency.

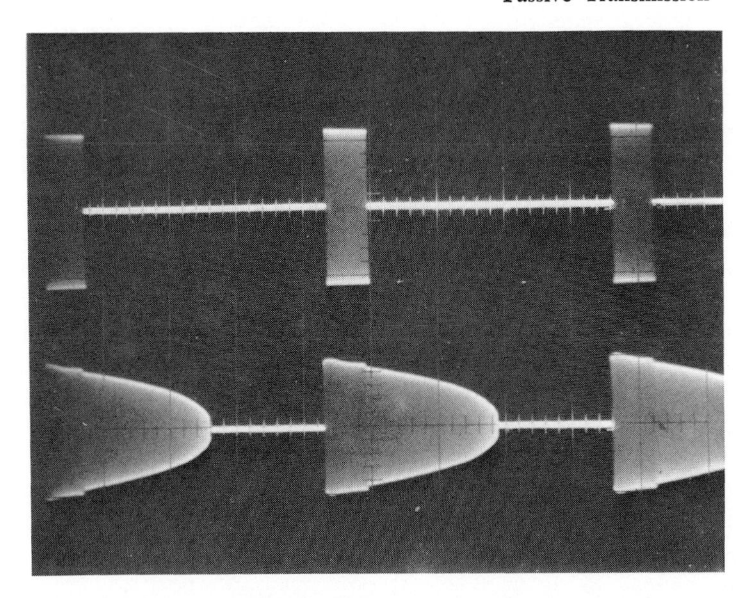

Fig. 13.10 The envelope of the applied wave form above and the overall response below. Without the tunnel diode, the response would effectively terminate immediately.

connected in series with a capacitor. This circuit was placed in a pulsed radio-frequency field having approximately the frequency of the tuned circuit. The induced voltage was rectified by the tunnel diode, now acting as an ordinary diode, to charge the other capacitor. With the end of the incoming radio-frequency pulse, the charge storage capacitor now acted as a power source for the negative resistance of the tunnel diode to produce oscillations in the resonant circuit, at the frequency to which the resonant circuit was tuned. Notice that the returned pulse does not involve any more voltage than simple ringing by a resonant circuit, but the duration of the signal is much longer. In a particular case with a small six-turn coil resonated with 20 pF, a 1N3712 diode, and 4.7 μF capacitor, the return pulse lasted a millisecond and so could be heard on a standard receiver. The incoming energy was at a frequency of 92 MHz and that returned was shifted in frequency to 110 MHz, so that the two signals could tune separately on the receiver. Representative wave forms are shown in Fig. 13.10. On the top line is the energizing pulse alone. The complete wave form is seen below, there being no noticeable after-signal if the tunnel diode is unsoldered. The trace appears very similar in relative proportions if the storage capacitor is reduced to 0.1 μF, the 1-msec interval being reduced to 20 μsec. Under one set of conditions of coupling between transmitting and receiving coils, the following conditions existed in the energized coil. When 6V peak-to-peak were applied, 0.2 V was observed during the return interval. There was some frequency variation during the return, but the tuning on a radio was rather sharp. Under these conditions, the storage capacitor had charged to 0.16 V. If the coupling was increased, then the storage capacitor charged to a considerably higher voltage. Following the initial signal, there was then a delay while the storage voltage gradually dropped until the diode was biased into the negative resistance condition. Only then do oscillations start, and the corresponding signal appears as in Fig. 13.11. The maximum voltage across the storage capacitor, and the delay with close antenna spacing before the return signal is generated, can be controlled by placing any silicon diode in parallel with the storage capacitor, thus regulating or limiting its initial voltage.

It was felt that such a circuit as this would be capable of transmitting physiological information by any of several means of modulation. The delay in the start of oscillation could be used to transmit an indication of an applied voltage or associated resistance, whereas frequency modulation could transmit diaphragm position. Notice that the returned frequency was

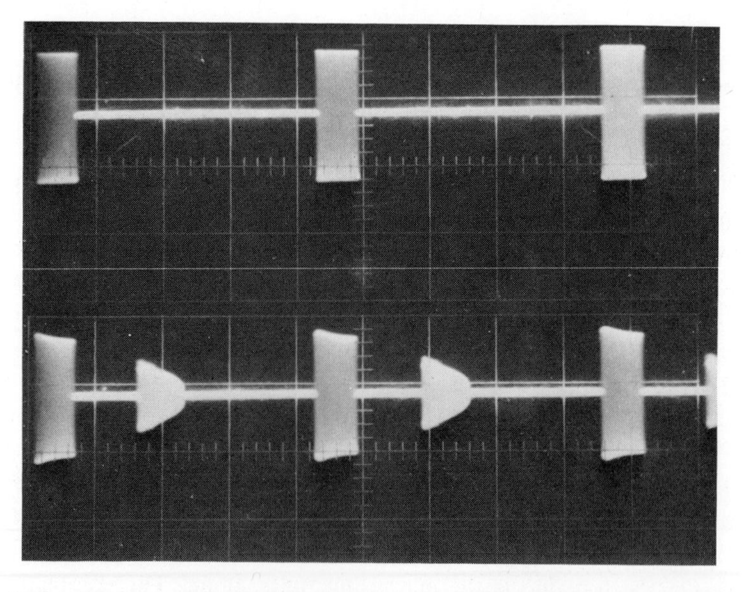

Fig. 13.11 If the capacitor is charged to an excess voltage, there is an interval before oscillations resume while discharge down to the negative-resistance voltage takes place. The magnitude of the delay can be used to telemeter resistance or other information or the excess voltage can be blocked to give immediate response.

purposely made slightly different from the incoming activation frequency. It was thought that such units as these might be employed in connection with radar sets. A radar set is a highly engineered piece of equipment that is mass produced to some extent. As mentioned (Chapter 10), although the frequency is very high, the near field extends for several hundred feet. Within this range the beam is well collimated, and the energy density is high. Thus this might be a convenient activator for suitable passive transmitters. One thing that confuses the displayed pattern on a radar screen is the clutter of reflections from various other objects in the scene. Thus a reflector that does not merely return the same frequency, but actually shifts it somewhat can be made extremely noticeable by detuning the radar receiver slightly off of its own transmitter frequency. The sweep is delayed until the end of the outgoing radar pulse. Preliminary experiments indicate considerable increase in range by this technique.

In ordinary use radar sets have some applicability. It is said that the Wallops Island 10 cm wavelength radar can track a honeybee to a distance of 10 km (Glover et al., 1966), though following an individual among a mass of flowers and other vegetation could become extremely difficult. With birds, both local activity and major migrations can be followed if a time exposure is used for identification of the tracks (Flock, 1968), with large birds being detected at ranges as great as about 100 miles.

In range determination by the usual methods, it is necessary that the returned signal be generated with no delay, unlike the devices mentioned in the previous paragraphs. This result can be accomplished by associating a diode mixer continuously driven by a suitable small oscillator with the remote antenna if a frequency shift is required. If not, then various types of reflectors can be employed. A resonant length of wire will reflect appreciable amounts of energy, and if a circularly polarized beam is employed it will do so for a variety of wire orientations. However, the signal from a passing seagull can be almost as great. There are several kinds of reflector which return a beam on themselves with rather good intensity. Thus three perpendicular surfaces, meeting like the walls in a corner of a room, have this property. The weight can be low but size must be fairly large so that each surface can be several wavelengths across. Similarly, a lens in front of a reflector has this retrodirective collimating property. In both cases the perfection of returned reflection can be periodically spoiled in order to communicate information. Thus, if a corner reflector is made with a flexible side upon which vibrations impinge, then the vibrations of this reflector will cause amplitude modulation in the returned beam, from which the frequency of the vibrator can be inferred.

There are a number of transmitter types which are passive in that they do not contain a battery, and yet they do contain an entire active trans-

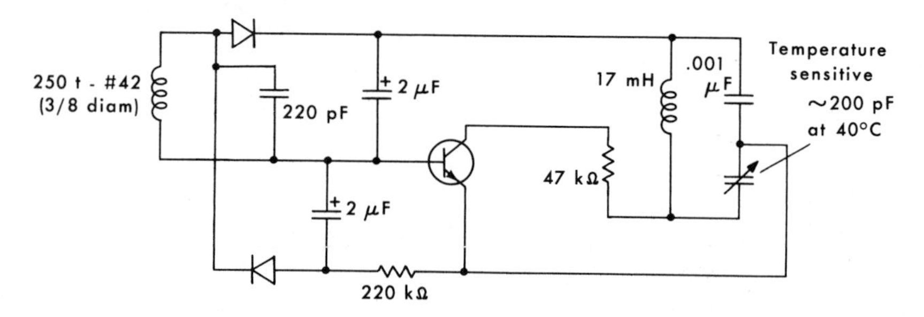

Fig. 13.12 Power induced into the voltage doubler circuit at 1 MHz supplies the oscillator to generate a frequency-modulated signal at 100 kHz to indicate temperature changes. Any transistor type is suitable, but precautions must be taken to prevent motion of the energizing source relative to the transmitter from causing voltage and frequency changes.

mitter of standard construction. An example is shown in Fig. 13.12, in which an ordinary oscillator is modulated in frequency by changes in temperature which cause changes in capacitance of the tuning capacitor. This oscillator can use almost any transistor type, with the signal being radiated at a frequency of approximately 100 kHz. Power is supplied to this oscillator by radio energy induced from outside into the coil on the left at a frequency of approximately 1.0 MHz. This coil is resonated by the 220 pF capacitor, and the resulting alternating voltage rectified by what is sometimes called a "full-wave voltage doubler." Positive half cycles charge the upper filter capacitor, while negative half cycles charge the lower filter capacitor; these two are discharged into the oscillator in series, thus providing a total voltage equal to twice the peak value of the applied alternating voltage. The frequency of the oscillator is somewhat responsive to changes in applied voltage. Even if Helmholtz coils or other arrangements are provided to give a uniform energizing field, changes in orientation can still produce problems. Thus the circuit is suitable if implanted at a particular position so that placement of an external coil at a particular body spot would always result in a known energizing field. The alternative is to actually regulate the supply voltage with a string of diodes in the way indicated in Fig. 2.2, although this somewhat reduces the overall available voltage. One transistor can both stabilize and rectify, and a voltage doubler is not essential. If such a circuit is used instead to transmit a pattern of voltage fluctuation (as opposed to absolute voltage levels), then this regulation problem is relieved somewhat. This can be done by replacing the temperature-sensitive capacitor with a voltage-sensitive diode pair connected in series opposition. The leads from the electrodes are connected across the lower of these varicap diodes, and a small inductance can be placed in series with one lead in order to keep the biological signal source from partially "shorting" the radio-frequency currents.

Instead of inwardly induced power being returned on a different frequency, it can be returned at a different time. Thus a pulse can be injected periodically, and in the intervening interval the returned oscillations can be monitored. This can be done at an extremely rapid rate if the energy-storage device is a capacitor, such as in the "echo capsule" mentioned in Chapter 3. If the energy-storage device is a small storage battery, then an extended period of inwardly induced power may be followed by hours or days of outward transmission. With some circuits it is possible to recharge a battery without interruption of the transmitted signal.

Some of these circuits that return a signal following receipt of an incoming signal are most usefully considered in another way. The subject of "transponders" will be taken up in Chapter 17. In longer range experiments, the length of time required for the receipt of a returning impulse following an outgoing one can be a direct measure of range to the subject. This aspect, plus battery-power conservation or elimination are two useful features of these circuits.

14

Field Work

Field work implies the study of relatively wild animals in their natural habitat. In some cases this means a study covering greater distances, but certainly not always. In many cases there is an implication of primitive surroundings without laboratory facilities for the care and repair of equipment. If the location is rather inaccessible, then a further premium is placed on reliability because of the difficulty of repeating a series of observations.

When there is a premium on lightness and portability, it will generally be true that battery operated transistorized receiving equipment will be employed. From a safety standpoint, such apparatus is also less likely to administer an electric shock. However, certain war-surplus receivers are rugged, inexpensive, and relatively small. Since these can be operated from the storage battery in a vehicle, they are sometimes applicable. In extended field work, it is well to remember that elevated temperatures cause a more rapid loss in the total energy contained in most battery types. Especially in desert work, direct sunlight can produce surprisingly high temperatures in various containers. It is also important to note that a cool evening does not compensate for a hot day with regard to deteriorating effects because of the generally nonlinear relationship involved. Many batteries will end up essentially dead before use on a long expedition if some precaution is not taken to maintain the average temperature with few fluctuations. In the case of the usual flashlight cells it is desirable that they be stored in a frozen condition, if possible.

Very low temperatures may not damage batteries, but will render them ineffective at the time. Even an external transmitter on an animal may go to extremely low temperatures, and in some cases receivers are required to operate at temperatures below freezing. It will often be found that nickel cadmium rechargeable cells are most effective in this case.

In many field studies it is quite difficult to prepare transmitters at the site, especially if the units are very small and compact. Thus in many cases it is desirable that the entire transmitter be constructed and closed in the laboratory before leaving. This leaves the problem of turning them on when it is desired to start an actual experiment. Obviously, the problem is simple with the relatively large externally carried transmitters, but with some of the smaller internal units this is more of a problem. For units that are not ultimately small it is convenient to incorporate a magnetic reed switch, as mentioned in Chapter 16, in series with the battery. A small permanent magnet beside the capsule effectively holds the switch open, and the entire capsule can be hermetically sealed; lifting the magnet away starts transmission.

There are other ways of switching on a transmitter which are perhaps simpler and more compact. A small circuit can be fabricated in its entirety with a pair of projecting wires constituting a break in one battery lead. This transmitter can then be cast into a solid block of plasic. A small disk of Teflon with a pair of holes just large enough to slip over the two wires is slid down them into contact with the plastic mass. The entire unit is then dipped in liquid silicone rubber and then removed. Since nothing sticks to Teflon, the disk can be slid away after the setting of the rubber is completed. This leaves a small indentation in one side of the transmitter, from which projects a pair of wires as seen in Fig. 14.1 at the left. The ends of the leads are tinned (coated with solder) for convenient later connection. To start the transmitter in the field, it is only necessary to twist the wires and solder them together. This is not quite as convenient as merely removing a magnet from the side of a package, but it can be done with an old nail heated in a campfire. The wires are then pushed down into the recess and some silicone adhesive rubber squeezed in to seal the entire package. This rubber sets by itself within approximately an hour after exposure to the moisture of the air, and is quite convenient. Such a transmitter in the operating condition is seen at the right of Fig. 14.1. This transmitter was recovered after a week in the gastrointestinal tract of a marine iguana on the Galapagos Islands. These transmitters are approximately 1 cm across, and are relatively indestructible against dropping or being bitten or walked on.

Transmitters of the above type were constructed by using a circuit rather like the one shown in Fig. 1.7. They were also fed to large tortoises on the Galapagos Islands in order to study the temperature regulation of these animals. A picture of such an experiment was given in Fig. 1.8, in which a 170-kg animal was being monitored. These animals were induced to eat

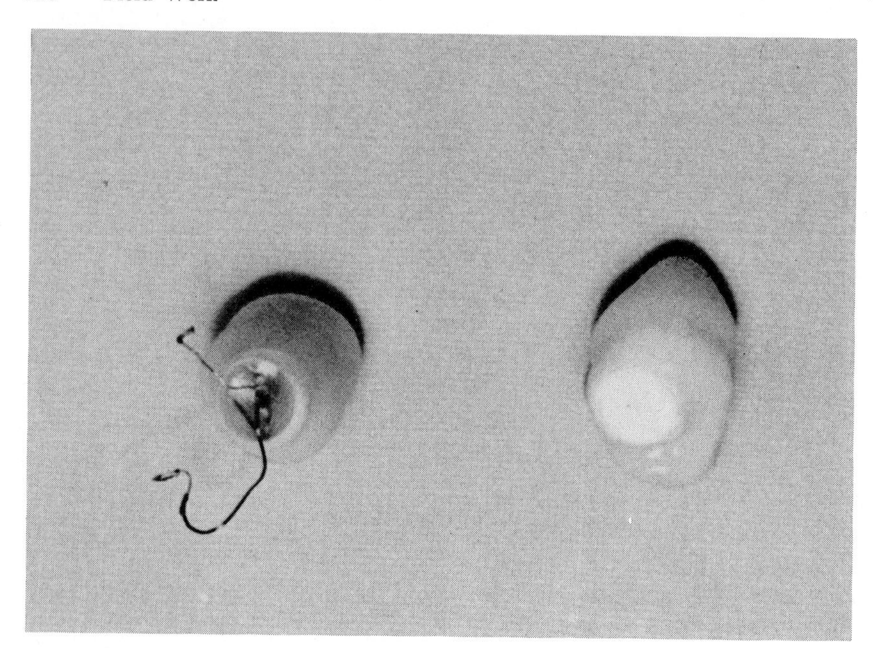

Fig. 14.1 Ingestible temperature transmitters of approximately 1 cm diameter, with projecting wires to be soldered together to start transmission. The unit on the right, closed off with a drop of self-hardening silicone rubber, was recovered from the gastrointestinal tract of a marine iguana on the Galapagos Islands.

a transmitter either by placing it in a banana, or by rolling them on their backs and tickling a foot, in which case a transmitter dropped into the upper jaw as they hissed was readily swallowed. In these experiments it was also important to measure the air and ground temperature in the vicinity of the animal, and thus they were periodically approached in order to make observations at short range; this did not seem to be at all disturbing to their normal pattern of activity. The data from such an experiment is given in Fig. 14.2, where it is seen that temperature deep in the body does not fluctuate much, and is generally higher than the ambient values. Thus the limbs did not serve as a radiator to rapidly bring body temperature to that of the ground. Shell temperature was seen to cycle about that of the body, with the body temperature having maxima delayed in time. The similarity of core temperature to that calculated below suggests that, in part, these large animals act as a single thermal time-constant system, with heat flowing in through a region of relatively high thermal resistance into a region of lower thermal resistance. It is tempting to associate these two regions with the shell, and with the rest of the body, the body perhaps being stirred to a uniform temperature by blood flow. In any case, this would seem to be a desirable biological arrangement in which a small increment of heat from the sun would cause the entire body to increase slightly in temperature uniformly, rather than starting a wave of heating moving inward (Mackay, 1964B). Similar observations were made on smaller tortoises, but fluctuation amplitude and mass were not compared, for the conditions the animals chose were not identical.

This has been an example of an experiment performed in the field under conditions where repair facilities were essentially nonexistent. Thus the extreme simplicity and reliability of click counting by ear with a stop watch was employed, in spite of the inconvenience of someone having to be present around the clock. At this isolated location there was very little interference from commercial radio stations, and thus there was little worry in advance over setting the endoradiosonde frequency. Also, because of the spread in transmitted frequencies from these blocking oscillators, the receiver could still pick up a signal if simply tuned away from any commercial station. This was occasionally a convenience at night when remote stations were heard. If this had proven to be a difficult situation, then the transmitter frequency could have been lowered by gluing a piece of ferrite to the outside of the transmitter, or raised by soldering a loop of wire (a shorted turn) around the transmitter and covering it with rubber. Before use, each transmitter was recalibrated with the help of a mercury thermometer taped in

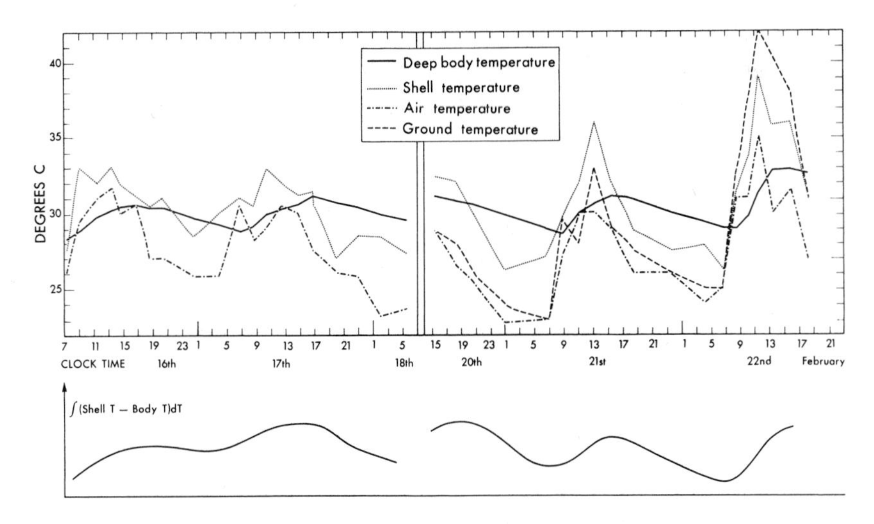

Fig. 14.2 Signals from within and around a freely moving 170-kg tortoise on the Galapagos Islands. The computed curve below suggests that, in part, heat flow within the animal takes place as a single time-constant system through a boundary region.

contact with the transmitter and rested in a can of salt water which was gradually warmed over a fire.

The above supplies an example of an experiment in which short-range transmission may not be convenient, but appeared acceptable. In other cases it may be difficult to place within an animal anything but a short-range transmitter, and a larger external transmitter may either be pulled off by the animal or interfere with his normal activities. It may also be true that periodic approach by the investigator will interfere with the normal pattern of activities. In that case, wires can be stretched over the ground in a pattern that will always leave an animal near a wire, with a resulting clear signal. For example, at the top of Fig. 14.3 is a row of adjacent single-strand "hairpin" loops. If these are each about a half meter wide, then an animal between wires and carrying the above transmitter type will always be successfully received, even if the loop has a length of hundreds of meters. Some irregularities are avoided if the length of a loop is less than a wavelength, and thus these systems are especially convenient with low-frequency transmitters. A small transmitting loop with plane vertical will induce an especially strong signal when right over a wire, while if the loop is in a horizontal plane there will be an especially strong signal to either side of the receiving wire and no signal when the transmitting coil is directly centered over the receiving wire. The signal may come from any of the loops within the area to be monitored and the investigator can either automatically record from all of the loops simultaneously, or he can cycle

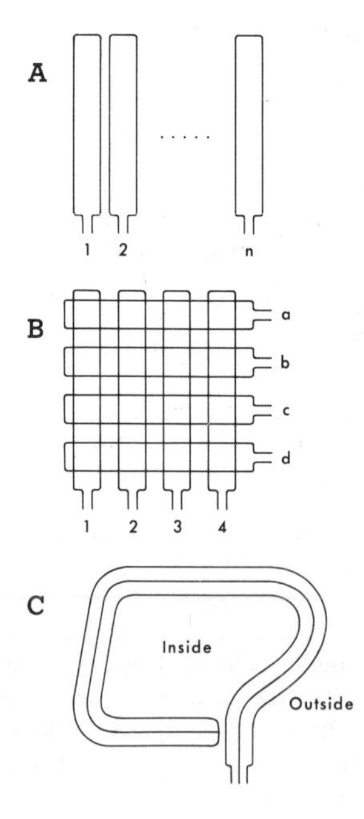

Fig. 14.3 *A:* A row of adjacent single-turn hairpin loops allows monitoring of the signal from a weak transmitter being carried by an animal near the surface of the ground. *B:* A matrix of crossed loops allows the localization of the position of the animal to a particular row and column. *C:* A long narrow loop allows the monitoring of the passage of a marked animal from inisde to outside a significant region. A double loop is used to indicate actual passage rather than mere approach and return.

from loop to loop by hand or automatically, or with strong signals he can combine the several signals directly through diodes into a single receiver.

Quite the same circuit arrangement that produced the tracing in Fig. 10.5 can be used to give continuous unattended recordings from a weak transmitter over a large area with such an array of antennas. Each antenna loop connects to a receiver through a separate diode (or relay) that is biased either on or off by a series of bistable multivibrators in a ring counter. Loss of signal in the receiver turns on a pulse generator to the multivibrators to cycle through the loops until a signal reappears in the output. The variable of interest is thus recorded "continuously" with an indication of frequency of activity shown by the drops to zero during switching.

With a set of crossed loops as in the center part of Fig. 14.3, it is possible to determine the instantaneous overall position of the animal by noting from which horizontal and which vertical loop the signal emanates. This can be especially convenient in dealing with small burrowing animals, whose movements can then be followed underground.

If position is not required, in some cases it is sufficient to loop a single wire back and forth. The exact pattern of sensitivity both to the transmitter and to unwanted disturbances then depends on the detailed layout. The symmetry and shielding of the receiver input transformer affect the ability of disturbance signals to be heard by causing a long loop of antenna wires to act together as a long antenna. It is difficult to achieve a sufficient balance to reject very strong signals such as lightning produces, even using a nearby comparison loop of the same area for further cancellation, and during storms signals can be lost.

In some cases the total activity pattern within a predefined region is not necessary, but merely the amount of time spent at points of special interest. Thus we may wish to know when an animal is in a feeding area, or a water area, or a nesting area, or not there. It is then sufficient to surround this area by a pick-up system so that we can say either that it is in the area of interest or not. If the area is relatively large, then a low-powered system can still be used by counting the number of crossings by the animal over the boundary of the area of interest. Thus, if he starts within the area of interest and has crossed the boundary an odd number of times, then he is certainly outside of the area. If a simple loop is placed around the area, then approach of an animal to the wire will cause the appearance of a signal, but he may not cross over. Thus a pair of concentric loops will indicate movement from inside a region to outside by the appearance of a signal at the first loop and then at the second loop, without a further appearance of a signal at the first loop indicating inward motion again. These loops of large area do pick up atmospheric disturbances such as lightning storms to an extreme degree, and this is minimized if a double wire, short-circuited at the end, is instead looped around the region of interest. This constitutes a very long thin hairpin loop in the sense of the previous discussion, which loop has been bent into a complicated shape. A transmitter right over this pair will induce a signal, but the reduced overall area minimizes irrelevant pick-up of other signals. (The input to the receiver must also be carefully balanced to prevent a monopole action.) A pair of these loops is depicted in use at the bottom of Fig. 14.3 in order to verify the completion of motion between the inside and outside of an interesting region.

In some cases the marking of the boundaries of an area can be done by using lengths of the double wire employed for carrying the signals from antennas to home television sets. If the investigator contemplates using kilometer lengths of such material, then it is important to pay special attention to the attenuation specifications and select some showing the lowest possible electrical losses in the plastic at the frequency of interest.

In principle, different animals could be coded by placing within them simple tuned circuits of different frequencies, or different combinations of magnets, conducting masses, and ferrite. However, such passive transmission of the crossing of animals near a double wire has not been as successful to date as the use of small active transmitters.

An intermediate-range transmitter that has proven rather reliable might next be mentioned. Figure 14.4 shows the circuit of a crystal-controlled

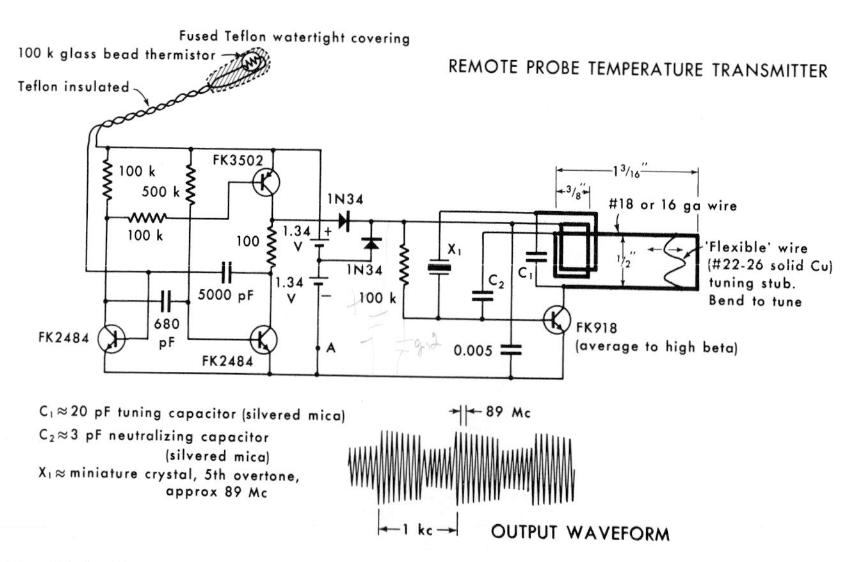

Fig. 14.4 Temperature transmitter with remote probe. The crystal-controlled output oscillator is cyclically switched between two amplitudes at a rate that carries the temperature information. The tuning stub is adjusted to give maximum output with a sloping wave form as shown, for spurious frequencies are generated if tuning is done for a square output wave form. The average current drawn is approximately 900 μA.

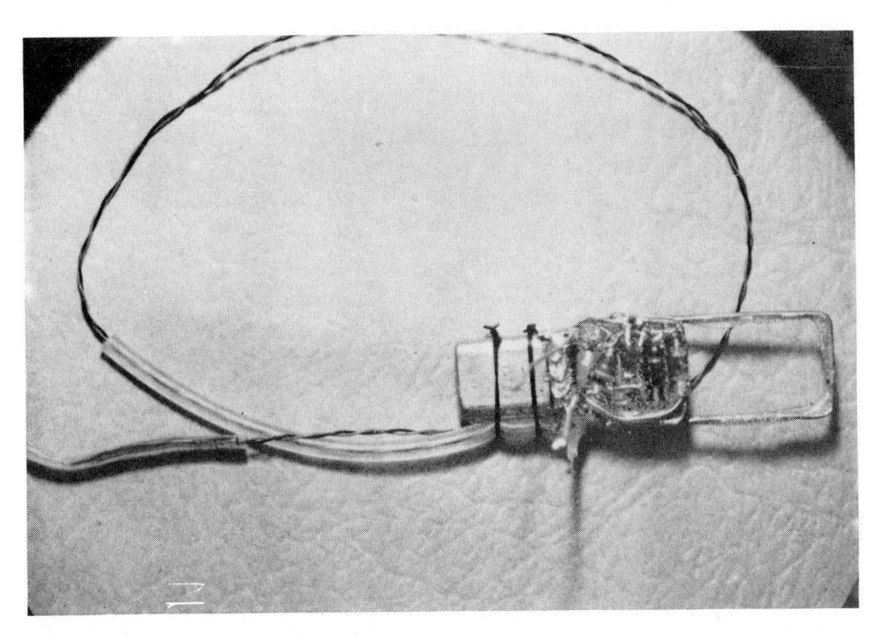

Fig. 14.5 The appearance of the circuit in Fig. 14.4 before being imbedded into a plastic block. To the left of the loop surrounding the components is the crystal can. The probe is at the far left. The signal is radiated from the loop in all horizontal directions if the plane of the loop is horizontal.

transmitter which puts out a signal from a small self-contained loop antenna in the frequency range near 100 MHz. Transmission is on an overtone of the basic frequency to which the crystal is ground, thus eliminating the need for output filters which would be required if a lower crystal frequency were used with frequency multipliers. This oscillator has an extra turn in the coil to provide for neutralization of the crystal capacitance (see Chapter 2). The oscillator is cyclically switched between two different amplitudes by the multivibrator at the left acting alternately to connect one or two batteries as a power source. The frequency of this subcarrier oscillator is controlled by the thermistor, which is enclosed in a sufficiently thick covering to prevent the interference by body fluids for a few weeks. If the circuit is adjusted so that the output wave-form envelope is quite rectangular, then spurious output frequencies can be generated. If the wave form is somewhat more slowly changing as shown, then action is generally more satisfactory. This circuit has been used for subcutaneous temperature monitoring, but other applications are obvious.

The actual construction of the circuit is shown in Fig. 14.5. All components except the battery and the can containing the crystal are placed within the small antenna turns. Because of the relatively high frequency of the oscillation, the antenna turns are painted with a layer of "Q-Dope," as mentioned in Chapter 4. This unit can then be potted into a solid block of almost any plastic without undue losses. The thermistor probe is shown at the lower left of Fig. 14.5, with its connecting wires looping around the top of the photograph.

Since the signal from this transmitter never ceases completely, the automatic frequency-control circuits in a standard receiver allow the receiver and transmitter to remain locked, in spite of any tendency of the receiver frequency to drift. If the plane of the transmitter loop is horizontal, then the signal can be picked up by a horizontal receiver loop, or on a standard receiver whose whip antenna has been oriented to be horizontal. The folded dipole array mentioned in Chapter 10 is also quite effective in receiving this signal. From this small transmitter antenna emanates a signal of approximately 50 μV per meter at 50 ft. Assuming an inverse first-power law of signal-strength reduction with distance, we should then expect useful signals of about ½ μV at approximately a mile. However, in the range of elevations above the ground from a dog to about a human, the useful range with ordinary receivers is more nearly 500 ft in relatively open terrain. The circuit is most conveniently turned on and off by breaking the connection at point *A* in Fig. 14.4. A similar transmitter with more power is described in Chapter 18.

Although one can track the signal from a transmitter carrying physiological information, in some studies the only parameter of interest is location. Thus one may wish to know the path of migration of a species, or the area over which a particular individual will normally roam (the home range). In these cases tracking alone is employed, sometimes at extreme ranges. The transmitter in that case can emit a continuous signal, but it is more noticeable if periodically pulsed. By the pulsing characteristics, a given individual subject can be identified, although each of several transmitters may be operating on the same radio frequency. This periodic activation also increases the life of the experiment by reducing average battery drain.

Examples of such studies are provided by the work of Drs. Frank and John Craighead, who have worked with several species of large animals. In much of their work, they have employed transmitters fabricated into collars placed around the neck of the animal. The collar also serves as the transmitting loop antenna, with the large area contributing materially to the great transmission ranges achieved. In Fig. 1.16 was seen the placing of the radio collar on an immobilized cow elk (*Cervus canadensis*). In placing transmitters on these animals it is generally necessary to administer an anesthetic with a syringe dart, to be discussed later in this chapter. They have used similar transmitting collars to study the movements of grizzly bears, as in Fig. 14.6.

Fig. 14.6 A grizzly bear with a transmitter collar around its neck. (Courtesy Frank and John Craighead.)

Their transmitters are 2.5 × 5 cm in size and vary in output from 90 to 200 mW in pulses to a resonant metal loop antenna which is reinforced and padded to serve as a collar. The transmitter frequency is 32.02 MHz. Seven Mallory mercury cells, type 1450, power the transmitter for sixty or more days. They report a reception range up to 15 or 20 miles under favorable conditions with a five-element receiving antenna on a 40 ft tower. With portable direction finders they note an operational range of 2 to 4 miles. By incorporating into the transmitter a circuit that turns the transmitter on for 10 secs and off for 20 secs, battery life is increased to almost one year. They feel this to be more satisfactory in field work than an off period of 15 min and an on period of 3 min. In some of their work they employed a cubex quad antenna, whose reception compared favorably with a 3-element Yagi antenna, because it could more readily be dismantled and carried on a back pack.

A representative result from such a study is shown in Fig. 14.7, which shows the summer and fall ranges of five different grizzly bears. Circles are nuclei of ranges, that is, areas where the bear spent most of his time. Such studies also indicate the fraction of time spent in various activities such as feeding, bedding, travelling, and standing.

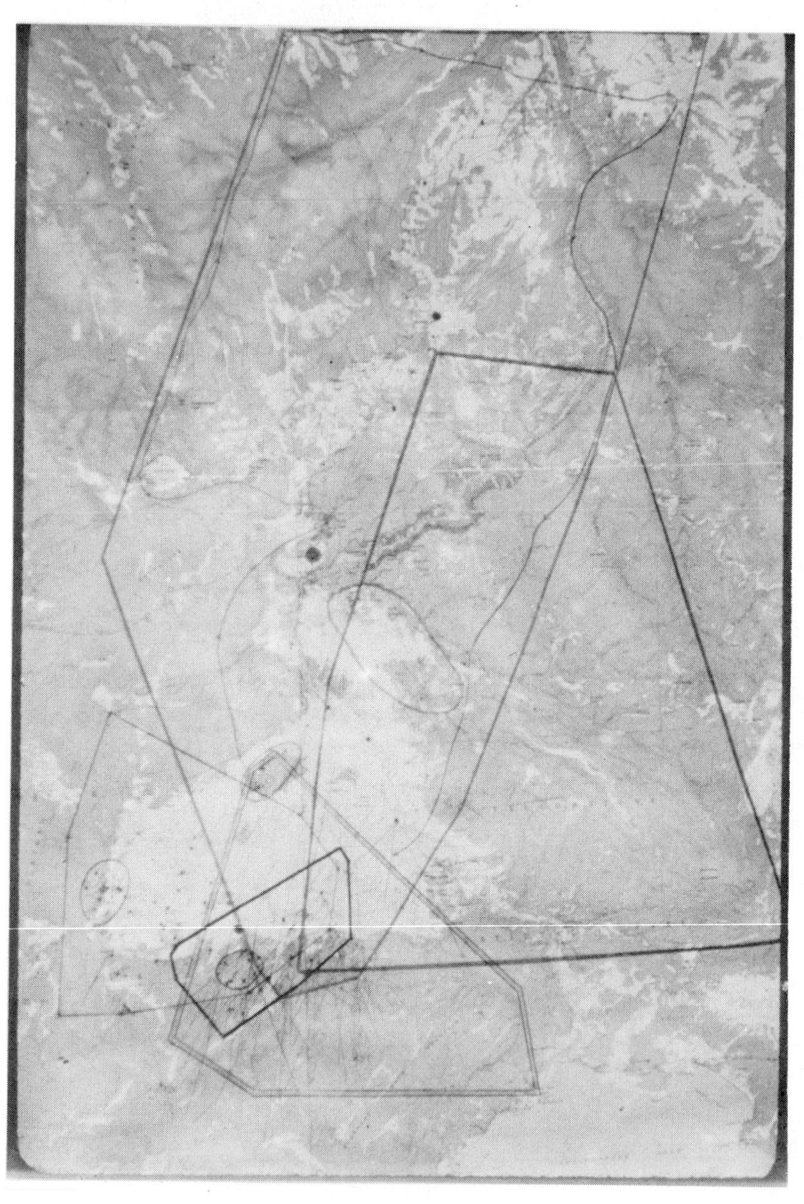

Fig. 14.7 Home-range data for several tagged animals plotted on a map of Yellowstone National Park, with special areas indicated by the rounded curves. (By Frank and John Craighead.)

The taking of tracking data over certain kinds of terrain can be made relatively automatic by continuously rotating a direction-finding receiving antenna at the top of each of a pair of towers. Whenever a signal null exists for a given animal transmitter, the momentary bearing of that receiving antenna can automatically be recorded. Thus the direction to the animal from two fixed points can be found at any time, to give the actual position of the animal. (Of course, the null need not appear at both antennas simultaneously, since an animal generally cannot move a significant distance in the time of an antenna revolution.) By having a number of receivers in parallel simultaneously monitoring signals from the antennas, it is possible to simultaneously follow the movements of a number of different animals, each carrying a transmitter on a different frequency. Such a unit has been set up at the Cedar Creek Natural History Area in East Central Minnesota, and has been described by Cochran et al. (1965). In this case radio signals from marked animals are received by rotating antennas supported on two towers placed 0.5 mile apart. Time and bearings for up to 52 animals are continuously recorded on 16mm film, with a mechanical accuracy determined to be $\pm 0.5°$. An operator takes bearings from the film as desired and determines locations by triangulation, although a digital computer program has been developed for automatic map construction. Tester and Siniff (1965) have discussed some aspects of such data. At the time of this writing, animal movements have been recorded with this apparatus for 35 red foxes, 15 white-tailed deer, 59 raccoons, 4 badger, 6 cottontail rabbits, 30 snowshoe hares, and 14 owls. They feel that analyses of such data are leading to better understanding of such behavioral and ecological aspects as predator-prey interactions, family group dispersal patterns, spacing of individuals in a population, home range and habitat utilization, and daily activity rhythms.

The above receiving equipment has been used with externally placed transmitters, generally in a collar. Construction of such collars for small animals such as rabbits and raccoons has been described by Mech et al. (1965).

The above observed angular uncertainty in a position measurement of 1° is not uncommon. This allows one to estimate the expected positional uncertainty at different ranges if added factors do not interfere, and also indicates the required mechanical stability of the receiving system.

Many systems will transmit to roughly 20% beyond the visible horizon, whose distance in miles over smooth terrain is given approximately by 1.22 times the square root of the altitude in feet. Also a radio signal moving along near the surface of the ground induces currents in the ground, thus dissipating power and diminishing the signal. Therefore an upwardly directed signal from a given transmitter type will carry to a greater distance.

Placing the receiving equipment in an airplane results in an increased detection range, though it is, of course, necessary then to have the additional information of the exact location of the plane during each reading used for position finding. As a specific example, one small transmitter was yielding a range of 0.05 mile to an antenna on the ground, while its signal was reliably received at a distance of two miles in a plane at 4000 feet. By considering the transmitter and receiver as interchanged, one can similarly understand why small transmitters can give somewhat greater ranges when affixed to a bird.

The same considerations apply to monitoring animals with the help of artificial satellites in the sky. Fairly small transmitters with suitably arranged antennas can be detected in the upward direction at appropriate distances which are in the general range of hundreds of miles. Some of these possibilities have been discussed (BIAC, 1966; Balmino et al., 1968). Physiological information could be transmitted as well as tracking data, and all parts of the globe would be covered in a cyclic fashion. Two of the schemes are described in Fig. 14.8. At the top is depicted the so-called IRLS system of NASA in which the subject would carry a small receiver-transmitter package (transponder) which would immediately return a signal upon receiving one. A satellite at position 1 would emit a signal and measure the time interval before receiving the returned signal. From this information the momentary distance to the subject would be indicated, and the subject could be said to be somewhere on circle number 1. Slightly later the satellite would be at position 2, and the subject could be said to be somewhere on circle number 2. Since the subject would not have moved significantly, the location must be either at A or at B. These two points would generally be sufficiently separated so that a knowledge of the original position of the subject, or its previous position, would be sufficient to eliminate this ambiguity. The lowest radio frequencies that could be considered here are in the "very high frequency" or VHF range (say, 150 MHz) where uncertainties in ionospheric propagation and multipath reflection from water can introduce some uncertainty. A number of subjects in one area could be handled, but each would have to carry a relatively heavy package.

A simpler relatively light transmitter would be all that need be carried by a subject involved in the method depicted in the lower part of Fig. 14.8. As the satellite sweeps by, the constant frequency being transmitted by the subject appears first of higher than normal frequency and then lower due to the Doppler shift. The character of the signals received is indicated at the right, there being an inflection point on the curves at the instant of closest approach by the satellite. The farther to one side of the subject is the path the slower is the frequency shift, and this can be used to accurately place the position of the subject. Assuming the transmitter maintains a

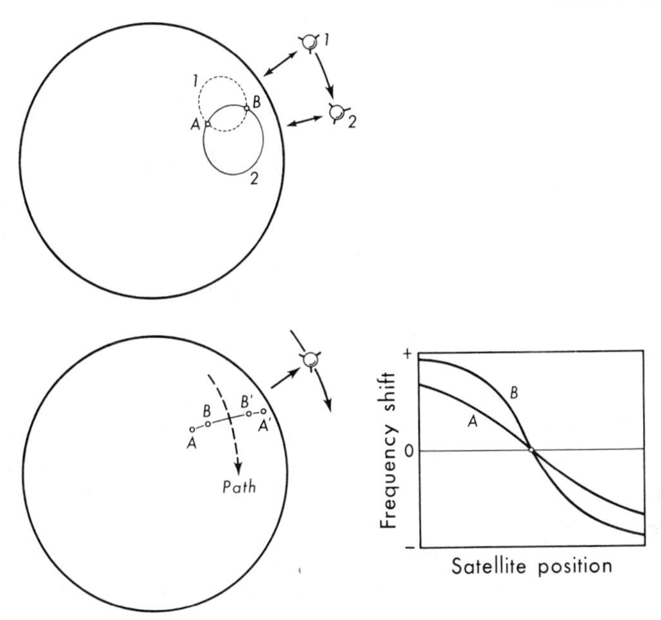

Fig. 14.8 Two methods by which an artificial satellite can be used to determine the position of a tracking transmitter on an animal. *Top:* A signal from the satellite is returned, with the transit delay fixing the animal position on successive circles for successive satellite positions. *Bottom:* The shift in frequency received at a satellite from a constant frequency transmitter indicates the position of the transmitter relative to the path of the satellite. In both systems there is an ambiguity regarding which side of the satellite path the subject is on; this is generally resolved by knowing where the subject was originally, or by the effect of earth rotation.

constant frequency to a few parts in 10^9 during the time of its passage (it can drift somewhat between passages), three (or generally more) frequency readings are sufficient to locate the subject. Once again there is the ambiguity, resolved by earth rotation giving a curve asymmetry. Best ambiguity resolution and greatest accuracy are achieved when the subject is not directly under the path of the satellite, and in general these methods should allow localization to within about two miles on the surface of the earth. The possibility of the method was demonstrated in reverse (a receiver on the ground) in the earliest satellite studies and it is used in the Transit navigation system (Kershner and Newton, 1962). The upper VHF range is suitable here with powers of the order of a tenth watt. The signal can be modulated to send physiological information. With this system, very simple weaker transmitters can be used, with several subjects being monitored on any one frequency in an area the size of a continent. Two-frequency transmission increases precision.

A specific unit that holds the promise of functioning from a bird when suitable satellites become available is based upon a thermally insulated crystal oscillator. This could be connected to the output stage for four seconds every twenty to conserve power. The actual pulse spacing could be modulated to send slowly changing or preprocessed biological information such as average heart or wing rate, temperature, orientation, altitude, etc.

These two methods both make use of, and require, the motion of the satellite with respect to the surface of the earth. All points on the globe could be covered by a single satellite in polar orbit if continuous information or position at any instant is not needed. A group of satellites in stationary or synchronous orbits would seem better able to continuously relay physiological information, and by "triangulation" to determine position, but they would be significantly more distant (approximately 500 vs. 20,000 miles). To monitor the entire earth continuously would require 8 times as many low level polar orbit satellites in rings as 24 hour period satellites (48 vs. 6), though the overall cost of the units and launch would perhaps be about the same. Approximately the same relative number is required if several satellites must be in view at the subject at all times. Given a choice, one should probably not consider synchronous units at this time for biological purposes, in view of the increased power requirements at both the subject and satellite which cannot be fully overcome by highly directive antennas.

As in the aquatic case of Chapter 15.4, the difference in time of arrival of a pulse (or wave train) at three receivers (satellites) indicates precise location. If the altitude of the subject in addition must be determined, a fourth receiver is again required. The system can be used in reverse, a receiver at the subject being used to time the difference in arrival of a signal from three or more satellites in order to find position. Such a passive navigation system is preferred by some groups for a number of reasons, including the ability to handle an indefinitely large number of subjects. Thus this may become one of the existing facilities. In animal work, however, it is not for the animal to know where he is, and thus the signal would either have to be recorded or relayed by transmission to one of the satellites. A transmitter at the subject then need only be active when returning a signal, thus prolonging battery life.

Such methods will be helpful in tracking various objects including life boats and weather balloons, and in direct biological applications will aid in monitoring the extended movements of some animals. There is special value where roads are not readily available. Thus the albatross or European White Stork and the caribou are subjects that could well be so studied. Some of the latter are perhaps vagrant in not returning to a given location, while the navigation of the birds is imperfectly understood.

As was mentioned in the chapter on passive transmission, birds and even insects give noticeable indications on radar sets, with large birds being detected at distances perhaps as great as 100 miles. Thus for some purposes such approaches should not be ignored.

A number of pieces of physiological information can be collected and transmitted from an animal if the investigator is willing to place a transmitter somewhat larger than the tracking units on the back of the animal. An excellent example of this is the cardiovascular observations made on baboons in Africa by Van Citters, Franklin, and Watson. Figures 14.9 to 14.11 are from some of their studies. The sketch in Fig. 14.9 depicts a baboon on whose back has been placed a box approximately 20 cm wide, 15 cm high, and varying in thickness from 3 to 7 cm, being curved to fit the back of the animal. Contained in this back pack was a high-frequency transmitter which was able to telemeter the signal from any of several physiological sensors, but not all at once. Various physiological sensors were surgically implanted as indicated. Doppler shift flowmeters were placed at several sites and a pressure transmitter was inserted into the aorta as indicated. Velocity could be monitored from the aorta, from one renal artery, from the superior mesenteric artery, and from the abdominal aorta.

Within this back package also was placed a receiver acting on a lower frequency. In this connection, heavy use was made of model airplane control equipment which is readily available, and some of which is quite effective. Signals from the investigator, through the action of a resonant reed relay (Chapter 3), were not only able to switch on and off the transmitter, thus conserving power, but were also able to select from among the various transducers which signal would be monitored. This same receiver was able

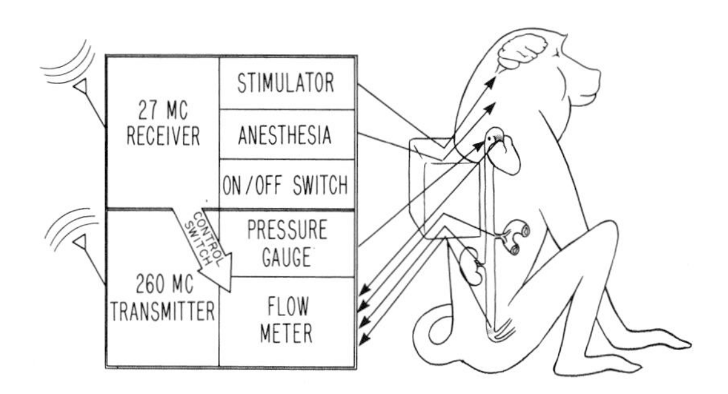

Fig. 14.9 Schematic of external telemetry system on a baboon. The receiver was able to activate an electrode implanted in the hypothalamus, to anesthetize the animal, to switch the transmitter on or off, and to select between the signals from the several implanted Doppler flowmeters. When activated, the transmitter sends information from the pressure gauge and selected flowmeter. (Figure courtesy R. L. Van Citters et al.)

to control the action of a brain-stimulating circuit. And finally, a suitable impulse into the receiver would cause the heating of a small wad of gun cotton in an implanted metallic cylinder, thus driving forward a piston to inject an anesthetic into the animal for recovery purposes.

A number of animals were successfully instrumented in this fashion. Transmission range was affected by the previously mentioned problem of changes in altitude. When animals came down from the trees to the ground, transmission that had extended to the horizon was reduced to a few hundred meters. Various transmitting antennas were employed. In some cases a loop antenna was placed within the box, and in other cases a base-loaded stub dipole was placed in the box. In other cases a length of wire extended from the top of the box (a base-loaded monopole).

A cause of some worry in all of these experiments is the effect on the activities of the animal and his interactions with his neighbors of any unusual appearing external device. Thus we might wonder in different cases if the subject animal might either become an outcast, or perhaps even a leader by virtue of his different appearance. In the present case, it was reported that there seemed to be no alteration in the behavior of the instrumented animal, or of his relationship with others. Indeed, it was noted that during mutual grooming, the other animal would not only groom the fur of the instrumented animal, but would indeed lift the back pack and reach up under it in order to groom that area. A report might be noted that the back-pack transmitter was best tolerated if the hair was not shaven from the region beneath the transmitter.

A sample tracing of cardiovascular parameters from this study is reproduced in Fig. 14.10. In this case flow and pressure in the aorta are recorded. In another study from this same expedition, using similar equipment, a few observations were made on freely roaming giraffes. In this case a pressure transmitter was implanted in the carotid artery just below the angle of the jaw. A relatively steady flow was noted at times of rather large pressure

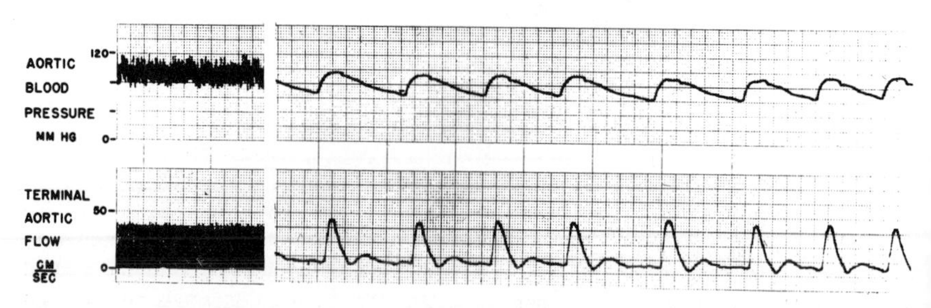

Fig. 14.10 Recordings of pressure and flow produced by the system in Fig. 14.9. (Courtesy R. L. Van Citters et al.)

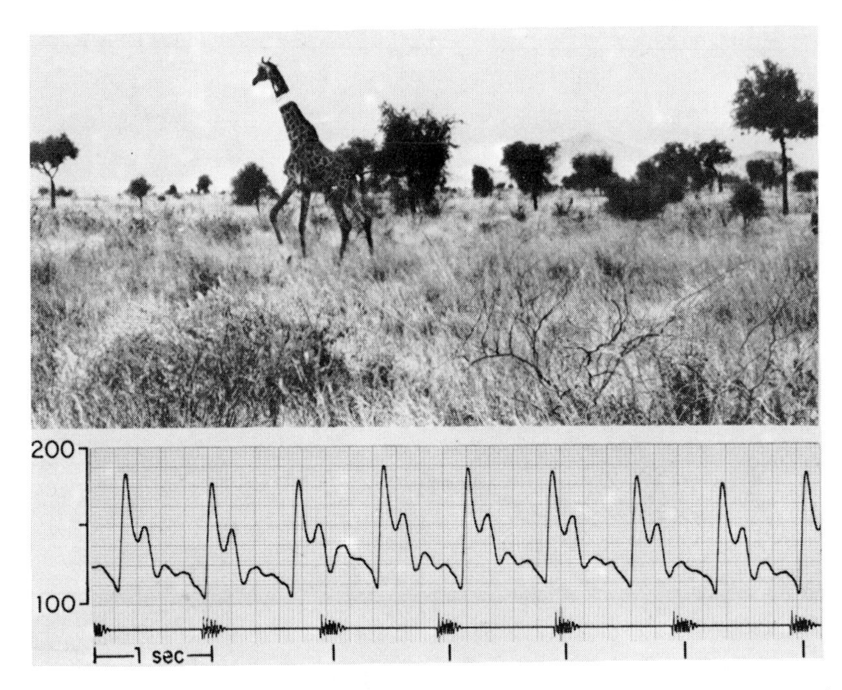

Fig. 14.11 Telemetry from the neck of a giraffe with a pressure recording from the carotid artery. (Figure courtesy R. L. Van Citters et al.)

oscillations. A sample recording is reproduced in Fig. 14.11 (Van Citters, Kemper, and Franklin, 1966).

In order to attach sensing and transmitting equipment to a wild animal, it must first be caught and immobilized. In some cases the animal can be trapped or noosed, and a transmitter tied into place without further problem. In the case of the larger species, this procedure can be fraught with difficulty. An anesthetic, of course, generally will be required if any surgical procedure is involved in the placement of the transmitter. In working with cold-blooded animals, packing in ice seems to be acceptable as an anesthetic. One often effective way of dealing with larger animals is to shoot them with a hypodermic syringe filled with anesthetic. A syringe can be propelled either by a crossbow, or else by the more sophisticated guns produced by the Cap Chur Company or the Pax Arms Company. These procedures can also be important in a zoo situation where not only are medicines thus sometimes applied, but where nervous animals are thus prepared for examination with minimum trauma.

Figure 14.12 is a photograph of a pistol in which a carbon dioxide cartridge propels suitable metal darts. At the top is the pistol, with the

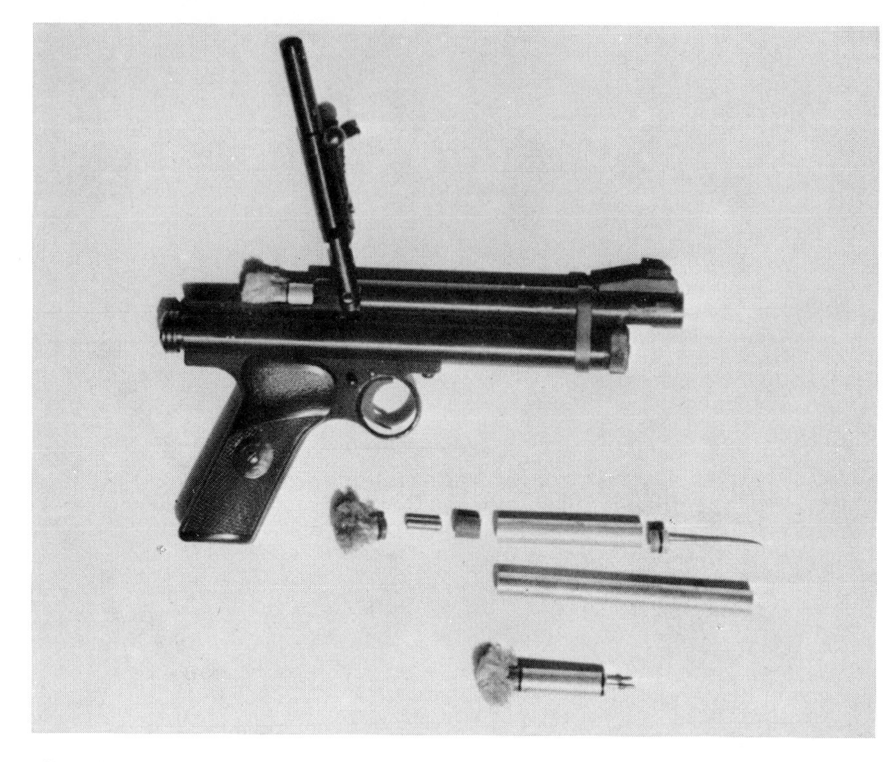

Fig. 14.12 Pistol for shooting a hypodermic syringe containing an anesthetic or other drug into an animal. The impact fires a small charge which causes the injection of the fluid. For details see text.

tuft of fibers at the end of the syringe showing. Below that is a dismantled syringe, with the tufted back cap seen at the left. Next is the small powder charge which fires on impact and forces forward the rubber plunger that is next in line. Next is the metal tube of the syringe, into which screws the metal needle at the right. This tube is initially filled with the anesthetic fluid with the plunger to the left, and upon hitting the animal the plunger is driven forward, thus injecting the anesthetic. Below is a larger capacity tube, and below that an assembled dart of smaller size with a slightly differently formed needle.

Investigators have been reluctant to use these methods in some rescue and study work because after being hit, an animal may run into dense bush before dropping. It can be impossible to find an unconscious and perhaps in-distress animal, and in some watery surroundings there is risk to the animal of drowning. It is possible to place a small short-lived radio transmitter within the dart to mark its location, assuming the dart stays with the animal. Figure 14.13 shows a transmitter which can be cast into a washer that screws between the tip and body of the dart, with the body serving as the transmitting antenna. One can also place the circuit and an antenna wire out on the tuft of fibers. A blocking oscillator in the 100 MHz frequency region drives a separate output stage, so that changes in the surroundings may shift the frequency but will not stop the action entirely. Sweeping in frequency assures a range of the receiver dial to choose from in case of interference, and also provides a signal to the output stage if it is detuned by ambient changes. The sweep in frequency produces a noticeable output from any FM receiver. It might be noted that a dart can similarly be used to monitor temperature, but its probe must go deep enough or else misleading indications result.

A number of immobilizing drugs can be employed, but one of the most useful is succinylcholine chloride. The drug is short-acting muscle relaxant that blocks nerve transmission at the myoneural junctions. A discussion of this chemical has been given by Pistey and Wright (1961). Average dose must be judged in terms of subject body weight, and it quite different for different species. It may even be that the sensitivity of a particular species for a given weight varies from season to season. Some representative information for various species has been collected by Dr. Clinton Gray of the United States National Zoo, and is summarized in Table 14.1. It is not possible to specify the exact doses, and it should be remembered that too

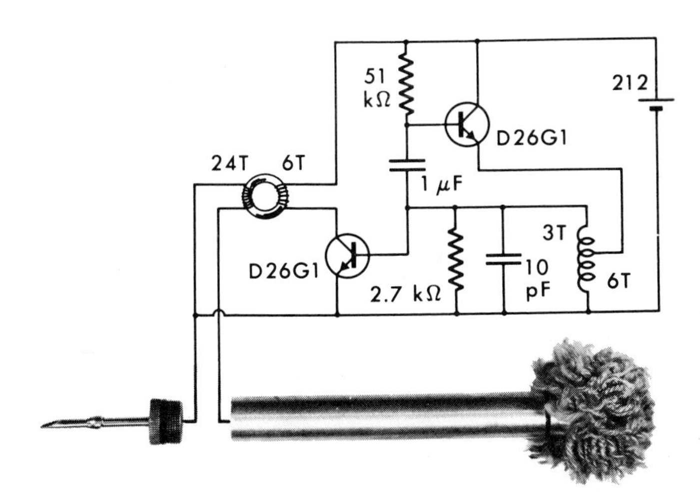

Fig. 14.13 Radio transmitter for marking position of an animal that has been darted and ran away before dropping. The blocking-oscillator coil can be wound on a matchstick, while the output transformer is on a small powdered-iron torus.

Table 14.1

Animal	Dose Succinylcholine Chloride	Comments
Sun Bear	14 mg	
Malay sun bear	14 mg	Female, 100 lb
Hybrid bear	100 mg	
Hybrid bear	100 mg	
		Female, 790 lb
Alaska brown bear	65 mg	785 lb
Alaska brown bear	100 mg	Male, 760 lb
Spectacled bear	15 mg	Male, 110 lb
Sloth bear	20 mg	Male, 130 lb
Sloth bear	15 mg	Female, 120 lb
Sitka bear	100 mg	Female, 425 lb
Peninsula bear	100 mg	Male, 740–750 lb
Peninsula bear	100 mg	Male, 850 lb
Black bear	150 mg	Male, 400 lb
Black bear	50 mg	Female, 85 lb
Black bear	90 mg	2 Females, 230 lb
Giraffe	21 mg	3/28/60; down in 8–10 min., immobile 30 min
Giraffe	25 mg	Down in 17 min, immobile 18 min
Giraffe	16 mg	Female
Giraffe	19 mg	Female
Nubian giraffe	10 mg	Adult male
Masai giraffe	45 mg	Adult male
Elk	8–9 mg	3 Females
Elk	12 mg	Male
Elk	15 mg	Adult male
Virginia deer	2.5 mg	Male, 100 lb
Virginia deer	2 mg	Female, 60 lb
Sika deer	4 mg	85 lb
Fallow deer	3.25 mg	40–50 lb
Fallow deer	4 mg	40–50 lb twice
Fallow deer	4 mg	Male, 40–50 lb
Fallow deer	4.5 mg	Male, 110 lb
Fallow deer	4 mg	Female, 30 lb
Fallow deer	4 mg	2 Females, 40–50 lb

Table 14.1 (*continued*)

Animal	Dose Succinylcholine Chloride	Comments
Fallow deer	8 mg	Adult
Fallow deer	5 mg	Spike
Fallow deer	5 mg	Doe
White-tail deer	8 mg	Male, adult
White-tail deer	5 mg	Female or young
White-tail deer	5 mg	Female, 92 lb
White-tail deer	6 mg	2 Females, 113 lb
Pere David deer	7 mg	Male, 3 years
Hog deer	1.5 mg	Male, 1–1/2 years
Eland	11 mg	2 Males, 1 year
Eland	14 mg	Male, 1000 lb—12 mg no effect
Eland	11 mg	Male, 700 lb
Eland	11 mg	6 Females
Zebra	30 mg	
Grant's zebra	55 mg	Male, adult
Grant's zebra	55 mg	Female, 700 lb
Bison	17.5 mg	Female, old
Congo buffalo	19–20 mg	Male, adult
Yak	15 mg	Male
Yak	4.5 mg	Female
Yak	10 mg	Female, large
Yak	10 mg	Female, old
Abyssinian ass	70 mg	Female, 400 lb—45 mg no effect
Onager	55 mg	Male, adult
Gaur	9 mg (12 mg)	Male
Gaur	16 mg	4 Males
Gaur	17 mg	Male
Gaur	15 mg	Female, young
Gaur	17 mg	Female, old
Gaur	18 mg	Female—dead in 26 min.
Mouflon sheep	9–11 mg	Adult
Peccary	3 mg	Male
Peccary	3 mg	Female
Sika deer	7 mg 5 mg	8 adults observed, the larger dose to males

Table 14.1 (*continued*)

Animal	Dose Succinylcholine Chloride	Comments
European roe deer	1.5 mg	Yearling
European roe deer	2.5 mg	Adult
Barasingha deer	7–8 mg	Male, 3 years
European red deer	11 mg	Male, 1–1/2 years
European red deer	12 mg	Male, 3 years
European red deer	14 mg	Male, adult
Eland	10–12 mg	2–1/2 years
Sitatunga	4 mg	200 lb
White-tailed gnu	7–10 mg	5–10 months
Grevy zebra	30–35 mg	850 lb
Hartmann zebra	50 mg	8 months
Hartmann zebra	90–105 mg	Adult
Przewalsky's horse	170–180 mg	2 years
Guanaco	25 mg	125 lb
Bengal tiger	90–95 mg	250 lb
	Sparine Pellets	
Dorcas gazelle	1 oz per day orally	30–40 lb
Grevy zebra	3 oz per day orally	800 lb
Grant's zebra	2 gm pellets per day orally	650–700 lb
	Acepromazine Maleate	
American bison	2.5 mg per 100 lb	7 years
	Sernylan	
Sooty Mangabey	10 mg	17 lb

Source of drugs: Anectine – Burroughs Wellcome
Sparine – Wyeth
Acepromazine maleate – Ayerst Laboratories

large a dose accidentally administered can kill a valuable animal. There is a continuing search for drugs displaying a wider latitude, with such things as the relative of morphine designated as M-99 holding out considerable promise at present (Harthoorn and Bligh, 1965). A dose is less critical if one can quickly reach the animal and breathe for it.

A few general comments about the use of these anesthetic guns might be made in closing this section. To maximize predictability all injections should be intramuscular, the effect being uncertain unless the needle penetrates a deep muscle mass. Following an injection, there is normally a lag of 4 to 5 min before the animal goes down. The variable time of the immobilization often ranges from 5 to 30 min. It is desirable that a source of oxygen be available. All ruminants should be placed on their sternum to avoid respiratory problems following any regurgitation. Drugs should be refrigerated, and it is best to use freshly diluted materials.

There is one further caution. One would expect behavioral changes in some animals subjected to painful manipulations while immobilized with certain drugs but not anesthetized. It might be mentioned that in humans, the administration of succinylcholine without anesthetic is frightening and uncomfortable even if nothing painful is then done. However, it is a useful agent in being relatively very fast acting, and more reproducible in terms of needed dose per body weight than agents that act on the central nervous system. In animals without plasma cholinesterase (e.g., dogs or porpoises, vs. cats and a majority of humans, which have it) the effect can be long lived, and prolonged artificial respiration perhaps necessary.

15

Aquatic Animals

Among the many species it is perhaps those that swim in water which are least susceptible to study by the usual laboratory methods. Thus to remove an aquatic animal from his liquid medium in order to make standard types of observations would produce significant psychophysiological changes. By being denied the opportunity to swim and dive in surroundings that are greatly different in their properties from air there are undoubtedly generally significant physical changes imposed also. Thus, for example, we should not expect the temperature of an animal to remain unchanged if removed from a liquid into air, and indeed, even the interruption of normal motion in the fluid changes the rate of heat exchange. In cases in which such an animal is confined to a small aquarium, it can be sufficient to extend wires from the animal to a central overhead point, and to allow him to swim around and around, perhaps without tangling. In that case, to prevent twisting of wires, a transformer can be constructed in which a primary winding is free to rotate in a coaxial fashion within a secondary winding. Even this can be considered as a form of short-range telemetry.

In considering telemetry systems, a distinction must be made between animals that spend all of their time in water, as opposed to those that spend part of their time in water. If the animal periodically surfaces for a breath of air (e.g., the aquatic mammals), then other possibilities are available. We must also distinguish those animals that live in the ocean from those that live in fresh water, with regard to the choice of telemetry methods. The choice between sound transmission and radio transmission of information through water can depend upon the frequency range of hearing of the subject animal, if he is not to be disturbed by the telemetry device itself.

In some cases, flashing lights or other telemetry methods may have special applicability.

The greatest interest in these methods relates to the monitoring of environmental and physiological parameters, whether they be transmitted from man (for example, in monitoring the condition of a diver using self-contained underwater breathing apparatus), or from fish or various aquatic mammals. The value of illumination, temperature, oxygen saturation, and pH preferred by a fish would be of interest, and can best be determined using telemetry methods for minimum disturbance. Various combinations of the present circuits allow the transmission of any of these. Depth might also be monitored as a pressure signal, although the first return of a periodic outward sound impulse could also be used to indicate the range to the surface or bottom. (If the return of an impulse caused the triggering of the next, then the frequency of impulses observed at a remote point would be a measure of range; similar considerations are taken up in the section on transponders in Chapter 17.)

The movement of various aquatic animals might also be studied in a normal habitat. Some displacements may be small. Thus it is suspected that perhaps abalone may move a few inches at night and then return to their original spot the next day. On the other hand, three species of king crab living on the continental shelf migrate for miles. It would be interesting to follow their movements in a more continuous fashion than tagging and possible recapture can provide. In terms of commercial fishing, as well as scientific studies, the movements of schools of fish could perhaps be followed using these methods. Thus one fish could be caught and be fed (or otherwise have attached) a tracking transmitter and then be thrown back to rejoin his fellows. A large group might thus be "kept in sight" by following this single signal.

Various special variables might be investigated. Thus the transmission of cochlear microphonics from a sea lion might be monitored during large excursions in pressure to give an indication of the rapidity of pressure equalization in the structure of the ear. We have been interested in some aspects of temperature regulation and find, for example, that the temperature of a dolphin seems somewhat more constant than that of a sea lion. A few examples of these observations will be given later.

1. ULTRASONIC TRANSMISSION

Sound signals carry very well through water, and can generally be more effective in transmitting information there than in air. The velocity of sound is rather independent of frequency, but the attenuation increases with frequency in a way that is discussed in all books on sonar systems, and most books on general ultrasonics (e.g., see Tucker and Gazey, 1967). A problem in the use of sound signals for the transmission of information is that a signal may arrive at a particular receiving point over several different paths. Since the paths will generally have different lengths, this means arrival over a range of times, and this can seriously garble a signal. Slow transmissions are less likely to become garbled due to multipath problems than fast ones. The magnitude of this problem generally increases with increasing range. Frequency-modulated signals are generally less disturbed than are amplitude-modulated signals.

Acoustic fish-tracking devices have been described in the literature (Trefethen et al., 1957). Baldwin (1965) has mentioned, among other things, the ultrasonic transmission of the electrocardiogram of several mammals. We have used the following few circuits for tracking and temperature transmission from fish. They are included for the sake of making examples specific, but it should be understood that they can be combined in various ways with the previously mentioned sensor circuits for the transmission of a variety of variables.

It is known that many fish and invertebrates do not respond to high-frequency mechanical disturbances, that is, to high-frequency sound signals. Thus it would seem that such methods as these would be most appropriate with these animals, where little disturbance to the animal would be expected. However, any nonlinearity in the pressure-sensing system of the animal could allow the "hearing" of the modulation envelope of the high-frequency sound signal. A sound wave impinging upon a surface exerts a small but steady force upon the surface, and this "sound pressure" effectively rectifies a sound signal somewhat to give changes in force in response to amplitude changes in the source. Thus an animal may not be totally unaware of a sound transmitter that is operating at a frequency beyond the normal range of hearing. Whether this can be disturbing to a normal activity pattern or pattern of physiological functioning can only be determined by tests. It is also true that the sound intensity at the transducer, which is very close to the animal, should not be high enough to produce burns or even significant heating. Intensities under 1 W/cm^2 seem safe.

In Fig. 15.1 is shown a periodically pulsing acoustic transmitter, whose pulsing rate is controlled by a variable resistance (such as a thermistor, cadmium sulphide cell, and so on). This is actually a "push-pull" version of the earlier blocking-oscillator circuits. The transformer containing the three coils is wound in a set of ferrite cup cores (Chapter 2). Because of periodic energizing at a slow rate suitable for timing by ear with a stop

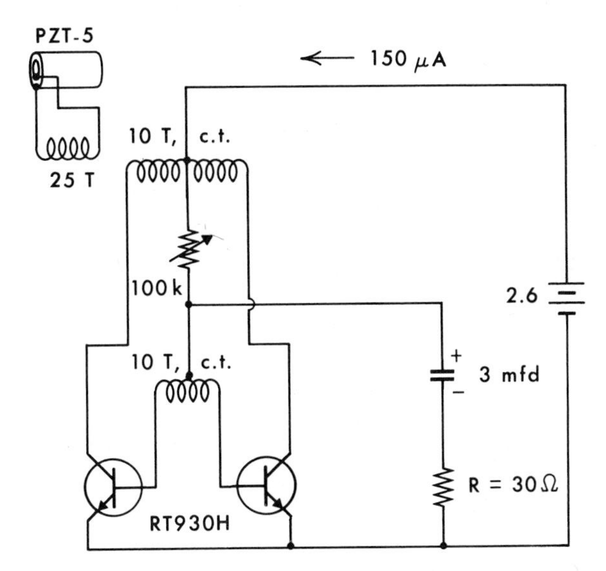

Fig. 15.1 A more powerful squegging oscillator with its rate controlled by the variable resistance. In this case the signal is coupled to a hollow cylindrical sound transducer of lead zirconate titanate through a ferrite-cored transformer. The rate is slightly affected by the immediate acoustic surroundings of the transducer.

watch, the average current is rather low. The electrical signal is converted into a high-frequency sound signal by the hollow ceramic cylinder of lead-zirconate-titanate which changes its diameter somewhat when a voltage is applied. These cylinders are formed in an electric field to polarize them, and if heated too much they can lose their piezoelectric properties through depolarization. However, it is generally possible to solder directly to a silver film coated on the inside and outside of the cylinder. The range of transmission of such a unit is a few hundred feet, depending upon factors to be discussed presently.

For somewhat greater precision of transmission, the circuit in Fig. 15.2 can be used. In this case a separate multivibrator is employed to switch on and off an output oscillator, with the pulsing rate again being determined by a variable resistance. Most of the circuit, except for the output transformer and batteries, can be fitted within the sound-transmitting cylinder, even though it is quite small. Connecting a relatively high capacitance capacitor across the batteries allows a greater range of transmission owing to the possibility of supplying greater momentary peak currents to the output oscillator. It was found that the very small 2 volt tantalum capacitors could be used in this application, even though the battery voltage was some-

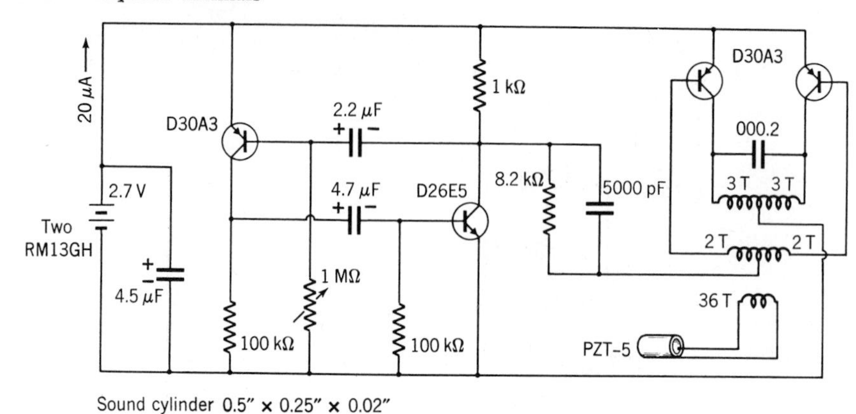

Fig. 15.2 An acoustic temperature transmitter that uses a complementary-pair multivibrator. Signals from other variable-resistance transducers can also be telemetered through water by this circuit.

what higher than that. Under these conditions their leakage current increased somewhat, but they were not observed to fail.

The astable multivibrator in this circuit employs a so-called complementary pair of transistors, that is, one is *PNP* and the other is *NPN*. The problem of generating a periodic short impulse separated by long intervals has been mentioned, and this type of multivibrator is essentially a pulse generator well suited to such an application. A complementary pair multivibrator has two specific advantages in this type of circuit over the usual multivibrator. Instead of the two transistors being on alternately, they conduct and go off together, that is, no current is being drawn part of the time. In generating these very asymmetric wave forms where the output may only be activated one one-thousandth of the time, there is an appreciable gain in battery life by having no current drawn during the longer part of the cycle.

The other advantage has to do with the generating of the asymmetric wave form itself. A large capacitor, which determines the long part of the cycle, must quickly be recharged to a standard initial condition during the short interval in order to again be ready to time the long interval of the next cycle. To supply the large recharge currents in an ordinary multivibrator, an emitter-follower connection, employing an extra transistor or some other special connection can be used in ways to be indicated elsewhere. But in this complementary pair configuration the recharging is done by the two transistors which switch on together to form a virtual "short circuit" through the low resistance of which rapid recharging takes place. The resulting circuit simplicity is an advantage. The duration of the generated output pulse is slightly more variable, being determined by a conducting active element, but this does not greatly affect the present application.

The components of such a transmitter are shown in Fig. 15.3. At the right are the three small tantalum capacitors which serve as an energy-storage element, and at the top is the pair of batteries. Below them is a piece of tubing that serves as a battery case, and whose end slides over the output transformer. The output transformer is seen to the left of the piezoelectric tube, into which all the rest of the electronic components are fitted, as seen at the bottom. The final assembly of components is covered with a piece of

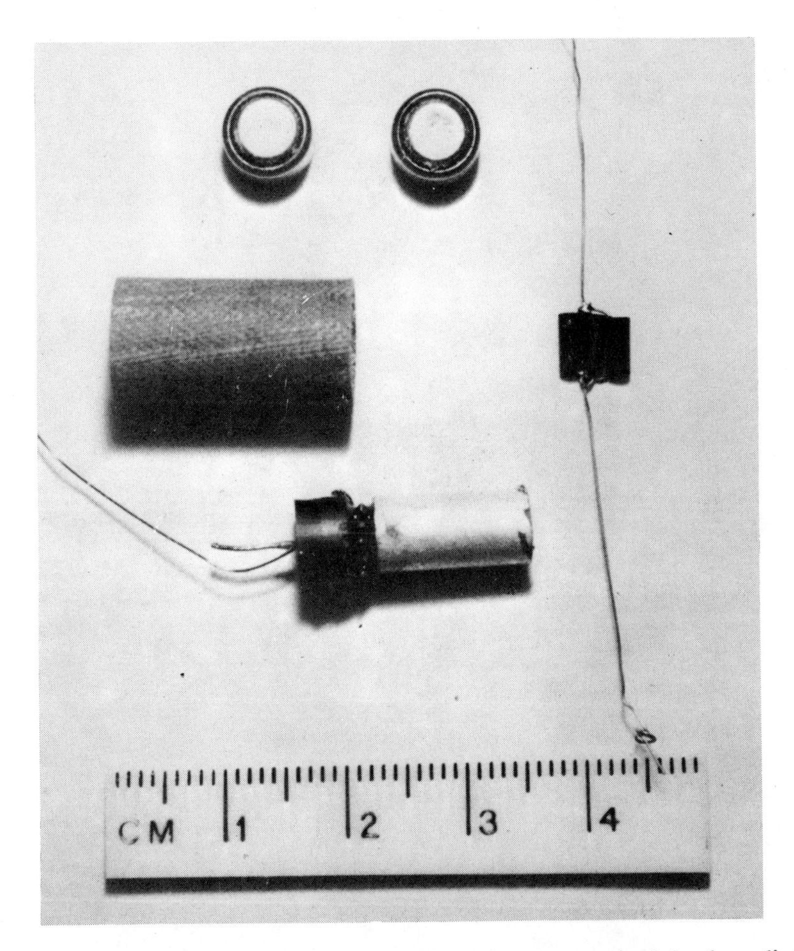

Fig. 15.3 Most electronic components in Fig. 15.2 are contained in the cylinder shown at lower center, with only the output transformer extending. To the right are three tantalum capacitors connected in parallel, and above are the two batteries.

heat-shrink tubing (Chapter 4) to give a unit which appears as in Fig. 15.4. The batteries are inserted into the left, and wax poured in to complete a water-tight seal. This provides an easy method for changing batteries. The unit shown was made for external attachment to the fin of a fish, and thus loops of wire tied around the body of the capsule were provided.

A few words should be said about mounting these sound cylinders. The ends must be plugged for effective acoustic action. If the entire transmitter is covered with shrink tubing, including both ends, then it is difficult to

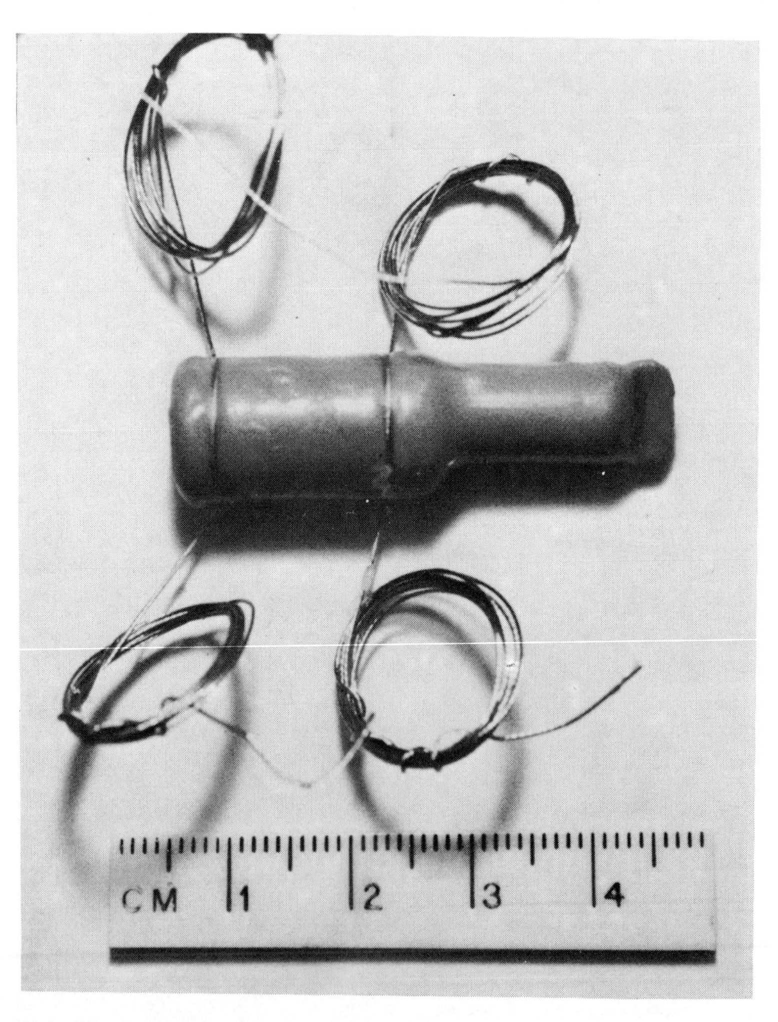

Fig. 15.4 The components of Fig. 15.3 are suitably placed and enclosed in a case of heat-shrink tubing. The wires are for use if external attachment is desired.

change batteries. A coating of wax alone over the cylinder provides little mechanical strength. Coating the cylinder with epoxy quickly results upon immersion in a decreasing resistance to values that are not acceptable in parts of some circuits. Therefore, we generally choose to mount the cylinder and its circuit in a hollow plastic cylinder. For this purpose, we machine a Teflon or Kel F cylinder closed at one end. The walls are made relatively thin, being in a range of 2 to 5 thousandths of an inch. The clearance between the outside of the ceramic cylinder and the inside of the plastic cylinder is made under 0.001 in. The plastic cylinder is filled with molten wax, and the whole circuit and ceramic cylinder pushed in, the wax serving as a lubricant. The circuit itself acts as a "backbone" for pushing the cylinder in, and it is useful to epoxy the circuit to the edge of the ceramic at one point. A hydrophone which is to receive such sound signals can be constructed in a similar way, or it can merely be covered with shrink tubing since there are no batteries to change. With the high frequencies presently being considered a little moisture leakage will not spoil the performance by reducing the resistance in general, but a hydrophone which is to pick up low-frequency sound signals must maintain an extremely high resistance.

These small ceramic cylinders have a characteristic frequency of mechanical resonance, being approximately 180 kHz for those with a diameter of 0.25 in. They tune relatively sharply even when damped by being placed in water. For transmission of signals in water at ranges less than approximately 500 ft the electrical oscillator should be tuned to this mechanical frequency. The inductance of the transformer can be chosen to resonate electrically with the capacitance of the crystal. If the frequency of oscillation is changed from 180 down to 50 kHz, there will be approximately one tenth the signal with a given voltage at the cylinder. However, for transmission ranges greater than approximately 500 ft it is better to use this lower frequency because of the decreased attenuation by the water over the transmission path. In such cases it can be desirable to go to a larger sized cylinder that has a mechanical resonance at a lower frequency, if other factors permit.

Other transducer shapes can be employed. The above cylinders were polarized in such a way as to cyclically change their diameter when an alternating voltage was applied. So that the fluid does not merely circulate from outside to inside of the cylinder, end caps in some form have to be provided. Instead, one can use a flat plate or disk of ceramic which will undergo changes in thickness in response to an alternating voltage, or perhaps a shear mode of oscillation, like a diaphragm, can be employed. Disks can be made which change their diameter in response to applied voltages. Such a disk glued across a hollow cylindrical plastic case causes the whole case to expand and contract, thus radiating a sound signal.

Actually, in this instance it is only the part of the plastic case near the disk which oscillates, most of the case generally being inactive at such high frequencies due to inertial damping.

Inexpensive receivers for this low range of frequencies are not readily available. However, it is possible to build a converter which will take these low-frequency signals and shift them up into the range where any inexpensive pocket, or other, receiver can be used to detect the signals. Thus a complete frequency shifting receiver system is depicted in Fig. 15.5. The sound signal from the test animal is picked up by a piezoelectric cylinder that can be quite similar to the one used in the transmitter. If the voltage generated therein is fed to an amplifier, then extra noise would be introduced. To some extent this is eliminated by increasing the voltage through resonance with the variable inductance shown. The voltage appearing across this inductance is fed to the emitter follower, which has a high input impedance and thus does not reduce the Q of the resonant circuit. This signal is combined with that from an adjustable frequency oscillator shown at the lower left. The transistor at the lower right is biased into a nonlinear range of operation so that the two signals applied to its base through resistors will mix in such a way as to generate sum and difference frequencies. By tuning the local oscillator one of these frequencies can be made to fall in the range of a standard radio receiver. The output from the collector of this mixer transistor can be coupled into a typical pocket receiver by wrapping ten turns of wire around the ferrite antenna rod, as indicated. This entire unit then becomes quite portable. The converter portion requires two voltage levels, and is conveniently powered by a pair of readily available flashlight batteries. As shown, the circuit covers the range of frequencies of 30 to 450 kHz, in two ranges. Extremely inexpensive components are employed, although others can be substituted; a transistor type 2N3391 is also inexpensive and shows low noise.

In the second part of Fig. 15.5, an alternative method of mixing the two signals is indicated using the improved performance of an inexpensive MOS field-effect device with double input. Assuming 9 volts is available from the receiver battery, the higher supply voltage is no inconvenience. The input impedance of this transistor is high enough to allow elimination of the emitter follower in the other circuit. The input signal is directly modulated by the local oscillator signal, as is required.

Feeding a signal directly into a tuned circuit as shown at the input to this converter requires a low-impedance source (under 25 Ω), if the resonant combination is to show a reasonably effective Q. If a signal is to be taken from a pair of electrodes in water (for the electromagnetic transmission to be mentioned in a later section), rather than from the ceramic transducer shown, then an amplifier stage must be placed before the tuned circuit. In

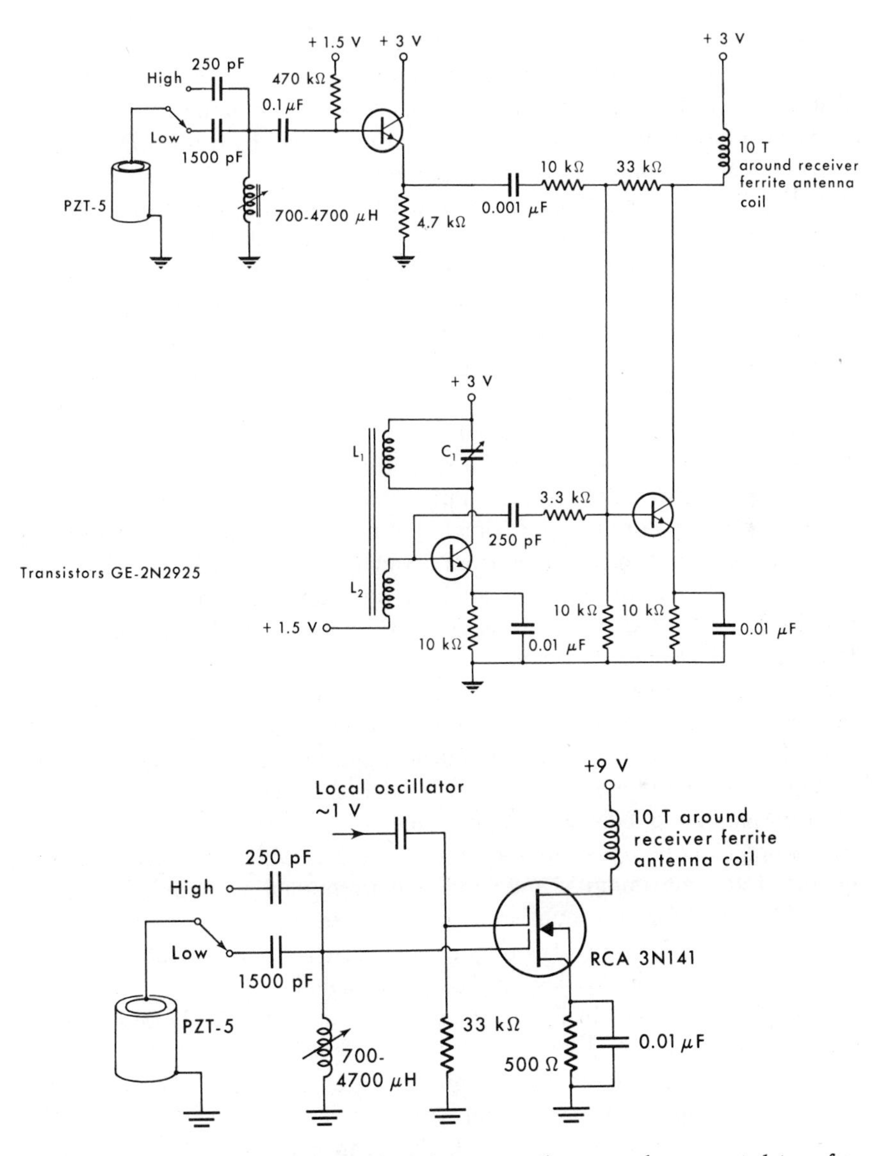

Fig. 15.5 The low-frequency signal of these transmitters can be converted to a frequency range suitable for a standard entertainment receiver with the help of the oscillator and mixer at the bottom of the upper figure. A sound signal is assumed to be received on the transducer at the upper left, but other signal modes can similarly be converted. The output coil can sit near the final receiver, but action is more reproducible if the coil is wound onto the receiver antenna coil. Below: An inexpensive double MOS transistor can be used for the efficient mixing of the two signals.

the situation illustrated, especially on the low-frequency range, the capacitance of the ceramic transducer combines with the series input capacitor to determine the resonant frequency of the input combination.

The overall circuit is shown coupled to the receiver by transformer action. Problems of a common ground connection between receiver and converter are simplified by direct connection of the converter output across approximately one tenth of the turns at the grounded end of the normal antenna coil in the receiver. In some receivers this requires placing a 0.1 μF capacitor in the lead wire connecting that coil to the receiver ground. The converter can instead be made to radiate its signal, in which case it need only be placed near a convenient receiver; the extra adjustment of intensity then must be made.

The oscillator at the upper left of Fig. 15.5 is simply a convenient circuit for generating a little power in the right range of frequencies. The coil L_1 is approximately 60 turns and L_2 approximately 20 turns, both wound on a single ferrite cup core of approximately 4 mm inside diameter. The tuning capacitor is one having a range from about 200 to 1200 pF.

The adjustment of this converter to an incoming signal is relatively simple. The standard AM receiver is set to an unoccupied frequency at the low end of the band. The capacitor C_1 is adjusted until a strong signal is received; this should be the reception of the local oscillator. To check, increase C_1 further until the next signal is received. This should be 455 kHz, the receiver's intermediate frequency, and the signal should be picked up equally over the whole broadcast band if the receiver itself is retuned. Note that in tuning to a steady input signal, a receiver with a beat-frequency oscillator should be tuned for zero beat, a receiver with a modulator should be tuned for maximum signal, and a conventional unmodified receiver should be tuned for maximum noise quieting. Having verified the dial setting which produces the 455 kHz frequency, then C_1 can be directly adjusted through a small range to pick up the desired signal.

With the equipment in hand, adjustment is simple and obvious, but an example might be useful here. To receive a signal on 70 kHz the receiver is set at some convenient low number such as 530 kHz. The converter oscillator is adjusted for maximum signal on the receiver. To check that the oscillator is indeed at 530 kHz, the capacitor is turned through a slightly increased value until 455 kHz is received; this is the only frequency that will be received over the entire band to which the receiver itself can be tuned, and can be checked by shifting the receiver between about 530 and 700 kHz. The oscillator is then returned to the setting previously found to give maximum signal. The converter tuning capacitor is then decreased enough to pick up the desired signal. In using the converter, a convenient reference is this 455 kHz intermediate frequency, which is the only setting picked up across the whole receiver band. Tuning the converter oscillator from this value results in signals appropriate to the local oscillator minus the desired frequency, then the local oscillator frequency, then the local oscillator plus the desired frequency. When heterodyning is taking place in such a way that a signal on the desired frequency is actually coming from the receiver, then only slight further adjustment will be required in the course of an experiment. When used with an actual transducer, the resonating inductance at the input circuit should finally be adjusted to yield the maximum signal.

The signal from this converter-receiver combination is then handled as would be the signal from any other receiver. Direction finding to a sound source has been accomplished in two ways. One is to place the receiver transducer in front of a reflector which can be aimed for maximum signal. A good reflector can be formed from a piece of foam plastic, very little sound energy passing from water into the trapped air. A transducer comparable in size or large with respect to a wavelength has intrinsic directional properties. These are generally to be avoided in transmitters, but can be useful in direction finding at the receiver. However, a pair of small omnidirectional receiving transducers can be spaced approximately a wavelength, and their signals added by connecting them either in series or in parallel. This combination does show direction-finding properties because of the difference in phase of a given wave at the two points. If the sound source attached to the subject has a carefully controlled frequency which is independent of such things as temperature changes, then changes in the received frequency give a measure of swimming velocity (or more precisely the component of that velocity towards and away from the observer). To make use of this the receiver system must be quite stable in its frequency conversions, and it is generally preferable to perform some more direct sort of frequency measurement on the incoming signal.

It might be noted, in closing this section, that ultrasonic transmission of information can be useful in other situations as well as those involving liquids; for example, such methods can be used to carry signals from within heavy metal reaction vessels from which it might be inconvenient to radiate radio signals. There is perhaps also an application for transmitters to be monitored in the body by direct listening through a stethoscope.

2. FACTORS AFFECTING LIMITING RANGE

The factors involved in the determination of the limiting range of an acoustic transmitter are perhaps best indicated by a rough calculation involving them. All numbers are approximate and indicate only the order-of-magnitude of the quantities involved. There can be great variations, especially due to variability in bottom conditions. Powers are specified in decibel notation and, as is also customary in underwater sound work, ranges are given in yards or feet. In referring to tables dealing with sonar systems it should be remembered that such factors as attenuation are often given for a round trip by the sound, whereas here we must consider the one-way passage.

The general type of transmitter might be one physically about half the size of a little finger. It might give out a 20-msec pulse every second to transmit temperature information for a month. The sound radiator might be a small hollow cylinder of lead-zirconate-titanate 1 cm long and ½ cm in diameter. Although the resonant frequency of these is approximately 170 kHz, it can be advantageous to use a lower frequency of 50 kHz; this is still above the hearing range of fish but is marginal for some cetaceans. During the pulse, the current drain on the battery is typically a few tenths of an ampere (although the average current is only one fiftieth of this amount), and the voltage across the transducer a few tens of volts. The circulating current in the transducer can be quite high, although the oscillator efficiency may not be optimum. Too much power in the transducer can heat it and cause a drift in frequency or even depolarization. Assume the power during the pulse to be about 4 W. A typical calculation follows.

Directional effects would be expected where the transducer size was a wavelength or more. At 50 kHz the signal should be radiated equally in all directions. If the projector (transmitting transducer) efficiency is 25 percent, there will be 1 W of omnidirectionally radiated power. Each watt of power gives 72 dB above 1 μbar at a range of 1 yd. In deep water there will be spherical spreading, with the familiar $1/R^2$ fall off. Thus the signal will be down 40 dB at 100 yds. Unless the water is highly aerated, thus increasing absorption, the tables indicate that the absorption loss in this path at this frequency will be about 1 dB. Thus the pressure level will be 31 dB above 1 μbar.

If the same small nondirectional transducer type is used as the receiving hydrophone, we might expect a characteristic of −120 dB relative to a volt per microbar, that is, 1 μbar would give 1 μV (this represents the ability of the unit to convert pressures to electrical signals in air or water or elsewhere). Thus the signal at 100 yds would be 31 dB above 1 μV or about 32 μV. Receivers often have input noise levels of about 1 μV in this range.

From tables (Knudson curves) we find the ambient noise spectrum level for quiet ocean in the frequency range of 50 kHz to be −80 dB relative to 1 μbar/cycle of bandwidth. If the width of the band employed were 4 kHz, the noise would be $-80 + 36 = -44$ dB, which is much below +31 dB. Thus, with a typical system of the above power, we might expect a range considerably greater than 100 yd, perhaps of the order of a mile, if there is small ambient noise. Traffic or rain or wind can increase noise by tens of decibels, and it appears that noise is usually considerably higher in shallow than deep water (say 10 dB) for a particular sea state. A 20 msec pulse requires a filter bandwidth of something over 50 Hz (the reciprocal of pulse length) for reasonable transmission. Thus, neglecting drift or Doppler shift problems, noise rejection might be accomplished by limiting the band of frequencies passed to a range of 100 Hz. In any such consideration, where a person listens to the final result, the ear will limit the range of final audio frequencies that interfere with the perception of a tone to something like 10 percent of the frequency of the tone.

If we go to a larger hydrophone (e.g., a vertical cylinder several feet long) the figure −120 dB could become −80 dB, which increase of 40 dB could contribute to greater range. If we go from 50 to 170 kHz, the one-way path attenuation per hundred yards goes from 1 to 5 dB. At 150 kHz the attenuation is 3.2 dB per 100 yds. At a range of ½ mile (1000 yd), changing from 150 to 50 kHz makes approximately 22 dB difference; this is extra loss over spreading loss. At the higher frequency, thermal noise in the water near the hydrophone would tend to dominate the ocean noise for low sea states, but the above would still hold.

In shallow water the sound cannot spread in all directions and thus a signal may not die out as rapidly with distance. This ducting results in a sound intensity which falls off somewhere between $1/R$ and $1/R^2$, that is, somewhere between the cylindrical case and the spherical case above. With a perfectly calm flat surface and a hard flat bottom the improvement in 100 yd would be 20 dB. This 20 dB would then extend the range to 1000 yd (not 10,000), with 10 dB going into further spreading and 10 dB into absorption (at 50 kHz) over the longer path. Sounds arriving over different paths may not all add up, but with a long pulse the above is essentially correct. Generally, however, there are losses at the interfaces, and there is not perfect reflection of the sound reaching them.

The velocity of sound in sand saturated with water is greater than in water. Thus there sometimes is a critical angle of about 25 degrees (the grazing angle) for total reflection over an appreciable range of frequencies. Over a mud bottom (fine silt) where the velocity is less than in water, there is no critical angle and so the beam can dive into the bottom. Organic sediments can serve as good reflectors due to the release of bubbles of methane gas. Some experiments have shown that at each bounce of a sound beam the backward signal drops 35 dB and the forward signal 10 dB, but this can be quite variable. Reflection from the surface depends on the wind, sea state, and the like, and is somewhat unpredictable. The effect of internal waves is not well known. If water depth is less than a wavelength, transmission can become zero because of a wave-guide effect.

Lakes and reservoirs often are isothermal in winter but display a temperature gradient from top to bottom at other times. The ocean often is warm on top and cool below. These effects can bend a sound ray down into the bottom. It is preferable to have the receiving hydrophone on the same side of any thermocline as the transmitter.

Avoidance of surface layers generally reduces back scatter from trapped air and reduces flow noise from a moving platform. A receiving hydrophone very near the surface may see no signal because of the combination of a direct ray with one reflected at the surface with a 180-degree phase shift.

These variable factors can cause the sound to diminish in shallow water with the familiar inverse square law, but the signal sometimes will fall off inversely with the first power of distance; in general, some intermediate situation will exist in which transmission in shallow water carries further than in deep water because some energy is reflected back into a path hitting the receiver.

3. ELECTROMAGNETIC TRANSMISSION

There are a number of ways in which the ultrasonic transmission of information is inconvenient. The multipath propagation problem was mentioned as sometimes garbling information. In some cases it is inconvenient or impossible to approach the body of water closely enough to dip a receiving hydrophone into it; this would be the case in field work from an airplane or with streams in some canyons. In other cases in which an animal may move in and out of the water ultrasonic transmission does not provide a continuous signal. And finally, in some cases the attachment of a noise-maker to an animal may produce psychophysiological effects. To monitor a pure performance variable such as depth of dive an acoustic transmitter should generally be acceptable, but the presence of such a transmitter might actually alter some variable such as heart rate. (In the former case an intelligent animal might actually become interested in noting frequency change associated with depth change, and this could perhaps even spur him on to greater diving efforts.) From the work of Scott Johnson it is known that the dolphin *Tursiops truncatus* responds well to frequencies up to approximately 150 kHz, and it would only be frequencies several times this that would be free of suspicion in this regard. In these various situations, perhaps in spite of somewhat lessened transmission range of information, it is desirable to consider the use of radio signals to carry the information.

Placing a transmitting loop antenna under water causes several changes. As was mentioned in Chapter 10, a radio signal is slowed by a factor of the square root of the dielectric constant or relative permittivity when travelling in a medium for which this factor is greater than unity. Thus wavelength is decreased for a given frequency, or the antenna acts as if it were physically larger. In terms of what is called the "radiation resistance" of the loop (the radiated signal being treated as a loss to the circuit in a resistance), the equations indicate a very significant increase in resistance due to the introduction of a medium such as water with its high dielectric constant (approximately 80). In part, it is for this reason that a signal being picked up by a receiver in air may actually increase if a transmitter is dipped below the surface of a pool of fresh water. The transmitters of both Fig. 3.2 and Fig. 10.3 have been employed in useful experiments on animals swimming in fresh water; an example is given in Fig. 18.10.

A signal rising from within a body of fresh water will partly enter the air at the surface, but because of the extreme impedance mismatch much of it will be reflected downward again. The part entering the air will be refracted parallel to the surface, to a large extent. If the plane of the transmitter loop is horizontal, thus producing horizontally polarized waves, the signal will most effectively cross the interface. In any case, the predominantly horizontally polarized signal above the surface of the water can be picked up well in the radiation field by a portable receiver with its whip antenna held horizontally. The matter of frequency preference will be treated in an approximate fashion in Appendix 2.

Satisfactory transmission can be achieved through many liquids. Thus in Fig. 15.6 is seen a short-range experiment with a mouse into whose peritoneum had been surgically implanted a very small temperature transmitter. This mouse has his lungs filled and is breathing an oxygenated fluorocarbon liquid, this being possible at one atmosphere pressure rather than the higher pressures necessary if the animal were required to breathe water (see Clark and Golon, 1966). Such animals should be relatively resistant to acceleration effects (if fluid density is adjusted), explosions (if free of intestinal gas), and decompression sickness (due to tissue supersaturation, and also the explosive form). The receiving antenna is seen at the top of the beaker. It was felt that the use of telemetry would provide the maximum reliability as the animal moved vigorously about in the fluid. As the mouse is introduced into the fluid and goes from air to fluid breathing, there is a relatively rapid cooling towards room temperature, due to contact of the liquid with the relatively vast surface area of the lungs. It is perhaps for this reason that there seem to be no aquatic mammals having gills. A typical recording of the cooling curve is reproduced in Fig. 15.7, where the temperature of the liquid T_L is also indicated at various times. When the mouse is removed considerably later, and returns to air breathing, he again warms, but with a characteristic time approximately three times as great. In this recording, note the somewhat nonlinear temperature scale, and the irregular upward deflections of the

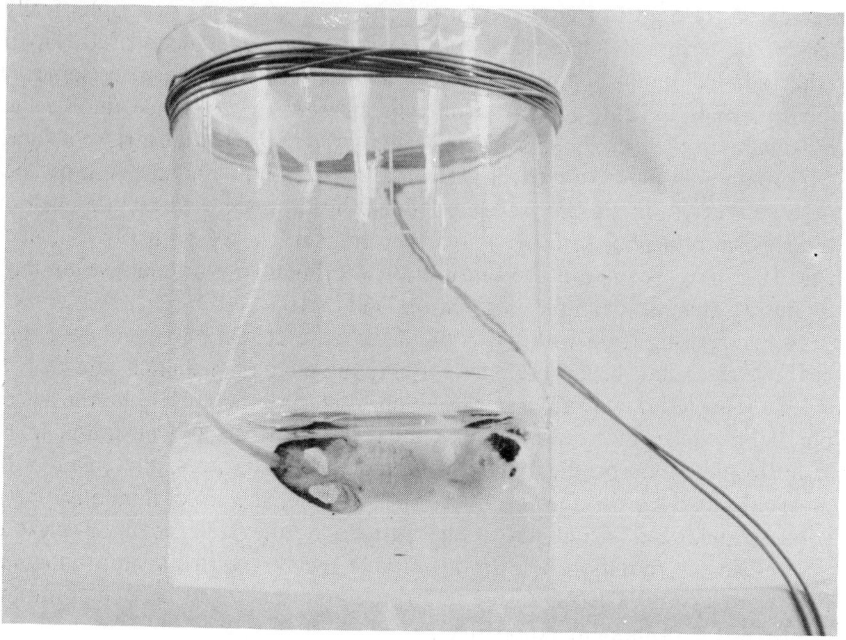

Fig. 15.6 A mouse immersed in and breathing an oxygenated fluid as core temperature is being telemetered. Brief prebreathing of pure oxygen was used to assure lung filling.

trace which denote momentary signal disappearance accompanying continuing activity. In this experiment the liquid was an inert commercial fluorocarbon coolant (3M Co., FC-75) having a dielectric constant of approximately 1.9 and a dissipation factor of approximately 0.0005; the telemetry aspect thus was not very different from being carried out in air.

Naturally occurring bodies of water show great differences in their electrical conductivity due to differences in the amount of dissolved minerals. The factors involved in the transmission of signals through relatively pure fresh water were indicated above. The opposite extreme has to do with animals that live in the ocean. In this case the partial conductivity of the water limits the passage of radio signals just as the metal walls of a shielded room prevent their passage. The higher the conductivity of the water, the more pronounced is the shielding effect. Also, the higher the frequency, the less able are signals to penetrate a given distance in a conducting medium.

In spite of this, some of the same methods can sometimes give useful information. Thus it was desired to study the marine iguana on the Gala-

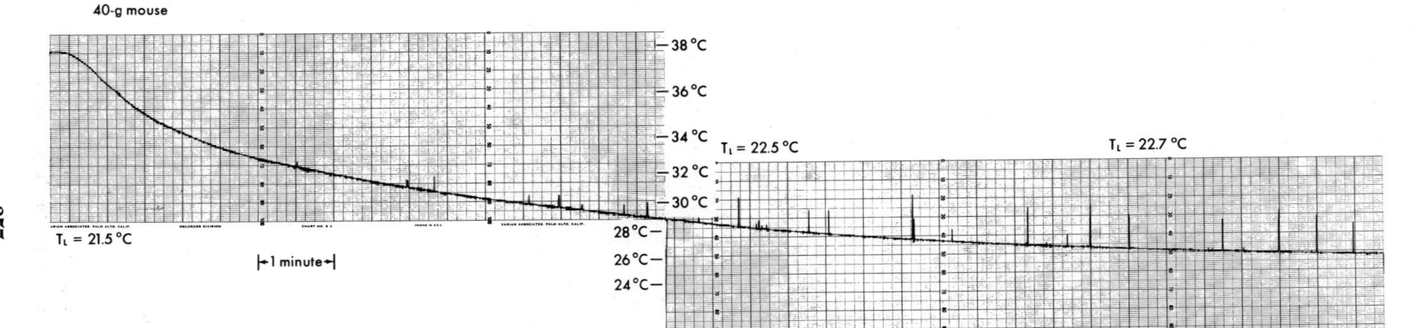

Fig. 15.7 When the fluid reaches the large area of the lungs of the mouse, his temperature drops drastically. The temperature T_L of the liquid increases slightly, as indicated.

pagos Islands with regard to temperature regulation, and without interfering with the normal pattern of activities which is involved in this control of temperature. This particular species periodically enters the cool Humboldt Current to feed on seaweed, and it apparently is the only lizard in the world which regularly enters the ocean. A photograph of such an animal with external transmitter attached was given in Fig. 1.15. A fairly large horizontal loop antenna was placed across the back of the animal, and a thermistor probe suitably inserted to monitor temperature. The transmitter itself was quite insensitive to temperature changes, and was in a covering giving neutral buoyancy in ocean water. Because it was desired to receive signals when the animal was on the land and also when in the ocean, the relatively low frequency of approximately 500 kHz was employed. It was found that, although the animal did warm considerably before entering the ocean, he did stay in and remained active until he had cooled all the way down to the temperature of the water. This animal has a metabolic process which is able to function under relatively wide excursions in temperature, as seen in Fig. 15.8. If it were not for the necessity to obtain transmission through an appreciable thickness of ocean water, a higher transmission frequency and smaller transmitting antenna would have been employed. With the unit shown, it was possible to track this animal through a mangrove tangle while monitoring his temperature and always remaining out of sight at ranges up to approximately 30 meters (Mackay, 1964B).

An interesting challenge is to obtain maximum range signals from within animals that are normally totally within ocean water. Some success along

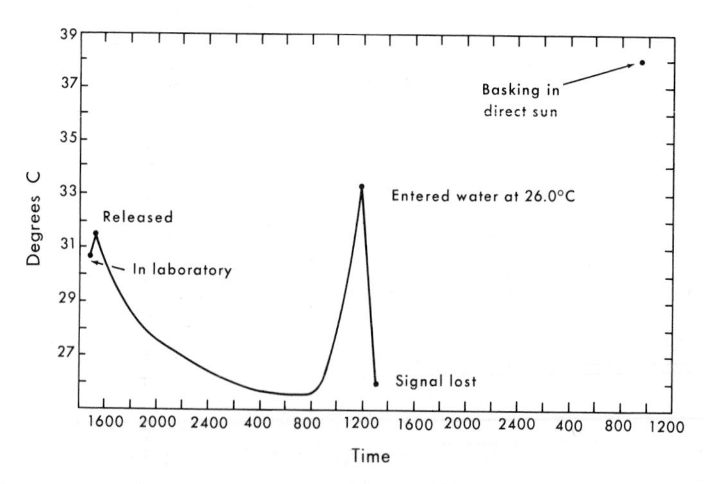

Fig. 15.8 Temperature of marine iguana on land and in the ocean near the Galapagos Islands. Activity was maintained under rather large excursions in temperature.

these lines has been achieved (Mackay, 1964C). It was found that a pair of exposed metal electrodes approximately 8 cm apart made an extremely effective transmitting antenna, and that useful transmission ranges could be achieved if the lowest possible frequencies were employed. Before going into these matters further it is desirable to briefly consider the transmission of signals through ocean water.

In a partially conducting or "lossy" medium, both velocity and penetration (or attenuation) of electromagnetic signals depend on frequency. Consider first the shielding or attenuation effect. In Chapter 10 there was mentioned the quantity "skin depth," or the distance of penetration through an extended conductor at which a signal is reduced to approximately one third of its initial value. Depending upon transmitter strength and receiver sensitivity, one should generally expect to be able to receive a signal at a range of approximately a dozen skin depths. The equation is indicated in Appendix 2, and depends upon electrical conductivity and frequency. The electrical conductivity of all ocean water is not the same, because of variations in salinity. But approximate characteristic numbers can be inserted into the equation to indicate that, for ocean water, the skin depth in meters is approximately 250 divided by the square root of the frequency. Physiological saline is somewhat less conductive, and thus the corresponding distance is a bit longer.

In a conducting medium both wavelength and velocity of propagation are definite functions of frequency. The wavelength in meters can be computed to be approximately 1.83×10^3 divided by the square root of frequency. Phase velocity measured in meters per second is approximately 1.83×10^3 times the square root of frequency. In making such calculations the equations developed in Stratton (1941) or Chapter 6 of Schelkunoff (1943) are valuable. Notice that a combination of the information of the previous paragraph with the present equations indicates that signals in sea water are attenuated at a rate of approximately 55 dB/wavelength.

We can compare electromagnetic signals and sound signals in the ocean, and note that, for both rate of attenuation and velocity, a 1 Hz electromagnetic signal is like an acoustic signal of 100,000 Hz. In ocean water the wavelength in meters of an electromagnetic signal with a frequency of 10 Hz is 585, whereas at 1 MHz the wavelength is 1.8 m. With an attenuation of 55 dB in such a distance it becomes clear why high-frequency radio signals are not very appealing for transmission.

The equations for the electric and magnetic dipole fields in an isotropic conducting medium also exhibit near and far fields, with a cross-over region again occurring at approximately one tenth of a wavelength from the source. Because the form of the equations is generally the same as in the air case, the shape of the fields is similar. The amplitude, however, changes drastically because of the attenuation factor and large phase changes occur. It is important to note that the same exponential attenuation factor appears in the field terms of the oscillating current dipole that apply to a uniform plane wave, and thus the same attenuation and velocity figures do apply to the dynamic fields of a dipole. The detailed mathematical treatments of signals in conducting media can become extremely complex, with almost every issue of the Institute of Electrical and Electronics Engineers Transactions on Antennas and Propagation containing a relevant article. The groups most generally interested in such problems are researchers attempting to determine accurate vector electrocardiograms, those interested in underwater and submarine communications, and those interested in long-range surface wave communication systems. Useful information is to be found in Hilliard (1962), Durrani (1962), Burrows (1962), Wheeler (1958), and all the papers in the May, 1963 issue of the *IEEE Transactions on Antennas and Propagation*.

Initial consideration of the transmission of radio signals from an ingested transmitter in an untethered dolphin suggested many problems. First of all, the transmitter would have to be quite powerful and of a very low frequency if any signal at all were to be transmitted. Also, *post mortem* studies on *Tursiops truncatus* indicated it to have an extremely long narrow gastrointestinal tract. Indeed, it appeared that a transmitter 1 cm in diameter would be unlikely to pass, although many subsequent observations have proven this to be untrue. Also, at that time there was no known way to anesthetize a dolphin, unconsciousness being accompanied by cessation of breathing. Thus an impacted transmitter could not have successfully been removed. Therefore an extremely thin transmitter was desired which would have a possibility of negotiating various turns of the intestine, yet a strong signal would have to be supplied by its antenna. There was at that time no information about the expected time of passage of an object through this species, and there were a number of other major uncertainties. A long thin ferrite core wound with the coil was considered for application as a transmitting antenna, but it might have had trouble turning corners in the body. Prior to this, a transmitter tested in the human body had conveyed its signal to the outside of the body by conductivity from a pair of exposed electrodes. The method had been successful, and thus the dolphin transmitter was constructed with a projecting length of springy wire with an exposed tip, the other electrode being at the transmitter. Currents flowing through the animal's body between the two electrodes would spread into the surrounding medium to provide useful signals at a receiver, and the method also provided satisfactory transmission when the animal was lifted from the water.

In the initial experiments core temperature was of interest, and thus a pulsing transmitter which could be timed by ear with a stop watch was employed. This allowed strong pulses to be generated while giving a battery life of at least 24 hr. The lowest possible radio frequency was desired, but it was known that an increase in frequency would reduce the likelihood of sensation or ventricular fibrillation in the subject. An initial series of experiments was performed in which the transmitter was applied to the human tongue. Individual pulses could be sensed if the frequency was 20 kHz, but not when the frequency was above that. Accordingly, a transmission frequency a little above 50 kHz was chosen. In preliminary experiments at Steinhart Aquarium in San Francisco this transmitter was tested from within the investigator's mouth. Various receiving antennas and probes were tested, which the dolphins playfully tried to eat out of the investigator's hands. It was found that a large receiving loop either in or out of the water made an effective receiving antenna, but that a pair of electrodes spaced a few feet apart and extending into the water were more effective. Grounding the receiver through a third electrode in the water minimized noise, and this is essentially the arrangement that has been used ever since.

The circuit of Fig. 15.9 was constructed with extreme care to make its diameter small. The output electrodes were pieces of platinum wire, while

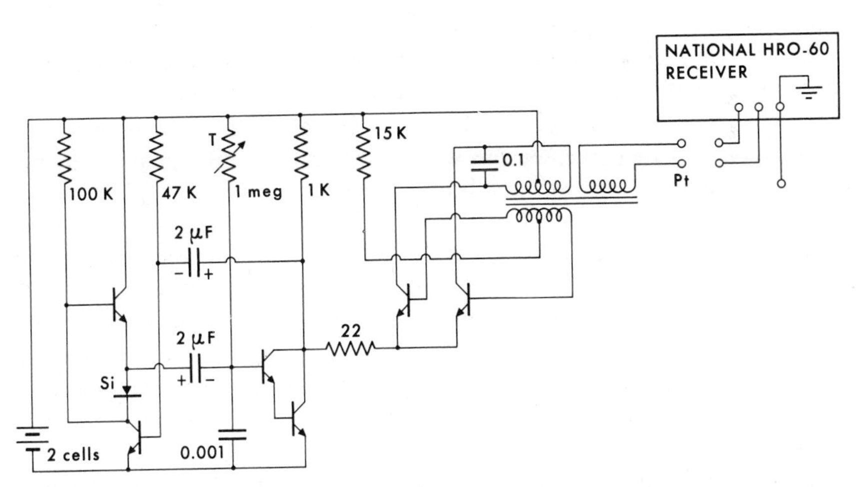

Fig. 15.9 Periodic impulses of electromagnetic energy come from the electrodes at a rate set by the thermistor in the multivibrator. The circuit was originally constructed by using an outdated transistor type MT101, but other types such as the RT930H are quite effective. The commercial receiver indicated tunes down into the range of 50 kHz. The signal is picked up from electrodes hanging in the water.

the receiver electrodes were merely small pieces of copper soldered to the ends of the wires. The circuit is seen to consist of a multivibrator which periodically applies power to the output oscillator in order to generate a short pulse. The problem of generating short pulses spaced by long intervals in this case was solved by the extra transistor and diode at the left in the drawing. As originally constructed, this circuit employed the now outdated transistor type MT101, but it has since been rebuilt a number of times using throughout other transistors such as the RT930H. For minimum diameter the original transformer coils were wound longitudinally through a ferrite tube, but more recently the output transformers have been formed on ferrite cup cores. A typical transformer would employ 18, 18, and 7 turns giving a peak-to-peak output voltage of 1.16 V when unloaded, and 0.85 V across 60 Ω. The commercial receiver indicated in Fig. 15.9 tuned down to this frequency band. As in the other cases, a beat-frequency oscillator was switched on for greater signal noticeabilty, and pulses of at least 20 ms duration were employed. The transmitter was inserted into the gill of a dead fish. The dolphin swallowed it whole without chewing, as shown in Fig. 1.17.

The studies on the dolphins have all been conducted at the United States Navy Marine Biology facility at Point Mugu, California. Successful transmission has been achieved in every experiment, with transmitters of the above type allowing strong reception of a signal by a receiver at pool side from an animal in any position within a pool 50 ft in diameter and 7 ft deep.

Sometimes a greater transmission range is achieved than expected when working in shallow water because some signal goes to the surface and is refracted along it. Some of the signal which travels through air with less attenuation can then reenter the water to give a useful signal at greater range. Although they have not proved necessary so far, omnidirectional systems related to those in Chapter 10 could also be devised for the present case. Corresponding to the use of three coils, four electrodes would be employed, not all in one plane. We can also produce electrode configurations that reject uniform fields in the water, such as might be produced by external disturbing sources. Thus, if two electrodes are each attached through a resistor to one terminal of the receiver, and the other terminal goes to a third electrode placed midway between the original two, then all fields of large extent tend to cancel. A generalization of this idea can employ an odd total number of electrodes arranged in a line, with the even ones in the line being each connected through resistors to one side of the receiver, whereas each of the odd ones is connected through a resistor to the other side of the receiver. This allows the covering of a large expanse, although there are a few dead spots in the array which must be separately monitored. In some cases, the limited range of radio transmission can be

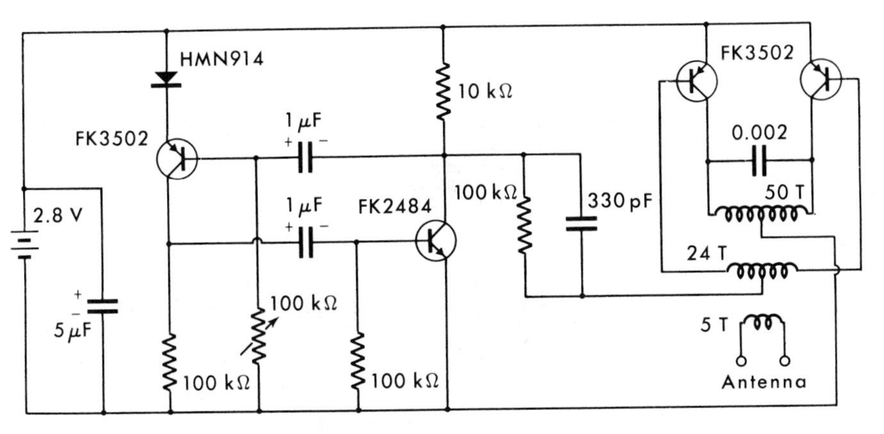

Fig. 15.10 A temperature transmitter using a different multivibrator type for generating short impulses at intervals long enough to time by ear.

overcome in trained animals by having them follow a receiving antenna in going about their tasks in the ocean.

Since the above exploratory observations (Mackay, 1964C), similar transmitters have been constructed in other forms. Thus a complementary pair multivibrator can be used to switch the output oscillator, as in Fig. 15.10. This circuit has a slightly higher repetition rate, for more convenient decoding by relatively rapid responding circuits, and a somewhat lower transmission range. The photograph of such a transmitter is seen at the bottom of Fig. 15.11. A pair of batteries is slid into the case under the contact, and the plastic cap is screwed on to start operation. The ring just in from the right-hand end of the case is one electrode, while the other is at the end of the projecting springy but somewhat flexible wire. Above this transmitter is seen a more complicated unit requiring three batteries, and being carried in a metal case covered by a plastic case. This unit is used for the continuous transmission of voltage fluctuations sensed by the electrodes themselves, which are converted into a radio-frequency signal reradiated from the same electrodes. Such a circuit is described later in this section. The complete electronic components are seen at the top of the picture before being slid into the case. The case has its watertight cap screwed off to the right. In this case, the other electrode is at the end of the screw cover for the battery compartment, and connection is automatic when this is put into place. After many experiments we have found it safe to feed an object 1 cm in diameter to *Tursiops truncatus,* although the initial transmitter was limited in diameter to 6 mm.

Some information about temperature regulation in this species has been

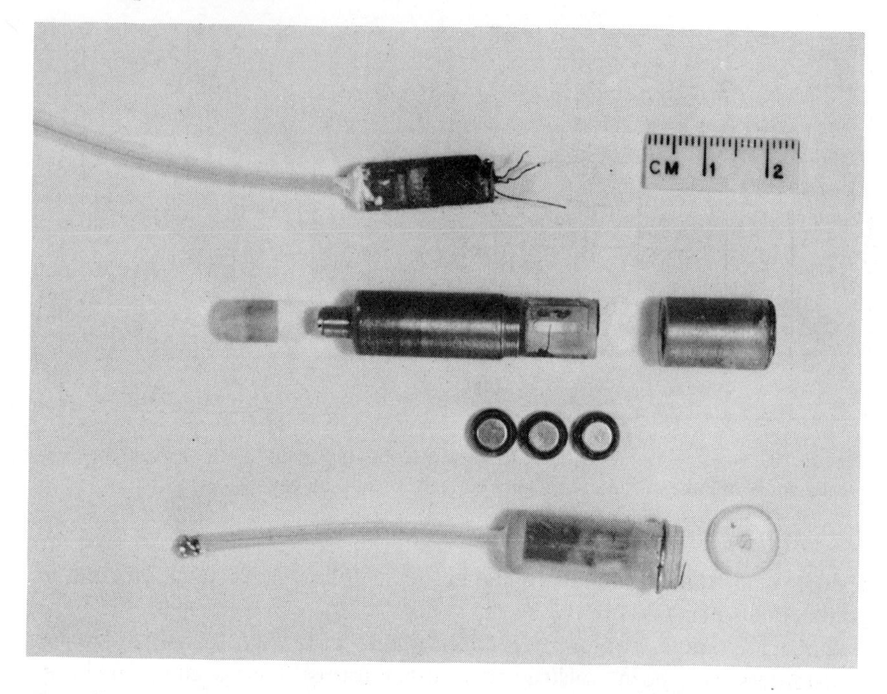

Fig. 15.11 At the bottom is a complete temperature transmitter for use in ocean water, and at the top are the components for a voltage transmitter. The upper transmitter employs three batteries, the one below, two. In both cases the signal appears between an electrode at the end of the case and at the end of the extending wire.

collected. Like other mammals, these animals show some fluctuation in temperature during the day and night. An example (Reid and Mackay, 1968) is given in Fig. 15.12. In this case digital timing circuits were used to record the interval between each pair of pulses, and a line through each point indicates the uncertainty in the observed temperature. Although the temperature regulator of these animals is quite effective when they are in the water, if they are removed into the air their temperature starts to rise, and their flippers and flukes begin to feel relatively hot when touched. Dumping the animal back into the water causes a large temperature transient which can appear oscillatory.

Some observations on the transient response of the dolphin temperature regulator have been made (Reid and Mackay, 1968). In this case a dolphin was suddenly transferred from water at one temperature to that at a somewhat different temperature, as shown in Fig. 15.13. In the second tank the animal was free to move about in a normal fashion while his temperature was recorded by telemetry. Several animals were studied, and it was observed that transference between pools at the same temperature did not cause a temperature transient within the animal. The result of a sudden cooling in the ambient temperature led to the transient in Fig. 15.14. A major part of the research here was in developing mathematical and computer methods for modelling observed data by relatively high-order differential equations. The fit to the overall data of two different mathematical models is indicated in Fig. 15.14. Model No. 1 involved a third-order equation while Model No. 2 is a second-order equation. From the actual form of the differential equation, we can state what kinds of feedback interconnec-

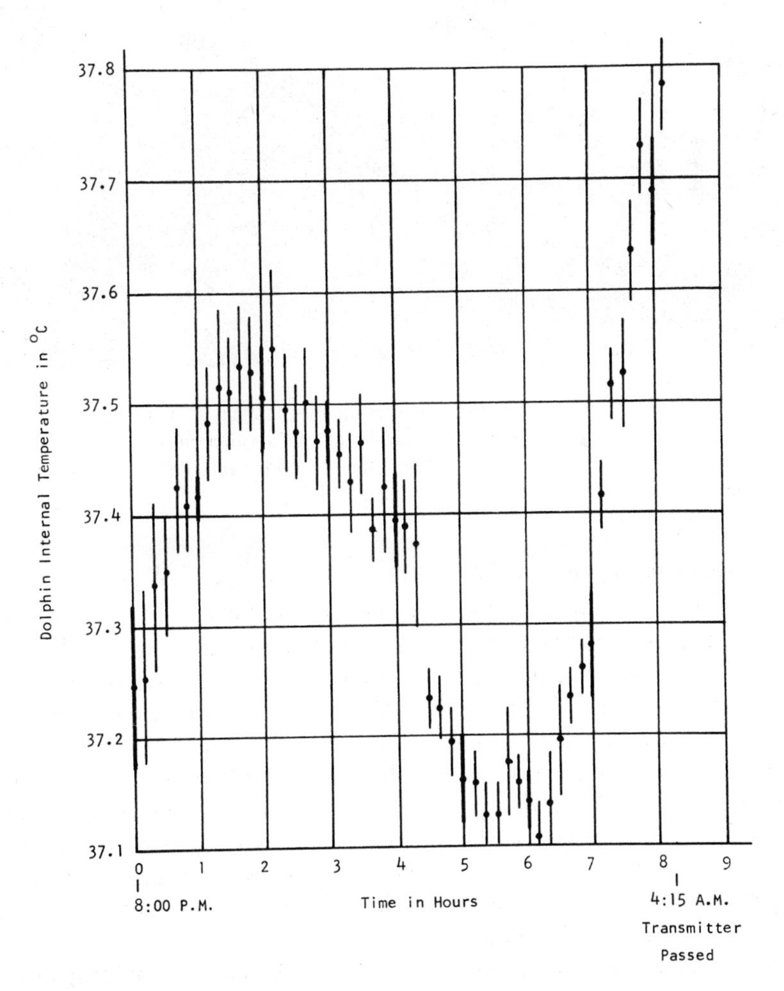

Fig. 15.12 Temperature in the gastrointestinal tract of an untethered dolphin during the night.

Fig. 15.13 To test the temperature-regulating mechanism of a dolphin it was suddenly transferred from a tank at one temperature to another at a somewhat different temperature. During free movement in the second tank core temperature was recorded.

tions might be functioning. Although such a process can give misleading results, in the present case, the plausible arrangement of Fig. 15.15 is suggested. This is an incremental model for dolphin temperature regulation, in which some rate-sensitive feedback seems to occur. From other lines of reasoning, the particular physiological components are denoted as possibilities. Such an arrangement could generate the overall form of the observed transient, but is not necessarily unique. With temperature gradients in the body, the effects of any vasoconstriction could contribute to such a transient. By implanting sensors at several positions, it will be possible to check such aspects now that anesthetics are known for these animals. In any case it is clear that these animals have a relatively effective temperature regulator if they remain in the water.

In some cases it is desirable to transmit rapidly changing information for which even frequently occurring pulses do not conveniently provide a sufficient information rate, in which case continuous transmission is desirable.

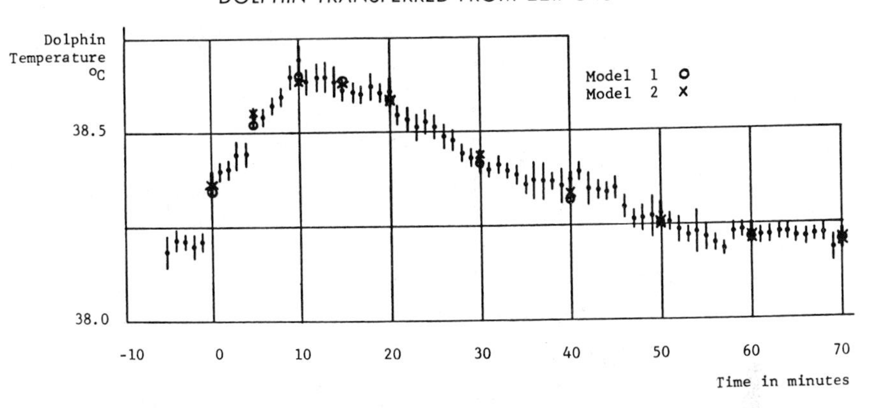

Fig. 15.14 Temperature transient in *Tursiops truncatus* following transfer between two different water temperatures. The apparent jump at the start of the curve is strongly dependent on the moment in the transfer called zero time. A digital computer was used to fit differential equations to the main part of this transient, and the perfection of the mathematical models is indicated.

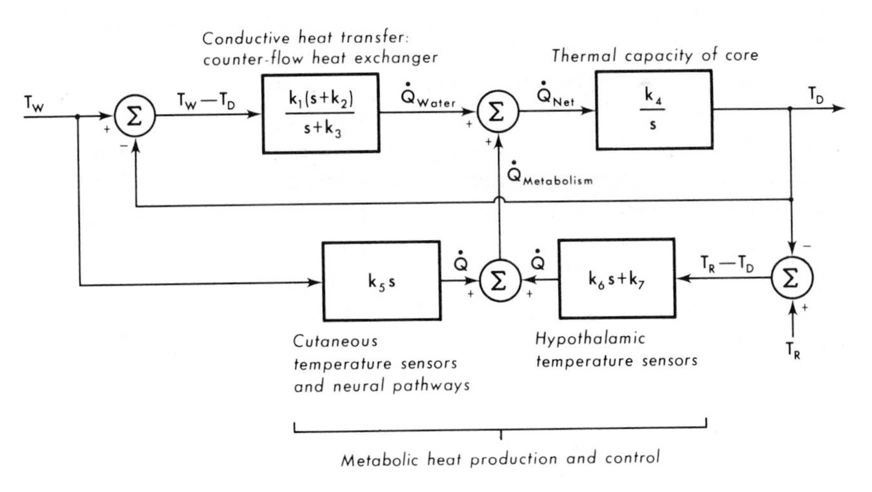

Fig. 15.15 A model that can explain the observed transient temperature response of the dolphin. Here T_W is the water temperature, T_R the dolphin "reference" temperature and $\dot{Q}$ is heat flow. The anatomical labels are added from other lines of reasoning.

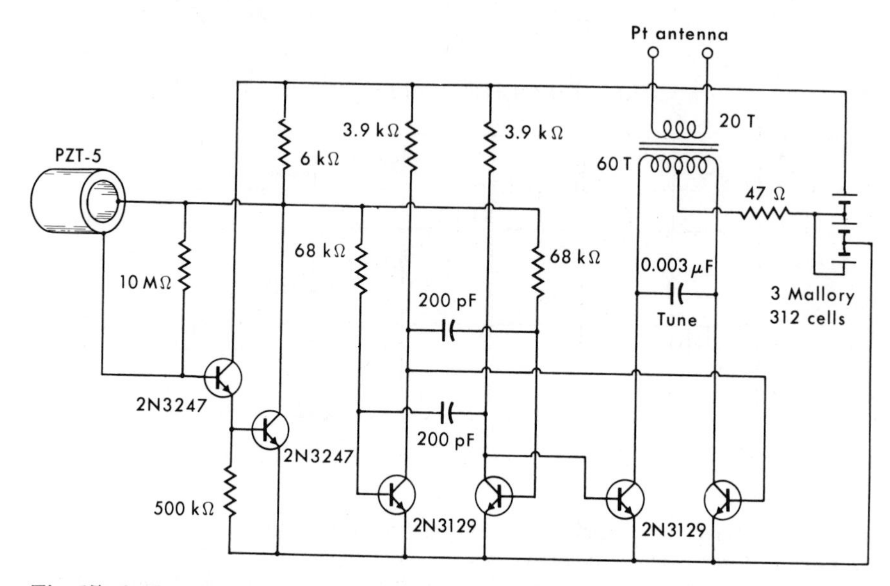

Fig. 15.16 Sounds sensed by the input transducer modulate in frequency the output electromagnetic signal coming from the antenna electrodes.

Thus Fig. 15.16 shows a circuit for the transmission of sounds picked up by the piezoelectric cylinder, and converted to a fluctuating voltage (under water such a sensor can be more effective than in air because of the better impedance match to the surrounding medium). To give a good low-frequency response, the signal from the transducer is fed into the high input resistance stage composed of the first two transistors, which has been shunted by a resistance of 10 MΩ. This signal is then communicated without an increase in voltage to control the frequency of switching in the multivibrator at the center of Fig. 15.16. The two transistors in this multivibrator control the switching of the transistors that comprise the output stage, which must be approximately tuned for satisfactory operation. Voltage in the circuit is that of two batteries in series, but one of them has a greater current drain and thus is comprised of two cells connected in parallel.

The basic frequency stability of this circuit is relatively high, although its sensitivity to applied voltages is also relatively high. The reason for this can be seen in terms of the approximately constant voltage between base and emitter of a transistor, which can be called "one diode drop." When a transistor in the multivibrator switches off, its collector tends towards the full positive battery voltage, but cannot go more positive than one diode drop because of being directly connected to the bases of the output transistors, whose emitters are grounded. The base of a conducting transistor is positive by one diode drop, and when switched off drops in potential by the downward swing of the other collector, which is one diode drop; thus the base drops to ground potential, from which it starts its upward voltage rise. The potential towards which it heads is positive by two diode drops because of the arrangement of the input-transistor pair, but switching of the multivibrator actually takes place at the halfway point when the voltage arrives at one diode drop. Since the voltage excursions involved are actually rather small, relatively small injected voltages cause significant changes in frequency. The alternative way of getting sensitivity to small voltage changes is to have switching occur far out on a slowly changing exponential voltage rise, but this leads to erratic performance. In spite of this sensitivity to small voltage changes, if the temperature changes, then the values of the diode drop all change by about the same amount, and the halfway point is reached at about the same time; that is, the frequency is unaffected. If the battery voltage changes, the currents through the various diodes change in somewhat the same way, and thus here also there is a tendency for compensation against unwanted frequency changes.

If moisture shunts the input circuit, thereby reducing its resistance, then the response of this circuit to low-frequency sounds will be limited. As shown, the transmitter is relatively effective for picking out frequencies as low as those of heart sounds.

Figure 15.17 shows a very similar circuit which takes in low-frequency

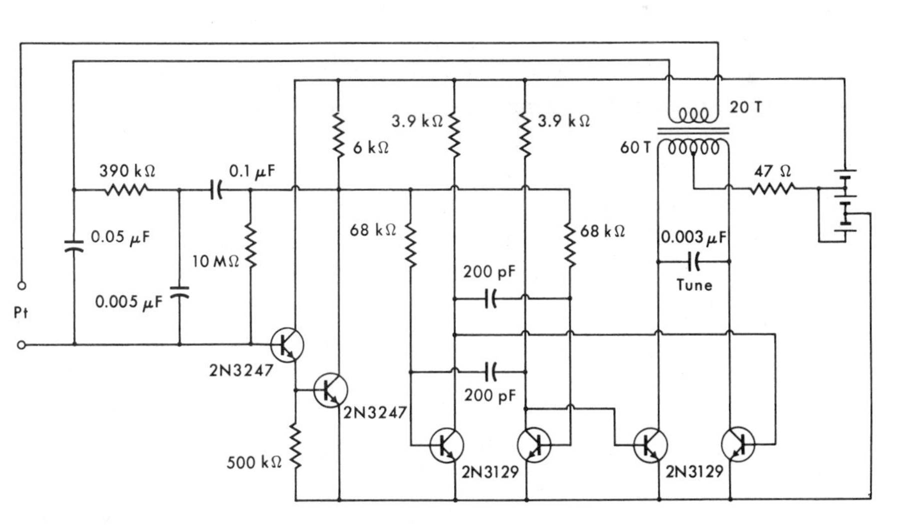

Fig. 15.17 Modification of Fig 15.16 in which low-frequency voltages picked up by the electrodes modulate the frequency of outgoing radio-frequency currents from the same electrodes.

bioelectric potentials at the input electrodes and converts them into frequency-modulated outgoing radio-frequency signals through the same pair of electrodes. The input filter arrangement allows for the effective circulation of radio-frequency currents in the electrode circuit, while passing only the low-frequency voltages of interest into the input amplifier circuit. This is the circuit that was seen packaged in the upper part of Fig. 15.11. Such circuits as these have been successfully swallowed and passed by *Tursiops truncatus*. Although continuous recordings of cardiac activity have not been obtained, indications have been obtained that when one of these animals, in a completely undisturbed state, holds his breath to dive and play with a toy, his heart rate drops by about one third. Diving bradycardia as observed in a laboratory can be augmented by emotional factors. Such a voltage-sensing circuit as was shown in Fig. 15.11 can be applied externally with suction cups, and the electrocardiogram transmitted from a human subject is shown in Fig. 15.18. The relationship between this circuit and that seen in Fig. 3.3 should also now be clear.

Similar circuits can be used with subcarrier oscillators, or they can be used directly with modulators of other sorts. In order to use the output stage previously shown, a large amplitude driving signal is useful. At the top of Fig. 15.19 is an oscillator into which has been inserted a resistance, the result of which is to give the large amplitude waveform shown. Although the frequency stability is somewhat degraded, this circuit can be frequency modulated by a variable inductance or variable capacitance. To drive a dual output stage, the arrangement shown below can be employed. If significant amounts of power are to be emitted, then it is still generally advantageous to add a capacitor to the primary of the output transformer in order to tune this circuit into the general range of the operating frequency.

Generally, when such a transmitter has been fed to a dolphin, it has been passed by the next day. Recovery of such a transmitter in murky water can be troublesome. However, one can rather accurately localize the position of the transmitter sitting on the bottom by attaching one of the receiver antenna wires to a long pole and moving it about until the loudest possible sound is

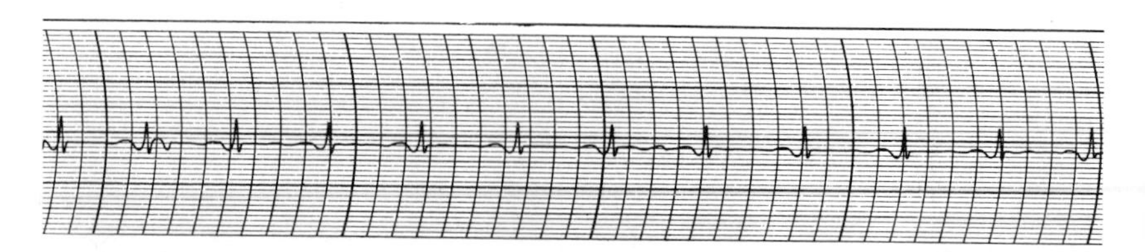

Fig. 15.18 Human electrocardiogram transmitted by the circuit in Fig. 15.17 when affixed with suction cups to the chest.

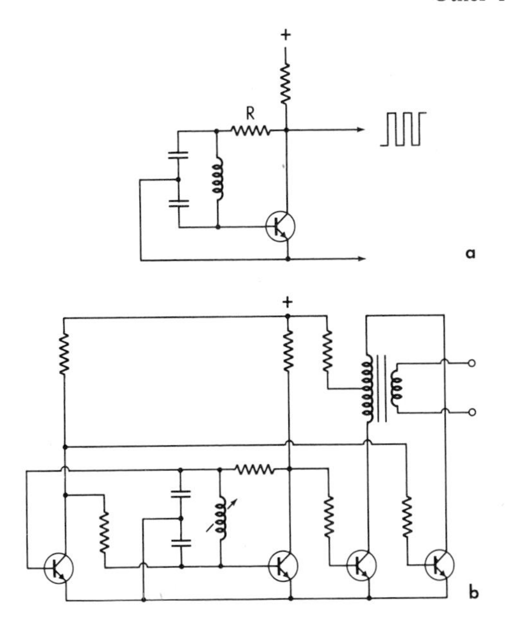

Fig. 15.19 Introduction of a resistor into an ordinary oscillator results in greater voltage excursion at *a*. At *b* is the corresponding configuration with a separate output stage.

heard. If a scoop is placed on the end of the pole, it will generally be found possible to retrieve the transmitter without actually being able to see it.

4. OTHER METHODS

In the case of the aquatic animals that periodically surface to take a breath of air, it is useful to consider the possibility of the radio transmission of tracking and physiological data during only these periodic intervals. In that case it would not be necessary for the radio signals to traverse the highly attenuating medium to study ocean-living forms. Most of these forms surface often enough to provide good continuity in direction-finding information. In the case of the transmission of physiological information, the situation is a bit more complicated, since all information generated since the previous breath would have to be stored, and then suddenly radiated in speeded-up form in a burst during the breath. This is probably most feasible at the present time by the use of preprocessing of the data, so that during the breath only the result of the experiment is transmitted, and not all of the raw data. Thus if we are interested in a possible diving bradycardia, instead of transmitting the complete electrocardiogram since the previous breath, we would merely transmit a signal proportional to the recent rate of change of heart rate.

There are some problems peculiar to the transmission of radio signals over the surface of water. There tend to be strong periodic nulls at the surface of the water owing to the combination of direct and reflected waves. Receiving antennas at several heights can be helpful, as can the use of a high-receiving antenna with an upward directed reflector near the surface of the water.

The velocity of a radio signal through air is affected somewhat by the moisture content, and humidity gradients over the ocean can trap an electromagnetic wave near the surface of the water. One can speak in terms of refractive index, and in a region where refractive index decreases with elevation, instead of the normal increase with elevation, there is the action of a duct that can propagate waves with very little attenuation over great distances in a manner similar to the transmission of waves through a wave guide. The depth of a duct over the surface of a body of water may be only 5 to 20 m, or it may be over 300 m deep. Ducts exhibit a low-frequency cut-off characteristic similar to a wave guide, with the frequency being determined by the depth of the duct and by the strength of the discontinuity and refractive index at the upper surface of the duct. The lowest frequency that can be propagated by such a duct is seldom below 50 MHz and usually will be greater than 100 MHz, even along the Pacific coast of the United States. A guided propagation through such a duct can extend ranges far beyond normal, although a receiving antenna outside of the duct would see little signal.

Schevill and Watkins (1966) have made some observations with 140 MHz transmitters of 1 mW power transmitting from the surface of the ocean with a quarter wavelength (50 cm) whip antenna. They note that with a receiving antenna at a height of 15 m, transmission range was limited to about 6 km. At 6.5 km small wavelets appeared to intercept the signal, and by 7 km the signal could only be received occasionally as the transmitter was carried on the crest of a wave. Local wave interference was pronounced at low angles of radiation, as when listening from a distance. Waves of 50 cm height at a transmitter-to-receiver distance of 30 km transform a steady transmitted signal into an unreadable one with a receiver in an airplane 300 m above the ocean's surface; with a calm sea and an altitude of 300 m, Schevill and Watkins found a good steady bearing to be received at the same distance. Range was generally shorter than anticipated from the curvature of the earth. In a zero sea-state, with the top of the quarter-wave whip antenna no more than 60 cm above the surface of the water, they report ranges for different receiver heights as reproduced in Table 15.1. They report that the distance at which the first bit of intermittent reception

Table 15.1

Receiver Altitude (Meters)	Range (Kilometers)
150	20
300	32
600	45
900	57
1200	70
1500	80

is noted is quite repeatable, and provides a means of estimating the distance from a transmitter.

These ranges appear to be useful, although there were some problems in actually applying these transmitters to the study of the migrations of whales. A small boat could perhaps be arranged with powerful booster transmitters and a somewhat elevated transmitting antenna to follow an instrumented animal, always remaining at a suitable range so as not to interfere with normal activities. Successes with various homing weapons suggest that this following action could be maintained automatically by an unmanned craft. Engines might simply be disengaged or turned off, for example, when the round-trip transit time to a transponder on the animal became less than a preset amount, with rudder constantly being directed for progress toward the animal.

Transponders are devices that return a signal upon receipt of one, and will be discussed in Chapter 17. They could be useful in the present applications in ultrasonic form, not only for locating the position of an animal, but also for communicating biological information as needed, thus conserving battery power. An "interrogating beacon" could be placed at a particular location so that any animal passing that particular spot and carrying a transponder would automatically have his transit recorded, perhaps in conjunction with certain kinds of information. A typical sonar set can provide the impulse to activate such a transponder, and by the delay time in the returning signal give an estimate of range. Because of the directional properties of the receiving system, it is also possible to determine the direction to such a transponder; direction and range give the exact location of the transponder. The range to two fixed stations, without direction information, is sufficient to indicate the position of a transponder, although there is an uncertainty as to whether the transponder is ahead of or behind the line joining the two receiving stations. This ambiguity can usually be resolved; for example, if the stations are at the shore line, the region behind is dry land. If an animal at the surface simultaneously emits a radio impulse in the air and a sound impulse in the water, then the difference in arrival time of these two signals at one receiving station gives range, which can be combined with direction information to uniquely fix the location of the animal. An animal fitted with a small sound transmitter that emits impulses at random times (perhaps transmitting physiological signals) can have its position uniquely fixed without requiring directional information. This can be an advantage since directional information can either be time consuming to collect, or else require relatively complicated arrays of receiving elements for instantaneous indication. The difference in time of arrival of an impulse at three receiving stations is sufficient to fix the location of the transmitter. If the animal is in addition carrying on significant diving activities, then the difference in reception time of an impulse at four locations, not all in one plane, is sufficient to fix the three-dimensional position of the transmitter.

To follow the long-term migration of an animal over great distances, it may prove most practical to periodically release floats, or batches of dye, or small explosive charges. To follow a major migration might require such a release only once per day. However, the actual difficulties in spotting an object or discoloration in the open ocean from the air makes the use of acoustic methods appealing. The well-known extremely long range of transmission of sound from a deep source makes such a method particularly interesting, especially when the high energy content stored in even a small explosive charge is considered. Of course, to prevent disturbing the animal, these charges would have to automatically be released from some external carrier, where they would drop away and fire after a significant delay time.

If it is expected that the animal under study will be seen at a later time, then it is probably most convenient for information to somehow be recorded (Chapter 17), and then later be collected for analysis. Thus the percentage of time spent at different depths can be inferred by placing a radioactive bead on the pointer of a pressure gauge, upon whose face has been placed a sealed photographic film. On recovery, degree of darkening at different positions corresponds to amount of time spent at these positions, although the actual temporal sequence is not indicated.

If a continuous indication of some parameter is desired, then it is sometimes possible to use other forms of transmission. Thus a light can be modulated to carry information through limited distances in water. In a tank with a captive animal such a procedure could be used, for example, to transmit heart rate, by attaching a small lamp to the animal and arranging that it flash with each beat of the heart. At night, the flashing of a small strobe light is quite noticeable, and might be used for tracking the position of an animal that periodically surfaces for a breath. In many cases it is possible to arrange that such a flash occur only upon surfacing, and with a suitable shade so that the animal is relatively unaware of the flashes.

With some of the larger aquatic animals it is interesting to contemplate various ways in which swimming activities might actually power various transmission systems. Changes in pressure upon diving could activate a bellows in order to do significant amounts of work, for example. Vibration of a small circular cylinder held perpendicular to the flow of water could be used to generate power as an oscillator, and would also give a measure of velocity since the frequency of vortex shedding is proportional to speed. The power involved in the flexing of the tail of the animal might be tapped more directly in the case of an implanted unit. The motion and efficiency of swimming might be monitored by an implanted accelerometer which, if self-generating, might directly power a small transmitter. Although it is probably impossible in general to tap the ambient sound sources in the ocean as a source of power because of the impossibility of producing a perfect diode (Chapter 2), it is perhaps possible to power certain small transmitters directly by the vocal activity of some aquatic animals. To work in a useful fashion with some of the large animals requires relatively large sources of power, although these same animals are also capable of carrying relatively large energy sources in the form of batteries. It is true, of course, that some experiments would probably require a powerful external booster transmitter, which would relay to a great range the low-power signals generated by some internal transmitters.

In working with various aquatic animals, different methods of equipment attachment can be employed. In some cases apparatus can be made self-detaching through the use of soluble washers, or metallic connections that corrode through electrolytic action when placed in salt water. Thus the above-mentioned recorder can be made self-detaching and recoverable, if it is not certain that the animal itself will be recovered. Objects can be attached to the shells of turtles with screws. Objects attached to the outside of a crab may well be lost following a molt. Some adhesives stick well to tissue, for example, Eastman 910, which retains some of its effectiveness under water. Dow Corning adhesive 269 is able to immediately stick an object to tissue completely immersed in water, although the strength of the bond is not extremely high. Silastic RTV891 liquid silicone rubber is able to glue materials under water, although its longer setting-up time suggests its use in combination with a more rapid adhesive type. The dorsal fin of some animals (e.g., most dolphins) is a rather inert structure as far as mechanical activity is concerned (although it can be involved in temperature regulation). Thus this is a convenient spot for placement of certain equipment, as well as being electrically appropriate for holding a transmitter with somewhat swept back vertical monopole antenna. At high speeds this placement introduces some drag but need not generate extreme drag by causing turbulence over the whole body of the animal.

In some investigations on various species it is desirable to inject various electrical, chemical, and other stimuli and then to record the resulting response. The same considerations that apply to carrying a signal *from* animals in the water apply to carrying a signal *to* an animal in the water. Because of propagation problems, there may be extra emphasis placed on the use of recorders with freely roaming animals. A magnetic recorder could be programmed to periodically inject a stimulus and then record the result. A single-channel recorder is sufficient for this entire operation since the recorded commands could be erased as used, and the responses recorded on the same tape somewhat "downstream," thus replacing stimuli signals by responses for later observation. These matters of remote stimulation are discussed in Chapter 16.

16

Inward Power and Telestimulation

1. REMOTE STIMULATION

The induction of electrical power into the body of an animal seems to have been an older concept than the outward transmission of radio signals carrying information. Chaffee and Light (1934) described a method for the remote control of electrical stimulation of the nervous system in which one or more primary transformer coils surrounded an animal cage and activated one or more secondary coils buried within the body of the animal, which was allowed free movement about the cage during stimulation. In this work, it was usual to discharge a capacitor through the primary transformer winding, thus giving very high currents, and there was little control over the wave form in the secondary coil. By going to higher frequency electromagnetic signals, a more efficient and flexible induction of power into an internal secondary winding can be made. Some of these matters have already been reviewed by Glenn et al. (1964).

A flexible scheme for the short-range transmission of electrical stimulating signals is depicted in Fig. 16.1. Within the animal is placed the receiving circuit shown, which comprises essentially a crystal-set receiver (Chapter 11). Near or around the animal is placed a transmitting coil which is activated by impulses of radio-frequency energy from some sort of pulsed oscillator. At the top of Fig. 16.1 is a set of wave forms appearing at several points within the circuit. If one wished to apply a periodic pulse to an electrode embedded in excitable tissue, then the wave form in the coil might be such as shown at *A*. In the implanted receiver at the left is a resonant circuit consisting of a receiving coil and its capacitor C_1, which combination is tuned to the basic radio frequency at which the signal is transmitted. Thus across the circuit at *A* appears the wave form directly

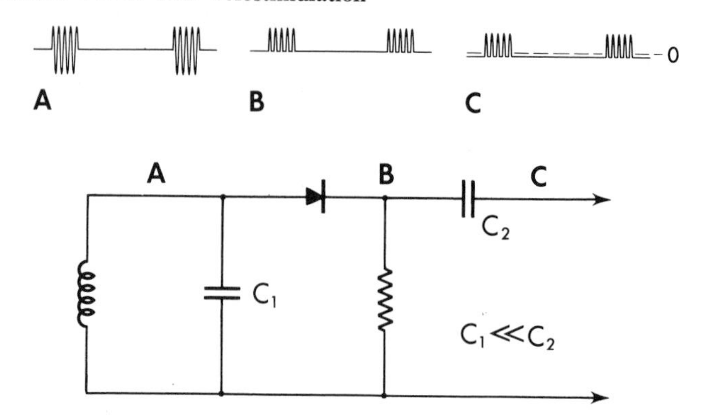

Fig. 16.1 Impulses that are to be applied to excitable tissue are used to amplitude modulate an external oscillator whose signal is picked up by the resonant circuit at the left of the implant. The induced signal is passed through a diode and to electrodes through any of several coupling configurations (see text).

above. This does ignore the effects of switching transients which generally do not much change these considerations. The voltage across the circuit at *A* is applied through a diode to give the wave form at *B*. These bursts of unidirectional impulses are able to exert a stimulating effect upon tissue, where the high-frequency wave form at *A* would not be. So to speak, radio-frequency positive half cycles are nullified in their effect by the immediately following negative half cycle, at *A*, although low-frequency alternating currents, of course, are able to stimulate tissue. A capacitor can be placed across the circuit at *B* to filter the wave form into square unidirectional impulses without fluctuations, but this is generally not essential. It will be noted that these impulses mimic the amplitude-modulation envelope of the input radio frequency, and thus various pulses of different frequency or overall wave form can be supplied. In the present case there is always a net flow of current in one direction, and this can cause problems at the electrodes. If a relatively large capacitance is placed in series with the output lead, for example, in the position of C_2, there can be no net flow of direct current. The wave form will actually be as at *C*, with large positive stimulating impulses, and a small negative voltage the rest of the time. A satisfactory stimulating effect can still be observed, with the small reverse flow of current over the longer interval merely preventing problems at the electrodes. In principle, for this back current to flow it is necessary to place a resistor as shown at position *B*, but it can generally be omitted in practice since the backward leakage through the diode and coil provides a corresponding path.

In one of the most ambitious applications of circuits of this general type,

Brindley and Lewin (1968) have implanted 80 stimulating receivers in the head of one human subject in attempting a prosthesis for blindness. Sensations of light and predictable patterns were reported.

These systems are convenient for the induction of power into the body over short ranges. In fact, grounding problems often require the use of some such high-frequency transformer arrangement even when wires are to be run directly from a stimulator to an animal. In such a case the signal to be applied modulates a radio-frequency oscillator, which is then coupled to the wires to the animal through a radio-frequency transformer having rather good isolation between windings. In some cases the transmitter coil and receiver are incorporated into the handle of a probe for stimulation of a neurophysiological preparation. Then with several probes in place at once, a stimulating current into one will not run into the others via the minimal capacitance to ground through the generator.

In the case of the induction of signals into a freely moving animal, the problem of amplitude control does exist, if this parameter is of importance. Thus the system is not omnidirectional, although with a trio of coils it can be made so. The energizing field must also be made uniform if reproducibility is to be achieved. In some cases it is feasible to place a string of silicon diodes or a Zener diode across this circuit to control the output voltage by limiting it to a particular value. Alternatively, a large overall feedback loop can be arranged in which a small transmitter senses the voltage across the internal terminals and broadcasts a signal to the stimulator of such a form as to maintain the electrode voltage (or current) quite constant. Notice that in most of these arrangements, near-field transmission is employed (Chapter 10), and thus these arrangements can be considered as transformers.

Perhaps the best known application of these methods is to the stimulation of the ventricles of the heart for pacemaker applications. Other investigators, however, have used them to help with the emptying of the neurogenic bladder, for control of incontinence, for the stimulation of the baroreceptors in the neck (at the carotid sinus) to help with the reduction of blood pressure, for phrenic nerve stimulation in connection with breathing problems for attempts at pain blockage, and also for bypassing a defective ear in cases of hearing loss. Many potential applications, ranging from the facetious to the important, suggest themselves. With such equipment, one could obviously rig a jumping-frog contest. Electrical ejaculation methods might employ radio-frequency techniques to advantage in certain cases where there were fertility or related problems. In using these techniques it undoubtedly would be possible to construct barless cages for animals, either by laying out a stimulating transmitting antenna over the limiting perimeter, or else by automatically administering some sort of graded shock as the animal wandered from some central point. Since these receivers require a high signal level for activation, there is little likelihood of accidental interference unless the subject is within a meter of certain electrocautery or diathermy machines, or extremely close to certain radio transmitters or neon signs or motors.

Another important application here is the remote control of the stimulating impulses to various sites within the brain of an animal. By using radio techniques it is possible to separately control a number of animals in order to make observations on modification of social behavior. In many cases it is most convenient to place at the animal a small radio receiver whose only function is to close a contact that activates a locally powered stimulator. In this case, wide changes in the intensity of the radio signals are tolerable and the circuitry of both the transmitter and receiver can be quite simple. It is also true that such methods allow the direct use of the readily available model-airplane control equipment. Delgado (1963) has discussed a number of these matters. He describes not only the application of electrical signals, but also the injection of various chemicals with the help of a so-called chemitrode pump. This pump consisted of two Lucite compartments separated by an elastic membrane. On one side was a synthetic spinal fluid or other solution to be injected, and on the other a solution of hydrazine. Current through the hydrazine solution released gas whose pressure pushed and injected the liquid from the other compartment. Current intensity was used to control injection rate between 1 ml every 10 min up to 1 ml every 10 hrs or more. A small radio receiver was used to switch on and off the current to the hydrazine compartment.

A similar pump arrangement with ultrasonic controls has been used by Baldwin (1965) under water to inject olfactory attractants into the nostrils of a shark. In that case the attempt was to perform studies dealing with the ability of the nurse shark to detect weak stimulants against a natural background, and to learn whether this animal employs its nostrils to localize the source of an odor.

Stimulation need not be electrical or chemical. For example, in studying some aspects of a heat receptor, a small resistance element could be placed in the vicinity of interest and activated by a remote signal to elicit a change in behavior. Refrigeration or cooling can also be obtained in a small electrical unit with a Peltier cooler.

At great ranges one can still use radio transmitters to activate receivers and stimulators being carried by test animals. Far-field transmission is then employed, and no new concepts are involved. In Fig. 14.9 there was a stimulating component to the experiment, with an electrode having been implanted in the hypothalamus of the brain of a baboon. Here again model-airplane control equipment was convenient. Electrical impulses to various sites can not only cause the activation of various motor functions but can also produce sensory illusions or temporarily alter the personality of the subject, or even induce sleep.

Delgado et al. (1968) have used intracerebral radio stimulation and recording in man. In this case a received signal at the subject was used to modulate an active constant-current transistor stimulator, with the subject being monitored by three-channel EEG telemetry. Their stimulations of the hippocampus and amygdala apparently produced a variety of sensations and responses, in one case eliciting behavior reminiscent of spontaneous crises that was important in orienting therapeutic surgery.

In employing such methods with a high frequency of transmission in a laboratory there can be certain problems. Radiated signals reflecting off walls and other objects can set up standing wave patterns displaying nulls in which little signal can be picked up. If an animal happens to be in one of these nulls when an impulse is transmitted, the expected effect may not be observed. Results can become especially irregular when several animals are simultaneously being handled on several frequencies. In this case it is desirable that the walls be covered with a highly absorbing material for radio waves so that significant reflections do not occur. There are highly absorbing materials with which walls can be coated, or one can place at one quarter of a wavelength before a conducting wall, at which a transmitter is aimed, a partly conducting sheet with the same impedance as free space. This number is approximately 377 Ω/square (for a square of any size, and thus no linear dimension is specified). Similar arrangements in some cases can make telemetry reception more reliable.

Inwardly induced signals can be used to turn on and off telemetry transmitters for purposes of conservation of battery life. This stimulation of a transmitter into activity is useful in short-term experiments in which it is desired to allow an experimental animal time to recover from surgery before observations are commenced and in longer term experiments in which continuous observations are not needed. The relatively long-range transmitter of Fig. 14.9 thus could be switched on and off at will by received signals on the lower frequency. The wave forms in Chapter 3 make it clear that a suitably biased Hartley oscillator can remain off if off, or will continue in oscillation if on. External impulses can switch a typical simple circuit between these conditions. This aspect has been commented upon by Lepri and Ramorino (1960) and by Lonsdale et al. (1966). A simple bistable multivibrator arrangement can always be placed so as either to supply power steadily to a small transmitter, or to switch it off, and this electronic switch can be activated by external impulses as picked up by any simple receiving arrangement. If the animal can be approached closely in order to switch on and off his transmitter, then in this last case a simple crystal-set receiver is sufficient. Of course, all of the passive transmitters considered in Chapter 13 have this property of activation only when needed, since they are active only when power is being induced into them. Also, the transponders to be mentioned in the next chapter have some of these properties.

2. MECHANICAL ACTIONS

In a number of cases it is useful to have the possibility of mechanical actions taking place, upon command, at various sites within the body of an animal. A number of examples are cited here. One of the most obvious is a continuation of the applications at the end of the previous section. Thus mechanical switches can be employed to turn on and off a transmitter. We have found it effective to incorporate within a transmitter circuit a small magnetically actuated switch that would serve this purpose. One convenient form of this switch is the so-called reed switch which is manufactured commercially by several companies. A small bar magnet placed parallel to the switch causes the in-line contacts to pull together, although a separate small magnet can be used to bias the switch so that the presence of an external magnet causes opening of the contacts. Closure of the contacts can either be used to connect the battery and start transmission, or else closure of the contacts can apply a cutoff bias to stop transmission. In this connection it might be mentioned that so-called class B or class C output stages are already biased so that they draw little or no power until an ac signal is applied to their input. It is extremely convenient to carry in the turned-off condition a transmitter incorporating one of these switches with a small magnet taped to the side. Transmission starts when the magnet is removed. Similarly, an implanted transmitter can be maintained in the off condition by a small magnet affixed to the skin of the animal. Lifting this away, perhaps following recovery from surgery, starts the transmission, with no battery life having been dissipated before that. Other switches (Fig. 16.2) are controlled by the sense of motion of a magnet, and remain in the state to which they are set.

Magnets have many other possible applications in inducing motions inside the body of an animal. If necessary, the clearing of the test volume in the bleeding-site detector was to be accomplished in this way. A pressure transmitter can be restrained at a particular point in the human gastrointestinal tract by a permanent magnet (using a conducting core for modulation and a piece of iron at the other end for restraint), but this method perhaps irritates the adjacent tissue. In the early studies on the dolphins (*Tursiops truncatus*), a number of capsules containing only a powerful permanent magnet (of Alnico 5) were passed through several subjects. By this means the momentary position of the capsule was determined with

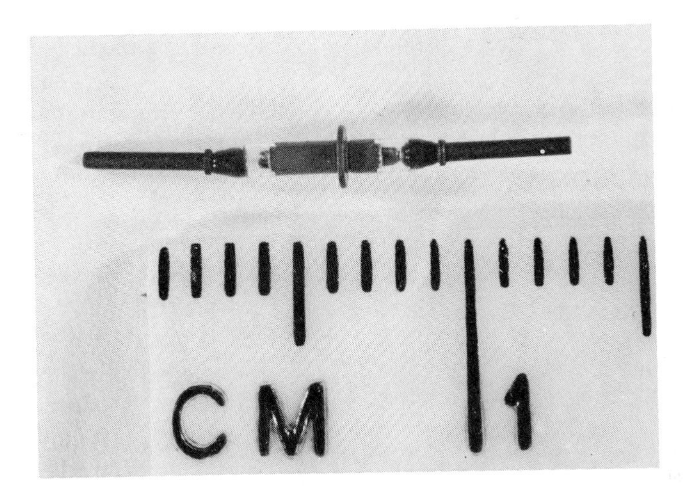

Fig. 16.2 Some magnetic switches employ a reed while this one has a moving armature. Movement of a magnet near this commercial unit opens one circuit and closes another, the new connection remaining after removal of the magnet (it is a latching single-pole double-throw switch). Applications include turning on transmitters after sealing or after implanting in an animal.

nothing more complicated than a magnetometer (in some cases a waterproof compass). But more important, it was felt that if the capsule did become caught in the gastrointestinal tract of the animal, an external powerful magnet could be used to jostle the capsule and perhaps work it by an obstruction. It was only after a series of these experiments with gradually increasing capsule size that free use was made of the somewhat larger transmitters described in Chapter 15. A spinning magnet outside the body can be used to rotate a directly underlying magnet implanted within the body, and thus provide some pumping action or other mechanical effect.

Various mechanical devices can be visualized in which it would be useful to have a single activation under control from outside of the body. In studies of the human gastrointestinal tract, it could be advantageous to have an adjustable capsule which would open and close upon command, thus taking a sample of the gut contents at the site in question. Similarly, such a capsule might be used to release a concentrated dose of a drug at a particular location, when the body would not withstand having a similar dose uniformly distributed. (It is known that organic derivatives of alginic acid are insoluble at neutral pH; such a covering on a capsule can prevent dispersal before leaving the stomach, but the present considerations aim at more specific sites.) Such a device might also be used for the remote performance of certain forms of minor surgery, including the taking of a biopsy. (An electromagnetic biopsy device with wires to the outside has been described by Driller and Neumann, 1967.) In some cases such actions could probably be caused to take place after an appropriate delay by the gradual solution or electrolytic decomposition of various restraining members. A spring-loaded capsule might be released to undergo its cycle by burning through a small metal loop restraining its cover; this could be accomplished from outside with little overall generation of heat with the help of a strong induction heater which could flash the internal shorted turn. In some cases an equivalent result might be achieved by having a small motor continuously run a filament of filter paper past an opening in order to continuously take up a sample. In other cases it may be sufficient to employ small geared-down motors, into which power is induced from an external coil into a suitable internal coil, to open and shut a lid.

One other method for producing a mechanical motion in the body by the induction of heat into a piece of metal is based on the "mechanical memory" properties of compounds composed of roughly 55% nickel and the rest titanium (e.g., deLange and Zijderveld, 1968). Such a material plastically deformed below its critical temperature will recover its original shape when heated above this temperature. The temperature for recovery varies from $-10°C$ to $100°C$ as a function of the Ni:Ti ratio around the stoichiometric composition. We have demonstrated the activation of a sample of such wire to open a cover within the body by an induction heating device.

A related, although perhaps simpler problem, is the restraining of a transmitter for an extended period within the stomach of an animal, and then after a relatively long delay to provide for its passage onward. This matter will be mentioned in Chapter 17, but it is mentioned here as an example of a similar type of consideration. From the stomach it may be possible to obtain information about heart rate or other parameters. It seems possible for many species to swallow things larger than they can pass. Thus the simple attachment of gelatin extensions on the transmitter could perhaps contribute to this result. An inflated bag around a transmitter would hold it in the stomach, and if sealed with a slowly dissolving string, later passage could be assured. Chemicals could probably be placed within a deflated bag to assure its partial inflation shortly after swallowing.

There are cases where it seems desirable to place within the body an electric motor in order to accomplish various functions. We have conducted a few experiments with a motor-activated clamp arrangement (Fig. 16.3) which was to be placed around an artery in order to shut off the blood supply to a particular organ at will. The artery in question would be passed through the spring-loaded clamp at the right, which is opened and closed by a motor suitably enclosed at the left. Enough power could be induced

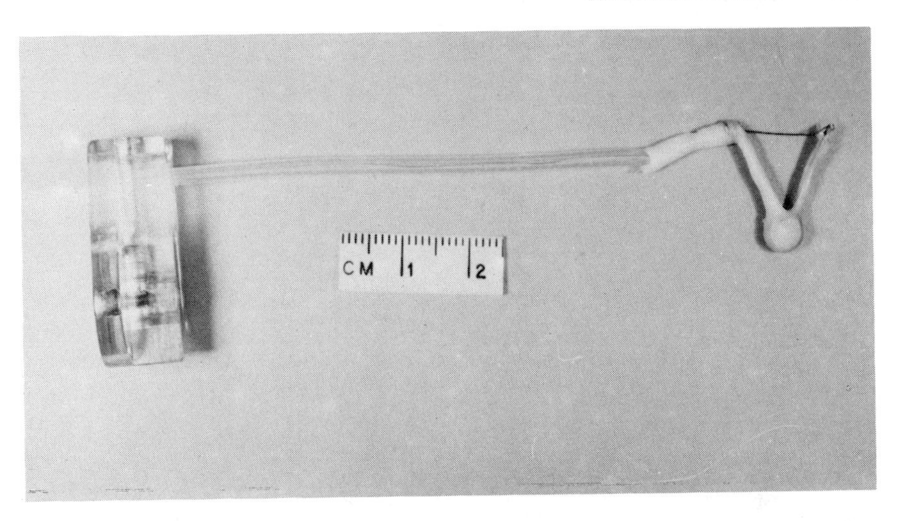

Fig. 16.3 Motor assembly at left used to open or close the blood vessel clamp at the right. External radio signals are used to activate the device.

into a subcutaneous pancake coil 8 cm in diameter to open or close this clamp on demand. As originally conceived, it was felt to be a useful way of checking the zero reading of an electromagnetic flowmeter. However, its applicability would obviously go beyond this in monitoring physiological effects produced by known input changes.

In an experimental animal a tubed skin flap can be produced through which is threaded either a loop of wire or the ferrite core of a transformer whose secondary winding is entirely within the body of the animal. A few external turns of wire can be looped through this core or a core through the coil, resulting in the transfer of significant power into the body of the animal without the problems associated with the penetration of the skin by wires or pneumatic tubes (Andren et al., 1968).

A group in Belgium has constructed a small motorized device that they have named an endomotorsonde. This is a small capsule containing an electric motor and a gear system which allows it to run up and down a dented nylon thread, while carrying various transducers and pulling catheters to desired locations within the gastrointestinal tract (Vantrappen et al., 1964). A photograph of such a unit is shown in Fig. 16.4. The horizontal filament is the nylon thread into whose dents fits a toothed wheel seen at the center of the forward face of the device. The thread is passed through the nose of a human subject and down the gastrointestinal tract. It is later hooked out through the subject's mouth. Direct introduction through the mouth and throat sometimes results in accidental biting upon the thread

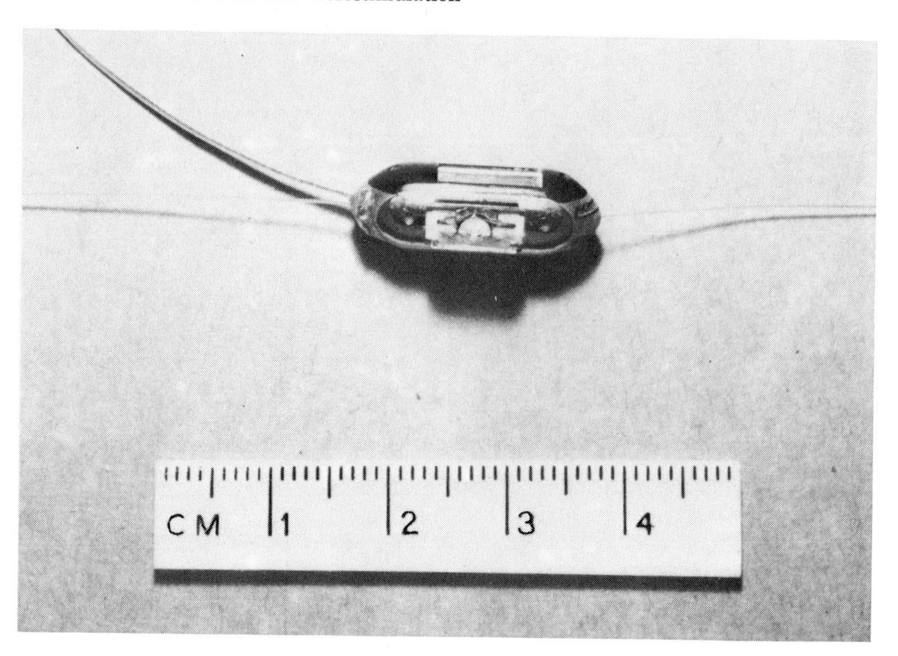

Fig. 16.4 An endomotorsonde made by a group in Belgium runs itself up and down the track of dented nylon thread swallowed by a human subject. The motor driven unit can pull various transducers or other devices to selected sites along the gastrointestinal tract.

which causes unfortunate nicks. The tiny motor is activated by the wires, which are seen trailing off up to the left. The little unit also contains a pressure sensor and antimony electrode. In the unit shown, the strength of the motor is increased by four stages of gear reduction. As built by these workers, such units constitute rather fine workmanship. Perhaps a commutatorless motor or a direct ratchet system could be used to expose fewer parts to body fluids and increase reliability. It is not clear whether inducing the power in for the motor and telemetering the signals out in order to do away with the trailing control cable would increase simplicity of operation.

To introduce the track upon which the device will run, a latex bag partially filled with mercury is attached at the far end. The day before an examination, it is introduced through the nasal cavity into the stomach. Once the mercury bag is sufficiently into the gastrointestinal tract, the study is started. Forceps are used to catch the thread in the pharynx and draw it in through the mouth, after which the endomotorsonde is attached and the motor started. The upper part of the jejunum is normally reached after about 10 min. Such a unit can introduce various objects to different parts of the human gastrointestinal tract.

Another group has worked for the last few years upon a class of devices which they call pod, and which accomplish some similar results in a different way (see Frei, 1966). These devices in general consist of a small permanent magnet with a moveable tail of silicone rubber. An alternating magnetic field causes a swimming action like a tadpole which propels the device forward through blood vessels. They note that a magnetic dipole in a liquid and in an oscillating field will vibrate and move about randomly. When encapsulated with a flexible tail, it will swim in a random manner if free, and in a forward direction if constrained in a tube. If a steady bias is added to the external field, the device will swim toward the source of the field with a degree of control that the steady field alone will not permit. This group has worked with pod magnets varying in size from 0.5 to 2 mm outside diameter. They have used pods to position catheters in otherwise inaccessible spots, and also electrodes. From pairs of X-ray images of a pod in a blood vessel, it is possible to judge the diameter of the vessel, or the rate of flow through it. A unit with two tails at opposite ends can be made to swim up or down stream by a change in applied frequency. Other applications for these devices will undoubtedly be forthcoming.

The above is a sampling of several ideas that relate to the general topic of this book. However, there are other methods which extend some of these related concepts much further. Thus there is a society which considers artificial internal organs. These organs might have their performance monitored by the present means, and several of these might have their performance improved with remotely powered internal pumps. An extreme example of this has to do with the attempts of a number of groups to design a complete replacement artificial heart which could be placed totally within the chest of a human, without having any sources of power mechanically perforate the chest. In that case a main consideration is efficient power transfer. A discussion of these considerations is felt to be a bit too peripheral to warrant inclusion here.

17

Related Methods

1. TRANSPONDERS

A transponder is a transmitter that radiates an impulse upon receipt of an impulse into an associated continuously active receiver. The reply to the incoming "interrogation" can be coded in such a fashion as to convey information. Such a transmitter conserves power since the transmitter, which is the main consumer, is only active when needed, such intervals perhaps being quite infrequent. This type of device has another very important application. If it is assumed that the returning impulse is generated without any delay in the circuit, which is generally effectively true, then the time between the transmission of the interrogation impulse and the return of a reply is a measure of the range to the transponder. This can be extremely valuable in localizing a subject, either on the ground or underwater. In the former case, it is often sufficient to assume that the radio signals travel with the velocity of light, and in the latter case an average velocity of sound (for an acoustic transmitter) can be employed with fair accuracy. When using radar methods, a transponder can give a stronger return trace than can a simple reflector for locating a subject, and a coded reply can distinguish one subject from another.

Various combinations of circuits already presented can be used to construct a transponder. If the most direct form, some small standard receiver will be used to activate a monostable multivibrator which can then turn on a transmitter for an instant. The start of the interrogation pulse can be used to trigger the multivibrator, which immediately leads to transmission. Thus the time delay in response can be negligible, although with some standard receivers there is a small delay in output following an input signal due to the finite rise time of signals passing through the intermediate amplifier filter. Arrangements by which a standard receiver can trigger a monostable multivibrator were given, for example, in Fig. 11.4. The control of a transmitter by a multivibrator has been indicated in a number of circuits, for example, Figs. 14.4, 15.2, 15.9, and 2.11. Any one of these can be changed to a monostable multivibrator, rather than being an astable multivibrator, for the above application. In each case, it is desirable that the outgoing frequency be set to a slightly different value from the incoming one in order to avoid the possibility of interaction directly between the output and input circuits.

There is another mode of operation of these range-indicating systems. The general scheme of one transponder type is indicated in Fig. 17.1. Here a transmitter is periodically triggered to radiate an outgoing pulse by some sort of pulse generator. This was precisely the situation, for example, in these last mentioned figures where a transmitter was periodically switched on by an astable multivibrator. In such periodic pulse generators there is always some sort of timing waveform which gradually increases until a critical point is reached at which sudden switching takes place to cause both the generation of an impulse and the resetting of the waveform to the initial condition. This action is depicted at the top of Fig. 17.1. In a multivibrator, this timing waveform is the change in voltage across a capacitor which is exponentially charging through a resistor. Whenever a capacitor is charged through a resistor and a suitable negative resistance is placed across the capacitor, this periodic action can occur. Notice that

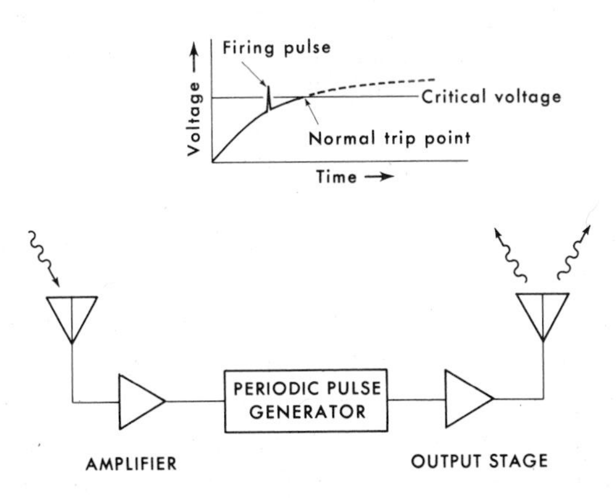

Fig. 17.1 A circuit that switches periodically can be drawn into synchronism with a more rapid train of impulses by injecting some of this signal into any sensitive part of the circuit. Among other applications, transponders can thus be made to return a signal upon arrival of one. The delay time in return to the investigator indicates subject range.

if a small added impulse is superimposed upon the timing waveform, then firing can be made to occur a little early, and in synchronism with the incoming impulse. Suppose then that a periodically pulsing transmitter is set to run at a low rate. A simple receiver circuit can then be connected into any sensitive part of the timing circuit, and the repetition rate will increase and lock into step with the incoming pulses. Suppose then that the investigator transmits a pulse, which results in a pulse being radiated back from the transponder. Suppose that, immediately on receipt of this returned pulse, a new interrogating pulse is transmitted. The time between successive pulses from the transmitter is then simply the round-trip signal passage time, and the pulse frequency of transmission is a direct measure of range.

This same type of operation could be achieved by setting the critical firing voltage up above the highest voltage achieved by the timing wave form, so that a pulse could never be generated without the arrival of an input pulse. Many circuits actually have this property because energy is gradually stored in a capacitor for generation of an intense short output impulse, and until this capacitor voltage builds up following a previous cycle, no new cycle can be generated. An input impulse then causes the discharge of the capacitor into the transmitter. However, the use of a device that normally pulses periodically was mentioned for several reasons. First of all, it illustrates the mechanism by which a slowly pulsing device can be locked into synchronism with a more rapidly occurring cycle by the injection of a little of the latter wave form into some sensitive spot on the circuit. Second, it depicts a general method by which an already existing periodically pulsing transmitter can rapidly be converted for range finding. And finally, in generalized tracking operations, it may be convenient for a transmitter to have a low intrinsic rate that will prevail whenever the transmitter at the investigator is inactive or has its signal lost.

It seems appropriate to mention at this point that a tracking transmitter having an intrinsic low blocking rate can be brought into synchronism with the signal from a low powered internal transmitter by this mechanism. It is no more difficult to track a transmitter containing information than one pulsing randomly, and so this can be a real advantage. If at any time the physiological signal becomes weak or unreliable, then the tracking transmitter will again take up pulsing at its own low rate so that the animal location need not become uncertain. In this application it is often sufficient for a tuned pickup loop to feed through a diode into the base connection of one of the transistors in the controlling oscillator, in order to produce this locking in or synchronism, since the tracking transmitter is usually quite close to the internal transmitter.

The circuits in the previous chapters are sufficient, in combination, to allow the construction of an underwater system. The speed of sound in sea water changes approximately 2 percent for a 10°C change in temperature, that is, approximately 3 m/(sec) (C°). An error of a meter in a kilometer of range is thus sometimes expected, although this depends upon conditions. With a source very close to the surface, there can be a "Lloyd's mirror" effect in which destructive interference between direct and reflected rays causes the signal to vanish. Periodic outgoing pulses can be displayed on an oscilloscope along with the returning pings, with the separation in the resulting pulse pairs being used as a measure of range. In some operations where a boat is maintained or periodically placed close to a marked animal, the same methods can be used to fix the position of the boat. In that case, one fixed transponder can be interrogated by a ship with three hydrophones, or a ship with one hydrophone can interrogate several transponders. To provide adequate base lines, the latter will generally prove the more effective.

Transponders on a subject animal can give a signal upon automatic activation at selected stations along a particular migration path. The reply signal can either carry information, or can merely signal the presence of the animal in the interrogation signal beam or region.

All of the passive circuits of Chapter 13 can be considered as transponders, and where their range is appropriate they can function in this role. The power-conservation aspects of transponders have been mentioned, and in some cases they are most effective in this sense. Alternate methods of power reduction, to repeat, include the use of clocks to turn on and off transmission at fixed intervals, the use of blocking oscillators, the use of implanted magnetic switches or bistable circuits, the use of external energizing, and the use of various circuits which are biased in such a way that an external pulse will cause them to start and remain in oscillation, or can similarly remove the energy stored in the resonant circuit by a properly phased pulse, thus permanently stopping oscillation (so-called oscillation hysteresis).

2. RECORDERS

In cases such as the monitoring of the welfare of a diver in the ocean or an astronaut in orbit, a continuous flow of physiological information is essential. Also, when the response to a stimulus is being monitored in order to decide upon the nature of the next applied stimulus, then immediate information is wanted about the various parameters being monitored. But in many cases where data is being collected, it is not necessary that it be available immediately and continuously, and in this case it is often useful to let the subject carry with him some sort of a recorder. The data stored therein can later be analyzed and can have the same value as if it were available instant by instant. This is certainly true when working with animals that are to be seen again at some later time, or with human subjects. The possible use of a self-detaching recoverable recorder in connection with aquatic animals was mentioned in Chapter 15. Self-detaching recoverable units could as well be made for terrestrial, or even aerial or arboreal animals.

Several companies produce relatively high-precision magnetic tape recorders about the size of a human fist. Availability and specifications change continuously so specific details are somewhat irrelevant. In general it can be said that these units can be employed to record the time course of some physiological variable for an hour, day, or other suitable period. To record all the details of a rapidly changing variable requires a higher tape speed and thus a larger reel of tape for a given period. In some cases this suggests the preprocessing of data to reject extraneous information before it is recorded. Thus in some cases a complete electrocardiogram is desired, while in others it is sufficient to record heart rate averaged over a few cycles; the latter requires considerably less of any given recording medium. If there is interest in the dependence of one variable upon another, then they can both be recorded as a function of time for later comparison.

In some cases a recorder need not run continuously, but can be switched on by the appearance of a significant value of the variable in question, for example, a critical value of temperature, heart rate, or the like. In cases of abnormal heart conditions it may be desirable to have a record of the conditions immediately preceding the critical ones. In that case, a pair of recorders can be employed. One recorder continuously runs a loop of tape through a closely spaced erase and record head, so that at all times the previous minute or more (depending upon tape speed) of information is stored along the loop. The other recorder contains a fresh roll of tape, and is inactive. When a critical event takes place, the first recorder is stopped, thereby preserving the immediately previous information, and the second recorder is started in order to store the subsequent time course of events. These functions can be made automatic if suitable switching circuits can be arranged to recognize the presence of a critical level in the variable under consideration.

Recording a variable as a function of time generally implies the progression of some medium past a recording region, but magnetic tape is not necessarily required. Thus an Accutron electric watch movement can slowly advance a plastic film into whose surface is scratched marks by a piezoelectric element related to a phonograph pick-up. Such a record can later be read under a microscope, and the overall size of the unit can be quite small. Companies have marketed electrolytic devices in which the electroplating of metal from one electrode to another causes an opening to gradually shift at a rate proportional to the current passing. In one version, a fine glass tube contains a column of mercury within a break in which is a droplet of mercury chloride solution; electric current through the ends of the tube causes the movement of this gap. This might be used as a time-base. If a radioactive light source were placed on one side and a piece of photographic film on the other, then blackening would depend inversely on current, and a microphotometer tracing could be used to infer the time course of some relatively slowly changing variable.

The time course of an occasional transient can be recorded without the continuous advance of some recording medium. Suppose that we have some two-dimensional recording medium such as a sheet of film, and we make deflections in the horizontal direction proportional to the variable in question, and deflections in the vertical direction proportional to the rate of change of this variable (its derivative). Then, when a transient occurs, a loop will be traced out. A constant frequency signal can be used to cyclically interrupt the trace, thus allowing for its later analysis.

When one speaks of recorders, it has become natural to think of magnetic tape recorders, but the intent of this section is much more general. Thus unusual means for leaving a record for later study or analysis are appropriate in different experiments. For example, with a diving animal one may not wish to know the time course of the animal's depth over a long period, but rather the percentage of time spent at different depths. In that case it can be sufficient to place a piece of photographic film on the face of a dial-type pressure gauge, and to attach a radioactive pellet to the pointer. Upon developing the film and reading it with a photometer, degree of blackening at a given position is an indication of amount of time spent at the corresponding depth. The amount of time a given temperature or humidity prevailed might similarly be recorded in other experiments when the time course of the variable was not needed. A radioactive light source can similarly be used. In that case, exposure to direct sunlight during assembly just prior to use should be avoided as the phosphor generally will then glow extra brightly for many minutes and give high readings. Such methods effectively plot a histogram of the variable being monitored.

A prime example of an unusual recorder for a special purpose is the "shark-bite meter" of Snodgrass and Gilbert (1967). The destruction of cables and moorings in the ocean lead to a need for quantitative data on the capacity of sharks to inflict damage. The so-called ganthodynamometer devised by these investigators essentially uses a standard Brinel hardness tester in reverse. Around a central soft aluminum bar are placed stainless-steel laths supported away from the aluminum by three stainless-steel ball bearings each. These units were placed in pieces of fish upon which sharks were allowed to bite. Under a microscope, the size of the indentations in the aluminum core could be evaluated, and from previous calibrations it was possible to evaluate the maximum forces produced. Maximum forces (not summations from several bites) of over 300 lb were recorded in this fashion. They also report that, from the pattern of indentations directly into a piece of polyvinyl chloride plastic, a loading on a single tooth as high as 30 kg/mm^2 was observed. Thus this recording system was able to yield previously unavailable data.

We have observed surprisingly high forces in the deformation of pieces of metal by the bite of a pet ocelot, in a similar fashion. Deformations in a washing machine indicate T.O.M. is able to produce forces over 75 kg using the neck muscles alone.

Wolff (1965) has discussed a number of compact recording schemes. He mentions a heart-beat totalizer based upon a so-called Solion electrolytic integrator. These useful chemical devices seem to have many applications both as transducers and for information storage (Hurd and Lone, 1957). In general the total amount of whatever is being measured can be recorded by any electrolytic integrator, which is then "read" by noting the current and time required to reverse electroplate back to zero. (Several electrochemical devices have also been discussed by Argue, 1965.) Wolff also discusses a recorder in which magnitudes as a function of time are stored by an electroplating process. In this, a thin copper wire is drawn slowly but steadily through a small plating bath by means of an electric watch movement. The length of wire exposed to the electrolyte was a fraction of a millimeter, and the speeds of travel a few millimeters per hour. The wire was made the cathode so that metal was plated onto it at a rate proportional to the magnitude of the variable to be recorded. To "replay" this variable thickness wire, it is drawn through a jet of high-pressure air, and either the flow through the jet or the pressure upstream from it are recorded. He notes a time resolution of 15 min. He also mentions a recorder system in which a multivibrator alternately electroplates radioactive nickel and ordinary nickel onto a single electrode in such a way that the proportion of time each nickel electrode contributes ions to the solution is determined by the magnitude of the variable to be recorded. Thus the composition of the "alloy" deposited on the cathode would vary as the quantity to be recorded varied. To replay this recording a steady stream of electrolyte would be allowed to flow past the receiving electrode above, which would now be subjected to a reversed current, thus dissolving the deposited material from it. A stream of electrolyte would then be monitored by a radioactivity rate meter, in order to reproduce the time sequence of events in reverse order. He discusses the possible problems associated with trying to construct such a recorder, and it seems probable that they can be overcome. However, reliable reversible plating is somewhat an art.

Electrolytic methods would seem to have many applications in connections with the topics being discussed. An electrolytic clock which might be used for causing an event to happen after a predetermined interval is often more accurate than an ordinary escapement clock, especially if long storage is to be involved before use. In such a device, a metal is slowly electroplated from one electrode to the other, and when the first is completely gone then the voltage across the cell suddenly rises; or alternatively, a mechanical action such as the snapping of a piece of wire can be made to take place. Two metals can be combined so that all of one will first slowly disappear under a given flow of electric current, and then the other will rapidly disappear to cause some discontinuous action in a sudden fashion at the end of a relatively long interval. Such a clock could be used to conrol the release of a harness or the starting of a transmitter after a desired delay (see Chapter 2, Section 10). As above, rather than a timing signal at the end of plating, a measure of degree can be had. A direct indication of a recorded time interval could be obtained by the electrophoretic migration of ions of one or more species down a color-changing indicator paper, the result depending upon time and voltage, and generally also on temperature.

In some cases it could be desirable to directly record some function of a set of variables. For example, there are many temperature-dependent processes, where the quantity of interest perhaps is a product of time and temperature. (Such a function would probably only retain its significance over a limited range of temperatures, and thus the temperature in question would be the number of degrees Centigrade above some specified lowest likely temperature.) We might achieve this result by combining a thermistor and battery with one of the previously mentioned mercury column devices, in which case the position of the gap would be a measure of the integral of temperature times time. With linearity of temperature response, such a unit is convenient for forming a true average, though in general the response should have the same temperature dependence as the process being monitored. In the technology of telemetry, such a recorder might also have direct application in monitoring the history of a battery, and thus allow for an estimate of the probable remaining life. In critical applications this could be important. Other functions of several variables might similarly be recorded, and it is here that the importance of deciding in advance what is significant to a particular experiment is seen.

3. BOOSTER TRANSMITTERS

There are situations in which a long range of transmission of a piece of information is desired, but in which only a weak signal is derived from the transmitter that is sensing the information. This might be the case, for example, with a small temperature transmitter which has been ingested by an animal. In that case it is sometimes possible to place near the animal a combination of receiver and transmitter which will pick up the signal and relay it to the greater distance. This can be done either by having the subject carry the transmitter in a harness on his body (in a pocket with human subjects), or by placing the booster transmitter at a location that the subject will periodically visit. Examples of this might be the placement of one of these "relay stations" near favored eating or drinking or resting sites. We might think that this result could be accomplished by attaching a receiving antenna to an amplifier whose output fed a transmitting antenna. However, if this is done, the strong output signal will enter directly into the input, and the device will simply oscillate in a fashion that generally does not transmit useful information about other signals. Attempts can be made to take some of the output signal and feed it back directly into the input circuit in such a phase as to cancel the effect of signals radiated from the output antenna back to the input antennas and circuit. Extremely careful cancellation can be made, but the slightest change in conditions (for example, expansions with temperature changes) will again lead to oscillation. This is because of the extreme strength of the radiated signal by comparison with the useful signal to be received, and hence the extreme precision which must be maintained in the cancellation. These same ideas relate to the use of what is termed a hybrid coil.

A more satisfactory general solution to the problem of overcoming a tendency towards oscillation is to take the received signal and shift it slightly in frequency before retransmission. Thus the outgoing signal cannot enter the receiver to set up a regenerative feedback path. Combining the incoming frequency with a local oscillator by heterodyning is an extremely effective way for shifting the frequency before retransmission. The frequency shifting circuit of Fig. 15.5 is quite effective in this application. It may be sufficient to use a frequency-doubling circuit in order to retransmit on a harmonic of the original frequency, but in some cases problems can arise due to the presence of harmonics in the original signal.

In the case of periodically pulsing transmitters it is not always necessary to shift the frequency, but it is still essential that switching transients in the output circuit not reactivate the input receiver. Thus a received impulse can activate a monostable multivibrator to turn on the output transmitter.

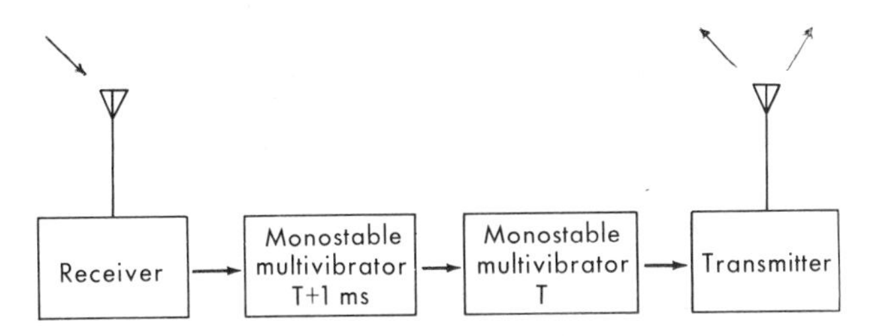

Fig. 17.2 Booster transmitters can be made to relay a weak signal from an internal or other transmitter to an extended range. In the case of periodically pulsing transmitters, the indicated arrangement can be used to prevent interaction from the output of the booster transmitter back to its own input.

A second monostable multivibrator can assure turning off of the transmitter before sensitivity is restored to further impulses at the input circuit. There are a number of ways of combining two multivibrators, one to control the duration of transmission, and the other to control duration of sensitivity following a received impulse. One way is depicted in Fig. 17.2. As discussed in the section on transponders, such an arrangement can easily be put together by combining previously described circuits in this book (e.g., Fig. 11.4 and Fig. 10.3 or 14.4).

Perhaps the greatest flexibility with the simplest use of available components results from receiving the weak signal on a complete receiver system which then modulates a complete transmitter on a different frequency. In Fig. 17.3 is seen a unit which was built into a standard pocket receiver, the final assembly being approximately the size of a cigarette package. With the receiver employed, an amplitude modulated signal at a frequency of approximately 600 kHz was converted to a crystal-controlled frequency-modulated signal at approximately 100 MHz. The unit was also effective for retransmitting the signals from transmitters that cyclically turn themselves on and off. The crystal was used at its fundamental frequency in a series resonant mode. The left transistor in the circuit drives the crystal through a capacitive divider to produce oscillations at approximately 30 MHz. The collector capacitance of the center transistor acts as a voltage-variable capacitor to produce frequency modulation (see Chapter 2, Section 8). The output transistor has no bias and thus its collector current flows in sharp pulses, the output circuit being tuned to the third harmonic of approximately 90 MHz. This frequency multiplication is relatively efficient, with changes in frequency being multiplied by the same factor as the frequency itself. A signal of 0.7 volt peak-to-peak produces a 50 kHz frequency deviation when the supply voltage is 9 volts, and a 100 kHz deviation when the supply voltage is 6 volts. Impedance matching is

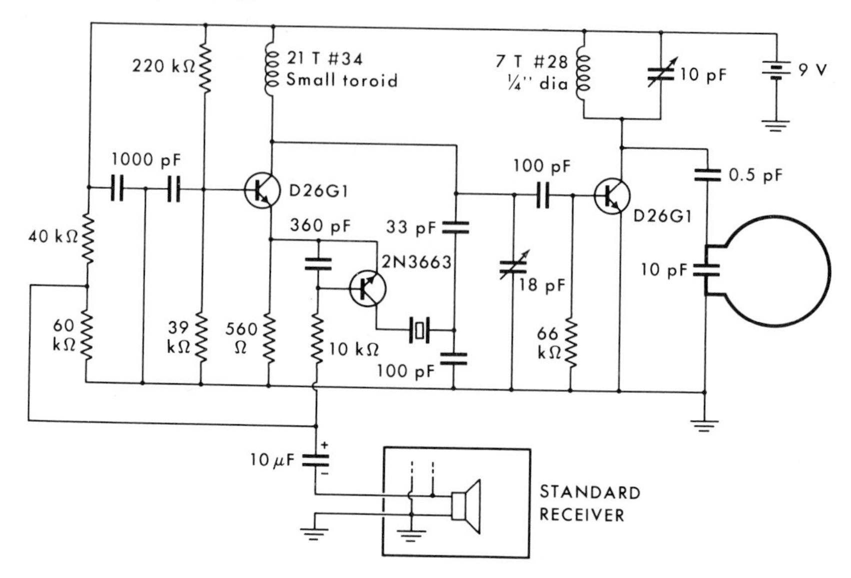

Fig. 17.3 Crystal-controlled frequency-modulated retransmission booster for relaying weak signals. Complete detection within a standard receiver allows for a change in the type of modulation as well as frequency. Connection is made elsewhere in the receiver than the speaker if a frequency range beyond the audible is required. The transmitting loop can be slightly deformed for tuning to resonance, and the two variable capacitors replaced by fixed ones yielding maximum signal. The loop can be replaced by a short wire (whip antenna) if its capacitor is replaced by a resonating coil.

supplied by the 0.5 pf capacitor to assure good power transfer into the loop antenna, which can radiate a strong signal. To prevent signal radiation during testing, the antenna was replaced by a suitably coupled resistor. With an 8 volt supply delivering 4 milliamperes, the harmonic power delivered into the resistor was 6 milliwatts. Doubling the bias resistor on the oscillator reduced the current drain to 2 milliamperes and the power delivered to 2 milliwatts. Transistor types differ greatly in their effectiveness as frequency modulators. When connected as shown, they provide considerable modulation sensitivity, though the emitter can be connected to the base to form a diode in series with the crystal; the signal is then inserted through a resistor to the connection between crystal and collector. The battery in the receiver is used as a power source. The signal is taken from the speaker, or elsewhere in the receiver if an extended frequency response is needed. If there is no direct electrical interaction between the circuits themselves due to their placement, the output of this unit will not feed back through the input because of the great difference in frequencies involved.

A superregenerative receiver locks into synchronism with any input signal, and these receivers tend to radiate a signal of their own. Thus in some cases a booster transmitter can consist simply of a suitably tuned and placed superregenerative receiver. This is an example of a situation where a self-oscillatory system can respond to local signals while radiating strong ones, although such a system is not extremely stable against changes in ambient conditions.

In Chapter 13 a passive transmission system was described in which a signal into one of a pair of decoupled coils was fed from the other. Tracking of the momentary frequency of a passive transmitter capable of coupling energy between the coils resulted, and the energizing signal was spread widely. In this case a long-range transmission action resulted, but small changes in ambient conditions which might result in a little of the strong output signal being coupled into the input circuit (Fig. 13.2) can again result in the radiation of signals unrelated to the parameter under study. Methods in which there is a shift of frequency or time generally provide greater stability.

4. MAGNETS AND RADIOACTIVITY

To produce their remote effects, radio transmitters must continuously be supplied with a source of energy such as from a battery. Radioactive materials and magnets are able to produce remote effects without further attention, and in some cases they can be used in experiments to produce results equivalent to those achieved by radio methods. Thus, although radioactivity has been mentioned in Chapter 9 and magnets in Chapter 16, they are again mentioned here and grouped together because of this property.

If we wish to follow the movements of a small animal at relatively short ranges, then probably the minimum encumbrance to the subject is had by using a radioactive marker. A circuit like that in Fig. 9.9 could be implanted in an animal at several sites. The movements of ectoparasites over this animal could then be remotely monitored if the parasites were rendered radioactive by the suitable placement of a small pellet, or the simple injection of a radioactive fluid. In 1964, Dr. Robert Stebbins was carrying on studies in the Galapagos Islands in which radioactive iodine had been administered to small lizards in connection with metabolism observations. As a separate matter, after dark these lizards would burrow under the ground, and their location could be found by walking over the general region with a scintillation detector. Thus it was possible for him to locate several individuals each evening so that he could observe the process by which they arose at dawn. In some cases the insertion of a radioactive tantalum wire into the body of an animal would be more effective than an injection of fluid.

In all cases the sensitivity of location is a matter of the strength of the radioactive source, the naturally occurring background activity level, and the length of time the investigator can spend in order to make a statistically significant observation in any one location. Locating a small animal at distances up to about a meter is generally relatively easy.

The energy associated with radioactivity can be used for longer range tracking through the generation of light. Thus a radioactive light source on a collar can be used to visually follow the movements of some nocturnal animals. This need not cause great radiation of the subject if an alpha or beta emitter is used since the penetrating power through most case materials will be low, leaving only secondary radiation.

Similar types of things can be done by placing within an animal a small permanent magnet. Somewhat greater ranges can be achieved in certain cases, and there is no possibility of radiation damage to either the subject or experimenter. Permanent magnets can be made from hard steel, with a material called Alnico-5 being especially good. In some cases a length cut from a needle is quite useful. Other types of material are somewhat more appropriate if a broad thin permanent magnet is to be formed with nearby faces carrying opposite poles, but this is generally not the case in the studies to be mentioned.

The strength of a magnetic field falls off with the cube of the distance to the source, much like the near field of a radio transmitter. Similarly, the direction of the field at a remote point depends upon the orientation of the source. Thus "direction finding" consists not of sensing a direction of a field, but, instead, the movement of a sensor in a general region to find positions of maximum strength. In work with the dolphins, for example, it was found possible to tell when a capsule containing a permanent magnet had left the stomach by moving a compass in the vicinity of the animal while looking for a maximum strength of response.

The range of efficient operation depends upon the weakest field that can be sensed, and this in turn depends upon the ambient magnetic field in which the investigator is working. The magnetic field of the earth at the North Pole is approximately ½ Oe, and at the Equator approximately ¼ Oe. (Being like a steady dipole, the field is twice as big at the poles as at the Equator.) At the poles the field is vertical and at the Equator horizontal. For comparison, the highly nonuniform field close in to a small permanent magnet may be as high as 5000 Oe. If a detector of magnetic field is stationary, then we can arrange to ignore (bias out) the steady part of the background field due to the earth. There are absolute changes in the earth's field of a few millioersted, and the so-called micropulsations are of the order of a few gamma (10^{-5} Oe) in a period of a few minutes. Although observations near limiting range are always a bit more uncertain and time consuming, motion of a magnet at a distance of several meters can readily be noted. A larger magnet can make a hidden animal noticeable at even greater distances, especially in areas free of masses of iron (Fig. 17.4).

Two magnetic detectors can have their outputs subtracted so that only differences in magnetic field at the two points are noted. The absolute limiting performance of such a gradiometer is somewhat superior to that of a single field detector in sensing the presence and motion of a permanent magnet. However, each of the two field-sensing elements must be several times more sensitive in order to accomplish the same result because of the subtraction of signals. Thus, although detectors can be made to sense either the field of a permanent magnet, or its gradient, it is generally quite sufficient to employ a single detector, which senses the field itself.

In some cases, instead of moving the detector to find the strongest field

Fig. 17.4 The large "dragon" on Komodo Island is seen swallowing the stomach from a goat carcass in which was a magnetic "transmitter." The experimental device also had a radio temperature capability. These animals swallow large portions whole.

in order to sense the position of an animal, we can fix the detector and note motions of a nearby magnet in response to movements of an animal. In some cases, activity of a particular structure can be noted. Thus a small magnet affixed to the wing of a hummingbird allowed the monitoring of the pattern of wing motion from the voltages induced in a coil placed near the feeder. With a plastic mouthpiece, a similar method applied to the lip of a bugler can answer questions about the tissue vibration frequency relative to that of the sound produced.

Activity in the gut of a human subject has been monitored with the use of an ingested permanent magnet and an external magnetometer (Wegner et al., 1961). The activity of an animal in a burrow might similarly be monitored by implanting a subcutaneous magnet into the back of the subject and placing around the run a suitable "search coil." If a transmitting magnetometer were placed in one animal and a small permanent magnet in another, then the amount of time the two animals spent near to each other could be followed, in connection with social studies. Magnet pairs on diaphragms can change quadripole moment to monitor pressure. The accidental loss of an intrauterine contraceptive might be noted with the help of an incorporated magnet and a compass, rather than by using projecting plastic fibers as is sometimes presently done.

The same magnetic techniques can be employed in other parabiological applications for identification. Thus a small door for a pet cat can remain closed to neighborhood feline friends who would enter for dinner, but are locked out without a suitable collar. The method can also be used to monitor unauthorized *post mortem* migrations from fur coat stores. The passive techniques using tuned circuits (Chap. 13) can be used here as well, while also being able to do such things as identify individual skiers buried in an avalanche.

As was mentioned in Chapter 9, there are a number of detectors of magnetic fields which can be employed to give high sensitivity with good ruggedness. A Hall effect sensor is satisfactory, but thin film or variable inductance detectors, or fluxgate detectors seem more effective.

5. FRAGMENTATION

A number of species seem capable of swallowing objects larger than they can pass. Thus a large transmitter might be employed to stay fixed for an extended period in the stomach where it might sense various forms of physiological information, and if it could later become smaller, then it would spontaneously pass. It was not possible to design a soluble transmitter, but it was possible to conceive of one that was held together at several places by electrical connections which would eventually dissolve. Thus such a transmitter would work for a period and would later break up

Fig. 17.5 A connection which, when swallowed, will break into separate pieces after a suitable time delay due to electrolytic action.

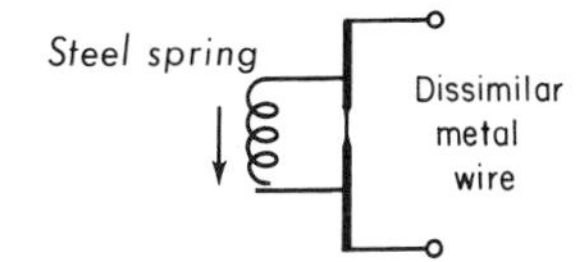

into small pieces which would then separately pass. An arrangement for a self-interrupting connection is shown in Fig. 17.5. Two dissimilar metals are connected in a loop, and when placed in a partially conducting fluid such as gastric juice, electrolytic action will destroy one. Several of these connections incorporating an aluminum wire were fabricated and fed to dogs, and in each case, after a suitable delay, the connections sprang apart. This type of system is related to the one sometimes used by oceanographers for the automatic release of equipment held to the bottom by expendable anchors. The delay time can be increased by placing a resistor in series with the steel spring, thus reducing the electric currents; this affects the direct connection only slightly.

Several observations on fragmented transmitters were made quite by accident. The dog in Fig. 17.6 has been used in a number of experiments in which she swallowed various radio transmitters. On several occasions,

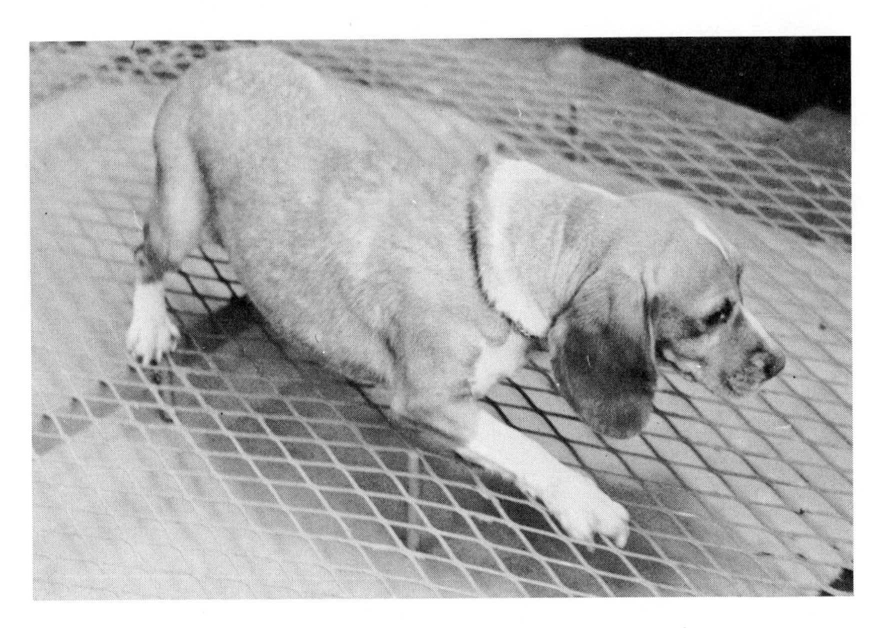

Fig. 17.6 Animal on an open-mesh floor through which transmitters can pass to prevent their being chewed and eaten again after initial passage from the animal. No discomfort is involved.

after they had reappeared she chewed them carefully into small pieces and ate them. It is interesting that a finely divided transmitter of rather great complexity was never observed to cause any harm to the animal. It is true that a mercury battery was never broken open, but all the other components were thoroughly exposed.

In order to avoid this problem, a raised screen was placed over the floor as shown. This is a kind of material known as expanded metal mesh. It was sandblasted to remove all possible sharp edges, and sprayed with a coat of epoxy paint. Any transmitter appearing would then fall through the openings out of reach; this procedure has proven completely effective. In spite of the rather large openings in comparison with the size of the metal strips, several animals have shown not the slightest signs of discomfort nor irritation to their paws when placed upon this shelf for several days at a time.

18

Zoo Exhibits

Workers in zoos have a special concern with these methods for several reasons. First of all, any research that is attempted in these locations is generally restricted by a requirement of minimum danger to the exhibit animals. Aside from the purely behavioral observations, the present techniques in some cases apparently provide the only acceptable study methods for minimum disturbance while taking scientific data. Second, such techniques can probably be used in monitoring incoming animals for disease. In some cases restraint or periodic recapture to measure temperature following a test injection has resulted in the self-destruction of nervous new animals. It would appear that temperature transmission by radio would minimize trauma of this sort. And finally, zoological parks in some cases can increase the interest value of their displays while providing a good instruction service by setting up exhibits incorporating some of these methods. It is to this last aspect that the following remarks are directed.

Any exhibit that is to hold the attention of the public, and be of maximum interest, should provide visitors with something to do. They must have a feeling of participation, and in general, the more that this is true, the better and more effective is the exhibit. Thus several of the following suggested displays provide a task that a visitor can perform, in some cases using his own radio receiver which he might happen to be carrying with him.

To date, there have been three exhibits of this sort, and they seem to have been relatively effective in spreading some of the ideas about the potentialities of these methods. The first such exhibit was held at the United States National Zoo in Washington, D.C., from August 10 to 13, 1966, in connection with the telemetry course and exhibit presented for the Smithsonian Institution. The second exhibit of this type was at the same location in October of 1966, in connection with the meeting there of the American Association of Zoological Parks and Aquariums. The third exhibit was held at the San Francisco Zoological Park in March of 1967, in connection with the annual North American Wildlife Conference. The following figures are taken from these exhibits.

A type of exhibit that gave visitors a chance to turn a knob and observe something employed a receiver with a direction-finding antenna, and a small external transmitter hanging from a collar on the neck of an animal. In two cases a pronghorn was used as the subject, and in another a llama. A three-element Yagi array was used for the receiving antenna. There were several reasons for doing this. It has relatively good sensitivity (gain), and distinguishes the forward direction from the backward one (Chapter 10). A relatively high frequency of transmission (in the 100 MHz range) was employed. This allowed all legal restrictions to be observed (Appendix 3). The relatively short wavelength also assured far-field operation and proper direction finding (Chapter 10), even if the animal wandered fairly close to the receiver. Of course, a signal radiated directly upward under a receiving antenna has no directional aspect. To be effective, this display requires a relatively large open area in which animals can roam.

A certain "James Bond" aspect seems to have made this exhibit somewhat more popular than otherwise would have been expected. In Fig. 18.1 is the sign that was prepared by the National Zoo for this display.

In setting up a display, two aspects were felt to be especially important, aside from the aspect of giving a visitor a feeling of participation. One desirable property of an exhibit is absolutely reliability. It is very discouraging to interested parties, and in some cases confusing, to perform an operation and find nothing happening. An improperly prepared exhibit can deliver nothing but static, which is most misleading and confusing. Thus every effort was made to assure proper operation at all times. The second factor is to protect the exhibit against occasional vandals and thieves. Some visitors find it a challenge to see what will fall apart under abnormal use. Although these persons are in a minority, their actions must be considered if an exhibit is to remain functional.

The layout of the tracking experiment is shown in Fig. 18.2. The rotating antenna is at the top of a piece of steel pipe where it is relatively inaccessible. The antenna is directed slightly downward since the effectiveness of such an array falls off somewhat when the transmitter is above or below the plane of the array. The antenna is rotated by a friction drive between two wheels. The driving wheel is activated by a long flexible shaft attached to the knob on the railing. Rotating this knob slowly turns the antenna, and sudden movements do no harm.

The receiver is a commercial entertainment FM portable unit, which was

Fig. 18.1 A display poster accompanying a direction-finder exhibit.

Fig. 18.2 The layout of a direction-finder exhibit, with a three-element Yagi antenna array shown at the top. This is rotated through a friction wheel arrangement by the flexible shaft connecting down to the knob. A standard entertainment receiver is seen in the box at the base of the upright support.

selected as discussed in Chapter 11. It was placed in a wooden box with a plastic front, as shown at the base of the upright in Fig. 18.2. Holes in the front of the box allowed the sounds from the speaker to be heard. A lock prevented unauthorized persons from having access to the receiver. This arrangement proved to be adequately weatherproof. Little heat is dissipated by one of these transistorized portable receivers, and so cooling is only a problem on very hot days when the direct rays of the sun heat the box somewhat in the fashion of a greenhouse. However, it is well for an attendant to pass by every few hours to retune the receiver for best action, since extreme changes in temperature do tend to cause small drifts.

Standard twin-lead television antenna wire carries the signal down from the antenna and its matching network, as is shown in Fig. 18.3. The size

Fig. 18.3 The appearance of a direction-finding antenna and matching network, with crossed arrows attached below to indicate to the public the direction in which the sensitivity lies.

of the antenna is adjusted to the transmitter frequency, and the matching network supplies an extra stage of tuning which helps to reject the signals from nearby commercial FM stations. Colored arrows are attached beneath the antenna to point to the transmitter when either a maximum or minimum signal is achieved.

The transmitter circuit itself is shown in Fig. 18.4. The output oscillator is crystal controlled. It is periodically switched on and off by the multivibrator at the left. When on, the second multivibrator at the center modulates its signal into a tone at approximately 1 kHz. Thus, in the receiver, no beat-frequency oscillator or modulator is required to make the signal extremely noticeable. It also means that any passing person who happens to be carrying a portable FM receiver will be able to pick up the transmitted signal on his own receiver. This seems to be a popular aspect with many visitors.

The circuit as shown is quite powerful. Thus care must be taken in the choice of the transmitting antenna if the display is not to exceed legal restrictions with a signal that carries too far. In most cases adequate range is achieved if the output loop, rather than going around the animal's neck, is merely made a few centimeters in diameter. In that case the loop can be placed in a horizontal plane under the animal's chin, while hanging from a simple collar. The arrangement of Fig. 14.4 is also quite adequate in this application.

In the other exhibits of a particular set, it is generally desirable to illustrate the transmission of physiological information using these methods. In one case the transmission and continuous recording of subcutaneous temperature from a freely moving mammal was displayed. In this case (Fig. 18.5) the subject was a llama, and the transmitter of Fig. 14.4 was employed. The thermistor probe was pushed sideways through a small puncture in the skin, and the transmitter tangled in the fur on the animal's back. Fluctuations in subcutaneous temperature of several degrees were observed as the animal went from within the building to outdoors and into the direct sunlight, and there were smaller rapid changes correlated with various stimuli.

This particular transmitter, with its self-contained small transmitting loop antenna, gave a field strength of approximately 50 μV at 50 ft. Although an effective range of a mile might be expected, proper action at 500 ft was more common. This transmitter draws a current of approximately 1 mA. Batteries type 675 last approximately a week, and type 625 last approximately two weeks.

This display was to depict the automatic decoding and recording of a continuous signal. A commercial receiver was employed with a suitable demodulator (Chapter 11) to directly drive a penwriter. A suitable length

TRACKING TRANSMITTER

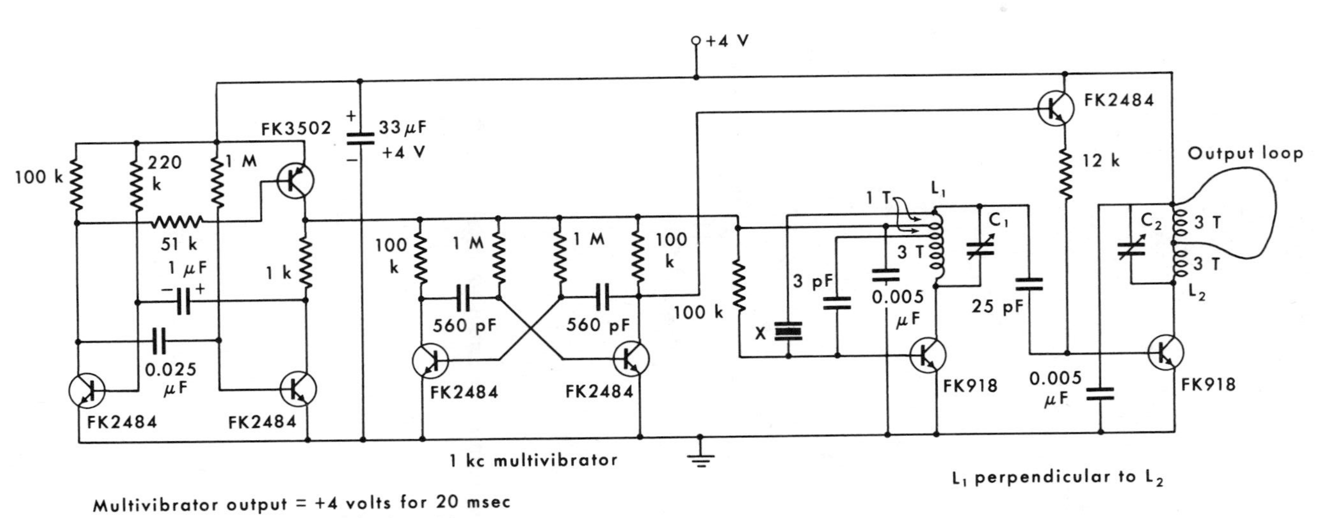

Fig. 18.4 **A pulsed crystal-controlled tracking transmitter for frequency of approximately 90 MHz.** The crystal is a fifth overtone series-resonant miniature unit; C_1 and C_2 are approximately 20 pF; L_1 and L_2 should not couple with each other and thus are mounted with axes perpendicular. The output loop is not tuned.

Fig. 18.5 A llama carrying a small transmitter of subcutaneous temperature in the fur on his back was free to move about in a large cage. The recorder and receiver were in a stainless-steel cage chained to a nearby tree.

of twin-lead wire was short-circuited at its ends to form a folded dipole; a break in the middle of one wire was soldered to another length of twin-lead to carry the signal down to the receiver. This receiving antenna was placed in one corner of the cage area.

The receiver and recorder were placed outside the cage area for ready observation by the public. They were placed within a veterinarian's sturdy glass-fronted stainless-steel cage, as shown in Fig. 18.5. This cage was locked shut with a padlock. A chain passing through an opening at the top of the door and through the side of the cage was looped around a nearby tree, and padlocked closed. This was effective in preventing disturbance to the equipment or its total removal. Nearby descriptive material is depicted in Fig. 18.6.

Some visitors find it interesting to compare the relatively constant temperature of a so-called warm-blooded animal with the rather variable temperature of a so-called cold-blooded animal. An indication from a poikilotherm is conveniently obtained by feeding a temperature transmitter

Fig. 18.6 A sample poster used in connection with the llama exhibit.

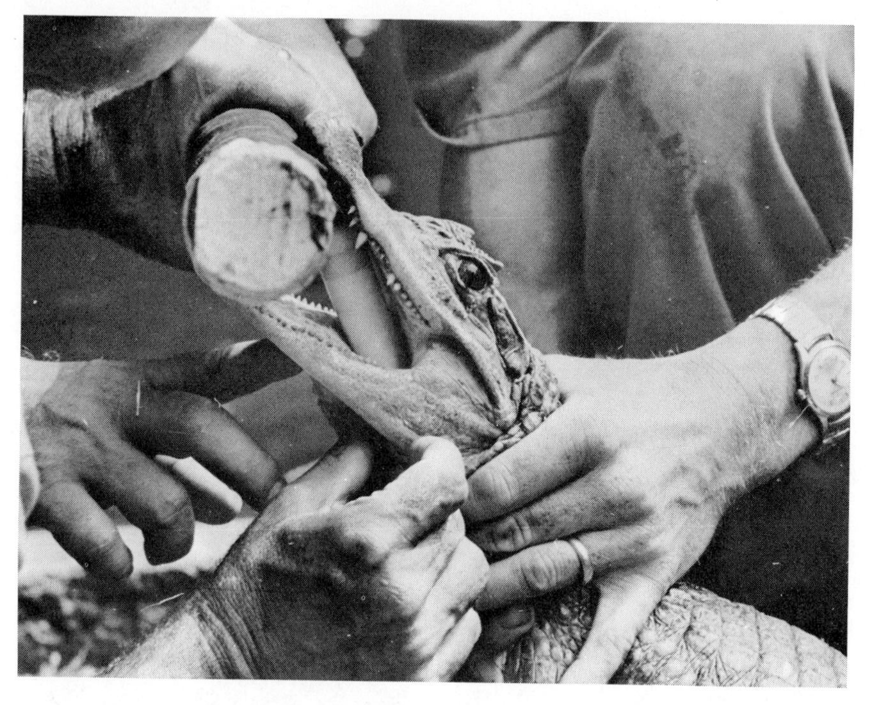

Fig. 18.7 One method of having a caiman swallow a transmitter is to allow him to bite on a stick and then slide the transmitter into the angle of the jaw.

to one of the large lizards. In three of the exhibits, a caiman was employed. With such an animal the transmitter reappearance time is quite variable, ranging from five days to three months in our observations. Feeding of the transmitter can be done in either of two ways. In Fig. 18.7, the animal has been allowed to bite upon a piece of wood over which has been placed a rubber covering to protect the teeth. The transmitter has then been slid in from the side, and is readily pushed down the throat. An alternative procedure is depicted in Fig. 18.8 where a padded plastic tube is first held in the open mouth of the animal. The transmitter, held in a pair of long forceps, is then slid through the tube and down the animal's throat. The first of these animals was about a meter long, and the second about twice that long. In neither case did swallowing the transmitter present any problem.

For maximum reliability in the continuity of signal reception, this transmitter was of the type employing a circularly polarized field, and was described in Figs. 10.3 and 10.4. After the circuit was cast into a block of plastic, it was inserted into a smooth tough outer case turned from a rod of Teflon. After attachment of the batteries, a mixture of paraffin and beeswax was melted and poured in, after which the covering was completed by screwing on a suitably shaped Teflon cap. In some of the displays two batteries were employed, and in other cases three cells were used to activate the transmitter for somewhat over a week.

This transmitter type pulses periodically at a rate which can be timed by ear. As indicated in Chapter 11, this arrangement is both very simple and extremely reliable. Radio interference causes negligible scrambling of the information when demodulation is done by the ear. But perhaps more important in the present case, any passing visitor can listen to the clicks and time them with his wrist watch, in order to obtain an indication of the temperature of the freely moving animal. Once again, this gives the visitor something to do, or a feeling of participation. By setting the transmission frequency at the low end of the commercial AM band, it is actually possible for a passing visitor to pick up the signal on his own portable receiver for decoding. This aspect seems quite appealing to many, and is extremely desirable.

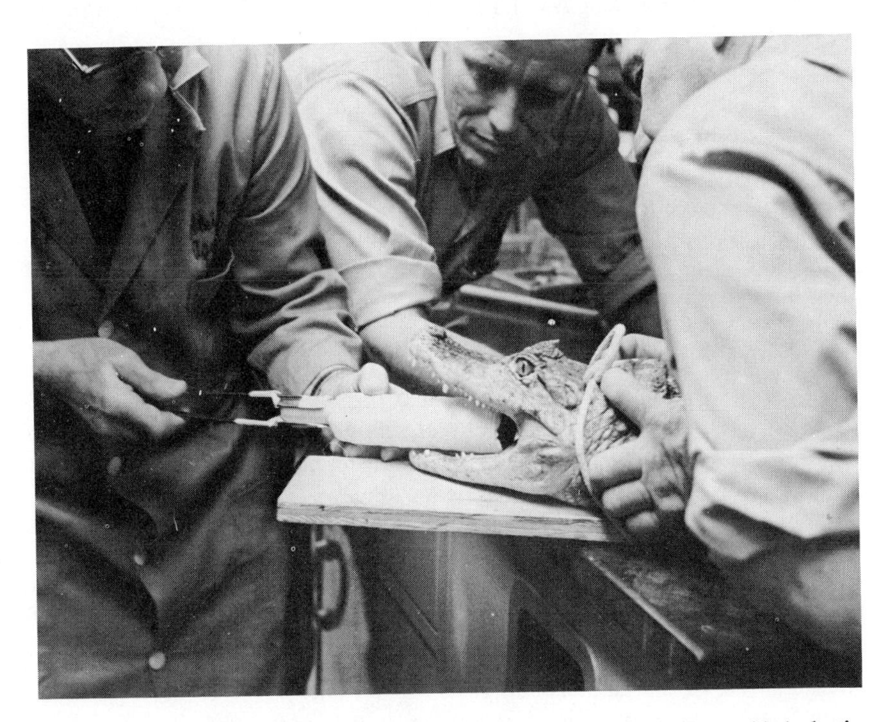

Fig. 18.8 A transmitter can be slid down the throat of a caiman if a padded plastic tube is held between the open jaws.

This transmitter fulfills legal restrictions, and yet a signal can be picked up by a visitor whether the animal is basking in the sun, swimming in the water, or hiding under a rock. Descriptive material to accompany such an exhibit is suggested by Fig. 18.9.

An unmodified portable receiver, selected according to the criteria of Chapter 11, should be placed beside the exhibit so that a signal will be available to any passerby not carrying a receiver of his own. Near this receiver should be placed a small radio-frequency oscillator, from whose coil will radiate some signal into the adjacent receiver. This oscillator is tuned until the signal beats with the transmitter frequency from the animal. This beat-frequency oscillator is in no way connected to the receiver, but merely couples in its signal by induction. Such an action makes the received signal much more noticeable, and actually significantly increases the effective range of transmission of a given transmitter. The received pulse from the receiver can instead be used to momentarily gate on an oscillator to put a strong constant tone from the loudspeaker, but it is felt that the present simple system is preferable because the small changes in intensity of the audible signal as the animals move about are quite interesting and instructive (as well as indicating which animal is actually carrying the transmitter).

The comparison between warm- and cold-blooded animals can be made even more direct if the same transmitter type is used in both parts of the exhibit. Many of the docile animals in a zoo are ruminants, and they will not pass an ingested transmitter for a very long time, if ever. In one exhibit the same transmitter type as used in the caiman was fed to a sea lion. In this case the transmitter was placed in a dead fish from which the viscera had been removed. A sea lion readily swallowed the fish whole. A sea lion, although warm-blooded, does show some variability in temperature. With these animals, a transmitter will generally pass in somewhat under 24 hours, and have to be refed to maintain the exhibit. Such an exhibit is shown in Fig. 18.10.

Figure 18.10 shows the box containing the receiver and beat-frequency oscillator. The signal was completely reliable wherever the animals went in an area approximately 15 m by 30 m, both in and out of the fresh water. In this case the signal from the battery-operated portable receiver was actually fed through a power amplifier to a pair of outdoor speakers in order that visitors could hear the signal over the noises produced by the sea lions themselves in calling to the visitors.

In this particular exhibit, the first passage of the transmitter through an animal resulted in it falling into a drain. Before it could be retrieved, during the night one of the animals had gotten it out and was next seen throwing it up in the air and playing with it on the island at the center of the pool.

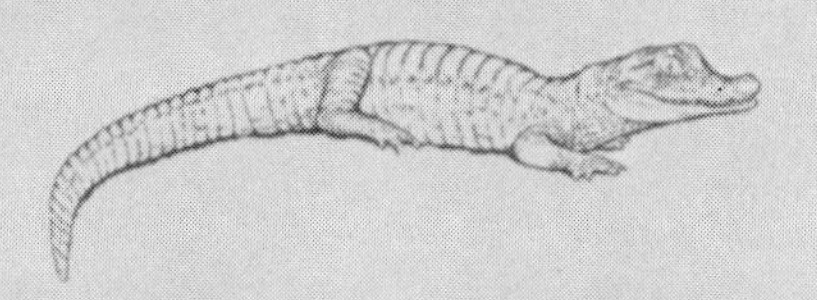

REPTILE TEMPERATURE BY RADIO

Like other cold-blooded animals, a caiman's temperature varies with his environment (and to some extent with his movements). If a caiman is too warm, he can regulate his temperature by moving into the shade or into the water; if cold, he can move into a warmer spot. Use of a thermometer would interfere with his movements.

This caiman has swallowed a one-inch by three-inch battery-operated radio transmitter that monitors his internal temperature and transmits it as a pulsating radio signal which is picked up by the National Zoo's radio receiver. The transmitter does not harm the caiman, and he will eventually pass it in his waste without being aware of it. The transmission frequency is about 540 kilocycles. If the caiman is near the edge of the caged area, you may be able to pick up the signal—a series of clicks—on a pocket radio; you can compute the caiman's temperature (Fahrenheit) by counting the clicks for two minutes, multiplying that number by .48, then adding 21.2. A similar computation is being done electronically by the Zoo.

This exhibit was prepared by Dr. R. Stuart Mackay of the University of California, together with the staff of the National Zoo, in connection with the Bio-Medical Telemetry course Dr. Mackay is presenting for the Smithsonian Institution. Related displays are at the llama pen and in the deer area.

Fig. 18.9 Poster used in connection with an exhibit of temperature regulation in cold-blooded animals.

Fig. 18.10 A portable entertainment receiver in a box successfully received the temperature-transmission signals from sea lions, whether they were in the water or on the island. An extra amplifier and speakers were employed to increase the volume of clicks so that they could be heard over the noise of the sea lions.

An hour later, another animal had swallowed it and was swimming around being monitored. The transmitter then spontaneously passed through three more animals, and the problem was how to terminate the experiment.

When three batteries are used in this transmitter, its life will be approximately ten days. If the receiver is occasionally retuned for maximum signal intensity, then there is no reason for a lack of signal at any time during this period. Figure 18.11 shows some poster material which has been used in connection with such a sea lion exhibit.

A suitable demodulating circuit can be used to indicate temperature directly on a meter or recorder. However, the activity of counting clicks is not only very reliable, but is quite appealing to many people, and thus it is an important aspect of any public exhibit. Counting the number of clicks falling in one or two minutes (the latter for greater accuracy) is a more natural and simpler operation for most people when using a wrist watch than is the timing of a specified number of clicks. Thus this generally less desirable procedure (Chapter 11) is indicated. A calibration curve can be posted from which a direct indication of temperature can be taken. In Fig.

SAN FRANCISCO ZOOLOGICAL GARDENS

ANIMAL TEMPERATURE BY RADIO

SEA LION

(*Zalophus californianus*)

CALIFORNIA COAST

WARM-BLOODED ANIMALS SUCH AS THIS SEA LION HAVE AN APPROXIMATELY CONSTANT TEMPERATURE, JUST AS DO HUMANS. CERTAIN ILLNESSES CAN BE DIAGNOSED BY FOLLOWING BODY TEMPERATURE CHANGES, PERHAPS AFTER AN INJECTION OF CHEMICALS THAT CAUSE A RISE IN TEMPERATURE IF HARMFUL BACTERIA ARE PRESENT.

TO MINIMIZE FRIGHTENING ANIMALS BY PERIODIC RESTRAINT TO MEASURE TEMPERATURE, THEY ARE FED A RADIO TRANSMITTER IN ORDER TO MONITOR BODY TEMPERATURE.

THIS ANIMAL HAS SWALLOWED A SMALL RADIO TRANSMITTER WHICH SENSES TEMPERATURE DEEP IN HIS BODY AND TRANSMITS IT AS AN F-M RADIO SIGNAL WHETHER HE IS IN OR OUT OF THE WATER. THE ENTIRE TRANSMITTER IS 1" IN DIAMETER AND 3" LONG, AND ITS BATTERIES RUN CONTINUOUSLY FOR ABOUT TWO WEEKS. IN LESS TIME, THE ANIMAL WILL PASS THE TRANSMITTER IN THE USUAL WAY, BEING COMPLETELY UNAWARE OF IT IN THE MEANTIME.

A LOOP ANTENNA SYSTEM IS INCLUDED IN THE TRANSMITTER PACKAGE. IT SETS UP A RADIO SIGNAL WHICH SPINS IN SPACE TO MINIMIZE THE CHANCE THAT THE SIGNAL CAN MOMENTARILY DISAPPEAR FOR ONE PARTICULAR ORIENTATION OF THE TRANSMITTER. THIS DOES NOT HAPPEN OFTEN ENOUGH TO SPOIL THE EXPERIMENT, WHICH IS INTENDED TO DETERMINE IN GENERAL HOW TEMPERATURE VARIES IN AN AVERAGE 24 HOUR PERIOD.

THE FREQUENCY OF TRANSMISSION IS APPROXIMATELY 500 KILOCYCLES, WHICH IS AT THE BOTTOM OF THE USUAL A-M RADIO STATION BAND. SOMETIMES WHEN THE ANIMAL IS CLOSE TO THE EDGE OF THE POOL, YOU CAN PICK UP THE SIGNAL ON YOUR OWN POCKET RECEIVER IF IT IS TUNED FAR TO THE LEFT.

THE SIGNAL CONSISTS OF A SERIES OF CLICKS. IF YOU TIME THESE WITH YOUR WATCH, NOTING HOW MANY CLICKS OCCUR IN ONE MINUTE, THEN THAT NUMBER PLUS 19 IS THE APPROXIMATE TEMPERATURE IN DEGREES FAHRENHEIT. THIS SAME TIMING OPERATION CAN ALSO BE DONE ELECTRONICALLY TO INDICATE OR RECORD TEMPERATURE DIRECTLY.

EXAMPLE: 80 CLICKS PER MINUTE
PLUS 19
99°F = BODY TEMPERATURE

THIS DISPLAY WAS PREPARED BY PROF. R. STUART MACKAY OF BOSTON UNIVERSITY IN COOPERATION WITH THE STAFF OF THE SAN FRANCISCO ZOO AND ZOOLOGICAL SOCIETY IN CONNECTION WITH THE NATIONAL MEETING OF THE NORTH AMERICAN WILDLIFE CONFERENCE. TWO OTHER DISPLAYS OF A DIFFERENT KIND ARE TO BE SEEN AT THE LLAMA PADDOCK AND THE CAIMAN EXHIBIT IN THE PACHYDERM BUILDING. AT THE LLAMA EXHIBIT, YOU CAN SEE A TRACKING EXPERIMENT SUCH AS IS USED TO FOLLOW THE MOVEMENTS OF ANIMALS; AND AT THE CAIMAN EXHIBIT, AN ALLIGATOR-LIKE ANIMAL HAS SWALLOWED A RADIO TRANSMITTER OF TEMPERATURE. HERE YOU CAN OBSERVE THE LARGER FLUCTUATIONS IN TEMPERATURE CHARACTERISTIC OF COLD-BLOODED ANIMALS.

Fig. 18.11 Sample of poster used in connection with the sea-lion exhibit.

18.12 is given a typical calibration curve for the circuit of Fig. 10.3. Large stop watches or stop clocks operated by the visitor can also be set up to time a given number of clicks, and calibrated directly in temperature. However, in many cases it is preferable that the visitor be able to make a simple computation in order to obtain temperature, rather than taking the result from a graph; taking information from a graph sometimes seems to give a feeling of disassociation in which any result is possible. If the calibration data falls approximately upon a straight line, then a simple computation indicates temperature in a way that is interesting to many people. If the

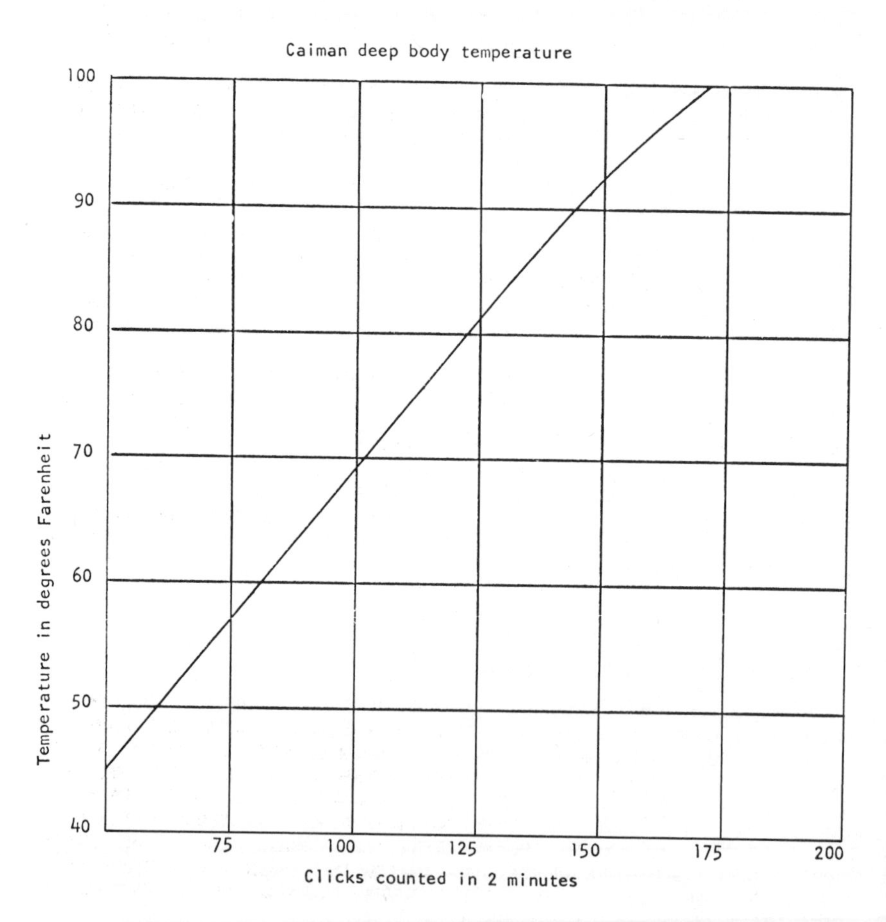

Fig. 18.12 Sample calibration graph used by the public in connection with these displays, along with an indicated simplified approximation that gave slightly less accuracy.

line passes approximately through zero counts at zero degrees, then the calculation is especially simple, requiring only multiplication of the number of counts by a specified factor. In some cases a line that is shifted slightly to pass through the origin will fit data adequately well, and allow this simplified calculation. In any case, the first step in preparing for such a computation is to plot the calibration data on a piece of graph paper, and draw through it a representative straight line. This is only possible if a straight line is an adequate representation of the data. The temperature is then the number of clicks in the specified period times the constant M plus another number B. These two numbers can initially be calculated from the representative straight line as follows. The slope of the line is M, which is a change in temperature divided by the corresponding change in number of clicks (taken from any pair of points on the straight line). The number B is the intercept on the temperature (vertical) axis, and is the temperature corresponding to zero clicks. The number B becomes zero if the line passes through the origin (zero temperature and zero clicks), with the simplification mentioned above.

In the United States the public is generally more familiar with temperatures measured in Fahrenheit than in Centigrade, so calibration in the former units is suggested.

These same methods can also provide displays for aquaria while monitoring the health of valuable swimming mammals. In the case of a large animal such as a killer whale, an ingested transmitter appears the most practical way of taking temperature, and a slightly buoyant unit can be used for convenient recovery.

Other displays can be constructed to generate interest in the possibilities of these methods. The continuous recording of the electrocardiogram from an unrestrained rabbit has proven of interest to many observers. With any animal, the most economical such display is either on a long persistence oscilloscope or simply direct listening to the heart signal. Either an external or an internal transmitter can be employed, with the latter being more effective. The relationship to the pacemakers implanted in humans removes any offensive aspect of the surgery in the minds of many. A little work has been done on preparing a wind-speed telemeter to be carried by a bird, based upon the use of the relative temperature of a heated and unheated thermistor. It remains to be seen how effective this will be, and whether undue power will be required. Other special exhibits can be designed to fit the interests of special groups.

19

Human Studies and Clinical Applications

It is appropriate to finish with mention of those applications that are in some ways the most urgent or important, namely, the care and study of man himself. Many examples from human studies have been cited in the past chapters, and all these methods are relevant to the affairs of man. But certain procedures can be emphasized as being especially significant in some circumstances.

1. HAZARDOUS ENVIRONMENTS

Monitoring the welfare of an individual in a hazardous environment or carrying on a dangerous activity is absolutely essential in many cases. The use of a portable recorder to take data for later examination is not an adequate substitute in these cases because presumably moment-to-moment information is required so that safety precautions can be taken, should the need arise. Presumably the individual in these cases is carrying out a significant activity or he would not be in this situation, and thus the use of telemetry to provide minimum interference with the activity also seems important. In all cases the parameter to be monitored should be that which is most rapidly responsive in indicating the potential onset of distress. Examples of several classes of activity falling in this general category will be mentioned, but there are certainly others as well.

Diving and swimming under the ocean involves a variety of potential hazards. Aquanauts living under the sea on saturation dives in many cases should be monitored, at least while they are swimming outside of their habitat. Since respiratory and gaseous exchange questions are among the dominant considerations of high pressure physiology, regularity of breathing would seem to be an important parameter. Distress, anxiety, loss or flooding of a mouthpiece, etc., might thus all be inferred by the resistance changes of a warm thermistor fixed in the mouthpiece. Other breathing sensors could similarly be employed. Extreme or prolonged irregularity might be cause for investigation by another diver.

Using modifications of the methods indicated in Chapter 15, it would also be possible to monitor such things as brain wave patterns or electrocardiograms of a diver. Electroencephalographic parameters are of importance because brain tissue is so sensitive to internal state changes, in the extreme case of oxygen poisoning there being convulsions. In some cases chilling (temperature changes) might also be observed, though this is a factor of which the swimmer himself is more aware. Monitoring during recompression treatment of decompression sickness can also be helpful. In certain cases special transducers might be developed to serve specific functions in guiding the activities of a diver. For example divers, tunnel workers, airplane pilots, and practitioners of hyperbaric medicine can all be afflicted with decompression sickness if they too rapidly go to a region of reduced ambient pressure. This is because of the inert component of their breathing gas coming out of solution in their blood and tissues in the form of bubbles. The ultrasonic detection of the bubbles of decompression sickness in animals has been demonstrated (Mackay, 1963B) as in Fig. 19.1 (from Mackay and Rubissow, 1970). In connection with his doctoral research, Rubissow seemingly sees preclinical bubbles in the human knee joint. If one assumes that any bubble which once forms may expand to cause problems if there is further decompression before a pause or a therapeutic recompression, then a non-scanning ultrasonic bubble sensor on a diver might give warning that his further upward progress should be modified. Other special transducer systems might similarly have applicability to special activities.

In space science research telemetry is always used. In some cases this is simply so that data is not lost if an orbiting capsule is not recovered. The heart beat of astronauts is generally monitored through the transmission of the electrocardiogram sensed by electrodes on the chest. In many cases the only usage that is made of this information is to note changes in heart rate, but it is of interest to have the entire electrocardiogram if something unforeseen does happen. Breathing rate can be transmitted by sensing resistance changes of the chest through the same electrodes. In at least one case, an astronaut was asked to take his own blood pressure by pumping up an armcuff by hand. The cycling of such a cuff can be made automatic, with placement perhaps not on the arm. (The method has been used on the tail

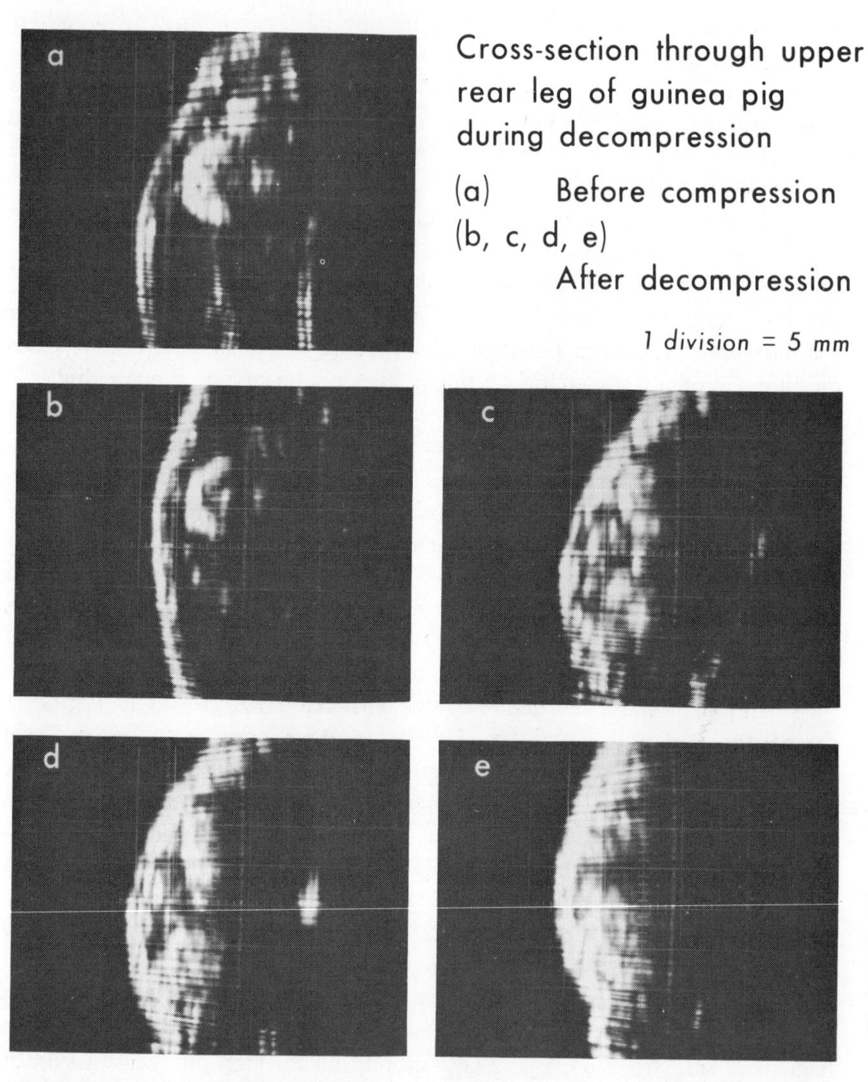

Fig. 19.1 Ultrasonic images showing the changing opacity of tissues in which bubbles form due to decompression. At a, the structure flattened at the left is the bone. The sound is from the left; and at d, bubbles can be seen forming on the inner surface of the leg through a region in which bubble formation is delayed. The last four images are at 1 minute, 1 minute 30 seconds, 1 minute 35 seconds, and 1 minute 50 seconds, after abrupt return to atmospheric pressure.

of a mouse, for example.) In the manned flights, the information about the astronauts is a small fraction of the total information being telemetered, a large part of it being computations and data about the vehicle itself. These many signals must be combined, and digital transmission is often used. The complexity that telemetry can achieve is indicated by the report released by NASA regarding the first manned landing on the moon. The voices of the astronauts were combined with 900 other signals and beamed to earth at a frequency of 2282.5 MHz. The signal picked up in California was relayed to the space flight center in Washington, D.C., where it was broken down into parts that went to Houston, Texas, the voices going to New York for rebroadcast to the world; the voices were also relayed back into space to the command ship 70 miles above the lunar surface. The second astronaut on the moon transmitted his voice to the first one on the moon and the combined signal went to the lunar module for transmission to earth. The best television reception seemed to be picked up at Parkes, Australia, from which it went overland to Sydney, flashed to the Moree Earth Station 200 miles to the north, beamed up to the Intelsat communications satellite over the Pacific Ocean, relayed to Jamesburg, California, passed by microwave ground signal and coaxial cable to Houston and finally transmitted to New York for distribution to individual television sets. This system is beyond the means of the individual investigator, but suggests what is possible.

Many sporting events are hazardous. Those who are interested in sports medicine or exercise physiology in many cases can study their athletes while actually engaged in their chosen activity only by the use of telemetry. Thus if one wishes to study impacts by accelerometers placed on a football player, or heart beat irregularities, it can perhaps best be done through the use of small transmitters placed on the belt or in the padding. The resulting signal can be used to collect data for future use or to immediately warn of impending danger so that the activity can be terminated.

In passing, it might be noted that a number of sports are scored in traditional ways that make no use of modern technology. In events ranging from fencing to the rope climb, telemetry with suitable sensors might be employed to make scoring more satisfactory without added encumbrance to the participant.

Humans are sometimes treated for leukemia by living in a radioactive environment of 1R/hr and their status could be monitored by these methods.

Miners work in a hazardous environment, and their health and position following a cave-in would be of great interest. Such an application is made difficult by the transmission of signals through great distances of earth, though a tough antenna wire strung through the mine with the power cables might remain intact following many disturbances, and part of it would always be near any miner. In any case, it might be helpful to monitor the participants in a rescue operation, where stresses can be high and conditions dangerous.

It is impossible to list all the activities which are potentially dangerous and where a warning of distress could be useful, but these few examples may be indicative.

2. CARE OF THE CRITICALLY ILL

Care of an injured or sick person often starts with a ride in an ambulance. Telemetry equipment in an ambulance can communicate various physiological parameters ahead to the hospital so that suitable preparations can be made. Someone in the ambulance must be qualified and make the gesture of affixing the sensors. Some ambulances have radio transmitters, and such a transmitter can simultaneously be used, with small modification, to transmit voice signals and also an electrocardiogram. Because certain regulations restrict mobile land FM to intermittent voice transmission, some commercially available transmitters will overheat if the continuous transmission of an electrocardiogram is required during a long ride. The question remains as to what is to be done if irregularity of heartbeat is observed. It can be debated whether an ambulance driver should be encouraged or allowed to apply closed chest massage or electric-shock ventricular defibrillation, but short time intervals are important in these cases. If only heart beat were of interest, then it might be simpler to simply fit each ambulance with an electrocardiograph unit, but presumably these considerations apply to other parameters as well.

During the first day after major surgery, or when a patient is very sick from any cause, he may be quite immobile with various catheters inserted into his body. Under these conditions, there may be little advantage to telemetry over systems where sensors communicate their signals via wires. But when the patient starts moving, that situation can change. Thus it may be desirable to monitor his heart during brief intervals when he first goes to the bathroom. At a much later stage, for example with a heart patient just before total release, it may be desirable to monitor electrocardiogram as he proceeds rather normally in a recreation area. It is quite possible to carry on monitoring by telemetry in areas ranging from intensive care units to visiting areas. Short range transmission from a subject no farther than to his mattress may be considered in some circumstances, while in others the signal must be received from a highly mobile subject. In the latter case a somewhat more powerful transmitter may be employed. Sometimes it can be observed that as a subject walks through a doorway the signal at the receiving antenna will disappear. This merely requires that there be another receiving antenna in the second room. In some cases, it is quite sufficient to stretch a long wire throughout the area where the subject may go (as in Chapter 14), in which case the transmitter need be no farther from the receiving antenna than the distance from belt buckle to floor, whether the subject is bathing or being wheeled down the hall on a gurney. On an even larger scale, one might imagine setting aside one bleacher in a stadium for the monitoring of ex-heart patients after their release.

Monitoring by telemetry may help assess the detailed effects and value of early ambulation, and thus settle some debates relating to its desirability in various cases.

As an example of a possible preventive measure, ventricular premature beats appear as a portent of trouble that can be averted. Potential problem subjects might thus be monitored routinely, especially those with coronary artery disease. Automatic processing of such signals would be required, including the several factors of prematurity and "compensating following pause" and differing shape from the average.

Progress is being made in the evaluation of irregularities in an expected pattern by computers, with the analysis of electrocardiograms in particular being fairly well advanced by a number of groups. Several companies have commercially available equipment that is to be placed at nurses' stations and which will sound an indication when a variable passes outside preset limits, for example, when heart rate becomes too low. There have been problems in the acceptance of such equipment, perhaps because of the relatively little training that goes with it by comparison with simpler devices such as X-ray machines. But with the increasing use of medical care, without corresponding increase in those who supply it, such equipment must be developed to reduce the load on overburdened staff.

It should be noted that some hospitals presently use their computers as little more than a digitizer and electric typewriter to produce orderly records, while much more is possible.

In monitoring a subject, some care should be taken to monitor those parameters that are significant and will be used, and not simply wrap the subject into a mass of sensors. Thus the only needed contact with a patient in some cases may be a set of electrocardiogram electrodes plus a small catheter in the femoral artery, the latter giving temperature, blood pressure, blood gases and electrolytes, heart rate, and cardiac output by automated dye dilution.

The variables of interest may change as new knowledge becomes available. Thus Rushmer's group has placed emphasis on dynamic variables such as rate of increase of ventricular pressure or initial emptying rate as an index of cardiac condition, rather than, say, traditional variables such as stroke volume. Simple measurements may retain interest, for example, a drop in toe temperature can be a bad sign for a subject in shock.

The decision as to whether to transmit the signal from the sensors by wire or by radio largely depends on whether the subject should move about or would otherwise tangle the wires and perhaps pull them loose. If lead wires are employed, it should be noted that these sometimes themselves fail.

In some cases telemetry is not indispensable, but very convenient. In this connection the reactions of the nurses and subject depicted in Fig. 6.3 are of interest. The subject was an epileptic, and thus a glass thermometer would not have been placed in his mouth. The nurses liked this system because it was faster, because they did not have to wake the patient if he was asleep, and because they did not have to use a rectal thermometer. The subject himself was interested, and was often seen holding a radio receiver in the vincinity of his head taking his own temperature.

3. ELECTRIC SHOCK HAZARDS

People occasionally are electrocuted in hospitals. The estimated number is subject to debate, but clearly with the increasing use of electronic monitoring equipment, sometimes from different and incompatible sources, this number will tend to rise. In some cases, equipment made by different manufacturers have incompatible grounding systems that lead to problems, while in other cases, the problem may simply be a worn power cord.

A special worry is for heart patients who are often observed by a catheter whose tip is extended into the heart itself. A very tiny leakage current can then cause ventricular fibrillation, which is normally not spontaneously reversible in humans. This problem is minimized by filling the catheter with a non-conducting solution if only pressure is to be measured. However, if atrial beat is to be monitored, then a conductor may traverse the catheter. Failure of some electrocardiographic units can pass dangerously high currents, even through electrodes that are placed outside the body.

Less serious inconveniences can sometimes result from the use of high frequency currents. In one case, a patient was routinely fitted with a rectal thermistor during surgery elsewhere on his body. Because of faulty grounding arrangements, during the use of the electrocautery apparatus, sparks jumped to the thermistor leads and severely burned his rectum. An explosion of the normal gases in the bowel also could have resulted in this case.

In general, small leakage currents can burn when using surface electrodes leading to monitoring equipment.

These problems are avoided if the signal is carried by telemetry. In this case, the transmission of the signal for an inch is quite sufficient, as the purpose is simply to supply electrical isolation even under conditions of malfunction.

4. SIGNALS FROM STERILE REGIONS

Wires and connectors can be sterilized in order to carry power and signals to and from sterile instruments, but in some cases, greater effectiveness and considerable convenience results from the use of telemetry. This is especially true if sound signals are involved.

Forceps are used to grasp kidney stones, gall stones, broken needles, and other foreign bodies in order to remove them from human subjects. The grasping may be somewhat of a problem, for example, kidney stones may shift to another part of the kidney and gall stones may escape. When the tip of a pair of forceps contacts a hard object, sound waves traverse its length and can be picked up at the handle end. Thus we fitted each of a set of forceps with a metallic projection to which could be clamped a sound transducer, and this provided a sort of eye in the tip of the grasping forceps from which hard objects could not escape (Fig. 19.2). In the case of stones, the large ones were obvious, and the small ones passed, but this device was helpful with the intermediate range of sizes. However, listening to an amplified signal was disturbed by the magnified sounds of the wires dragging on the sterile drapes. The device became much more satisfactory when fitted with the unit of Fig. 9.2 so that the signal was carried by radio, it being turned on by inversion of a mercury switch. The interchangeable forceps were heat sterilized in the usual way and the transmitter gas

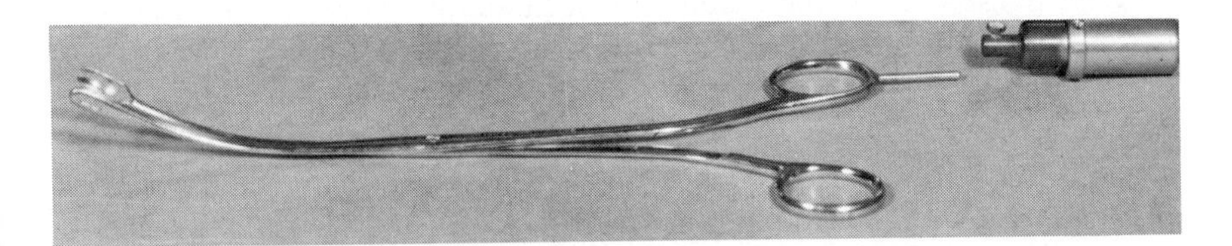

Fig. 19.2 A set of forceps can each be fitted with a small projection to which can clamp a sound sensor and transmitter. Urinary calculi and other hard objects are then less able to elude the grasping tip during removal due to the sounds of contact that are propagated along the tool. The sound unit is separately sterilized to avoid overheating.

sterilized. (Some suitably encased batteries can transmit during heat sterilization, but generally little life will remain.)

One other similar example is the detection of fetal heart sounds, during delivery, by a microphone affixed to the abdomen of the mother. A significant drop in fetal heart rate during birth can be cause for alarm and drastic action. The signals can be carried by wire, but there is interference by induced sounds as above. Under actual hospital conditions, the procedure becomes simpler if the signals are carried by radio. The attending obstetrician can wear a portable receiver and listen to the sounds directly, the human ear being very good at distinguishing such signals. In the latter case, gross changes in rate are noticeable, while more subtle changes must be indicated by electronic circuits.

5. HIGHLY EMOTIONAL SITUATIONS

Human situations are sometimes dominated by emotional factors, and when this is the case, monitoring by the traditional methods can introduce serious alterations. An extreme case is the investigation of some matters sexual. Masters has suggested that perhaps this aspect of reproductive physiology is the last unexplored medical area. In any case, a number of gynecologists have expressed difficulty in obtaining reliable information on sexual activity in connection with the fertility of their patients. Studies in this area should leave the average subject relaxed and as unwatched as possible if representative responses are to be recorded. For some physiological measurements, a pressure or temperature transmitter may be placed in the female. However a simple domeless diaphragm whose rim serves as the antenna and tuned circuit in a Colpitts oscillator can be cyclically deformed by intercourse, the resulting frequency modulation supplying the desired record. A 2 mm loop deflection typically produces a frequency deviation of 40 kHz at 100 MHz.

In psychological studies of the interaction between mothers and infants, a sound transmitter can be useful. One location that is out of the way is on the sole of the foot of the infant. From this one can record, for example, the sound of crying and character of the response of the mother under normal home conditions. Recorders of sufficient frequency response are suitable for storing this data, but if the baby is to be lifted and carried between rooms while his sounds are monitored, a small transmitter is clearly less intrusive than an attached recorder because of size. Many psychological studies can similarly be divested of interference by the presence of the investigator.

In some cases, psychiatric interviews might be more productive if the subject were fitted with electrodes which, by sensing electromyographic potentials, would indicate small changes in pattern of tension and alertness which might otherwise by missed by the therapist. Similarly, patients who report that their pains come or go as they enter or leave a hospital or church might be monitored by these methods for otherwise unnoticeable patterns of muscle tension.

Brain wave patterns can be monitored for a number of purposes, some recordings being sensitive to emotional or distraction factors. One application is to detect suspected latent seizure bursts in persons with behavior disturbances during normal activity and during periods of increased physical activity or other stress. In some cases, telestimulation has also been employed (Chapter 16). Special voltage changes in normal humans are also observed if the conditions of the experiment do not interfere. Thus expectancy apparently makes the mastoid negative with respect to the top of the head, and such potentials sensed with non-polarizing electrodes have been telemetered (Walter et al., 1967). They state that this "contingent negative variation" appears attenuated by distraction, anxiety, a sense of constraint or surveillance, and uncertainty, while concentration, reassurance, exhortation, and sometimes competition, augment it. In exploring any such effects, telemetry can be useful.

Various indices of emotion itself are employed in the so-called lie detectors. Such methods are always involved with subtle questioning and interpretation, and in some cases the observation might be more effective if the subject were not distracted by the presence of unfamiliar recording equipment. Changes in the electrical resistance between the palms of the hands are an often employed attempt at an index of the meaningfulness of a topic to a subject. These changes may be associated with changes in the pattern of blood flow. This is suggested by changes in subcutaneous temperature within a few heart beats observed in response to unexpected stimuli. An extreme example is the case of a blush, now less common, which can produce a change in temperature of a few degrees. The red neck of a courting male ostrich undoubtedly can undergo a similar temperature change under some ambient conditions. Changes in heart rate and in peristaltic patterns have already been mentioned, and other possible indices can be imagined.

6. DRUG DOSE AND RELATED MATTERS

Much of our information about drugs, though certainly not all, is based on experiences with anesthetized or immobilized animals. In some cases, the degree of restraint causes an effect that introduces an uncertainty into the overall observation. The present approaches are for quantitative studies on more nearly normally behaving animals. But beyond that is the question of individual differences among patients in response to a given drug, quite aside from species differences between test animals and humans. Thus pharmacologists express concern over the factor of differences in absorption and metabolism. With the help of these techniques, in some cases it should be possible to monitor the circulating level of a drug in an individual, or to match the dose of a drug to the eventual response of a given individual. Thus the pressure transmitters of Chapter 5 should allow adjusting the dose of a spasmolytic to an individual, while the pH transmitters of Chapter 7 should allow for the individual fitting of the proper dose of an antacid preparation.

Subjects of the same sex and size do not necessarily absorb a drug into the blood stream at the same rate, and there can be differences among patients in their reactions to drugs caused by race, individual heredity, personal idiosyncrasy, or allergic reaction. Not all of these require telemetry for their observation, but in some cases the response is best observed using these techniques to avoid corruption by extraneous influences.

One can judge human interaction with non-chemical therapies also. Thus, ECG and accelerometers should aid evaluation of blind guidance devices; subject tension and movement ease seem measurable by the a.c. signal of the latter.

Many other affairs that are somewhat characteristically human can be guided in a similar fashion by telemetric observations. As a final example, one might consider the matter of teeth brushing. Acidity values corresponding to a pH of less than about 5.5 can be damaging to the dental enamel. Bacterial aggregations or plaques, breaking down dietary carbohydrates and producing acid, are disturbed by studies involving the insertion of a pH electrode, and the effects of chewing and saliva stimulation are not present when a subject's mouth is held open for such an observation. Thus Graf (1970), using circuits rather like those presented here, constructed a glass electrode and transmitter combination in a removable partial denture. A representative finding was that eating an apple just prior to bedtime resulted in the reduction of pH into the dangerous range for approximately two hours. Brushing could correct this, and some foods did not produce this problem.

In these discussions, we have seen how information can be transmitted from subjects, perhaps from within, with minimum interference to their normal patterns of either localized or overall activity. The sensing of the information is at least as important as its transmission, and it is hoped that from the selected examples of various sensor-transmitter combinations it will be clear how other variables could be monitored and other studies carried out.

Appendix 1
Antenna Turns

We want to maximize the field produced by a magnetic dipole, that is, by the small coil acting as the transmitting antenna in an endoradiosonde. If the oscillator or driving circuit produces a constant current I, in a loop of N turns and area A, then the problem is simple. It is apparent that the magnetic field will be a maximum at a given distance for a maximum value of dipole moment, NAI. This is independent of frequency and requires a maximum number of turns over the greatest possible area, while still giving a usable value of inductance.

A more often encountered and hence more practical case involves the voltage-limited oscillator. If an antenna coil has considerably more inductance than resistance at the frequency of transmission, then we achieve the greatest flux per volt with a minimum number of turns. This result can be stated in more specific form with the help of a few equations. Suppose that in an oscillator, a parallel connected RLC circuit is caused to oscillate at a fixed voltage amplitude E_0. In a typical transistor oscillator, this would be the collector supply voltage. For an optimum magnetic near field, a maximum NAI is required, but in this case I is determined by the inductance, which depends upon the number of turns N, the diameter d, and the "form factor" F. The inductance of a typical coil can be expressed as $L = FN^2d$ μH, where d is expressed in inches. In metric units where d' is in meters

$$L = 4 \times 10^{-5} FN^2d'.$$

It is known that the form factor is a function only of the ratio of diameter to length of coil. Thus for a typical coil having a diameter-length ratio of 2 the form factor is 0.027; for a ratio of 3 the factor is 0.032. A table of form factors can be found in the IT&T Co. reference (1963). The current I

through the coil is $I = E_0/\omega L$, where ω is 2π times the frequency. When this current is substituted into NAI, and the above value for L used, the result is

$$NAI = \frac{\pi d' E_0}{16 \times 10^{-5} \omega F N}$$

This shows that the maximum field is attained with a fixed E_0 if the number of turns is decreased and the diameter increased. In some cases it is necessary to reformulate this expression by considering the added frequency constraint that $\omega = 1/\sqrt{LC}$.

For a given voltage E_0 available from an oscillator, the output field may be still further increased by utilizing tapped coils, as was shown, for example, in Fig. 5.1 at the output stage. In such a circuit the total number of turns N is bridged by the resonating capacitor C. The coil is connected to the circuit at a tap so that only Na turns are presented to the circuit, where a is the turns ratio. If the voltage across the Na turns is again limited to E_0, the total magnetic moment is given by

$$NAI = \frac{\pi d' E_0}{16 \times 10^{-5} \omega F N a}$$

The output power can be increased greatly by reducing a, as well as by reducing N. A tapped coil arrangement generally enables construction of a higher Q coil, insuring greater stability and higher efficiency.

If a given coil in a voltage-limited oscillator, with N turns, is extended and the capacitor connected across the new total inductance is simultaneously decreased to maintain the same resonant frequency, it will be found that no improvement in power output is obtained. The expression for NAI is precisely the same as the original expression, after the addition of some new number of turns N'. For a specific example see pages 239–240.

It is generally necessary to consider coil losses in the oscillator resonant circuit. These may be represented as an equivalent resistance R connected in parallel with the parallel connected inductance L and capacitance C. The ratio of peak magnetic energy in the inductance to the average power lost in the resistor per cycle is well known to be given by

$$\frac{\frac{1}{2} L I_L^2}{\dfrac{\pi E_0^2}{R\omega}} = \frac{R}{2\pi\omega L} = \frac{Q}{2\pi}$$

Thus the efficiency of the circuit is strongly affected by the Q of the resonant circuit.

It is informative to consider the case of a voltage-limited oscillator circuit in which the maximum power supply is also limited. This is a case met in practice when a transmitter must operate on low currents for prolonged periods. In this case, both E_0 and R are specified. The power dissipated in the circuit at resonance is $E_0^2/2R$. At resonance $\omega L = R/Q$, and since $I = E_0/\omega L$, $I = E_0 Q/R$. The dipole moment then is

$$NAI = \frac{AE_0}{R} NQ.$$

Since the factor AE_0/R is fixed, optimization of power output consists of maximizing NQ. A general theoretical solution is difficult since changing the number of turns does not change Q in a simple way. This is because Q involves not only series wire resistance but also adjacent wire interaction, skin-depth losses, radiation losses, and so on, but the requirement is instructive and not unexpected.

The known equations that are applicable for far-field transmission again show a requirement for a maximum dipole moment, with a radiated field strength that increases with the square of the frequency.

Related considerations apply to other antenna types. Thus a strip transmission-line ring antenna can be useful, and its impedance can be altered by selection of the tap or driving point.

Appendix 2
Frequency Choice

A number of factors affecting the choice of frequency were mentioned in Chapter 10. A signal passing through a partially conducting medium is attenuated by a sort of shielding action which becomes stronger with increasing frequency. For penetration through a lossy medium, this suggests going to lower frequencies, but there are factors which indicate increased effectiveness with increasing frequency. Thus in the far field of a dipole of fixed amplitude (not power) the field strength goes up with the second power of frequency in the magnetic case and the first power in the electric case. Also, a wave reaching a receiver will be able to induce in a loop a voltage that increases with frequency. Thus we should expect an optimum intermediate frequency which would depend upon the conductivity of the intervening path, and the length of this path. Solving this problem exactly in any given case is extremely complicated, but the following may make the considerations a bit more specific while indicating the effect of the different factors. It should, however, be emphasized that the range of validity of the resulting expressions depends on the specific case involved.

If $\omega = 2\pi f$, where f is frequency, μ = permeability, and σ = conductivity, then, for the case of an isotropic medium, it is known that the field expressions include an attenuation factor with distance χ of

$$e^{-\alpha\chi}$$

where

$$\alpha = \sqrt{\frac{\omega\mu\sigma}{2}} \text{ nepers per meter}$$

Alpha is the reciprocal of skin depth and can be shown to be 2π divided by the wavelength.

An expression for the compromise between penetration and coupling suggests an optimum frequency f_o, as follows. If the signal transferred is

$$S = Kf^n e^{-K'\sqrt{f}}$$

where n has to do with the way the radiation process increases in effectiveness with frequency, then

$$\frac{dS}{df} = K\left[nf^{(n-1)} e^{-K'\sqrt{f}} + f^n\left(\frac{K'}{2\sqrt{f}} e^{-K'\sqrt{f}}\right)\right]$$

For a maximum, set $dS/df = 0$.
Then

$$f_o = \left(\frac{2n}{K'}\right)^2 = \frac{4n^2}{\pi\mu\sigma\chi^2}$$

The value of n depends on the circuit requirements and limitations. It is difficult to make the impedance of an oscillator circuit less than the very low radiation resistance (e.g., Stratton, 1941) of the small transmitter antennas. An increase in frequency increases this, thus producing greater output power. Penetration of a signal through an interface, if involved, is also a somewhat frequency-sensitive process. In some cases it appears that n can be approximated by a small integer. In the near field, however, the effectiveness of coupling need not change with frequency. Then $n = 0$, and the lowest possible frequency is best.

Appendix 3
Legal Factors

Legal restrictions upon transmitters occasionally change, and thus it is impractical to make any general statements about them here. However, it is useful to mention some kinds of restrictions presently in force. This will give an impression of the kinds of factors to be considered. Of course, different groups have special licensing arrangements, but the following extracts from Part 15 of the United States Federal Communications Commission regulations apply to devices of low power.

A low-power communication device is defined to be a restricted radiation device, exclusive of those employing conductive or guided radio-frequency techniques, used for the transmission of signs, signals (including control signals), writing, images, and sounds or intelligence of any nature by radiation of electromagnetic energy. Examples included are wireless microphone, photograph oscillator, radio-controlled garage-door opener, and radio-controlled models. Under subpart E, it is noted in Section 15.201 that a low-power communication device may be operated on any frequency in the bands 10 to 490 kHz, 510 to 1600 kHz and 26.97 to 27.27 MHz. Section 15.211 lists requirements for operation at frequencies above 70 MHz, and Section 15.212 lists requirements on telemetering devices and wireless microphones in the band 88 to 108 MHz.

Although limitations are numerous, some of the most relevant to the present applications are the following. Section 15.203 notes that a low-power communication device operating on a frequency between 160 and 190 kHz must have a power input to the final radio-frequency stage (exclusive of filament or heater power) which does not exceed one watt. Emissions outside of this frequency range must be suppressed by at least 20 dB below the unmodulated carrier. In a range of frequencies between 510 and 1600 kHz, the power input to the final radio stage (exclusive of filament or heater power) must not exceed 100 milliwatts. Again frequencies outside this range must be suppressed by 20 dB. In the frequency band 26.97 to 27.27 MHz, the power input to the final radio stage must not exceed 100 milliwatts, and the antenna shall consist of a single element that does not exceed 5 ft in length.

Of special interest are the provisions of Section 15.212 which applies to telemetering devices and wireless microphones in the band 88 to 108 MHz. In part C it notes that, "Emissions from the device shall be confined within a band 200 kc/s wide centered on the operating frequency. Such 200 kc/s bands shall lie wholly within the frequency range 88–108 mc/s." In part D, it notes, "The field strength of emissions radiated within the specified 200 kc/s band shall not exceed 50 μv/m at a distance 50 ft or more from the device." In part E, it notes that, "The field strength of emission radiated on any frequency outside the specified 200 kc/s band shall not exceed 40 μv/m at a distance of 10 ft or more from the device."

Other sections relate to further United States restrictions, and for complete information the reader is referred to the documents listing the currently applicable rulings.

It should be noted that the regulations are not presently international in scope, and thus different rules hold in different countries. For example, the United Kingdom has recently formulated regulations setting aside the band of frequencies from 102.2 to 102.4 MHz for bio-medical purposes exclusively. Transmitters unstabilized in frequency and with low power (10 μV/m at 15 m) are to be within the range 102.2 to 102.35 MHz. Two bands 10 kHz wide, centered within 3 kHz of 102.36 or 102.39 MHz, are for crystal stabilized transmitters of higher power (10 μV/m at 60 m). A channel at 102.375 MHz is to be used for the remote control of the telemetry transmitters on either of the preceding channels. (These frequencies are all within the commercial FM band of the United States). A transmitter completely in an animal is limited to the range of frequencies between 300 kHz and 30 MHz, and the mean dc power input must not exceed 5 mW with continuous oscillation or 1 mW if pulsing is used. In all cases limits are specified on accidental radiation and spurious response by the associated receiver.

Appendix 4
Laboratory Experiment on Transmitter Construction

This section outlines the construction of a simple temperature transmitter for use with any AM receiver, and which has a range of 1–2 meters. It can be suitable for carrying a signal out from within a person or animal who has swallowed it, or from a gopher in a burrow, or from a fish in a tank, etc. It is on intermittently at a rate which carries the temperature information, and since it is off the larger fraction of the time, the average drain on the battery is quite low.

The coil in Fig. A4.1 is wound on a length of soda straw (approximately 0.5 cm diameter). Wind 50 turns along about 1 cm, take a loop to the side for the tap, and continue winding another 100 turns in the same direction on top of the first turns. Secure the turns with sticky tape or wax or fingernail polish, avoiding kinks. Insulation should be removed from the ends by chemicals to minimize nicking and breakage; with some wire it is sufficient to pass the ends into a blob of molten solder.

Lightly sandpaper the battery and wipe the surface clean. Affix battery to table with double-sided sticky tape. Place a drop of 4N hydrochloric acid on surface to be soldered. (In other soldering operations rosin core solder is used, and no other flux.) Clean the tip of the soldering iron by rubbing on a cloth. In one sequence, melt a droplet of solder on the tip of the iron and rub the flat of the iron on the battery for one second to transfer the solder. Blow on the battery to cool it fully and then rapidly solder copper leads, previously coated with solder, to the battery. Check the voltage.

Connect the transistor, thermistor and capacitor as a subassembly to be inserted into the coil and soldered to it, after cutting all leads to a length of 3–10 mm except for the collector lead as shown. The transistor here is a D26E1 or D26E5, but others can be substituted.

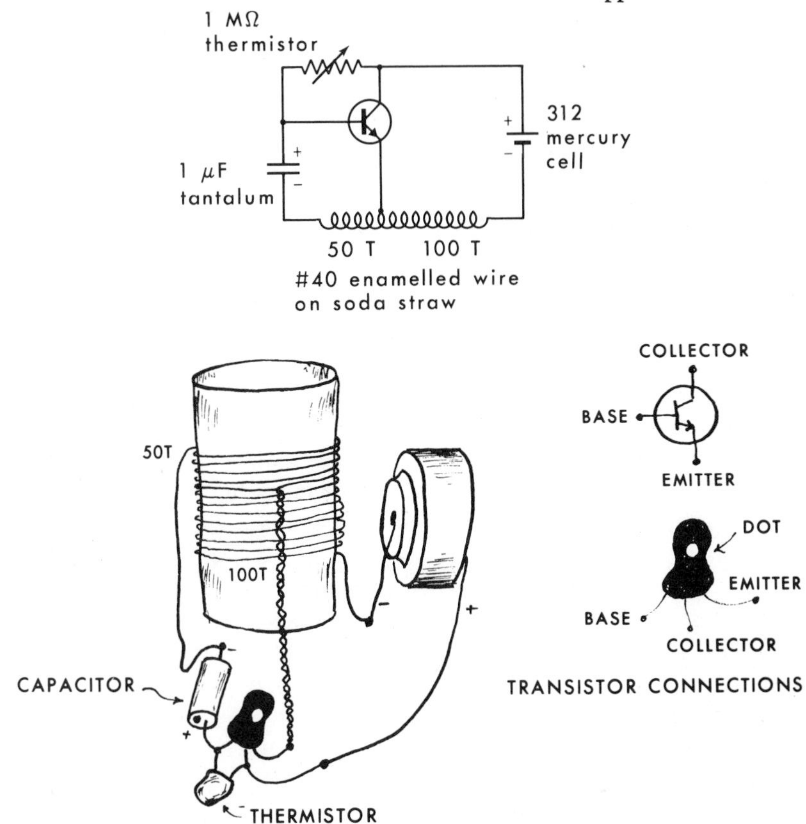

Fig. A4.1 Schematic and corresponding pictorial diagram of simplified temperature transmitter construction. The circuit is similar to that of Fig. 1.7, but the tuning capacitor has been eliminated and the turns reduced for increased transmission distance.

For economy during construction in a class, it is quite practical to hold the components with masking tape placed on the table sticky side up, and to manipulate them by pushing with toothpicks. No tuning capacitor is employed, the capacitances incidentally associated with the transistor and coil largely serving this function. All components can go in the coil except the battery which should go beside the coil or at its end. Battery placement beside the coil generally gives the stronger signal. Be certain bare wires are not touching each other before connecting the battery.

A click will be heard from a nearby AM (standard entertainment) receiver about once per second over a considerable range of tuning. Start with a separation of about a foot. An oscilloscope connected across the entire coil should display the periodically interrupted waveform. If there should be an apparent lack of radio function, it can be due to continuous oscillation or else lack of oscillation, noticeable this way and perhaps corrected by redoing the coil.

If a PNP type transistor were instead used, the circuit would be the same except the battery and capacitor would be reversed. A buzz rather than a periodic click will come from the receiver if the capacitor is replaced by one of 0.001 microfarad, in which case there will probably be no preferable polarity on this component.

Transmission distance is usually 1–2 meters. Enclose the transmitter in a plastic bag and note the transmission distance through fresh water, salt water, sand, and from within the open and closed mouth. Aquaria and bags of soil should be available, within or between which the transmitter can be moved. Note the effect of orientation changes by the transmitter and receiver.

For extended use, a wax potting or coating is desirable, perhaps with an outer cover of silicone rubber. If the transmitter is to be used inside an animal, sensitivity calibration should be with the unit in a beaker of 1% saline. The click rate is timed with a watch during exposure to different known temperatures. Long term stability depends upon the stability of the actual components obtained, especially the capacitor in low rate units.

Other variable resistance elements can replace the thermistor to transmit light intensity or humidity, or etc.

If sealed functioning transmitters might be carried home on an airplane, suitable shielding should be achieved by tightly wrapping in a square foot of aluminum foil.

Appendix 5
Phaselock Loop Receiver

A phaselock loop receiver (Chapter 11) can be formed from standard integrated and other components. Such a circuit has been prepared by Keith Brocklesby and Richard Barwick of Canberra, and is depicted below. The block diagram gives the scheme of the interconnections and the overall details of operation, while the complete diagram indicates a set of suitable commercial components. The operating frequency is 27.05 MHz and the sensitivity at the input is −148 dB which corresponds to 0.012 microvolt across 50 ohms. Maximum input is −110 dB at maximum sensitivity before distortion appears, beyond which an attenuator is employed. The noise figure is 9.7 dB. The loop bandwidth is 10 Hz, and the control range is 3 kHz.

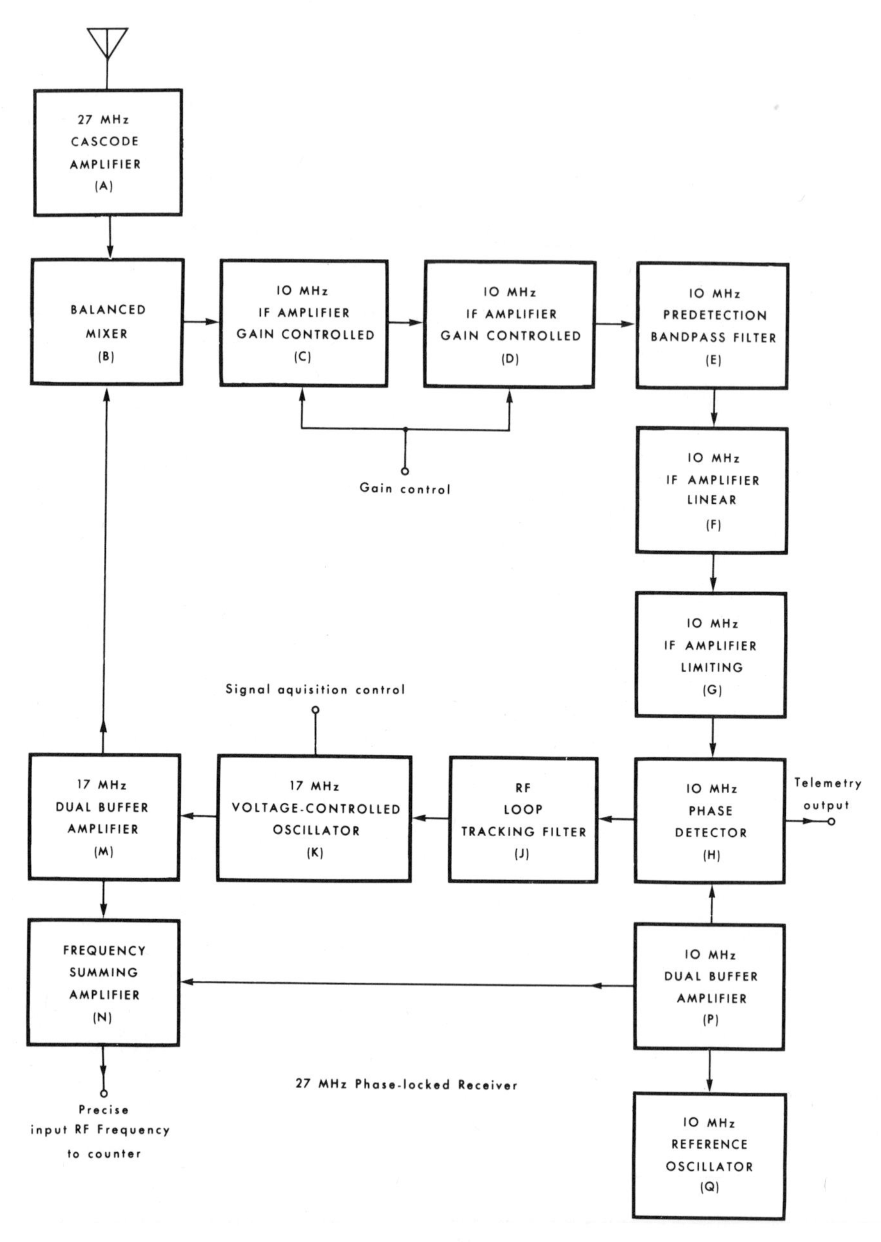

Fig. A5.1 Logic of interconnections of a phaselock loop receiver.

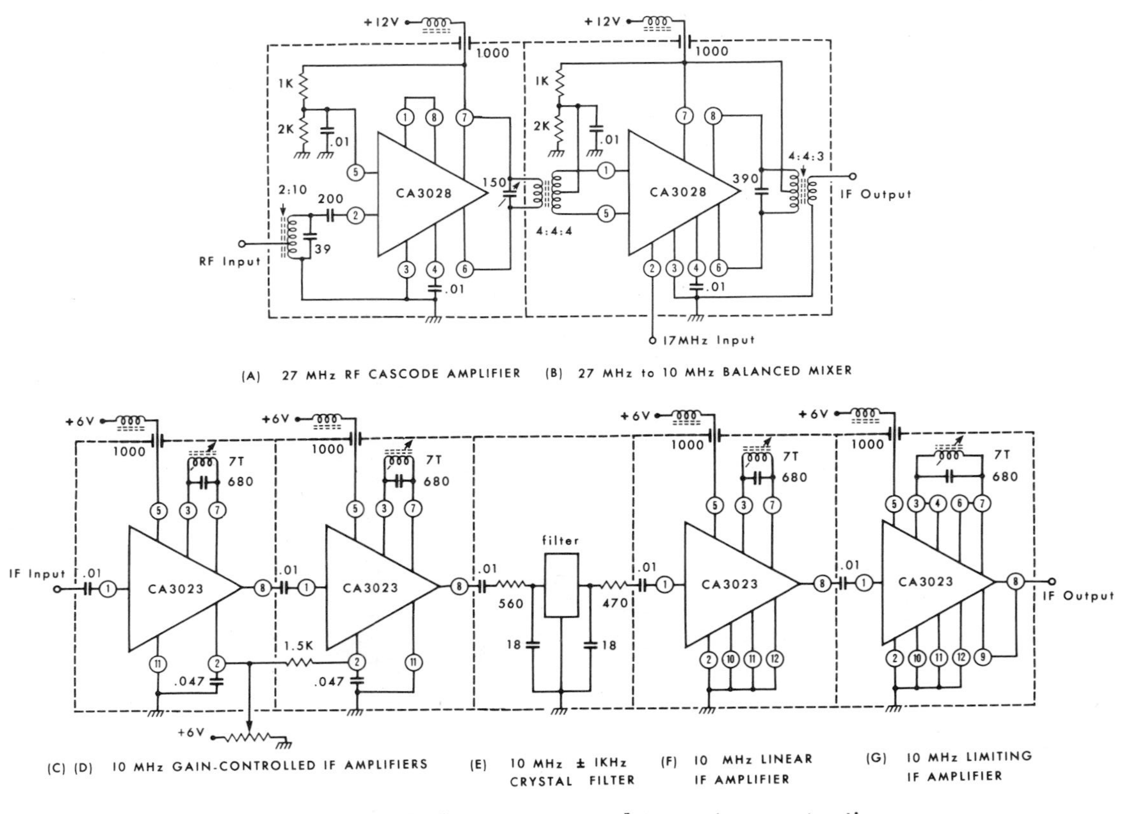

Fig. A5.2 Standard components used in receiver construction.

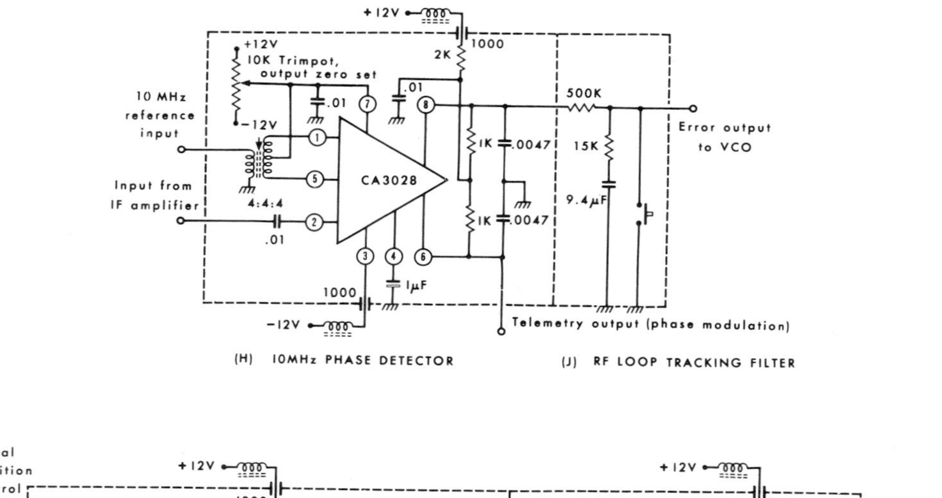

(H) 10MHz PHASE DETECTOR (J) RF LOOP TRACKING FILTER

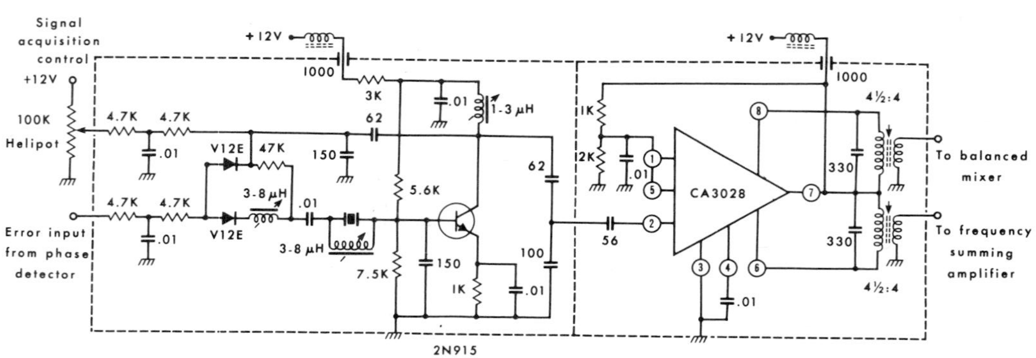

(K) 17 MHz VOLTAGE-CONTROLLED OSCILLATOR (M) 17 MHz DUAL BUFFER AMPLIFIER

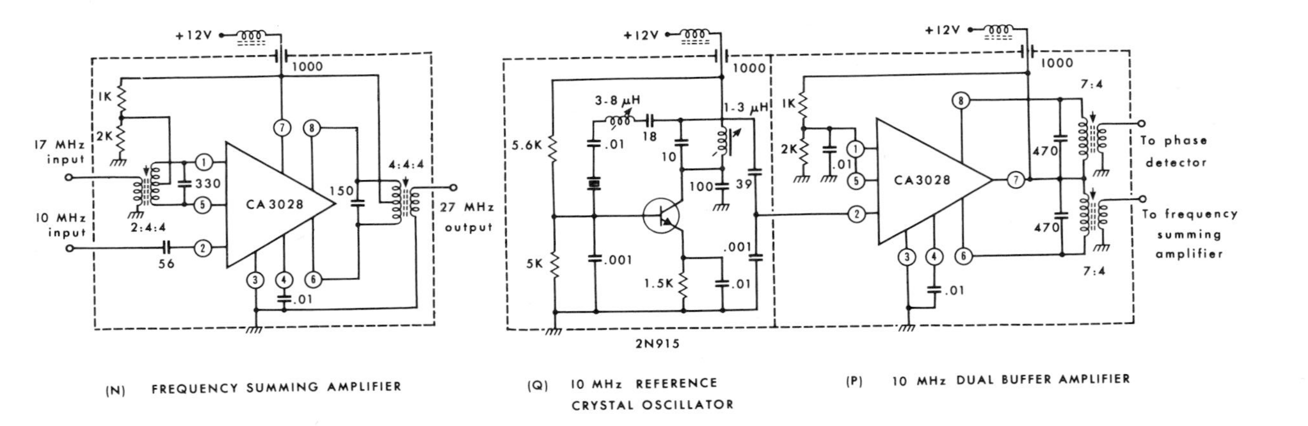

Fig. A5.2 (*contd.*) The large triangles here represent linear integrated circuits, in this case monolithic silicon amplifiers manufactured by R.C.A.

References

Adam, G. (1963), *IEEE Trans. Electron Devices Ed-10,* 51–58.

Andren, C. F., M. A. Fadali, V. L. Gott, and S. R. Topaz (1968), *IEEE Trans. on Bio-Med. Eng.,* **BME-15**, 4, 278–280.

Argue, G. (1965), *Ind. Res.,* 7, 85–90.

Aukland, K. (1967), in *Blood Flow Through Organs and Tissues* (Eds: W. H. Bain and A. M. Harper), Williams & Wilkins Co., Baltimore, 157–162.

Baghdady, E. J. (1961), *Communication System Theory,* McGraw-Hill, New York.

Baker, L. E., L. A. Geddes, and H. E. Hoff (1966), *Med. Biol. Eng.,* **4**, 371–379.

Baldwin, H. A. (1963), Technical Documentary Report No. SAM-TDR613–36, USAF School of Aerospace Medicine, Brooks Air Force Base, Texas.

Baldwin, H. A. (1965), *BioScience,* **15**, 95–97.

Balmino, G., S. J. Criswell, D. L. Fernald, E. H. Jentsch, J. Latimer, and J. C. Maxwell (1968), Spec. Report 289, *Animal Tracking from Satellites,* Smithsonian Astrophysical Observatory, Cambridge, Mass.

Bárány, F., and B. Jacobson (1964), *Gut,* **5**, 90–93.

Bassen, H. I. (1967), *Suppl. to IEEE Trans. on Aerospace and Electron. Systems,* **AES-3**, 6, 530–533.

BIAC (1966) Special Bioinstrumentation advisory council report S-1 from American Institute of Biological Sciences.

Birzis, L., and S. Tachibana (1962), *Life Sciences,* Pergamon Press, Vol. 11, pp. 587–598.

Brindley, G. S., and W. S. Lewin (1968), *J. Physiol.,* **196**, 479–493.

Burrows, C. R. (1962), *IRE Trans. Antennas and Propagation,* **AP-10**, 328–334.

Burrows, C. R. (1966), *IEEE Trans. Antennas and Propagation,* **AP-14**, 386–388.

Busser, J. H., and M. Mayer (1957), *ONR Res. Rev.,* 9–13.

Caceres, C. A. (1965), *Biomedical Telemetry,* Academic Press, New York.

Chaffee, F. E., and R. U. Light (1934), *Yale J. Biol. Med.,* 7, 83–128 and 441–450.

Chang, J. J., J. H. Forster, and R. M. Ryder (1963), *IEEE Trans.* on *Electron Devices,* **ED-10**, 281–287.

Chen, A. S. (1963), "Passive Telemetering System Feasibility Study," M.S. Thesis, Electrical Engineering, University of California, Berkeley.

Chow, W. F. (April 1, 1957), *Electronics,* **30**, 180–182.

Clark, L. C., and F. Gallan (1966), *Science,* **152**, 155–156.

Clark, L. C., R. Wold, D. Granger, and Z. Taylor (1953), *J. App. Physiol.,* **6**, 189–193.

Clifton, J. S., and D. Parker (1970), in *Proc. of the 2nd Int'l. Conf. on Med. Physics,* Int'l. Conf. on Medical Physics, Inc., Boston.

Cochran, Wm., and R. D. Lord (1963), *J. Wildl. Man.,* **29,** 898–902.

Cochran, Wm., D. W. Warner, J. R. Tester, and V. B. Kucchle (1965), *BioScience,* **15,** 98–100.

Cochran, W. W. (1966), "Some Notes on the Design of a Directional Loop Antenna for Radio Tracking Wildlife," A Bioinstrumentation Advisory Council Information Module, Distributed by the American Institute of Biological Science.

Cole, L. C. (1963), Radiotelemetering of environmental conditions for ecological field studies, in L. E. Slater (Ed.), pp. 203–209.

Collins, C., and R. S. Mackay (1960), *IRE Trans. Med. Electron.,* **ME-7,** 349–350.

Collins, C. C. (1966), "A Study of the Mechanism of Sensory Evoked Pressure Response in the Rabbit Eye," Doctoral Thesis, University of California, Berkeley.

Craighead, F. C., and J. J. Craighead (1965), *BioScience,* **15,** 88–92.

Daniel, E. E., and K. M. Chapman (1963), *Am. J. Digest. Disease,* **8,** 54–60.

deBretteville, A. P. (1953), *Ceramic Age,* **61,** 18–24.

deLang, R. G., and J. A. Zijderveld (1968), *J. Appl. Physics,* **39,** 2195–2200.

Delgado, J. M. (1963), in *Biotelemetry* (Editor: L. Slater), Pergamon Press, New York, pp. 231–249.

Delgado, J. M. R., V. Mark, W. Sweet, F. Ervin, G. Weiss, G. Bach-y-Rita, and R. Hagiwara (1968), *J. Nervous and Mental Dis.,* **147,** 329–340.

Dewhurst, D. J., and R. L. Kirsner (1969), *Proc. 8th Int. Conf. on Med. and Biol. Eng.,* Inst. Electrical and Electronics Eng., New York, p. 28.5.

Doyle, J. M. (1966), *Thin-Film and Semiconductor Integrated Circuitry,* McGraw-Hill, New York.

Driller, J., S. K. Hilal, W. J. Michelsen, B. Sollish, L. Katz, and W. Konig, Jr. (1969), *Med. Res. Eng.,* Aug.–Sept., 11–16.

Driller, J., and G. Neumann (1967), *IEEE Trans. Bio-Med. Eng.,* **BME-14,** 52–53.

Durrani, S. H. (1962), *IRE Trans. Antennas and Propagation,* **AP-10,** 524–528.

Edge, G. M., G. Horn, and G. Stechler (1969), *Proc. Physiol. Soc.,* June, 2–4.

Eklund, C. R., and R. E. Charlton (1959), *American Scientist,* **47,** 80–86.

England, S. J., and B. Pasamanick (1961), *Science,* **133,** 106.

Falb, R. D., G. A. Grode, L. M. Luttinger, and R. I. Leininger (1966), in *Proc. Annual Conf. on Eng. in Med. and Biol.,* Inst. Elec. and Electronic Engineers, New York, 261–271.

Farrar, J. T., C. Berkeley, and V. K. Zworykin (1960), *Science,* **131,** 1814.

Farrar, J. T., V. K. Zworykin, and J. Baum (1957), *Science,* **126,** 975.

Findlay, A. L. R., G. Horn, and G. Stechler (1969), *Proc. Physiol. Soc.,* June, 4–6.

Fischmann, E. J., and M. R. Barber (1963), *Amer. Heart Jour.,* **65,** 628–637.

Flock, W. L. (1968), *IEEE Spectrum,* June, 62–66.

Fox, R. H., R. Goldsmith, and H. S. Wolff (1961), *J. Physiol.,* **160,** 22–23.

Franklin, D. (1970), in *Proc. of the 2nd Int'l. Conf. on Med. Physics,* Int'l. Conf. on Medical Physics, Inc., Boston.

Franklin, D. L., N. W. Watson, K. E. Pierson, and R. L. Van Citters (1966), *Am. J. Med. Electron.,* **5,** 24–28.

Franklin, D. L., D. W. Baker, and R. F. Rushmer (1962), *Inst. Rad. Eng. Trans. Bio-Med. Electron.,* **BME-9,** 44–49.

Frei, E. H., J. Driller, H. N. Neufeld, U. Barr, and L. Blieden (1966), *Proc. Annual Conf. on Eng. in Med. and Biol.,* 65.

Fryer, T. (1965), *Proc. 18th Annual Conference on Engineering, Medicine, and Biology,* Institute of Electrical and Electronics Eng., New York.

Fryer, T., and G. Deboo (1964), *Proc. 17th Annual Conference on Engineering, Medicine, and Biology,* Institute of Electrical and Electronics Eng., New York.

Fryer, T. B., H. Sandler, and B. Datnow (1969), *Med. Res. Eng.,* Mar.–Apr., 9–15.

Galler, S. R. (1965), in *Biomedical Telemetry* (Ed. C. Caceres), Academic Press, New York, 237–254.

Gardner, F. M. (1966), *Phase Lock Techniques,* John Wiley, New York.

Garrels, R. M., M. Sato, M. E. Thompson, and A. H. Truesdale (1962), *Science,* **135,** 1045–1048.

Geddes, L. A., and L. E. Baker (1967), *Med. Electron. and Biol. Eng.,* **5,** 271–293.

Geddes, L. A., and H. E. Hoff (1965), *Digest of the 6th International Conference on Med. Electron. and Biol. Eng.,* p. 64.

Geddes, L. A., and L. E. Baker (1968), *Principles of Applied Biomedical Instrumentation,* J. Wiley & Sons, Inc., New York.

Glenn, W. L., J. H. Hageman, A. Mauro, L. Eisenberg, S. Flanigan, and M. Harvard (1964), *Annals of Surgery,* **160,** 338–350.

Glover, K. M., T. Konrad, K. R. Hardy, and C. R. Landry (1966), in *Proc. 12th Conf. on Radar Meteorology,* Amer. Meteorological Soc., Norman, Okla., 254–258.

Goodman, B. (1966), *The Radio Amateurs Handbook* (43rd. Ed.), American Radio Relay League, Newington, Conn.

Gott, V. L., J. D. Wiffen, and R. C. Dutton (1963), *Science,* **142,** 1293.

Gott, V. L., D. Wiffen, D. E. Koepke, R. L. Daggett, W. C. Boake, and W. P. Young (1964), *Trans. Amer. Soc. Artificial Internal Organs,* **10,** 213–217.

Graf, H. (1970), in *Proc. Fourth International Congress on Pharmacology* (Basel), Vol. 5, Schwabe and Co., Basel.

Greatbatch, W. (1965), *Med. Electron. and Biol. Eng.,* **3,** 305–306.

Gualtierotti, T., and P. Bailey (1968), *Electroenceph. Clin. Neurophysiol.,* **25,** 77–81.

Harthoorn, A. M., and J. Bligh (1965), *Research in Vet. Sci.,* **6,** No. 3, 290–299.

Haugen, M. G., W. R. Farrell, J. F. Herrick, and E. J. Blades (1955), *Proc. Nat. Electron. Conf.,* 465–475.

Haynes, H. E., and A. L. Witchey (1960), *RCA Engineer,* **5,** 52–55.

Hilliard, E. J. (1962), *Electronics,* **35,** 52–56.

Hurd, R. M., and R. N. Lane (1957), *J. Electrochem. Soc.,* **104,** 727–730.

IT&T Co. (1963), Reference Data for Radio Engineers, 4th Ed., IT&T Co., Publishers.

Jackson, B. T., and R. H. Egdahl (1960), *Surgery,* **48,** 564–570.

Jackson, B. T., and R. H. Egdahl (1962), *Surgery,* **52,** 165–173.

Jacobson, B., and B. Lindberg (1963), *Electronics,* **36,** 58.

Jacobson, B., and R. S. Mackay (1957), *Lancet,* 1224.

Jacobson, B., and R. S. Mackay (1958), *Advances in Biological and Medical Physics, Vol. 6* (Eds. J. Lawrence and C. Tobias), Academic Press, New York, 201–261.

Jasik, H. (1961), *Antenna Engineering Handbook,* McGraw-Hill, New York.

Kahn, A. R., and S. Koller (1966), in *Proc. 19th Annual Conf. on Eng. in Med. and Biol.,* p. 19.7.

Kalmus, H. P. (1955), *Proc. IRE,* **43,** 698–700.

Katchalski, E. (1962), *Polyamino Acids, Polypeptides, and Proteins* (Ed. M. Stahmann), University Wisconsin Press, pp. 283–288.

Kershner, R. B., and R. R. Newton (1962), *Inst. of Navigation Jour.,* **15,** 129–144.

Kimoto, S., T. Watanuki, M. Hori, K. Suma, J. Nagumo, A. Ouchi, T. Takahashi, M. Kumano, and H. Watanabe (1964), *Med. Electron and Biol. Eng.,* **2,** 85.

Ko, W., and L. Slater (1965), *Electronics,* **38,** 89–96.

Ko, W., W. Thompson, and W. Yon (1963), *Med. Electron and Biol. Eng.,* **1,** 363–366.

Kolin, A., and R. Wisshaupt (1963), *Inst. Electrical and Electronics Eng. Trans. Bio-Med. Electron,* **BME-10,** 60–67.

Lavallee, M., and G. Szabo (1965), *Digest of the 6th International Conference on Med. Electron. and Biol. Eng.,* pp. 568–569.

Leininger, R., M. Epstein, and R. D. Falb (1965), *Digest of 6th International Conf. on Med. Electron. and Biol. Eng.,* pp. 349–350.

Lepri, F., and M. L. Ramorino (1960), *Proc. 3rd Intl. Conf. on Med. Electron.,* 121. Institution Electrical Engineers, London.

Linder, R. (1962), *Bell System Technical Journal,* **16,** 803–831.

Lion, K. S. (1959), *Instrumentation in Scientific Research: Electrical Input Transducers,* McGraw-Hill, New York.

Lonsdale, E. M., J. W. Steadman, and W. L. Pancoe (1966), *IEEE Trans. on Bio-Med. Eng.,* **BME-13,** 153–159.

Lyklema, J. (1964), *Med. Electron. and Biol. Eng.,* **2,** 256–280.

Mackay, R. S. (1958), *Am. J. Physics,* **26,** 60–69.

Mackay, R. S. (1959), *IRE Trans. Med. Elec.,* **ME-6,** 100–105.

Mackay, R. S. (1960A), *IRE Trans. Med. Elec.,* **ME-7,** 67–71.

Mackay, R. S. (1960B), *Rev. Sci. Inst.,* **31,** 1241–1242.

Mackay, R. S. (1961), *Science,* **134,** 1196–1202.

Mackay, R. S. (1962), *Med. Electron. and Biol. Eng.,* **1,** 109–111.

Mackay, R. S. (1963), *New Scientist,* **19,** 650–653.

Mackay, R. S. (1963B), in *Proc. 2nd Symp. on Underwater Physiol.,* Publ. 1181, Nat'l. Acad. Sci. and Nat'l. Res. Council, Washington, D.C., 41.

Mackay, R. S. (1964A), *Med. Electron. and Biol. Eng.,* **2,** 3–18.

Mackay, R. S. (1964B), *Nature,* **204,** 355–358.

Mackay, R. S. (1964C), *Science,* **145,** 296–297.

Mackay, R. S. (1965), *Biomedical Telemetry* (Ed. C. Caceres), Academic Press, New York, 147–235.

Mackay R. S. (1968), *Copeia No. 2,* 252–259.

Mackay, R. S., and B. Jacobson (1957), *Nature,* **179,** 1239–1240.

Mackay, R. S., E. Marg, and R. Oechsli (1960), *Science,* **131,** 1668–1669.

Mackay, R. S., and G. Rubissow (1970), in *4th Symp. on Underwater Physiol.* (Ed: C. Lambertsen), Academic Press, New York.

MacKereth, F. J. (1964), *J. Sci. Instrum.,* **41,** 38–39.

Magondeaux, L. (1957), U.S. Patent 2,812,427.

Marchal, M. M., and M. T. Marchal (1958), *C. R. Acad. Sci. Paris,* **246,** 3519–3520.

Marchal, M. M., and M. T. Marchal (1959), French Patent 1,200,043.

Marey, M. (1869), in *Smithsonian Annual Report,* Washington, D.C., 226–285.

Marshall, W. H. (1965), *BioScience,* **15,** 92–94.

McElligott, J. G., J. R. Zweizig, and R. T. Kado (1969), in *Proc. Nat'l. Telemetering Conf. '69,* Institute of Electrical and Electronics Engineers, Inc., New York, 207–210.

McGinnis, S., and C. Brown (1966), *Herpetologica,* **22,** 189–199.

McLeod, F. D. (1967), in *Digest of the 7th Int'l. Conf. on Med. and Biol. Eng.,* p. 213.

Mech, D. L., V. B. Kuechle, D. W. Warner, and J. R. Tester, *J. Wildl. Man.*, **29**, 898–902.
Meindl, J. D. (1969), *Micropower Circuits*, J. Wiley & Sons, Inc., New York.
Nagumo, J., A. Uchiyama, S. Kimoto, T. Watanuki, M. Hori, K. Suma, A. Ouchi, M. Kumano, and H. Watanabe (1962), *IRE Trans. Bio-Med. Electron.*, **9**, 195.
Nelson, R. S., W. C. Dewey, and R. G. Rose (1963), *Jour. Nuc. Med.*, **4**, 206.
Noble, F. W., T. C. Goldsmith, C. J. Waldsburger, and T. C. Cook (1962), *Digest of 15th Conf. on Eng., Med. and Bio.*, The Conference Committee, Chicago, U.S.A.
Noble, F. W. (1966), *Proc. Inst. Elect. Electron. Eng.*, **54**, 1976–1978.
Noller, H. G. (1960), *Medical Electronics* (Ed. C. N. Smith), Iliffe, London, pp. 296–299.
Norgaard, D. E. (1956), *Proc. IRE*, **44**, 1735–1743.
Olsen, E. R., C. C. Collins, T. R. Altenhofen, J. E. Adams, and V. Richards (1968), *Amer. J. Surg.*, **116**, 3–7.
Orr, W. W. (1966), *The Radio Hand Book*, 16th ed., Editors and Engineers, Ltd., California.
Owens, B. B., and G. R. Argue (1967), *Science*, **157**, 308–309.
Pacela, A. F. (1966), *Med. and Biol. Eng.*, **4**, 1–15.
Perrenoud, J. (1957), *Gastroenterologia*, **87**, 349.
Pienkowski, E. C. (1965), *BioScience*, **15**, 115–117.
Pistey, W. R., and J. J. Wright (1961), *Canadian J. Comp. Med.*, **25**, 59–68.
Pye, J. D., and M. Flinn (1964), *Ultrasonics*, **2**, 23–28.
Rechnitz, G. A. (1969), *Analytical Chemistry*, **41**, 109A–113A.
Reid, M. H. (1966), "A Mathematical Method for System Identification with Application to Temperature Regulation in the Dolphin and Pressure Regulation in the Eye," Ph.D. Thesis, University of California, Berkeley.
Reid, M. H., and R. S. Mackay (1968), *Med. and Biol. Eng.*, **6**, 231–290.
Reynolds, L. W. (1964), *Aerospace Medicine*, **35**, 115–117.
Salzman (1957), *Surgery*, **61**, 1–10.
Schelkunoff, G. A. (1943), *Electromagnetic Waves*, D. Van Nostrand Co., New York.
Schevill, W. E., and W. A. Watkins (1966), "Radio-Tagging of Whales," Woods Hole Oceanographic Institution Technical Report No. 66-M.
Schuder, J. C., and H. Stoeckle (1962), *ASAIO Trans.*, **8**, 90–93.
Schwan, H. P. (1965), *Digest of the 6th International Conference on Med. Electron. and Biol. Eng.*, pp. 556–559.
Schwartz, N., R. S. Mackay, and J. Sackman (1966), *Bulletin Mathematical Biophysics*, **28**, 585–643.
Schwartz, R. J., and R. C. Spindel (1967), *IEEE Student J.*, **5**, 27–32.
Severinghaus, J. W., and A. F. Bradley (1958), *J. Appl. Physiol.*, **13**, 515–520.
Shimizu, A., and J. Nishizawa (1961), *IRE Trans. on Electron Devices*, **ED-8**, 370–377.
Singer, J. R. (1960A), *J. Applied Physics*, **31**, 125–127.
Singer, J. R. (1960B), *Proc. 3rd Intnl. Conf. on Med. Electron*, "Nuclear Magnetic Resonance Blood-Flow Meters," 25th July, 227–229.
Slater, L. (1963), *Biotelemetry*, Pergamon Press, New York.
Smith, N. T., and H. O. Schwede (1969), *J. Appl. Physiol.*, **26**, 241–247.
Smyth, C. N. (1957), *J. Obstet. Gynaec. Brit. Emp.*, **64**, 59–66.
Smyth, C. N. (1967), *Med. & Biol. Eng.*, **5**, 69–73.

Smyth, C. N., and H. S. Wolff (1960), *Lancet,* p. 412.

Snodgrass, J. M., and P. W. Gilbert (1967), *Sharks, Skates and Rays* (Eds: P. W. Gilbert, R. F. Mathewson, and D. P. Rall), The Johns Hopkins Press.

Spencer, M. P., and A. B. Denison (1959), *Inst. Rad. Eng. Trans. Med. Electron.,* **ME-6**, 220–225.

Stow, R. W., and B. F. Randall (1954), *Amer. J. Physiol.,* **179**, 678.

Stratton, J. E. (1941), *Electromagnetic Theory,* McGraw-Hill, New York.

Tausworthe, R. C. (1966), Tech. Report 32-819, *Theory and Practical Design of Phase-Locked Receivers,* Vol. 1–2, Jet Propulsion Lab., Cal. Inst. Tech., Pasadena, Calif.

Tester, J. R., and D. B. Siniff (1965), *Trans. N. Amer. Wildl. Conf.,* **30**, 379–392.

Togawa, T., and K. Suma (1965), *Digest of 6th Intnl. Conf. on Med. Electron. and Biol. Eng.,* 48–49, Japan Society of Medical Electronics and Biological Eng.

Trefethen, P. J., J. W. Dudley, and M. R. Smith (1957), *Electronics,* **30**, 156–157.

Tucker, D. G., and A. B. Gazey (1967), *Applied Underwater Acoustics,* Pergamon Press, New York.

Vantrappen, G., J. D'Haenes, S. Verbeke, and J. Vandenbroucke (1964), *Gut,* **5**, 96–98.

Van Citters, R. L., W. S. Kemper, and D. L. Franklin (1966), *Science,* **152**, 384–386.

Van Citters, R. L., and D. L. Franklin (1966), *J. Appl. Physiol.,* **21**, 1633–1636.

von Ardenne, M. (1960), *Medical Electronics* (C. N. Smyth, Ed.), Iliffe, London, 268–280.

Walawender, W. P., C. Tien, and L. C. Cerny (1970), in *2nd Int'l. Conf. on Hemorheology 1969,* Springer-Verlag, New York.

Walter, W. G., R. Cooper, H. J. Crow, W. C. McCallum, W. J. Warren, and V. J. Aldridge (1967), *Electroenceph. Clin. Neurophysiol.,* **23**, 197–206.

Webb, G. N., and R. E. Rogers (1966), *IEEE Spectrum,* **3**, 77–87.

Weinmar, J., and J. Mahlar (1964), *J. Med. Electron. and Biol. Eng.,* **2**, 299–310.

Wenger, W. A., B. T. Engel, T. L. Clemens, and T. D. Cullen (1961), *Gastroenterology,* **41**, 479.

Wheeler, H. A. (1958), *Proc. IRE,* **46**, 1595–1602.

Whitehead, J. R. (1950), *Super Regenerative Receivers,* Cambridge U. Press, England.

Wolff, H. S. (1961), *New Scientist,* **12**, 419–421.

Wolff, H. S. (1965), *Digest 6th Intnl. Conf. on Med. Electron. and Biol. Eng.,* Japan Society of Medical Electronics and Biological Eng., Tokyo, 309–312.

Wolff, H. S., J. McCall, and J. A. Baker (1962), *Brit. Commun. Electron.,* **9**, 120–123.

Woldring, S., G. Owens, and D. Woolford (1966), *Science,* **153**, 885–887.

Wyatt, D. G. (1961), *Phys. in Med. Biol.,* **5**, 289–320.

Zarnstorff, W. D., C. A. Castillo, and C. W. Crumpton (1962), *Inst. Rad. Eng., Trans. Bio-Med. Electron.,* **BME-9**, 199–200.

General Reading

Aagaard, J. S., and Dubois, J. L. (1962), Telemetering impact data from the football field, *Electronics,* 35(14): 46–47.

Abrams, L. D., Hudson, W. A., and Lightwood, R. (1960), A surgical approach to the management of heart-block using an inductive coupled artificial cardiac pacemaker, *Lancet,* 1: 1372–1374.

Abrams, L. D., and Norman, J. C. (1964), Experience with inductive coupled cardiac pacemakers, *Ann. N.Y. Acad. Sci.,* 111(3): 1030–1040.

Adams, L. (1965), Progress in ecological biotelemetry, *BioScience,* 15: 83–86.

Adams, L. (1965), Bibliography: Biotelemetry, *BioScience,* 15: 155–157.

Adams, L., and Smith, W. C. (1964), Wildlife telemetry, *Proc. 1964 Natl. Telem. Conf.,* Sec. 2-3: 1–7.

Adams, R. M., Fromme, G. L., Dizon, D., and Anstadt, G. L. (1963), A microwatt VHF telemetry system for implantation in animals, *Proc. 1963 Natl. Telem. Conf.,* Session 5-2, Albuquerque, N.M.

Adams, S. H., and Zander, H., Functional tooth contacts in lateral and in centric occlusion, *Jour. Am. Dent. Assn.,* 69: 465–473.

Adey, W. R. (1963), Potential for telemetry in the recording of brain waves from animals and men exposed to the stresses of space flight, in L. E. Slater (Ed.), *Biotelemetry,* Pergamon Press, New York, pp. 289–302.

Adey, W. R., Hanley, J., Kado, R. T., and Zweizig, J. R. (1966), Design considerations in multichannel telemetry, *Proc. 19. Ann. Conf. Eng. Med. Biol.*

Adey, W. R., Cockett, A. T., Mack, P. B., Meehan, J. P., and Pace, N. (1969), Biosatellite III: Preliminary Findings, *Science,* 166: 492–493.

Akulinichev, I. T. (1964), Use of radiotelemetry in space medicine, *Vest. Akad. Med. Nauk SSSR,* 19: 60–66.

Akulinichev, I. T., Baevskij, R. M., Denisov, V. G., and Jaxdovskij, V. I. (1963), Biotelemetering systems in astronautics (in Russian), in V. V. Parin, *Radiotelemetry in physiology and medicine,* pp. 10–13.

Akulinichev, I. T., and Baevskij, R. M. (1962), Importance of medical electronics for the security of space flights of animals and men (in Russian), *Tex. Dokl. II Vses. Konf. po Primeneniju Radioelektroniki v Biologii i Medicine,* NIITEIR, 61–62.

Akulinichev, I. T., Baevskij, R. M., and Gazenko, O. G. (1965), Means and methods of medic-biological experiments in space flight, *Proc. 1st Intern. Symp. on Basic*

Environmental Problems of Man in Space, Paris, Oct. 29–Nov. 2, 1962, Springer-Verlag, 425–451.

Akulinichev, I. T., Baevskij, R. M., Zazykin, K. P., and Frejdel, V. R. (1964), Radio-electronics in space medicine (in Russian), *Moskva–Leningrad,* 48 pp.

Akulinichev, I. T., Baevskij, R. M., Bajkov, A. E., Zazykin, K. P., Maksimov, L. G., and Siadrincev, I. D. (1963), The system of physiological measurements in the spacecraft "Vostok-5" and "Vostok-6" (in Russian), in V. V. Parin, *Radio-telemetry in physiology and medicine,* pp. 27–30.

Akulinichev, I. T., Emelyanov, M. D., and Maksimov, D. G. (1965), Okulomotor activity in cosmonauts in orbital flight, *Izv. Akad. Nauk SSSR Ser. Biol.,* 30(2): 274. Translation: *Fed. Proc. Transl. Suppl.,* 25 (1966), T31–T33.

Allard, E. (1962), Telemesure en medecine sportive, *Electronique Medicale,* No. 19–20: 37–40.

Allen, R. T., Hanson, M. L., and Kresge, D. J. (1964), Biotelemetry in medicine, *Bio/Med. Instrum.,* New York, 1(3): 15–19.

Allington, R. W. (1959), What's the stomach pressure?, *Univ. of Nebraska Blue Print,* 10–11.

Almond, J. A. (1965), Personal telemetry transmitter system, *USAF Tech. Doc. Rep.,* TR-65-87: 1–24.

Almond, J. (1966), A personal telemetry system—a production model, *Proc. Nat. Telem. Conf.*

Alnutt, R. (1963), *Techniques of physiological monitoring,* Vol. II, *Components,* A,RL-TDR-62-98 (II), USAF, 1–256.

Althouse, J. (1967), Build hydronic-radiation transmitter, *Radio-Electronics,* 38(5): 37–38.

American Aviation Inc. (1963), *Telemetering systems for the life sciences, a survey,* Rept. SID, 63–284.

Amlinger, P. R. (1969), Routine telemetry of electrocardiograms and computer analysis, *National Telemetering Conference,* 222.

Anderka, F. W., and Dyer, M. I. (1967), A design for a miniature biopotential radio transmitter, *J. Appl. Physiol.,* 22(6): 1147–1148.

Anon. (1957), Thermometer that transmits, *Radio-Electronics,* 10: 52.

Anon., *Discoverer III biomedical data report,* Directorate of Bioastronautics Projects, A.F. Ballistic Missile Division, Hq.A.R.D.C., Los Angeles, Calif., WDZP Report No. 2.

Anon. (1958), Few physiological changes noted in monkey's weightless flight, *Aviation Week* (Dec. 22, 1968), 69(25): 23.

Anon. (1958), Bibliography on medical electronics, *IRE Prof. Group Med. Electron.,* New York, pp. 1–91.

Anon. (1959), *Army biological experiment 2 B on Jupiter IRBM AM-18 (U),* Missile Systems Engineering Branch, Structures, and Mechanics Lab., DOD, ABMA, Redstone Arsenal, Huntsville, Ala., Rept. No. DSL-TN-16-59 (Secret), May 11, 1959.

Anon. (1959), Bibliography on medical electronics, Suppl. I, *IRE Prof. Group Med. Electron., New York,* pp. 1–68.

Anon. (1960), Development and use of short wave radio transmitters to trace animal movements, *Prog. Rep. Univ. Minnesota,* 1–26.

Anon. (1960), Bibliography on medical electronics, Suppl. II, *IRE Prof. Group Med. Electron.,* New York, pp. 1–76.

Anon. (1961), Biophysical telemetering system tracks pigeons in 25-mile flight, *Electron Prod.*, pp. 21W, 111W, 113W.
Anon. (1961), Radio aids sheep research, *New Scientist*, 236.
Anon. (1961), Radio rabbits, *Illinois Wildl.*, 16(4): 12.
Anon. (1961), While patients exercise, *Electrical Engineering*, 80(7): 556–557.
Anon. (1961), *Manhigh III USAF manned balloon flight into the stratosphere*, U.S. Dept. Comm., Off. Techn. Serv., AFMDC-TR-60-16.
Anon. (1962), *Results of the second U.S. manned orbital space flight*, May 24, 1962, NASA SP-6, pp. 55–62.
Anon. (1963), Development and use of short wave radio transmitters to trace animal movements, *Univ. Minnesota* (1962), 1–18 mimeo.
Anon. (1963), First manned space flights (in Russian), *Moskva.*
Anon. (1963), Studies of movements, behavior, and activities of ruffed grouse using radio telemetry techniques, *Univ. Minnesota*, 1–30 mimeo.
Anon. (1963), Application of telemetry for the management of sportsmen regarding their cardiovascular function, *Jap. Circ. J.*, 27:773.
Anon. (1963), *Telemetry transducer handbook*, Off. Techn. Serv., Washington, D.C., WADD-TR-61-67.
Anon. (1963), Rat walkie-talkie, *Illinois Wildl.*, 19: 8.
Anon. (1963), Biotelemetry: Prime space needed, *Electronics*, 36, Feb. 22.
Anon. (1963), Telemetry on muscle power, *Electronics*, 36, Sept. 20.
Anon. (1964), First team flight in space (in Russian), *Moskva.*
Anon. (1964), *Operating instruction, physiological telemeter*, AF-04(611)-9086, Flight Test Division, Hughes Aircraft Company, Culver City, Calif., July.
Anon. (1964), *Telemetry transducer handbook*, Air Force Flight Dynamics Lab., Washington, D.C., U.S. Dept. Comm., OTS, Publ. No. AD 421 951 (50 080).
Anon. (1964), Micro-telemetry aids for research, *Electronics Weekly*, 184: 4.
Anon. (1964), From Astronautics—to volleyball (in Russian), *Sport. Igry, Moskva*, 10: 4, 5.
Anon. (1964), Telemetric observation on exercise test of normal subjects, *Jap. Circ. J.*, 28: 375.
Anon. (1964), Cardiac telemetry and monitoring, *J. Amer. Med. Assn.*, 187, Suppl. 31–32.
Anon. (1965), Telemetric study on ST-T segment in exercise electrocardiogram, *Jap. Circ. J.*, 29: 753 (Abstract).
Anon. (1965), *Medical and biological applications of space telemetry*, U.S. Govt. Print. Off., Washington, D.C., Catalog No. NAS 1.21:5023, 66 pp.
Anon. (1966), *Development of a small animal payload and integration with a sounding rocket*, U.S. Govt. Print. Off., Washington, D.C., Catalog No. NAS 1.21:109, 98 pp.
Anon. (1966), One-channel telemetry for physiological signals, *World Med. Electronics*, 4(1): 13–14.
Anon. (1966), Telemetry in medicine, *World Med. Electronics*, 4(1): 20.
Anon. (1966), Telemetering instruments for medicine. Breathing and pulse changing an electromagnetic field (in German), *VDI-Nachr.*, 10: 11.
Anon. (1967) *Current books in bioinstrumentation*, AIBS/BIAC Information Module M11.
Anon. (1967), Biotelemetry equipment sources directory, *AIBS/BIAC Information Module M12.*

Anon. (1967), *A sampling of some current projects in implant bio-telemetry,* AIBS/BIAC Information Module M3.

Antipov, B. I. (1965), Measurement of heart rate in pole-jumping by telemetry (in Russian), Teor. prakt. fiz. kult., *Moskva,* 28(7): 38–39.

Apeloig, D., Frei, E. H., Yerushimi, S., Barr, I. M., and Blieden, L. (1966), Measurement of interelectrode tissue resistance of radio pacemaker, *Conf. Eng. Med. Biol. Proc.,* 8:164.

Ardenne, M. von (1958), Der verschluckbare intestinalsender, *Technik.,* 13:614.

Ardenne, M. von, Mielke, H., Rackwitz, H., and Volland, F. (1959), Die Technik des verschluckbaren intestinalsenders, *Nachrichtenntecknic,* 9: 449–456.

Ardenne, M. von, Mielke, H., and Reitnauer, P. G. (1964), Der gegenwartige Stand von Methodik und Anwendung der verschluckbaren intestinalsender, *Dtsch. Gesdwes.,* 19: 810–816, 876–881.

Ardenne, M. von, and Sprung, H. B. (1958), Uber einen verschluckbaren intestinalsender, *Z. Ges. Inn. Med.,* 13: 269–274.

Ardenne, M. von, and Sprung, H. B. (1958), Uber den verschluckbaren intestinalsender fur pH-Wert-Signalisierung, *Naturwissenschaften,* 45: 546–565.

Ardene, M. von, and Sprung, H. B. (1958), Uber die gleichzeitige registrierung von druck anderungen und lageanderungen bei dem verschluckbaren intestinalsender, *Z. Ges. Inn. Med.,* 13: 596–601.

Ardenne, M. von, and Sprung, H. B. (1958), Uber versuche mit einem verschluckbaren intestinalsender, *Naturwissenschaften,* 45: 154–155.

Artmann, J. W. (1967), *Telemetric study of the pocket gopher* (*geomys bursarium*), M.S. Thesis, University of Minnesota, Minn.

Artuso, G., and Zannini, D. (1961), A remote pulse-counting apparatus for the study of cardiac rate during work (in Italian), *Med. Lavoro,* 52: 29–32.

Asa, M. M. (1964), High fidelity fetal radioelectrocardiography, *Amer. J. Cardio.,* 14: 530–532.

Asahina, K., and Mono, A. (1963), Telemetry of electrocardiograms during exercises, in *Symp. Appl. Telemetry Syst. Med.* Ref.: *Med. Electron. Biol. Eng.,* 1: 578.

Askanas, Z. (1965), Telemetric registration in the application of cardiography, *Postepy Hig. Med. Doswiad.,* 19: 787–788.

Atkins, A. R., and Lock, E. B. (1965), A portable electrocardio-transmitter and receiver, *South African Electron. Eng.,* 37: 806–808.

Augee, M. L., and Ealey, E. H. (1968), Torpor in the echidna *Tachyglossus aculeatus, Jour. Mammalogy,* 49(3): 446–454.

Austim, W. T. S., and Harris, E. A. (1957), Measurement of heart rate in exercise, *Quart. J. Exp. Physiol.,* 42: 126–129.

Ax, A. F. (1964), *Validation of the aerospace medical research laboratories 3-channel personal telemetry system,* AMRL-TR-64-124, USAF, 1–27.

Ax, A. F., Andreski, L., Conrter, R., DiGrovanni, D., and Herman, S. (1964), The artifact problem in telemetry of physiological variables, *Bio-Med. Sci. Instr.,* 2: 229–233.

BIAC Staff Report (1967), Biotelemetry equipment sources directory, *AIBS/BIAC Information Module M12,* 19 pp.

Babskii, E. B., and Uljaninskij, L. S. (1961), Present problems of the physiology and pathology of the circulatory system (in Russian), *Moskva.*

Babskii, E. B., et al. (1964), Use of radiotelemetry for the study of evacuation activity of the stomach (in Russian), *Dokl. Akad. Nauk SSSR,* 156: 719–720.

Babskii, E. B., Baranovsky, A. L., Ganelin, G. Z., Oulianisky, L. C., and Ouchakowa, I. A. (1962), Electric stimulation of heart by means of radio frequency transmission of impulses (in Russian), *Reports USSR Acad. Sci.,* 147: 255–258.

Babskii, E. B., Sorin, A. M., Beboussow, A. S., Goukov, U. S., and Dimanis, V. (1963), Radiotelemetric survey of temperature in the human digestive tube (in Russian), *Reports USSR Acad. Sci.,* 149: 1213–1216.

Babskii, E. B., Sorin, A. M., Belousov, A. S., Dimanis, V., and Malkiman, J. J. (1964), Radiotelemetric study of the pressure in the human alimentary canal (in Russian), *Dokl. Akad. Nauk SSSR,* 158: 993–996.

Babskii, E. B., Botchal, B. E., and Belousov, A. S. (1965), Use of radiotelemetry for studying the temperature in the human stomach and intestine (in Russian), *Therap. Arck,* 37: 96–100.

Bach, L. M. N. (1963), Telemetering neurophysiological information from animals, in L. E. Slater (Ed.), *Biomedical telemetry,* pp. 321–326.

Baevskij, R. M. (1961), Application of radio telemetry in space medicine (in Russian), *Zarubeznaja Radioelektronika,* No. 1: 82–95.

Baevskij, R. M. (1962), Method of integral phonocardiography (in Russian), *Problemy Kosmiceskoj Biologii, Moskva,* tom 1: 412–414.

Baevskij, R. M. (1963), Biological telemetry and space flights (in Russian), *Problemy Kosmiceskoj Biologii, Moskva,* tom 2: 25–39.

Baevskij, R. M., Zazykin, K. P., Sazonov, N. P., and Frejdel, V. R. (1963), Design problems of dynamic telemetry systems in relation to space medicine tasks (in Russian), in V. V. Parin, *Radiotelemetry in physiology and medicine,* pp. 31–34.

Baevskij, R. M. (1965), Physiological methods in astronautics (in Russian), *Moskva.*

Baevskij, R. M., and Volkov, Yu. N. (1965), Clinicophysiological evaluation of seismocardiographic data obtained during space flights of Vostok V and Vostok VI, *Klinicheskaga Meditsina,* 43(2): 6. Translation: *Fed. Proc. Trans. Suppl.,* 24 (1965), T953–T956.

Bagno, S., Liebman, F., and Cosenza, R. (1960), Detecting muscle potentials in unanesthetized animals, *Electronics,* 33: 58.

Bailey, D. A., and Orban, W. A. R. (1963), The physiological response of athletes during all-out sports performance as monitored by radiotelemetry, *Progress Report,* U. of Saskatchewan, Saskatoon, Nov.

Bakirova, R. M. (1963), Results of radiopulsimetry by tests and during exercises in so called health-groups (in Russian), in V. V. Parin, *Radiotelemetry in physiology and medicine,* pp. 157–163.

Balding, H. A. (1965), Radio tracking a white-tailed deer, *The Ohio J. Sci.,* 67(6): 382–384.

Baldwin, H. A. (1964), A remote control technique for the study of olfaction in sharks, *Biomed. Sci. Instrum.,* 2: 217–227.

Baldwin, H. A., Ingle, G. F., and Hsiao, S. (1967), Remote interpretation of temperature transients, *Digest of 7th Intern. Conf. Med. Biol. Eng.,* 88.

Balin, H. (1964), Radio telemetry system for the study of ovarian physiology, *Obst. Gynec.,* 24(2): 198–207.

Balin, H. (1965), Ovarian biotelemetry, *Surg. Forum,* 16: 404–406.

Balin, H. (1966), Telemetry and reproductive biology, *J. Assn. Adv. Med. Instrument.,* 1: 17–21.

Balin, H., Fromm, J. H., and Platt, J. M., et al. (1965), Biotelemetry as an adjunct to the study of ovarian physiology, *Fertil. Steril.,* 16: 1–15.

Ballingtijn, C. M. (1965), Electric recording from unnarcotized, unrestrained small animals, *Med. Electron. Biol. Eng.,* 3: 71–73.

Bárány, F., and Jacobson, B. (1963), Endoradiosonde study of the effect of prostigmin and diphenoxylate on propulsive motility and pressure curves of the small intestine, *II World Congress Gastroenterology Munchen,* 767–771.

Barker, P. R., and Bredon, A. D. (1962), *Intra-arterial blood pressure sensing system feasibility study,* prepared for MMSCV Directorate Space Systems Division, Air Force Systems Command, Los Angeles, Calif. 41 pp.

Barr, N. L. (1954), The radio transmission of physiological information, *Military Surgeon,* 114: 79–83.

Barr, N. L. (1958), Long-distance telephone technic is latest aid to diagnosis, *Mod. Hosp.,* 91: 64.

Barr, N. L. (1959), The transmission of physiological responses from air to ground by electronic methods, *IRE Transact. Med. Electr.,* **ME-7.**

Barr, N. L. (1959), An account of experiments in which two monkeys were recovered unharmed after ballistic space flight, *Aerospace Med.,* 30: 871.

Barr, N. L., and Voas, R. (1960), Telemetering physiological responses during experimental flights, *Amer. J. Cardiol.,* 6: 54–61.

Barry, W. (1964), A radio telemetering device, *J. Appl. Physiol.,* 19: 528–530.

Barsukov, F. I., and Maksimov, M. V. (1962), Radiotelemetry (in Russian), *Moskva.*

Bartholmew, R. M. (1967), A study of the winter activity of bobwhites through the use of radio telemetry, *Occas. Papers Adams Ctr. Ecol. Studies,* Western Michigan University, No. 17: 25 pp.

Bartlett, Jr., R. G. (1963), Pulmonary function evaluation in air and space flight, *Industr. Med. Surg.,* 32: 2–8.

Basan, L. (1955), A new method for studying physiological processes during work, *Fiziol. Zh. SSSR,* 41: 95–96.

Basan, L., and Lovdzhiyev, I. (1958), Procedure for radio investigation of respiration during work and sports exercises (in Russian), *Fiziol. Zh. SSSR,* 44: 773–775.

Bassan, L. (1964), Radiotelemetric examination of heart rate during rowing-competitions (in German), *Theorie und Praxis des Leistungssports,* 112–148.

Bassan, L. (1964), Radiotelemetric examination of heart rate during row-training (in German), *Theorie und Praxis des Leistungssports,* 148–168.

Bassan, L. (1965), Radiotelemetric functional diagnosis (in German), *Schriftenreihe Aerztl. Fortb.,* 24: 31–48.

Bassan, L., and Haemer, W. (1963), Radiotelemetric examination of heart rate during swimming (in German), *Theorie und Praxis des Leistungssports,* 138–158.

Bassan, L., and Stoeckel, J. (1965), Radiotelemetric examination of heart rate during axe-working (in German), *Arch. Rorstwesen,* 14: 1061–1077.

Battye, C. K. (1962), Telemetering electromyograph with a single frequency modulated channel, *Electronic Engineer.,* 34: 398–399.

Battye, C. K., and Joseph, J. (1966), An investigation by telemetering of the activity of some muscles in walking, *Med. Biol. Eng.,* 4: 125–135.

Bauer, M., Lishshak, K., and Madaras, S. (1957), New method of recording of the secretion of the salivary glands during free movement of experimental animals (dogs) (in Russian), *Fiziol. Zh. SSSR,* 3: 132–135.

Bayevskiy, R. M. (1961), Use of radiotelemetry in space medicine (in Russian), *Zarubezhn. Radioelectronika* (Foreign Electronics), 1: 82–95.

Beal, R. O. (1967), Radio transmitter-collars for squirrels, *J. Wildl. Mgmt.,* 31(2): 373–374.

Beaupre, M. A., and Evand, H. A., Internal biological telemetry system feasibility study, *North American Aviation Rept.*, N60-1986, Contract No. AF04(647)585.

Beechey, P., and Lincoln, D. W. (1969), A miniature FM transmitter for the radiotelemetry of unit activity, *J. Physiol.*, 5–6.

Beenken, H. G., and Dunn, F. L. (1958), Short distance telemetry of physiological information, *IRE Trans. Med. Electron.*, **PGME 12:** 53.

Beenken, H. G., and Dunn, F. L. (1959), Short distance radio telemetering of physiological information, *IRE Trans. Med. Electron.*, Telemetry SET 2: 82.

Belford, J. R., Meyers, G. H., Parsonnet, V., and Sucker, I. R. (1964), Transducers for producing biological energy, *Conf. Eng. Biol. Med., Proc.*, 6: 17.

Bell, E. C., and Robson, D. (1967), Use of multivibrators in small telemetry systems, *Proc. Inst. Electr. Engineers,* 114(3): 327–332.

Bellet, S., Deliyannis, S., and Eliakim, M. (1961), The electrocardiogram during exercise as recorded by radioelectrocardiography, *Amer. J. Cardiol.*, 385–400.

Bellet, S., Deliyannis, S., Eliakim, M., and Figallo, E. M. (1962), The electrocardiogram during exercise as recorded by radioelectrocardiography: Comparison with the post-exercise electrocardiogram (master two step test), *Am. Int. Med.*, 56: 665.

Bellet, S., Eliakim, M., Deliyannis, S., and Figallo, E. M. (1962), Radioelectrocardiographic changes during strenuous exercise in normal subjects, *Circulation,* 25: 686.

Bellet, S., Eliakim, M., Deliyannis, S., and Figallo, E. M. (1962), Radioelectrocardiography during exercise in patients with angina pectoris, *Circulation,* 25: 5–14.

Bellet, S., Muller, O. F., Herring, A. B., and La Van, D. W. (1963), Effect of erythrityl tetranitrate on the electrocardiogram as recorded during exercise by radioelectrocardiography, *Am. J. Cardiol.*, 11: 600.

Bellet, S., Eliakim, M., Deliyannis, S., and La Van, D. (1964), Radioelectrocardiography during exercise in patients with angina pectoris. Comparison with the post-exercise electrocardiogram, *Circulation,* 29: 366–375.

Bellet, S., and Muller, O. (1965), The electrocardiogram during exercise. Its value in the diagnosis of angina pectoris, *Circulation,* 32: 477–487.

Belousov, A. S., Malkiman, J. J., and Sorin, A. M. (1964), Radiotelemetric Investigation of the functions of the digestive tract, *Vestnik Akademii Meditsinskikh Uank SSSR,* 19: 71–78. Translation: Clearinghouse for *Federal Sci. and Techn. Information,* S 520–4.

Below, W. C., Ellis, A. B., and Klein, E. A. (1961), A personalized microminiature biomedical data acquisition and telemetering system, *Digest of the 1961 International Conf. Medical Electronics,* New York City, str. 124.

Benazet, P., Bordel, R., Bnon, A., Fontaine, M., and Sevestre, J. (1964), Etude telemetrique de l'electrocardiogramme du cheval de sport, *Rec. Med. Vet Ecole. Alfort.,* 140: 449–459.

Bengulescu, D., and Nestianu, V. (1963), A semiconductor system for remote transmission of four biological currents (2 EEG and 2 ECG) (in Roumanian), *Studii si Cercetari de Fiziologie* (Bucuresti), 8(4): 663–671.

Bengulescu, D., and Nestianu, V. (1964), A semiconductor system for remote transmission of four biological currents (2 EEG and 2 ECG) (in French), *Rev. Roumaine Physiol.,* 1(2): 195–203.

Bengulescu, D., Nestianu, V., Avramescu, C., and Tiron, V. (1963), Transistorized portable set for the transmission of two EEG and two ECG channels. *3rd*

Annual Meeting of the Rumaniam Group of Electroencephalography and Clinical Neurophysiology (Bucharest), May 25–26, 1962, *EEG Clin. Neurophysiol.,* 15: 530.

Berg, A. I. (Ed.) (1960), Electronics in medicine (in Russian), *Moskva-Leningrad.*

Berktay, H. O., and Gazey, B. K. (1967), Communications aspects of underwater telemetry, *Radio Electron Engineer,* 33(5): 295–304.

Bernstein, A., Rothfeld, E. L., Parsonnet, V., and Zucker, J. R. (1965), Telemetric monitoring of the electrocardiogram in acute myocardiac infarction (abstract), *Am. Int. Med.,* 63(5): 908.

Berson, A. S., and Pipberger, H. V., (1965), Errors caused by inadequacy of low frequency response of electrocardiographs, *Digest 6th Intern. Conf. Med. El. Biol. Eng.,* Tokyo, 13–14.

Bert, J., and Collomb, H. (1965), Effet de l'immobilisation sur l'EEG du babonin. Comparison avec les résultats obtenus per télémetric chez animal en liberté, *C.R. Soc. Biol.,* 159: 1202–4.

Bert, J., and Collomb, H. (1966), L'électroencéphalogramme du sommeil nocturne chez le Babouin Etude par télémetrie, *J. Physiol.,* 58: 285.

Biryukivick, A. A., and Korol, V. M. (1964), *Radiotelemetric studies of school-children during intensive muscular activity.* Abstracts of papers presented at International Congress of Sport Sciences, Tokyo, 90–91.

Bland, D., Lafontaine, E., and Medvedeff, M. (1966), Radiotelemetric recordings of the electroencephalograms of civil aviation pilots during flight, *Aerospace Med.,* 37: 1060.

Bligh, J., and Beadle, M. (1963), The continuous radio-telemetric recording of deep body temperature of some African mammals, *Proc. Biochem. Soc. J.,* 89: 72 pp.

Bligh, J., and Harthoorn, A. M. (1965), Continuous radiotelemetric records of deep body temperature of some unrestrained African mammals under near-natural conditions, *J. Physiol.,* 176: 145–62.

Bligh, J., and Hartle, T. C. (1965), The deep body temperature of an unrestrained ostrich 'struthio camelus' recorded continuously by a radio-telemetric technique, *Ibrs.,* 107: 104–105.

Bligh, J., Ingram, D. L., Keynes, R. D., and Robinson, S. G. (1964), The thermo-regulatory performance of an unrestrained Welsh mountain sheep under field conditions recorded over a 12 month period by radio-telemetry, *J. Physiol.,* 175: 62 pp.

Bligh, J., Ingram, D. L., Keynes, R. D., and Robinson, S. G. (1965), The deep body temperature of an unrestrained Welsh Mountain sheep recorded by a radio-telemetric technique during a 12 month period, *J. Physiol.,* 176: 136–144.

Bligh, J., and Lampkin, G. H. (1965), A comparison of the deep-body temperature of Hereford and zebu cows recorded continuously by radio-telemetry under similar field conditions, *J. Agric. Sci.,* 64: 221–227.

Bligh, J., and Robinson, S. G. (1963), Continuous telemetry of the deep body temperature of sheep under field conditions, *J. Physiol.,* 165: 1 p.

Bligh, J., and Robinson, S. G. (1965), Radiotelemetry in a veterinary research project, *Medic, et Biol. Illustr.,* 15: 94–99.

Bodenlos, L. J. (1966), Transmitter back-pack for free roaming animal, *Lab. Animal Care,* 16(5): 454–458.

Bokser, O. J., and Klevcov, M. I. (1961), Improved radio methods for the investigation of reflexes (in Russian), *Bjull. Eksper. Biol. Med.,* No. 2: 111–113.

Bolie, V. W. (1944), Radiation resistance of a loop antenna in a conducting medium, with reference to biomedical radiotelemetry, *Proc. Nat. Telemetering Conf.,* 2–5, 4 pp.

Boone, J. L. (1965), Silicone rubber insulation for subdermally implanted electronic devices, *Proc. 18 Ann. Conf. Eng. Med. Biol.*

Boreen, H. I., Shandelman, F., and Berman, R. (1961), Biomedical telemetering, *Proc. Nat. Telem. Conf.*

Bornert, D., and Schicketanz, W. (1963), Zur anwedung der intestinal methode in der veterinamedizin, *Monatsh. Vet. Med.,* 18: 647–653.

Bornert, D., and Seidel, H. (1964), Drahtlos ubertragene EKG-ableitungen vom freibeweglichen pferd., *Arch. Exp. Vet.-Med.,* 18: 1217–1223.

Bornert, D., Seidel, H., Maiwald, D., and Bornert, G. (1964), Drahtlos ubertragene EKG-ableitungen vom freibeweglichen rind., *Arch. Exp. Vet.-Med.,* 18: 702–712.

Borovicka, J. (1964), Equipment for radiotelemetric transmission of physiological parameters from aircraft (in Czech.), *Cs. Fysiol. (Praha),* 13(3): 195–196.

Bosman, J., Boter, J., and Kuiper, J. (1965), Improvement of the measurement of physical load by wireless transmission of the ECG, *World Med. Electronics,* 3(6): 227.

Bostem, F. H. (Ed.) (1963), Medical electronics, *Proceedings of the 5th International Conference on Electronics,* Liege, July.

Botsch, F. B. (1966), A digital telemetry system for physiological variables, *Nat. Telemet. Conf. Proc.*

Botsch, R. W. (1961), Current methods of temperature telemetry in physiological research, in *Temperature Its Measurement and Control in Science and Industry,* Reinhold Publ. Co., New York, p. 19.

Botsch, R. W., Powers, J. J., and Koch, A. A. (1959), *Environmental protection research by means of radio telemetry, Part I.* Telemetric instrumentation and examples of field use in studying the man clothing environment system. Headquarters Quartermaster Research and Engineering Command, U.S. Army Technical Report EP 119 Octl., Project Ref. 7-83-01-006.

Bouchard, G. O. (1966), Telemetry for astronaut maneuvering unit, *Proc. 19th Ann. Conf. Eng. Med. Biol.*

Boyd, J. C., Sladen, W. J. L., and Baldwin, H. A. (1967), Biotelemetry of penguin body temperatures, 1966–1967, *Antarctic Journal,* Vol. II, No. 4: 97–99.

Brach, E. J., and Wilner, J. (1967), Data acquisition of physiological characteristics of plants, *Digest of 7th Intern. Conf. Med. Biol. Eng.,* 83.

Bradfute, G. A., Wright, J. L., and Burns, R. (1965), Measurement of intro-arterial pressure and other physiologic variables in the ambulatory human with UHF telemetry, *18th Ann. Conf. Eng. Med. Biol.*

Bradley, W. E., Wittmers, L. F., Chou, S. N., and French, L. A. (1962), Use of radio transmitter receiver unit for treatment of neurogenic bladder, *J. Neurosurg.,* 19: 782–786.

Bradley, W. E., Chou, S. N., and French, L. A. (1963), Further experience with the radio transmitter receiver unit for the neurogenic bladder, *J. Neurosurg.,* 20: 953–960.

Bradshaw, H., and Bradshaw, V. (1968), Tuning in on wildlife secrets, *Pop. Mech.,* 129(1): 115–116.

Brander, R. B. (1965), Factors affecting dispersion of ruffed grouse during late winter and spring in the Cloquet Forest Research Center, Minnesota, Ph.D. Thesis, Univ. Minnesota, Minn., 180 pp.

Brander, R. B. (1967), Movements of female ruffed grouse during the mating season, *Wilson Bull.,* 79(1): 28–36.

Brannick, L. J. (1963), Recording physiological functions from unrestrained dogs, in L. E. Slater, 303–309.

Bratt, H. R. (1965), Biomedical aspects of the X-15 program, *Military Med.,* 130: 445–459.

Bratt, H. R., and Curamoto, M. J. (1963), Biomedical flight data collection, *ISA-Journal,* 10(10): 57–62.

Bratt, H. R., et al. (1963), Review of in-flight biomedical data collection systems at AFFTC, May 1960 to Dec. 1962, *Biomed. Sci. Instrum.,* 1: 445–459.

Breakell, C. C., and Parker, C. S. (1949 1), Radio transmission of electrophysiological processes, *Lancet,* 167–168.

Breakell, C. C., and Parker, C. S., and Christopherson, F. (1949), Radio transmission of the human electroencephalogram and other electrophysiological data, *Electroenceph. Clin. Neurophysiol.,* 1: 243–44.

Breakell, C. C., and Parker, C. S. (1956), Lecture about radio-EEG, *Europ. EEG Congr.,* London, May.

Beckler, A., Kaeburn, L., Ettlesch, B. L., and Douglas, D. W. (1959), Space canaries: Implicit biological monitoring, *14th Ann. Meeting ARS* (Washington, D.C.), Nov. 16–20.

Brewer, A. A., and Hudson, D. C. (1960), Application of miniaturized electronic devices to the study of tooth contact, *J. Dental Res.,* 39: 758. *School of Aviation Med. Rept.,* 61-12, Oct. 1960.

Brewer, A. A., and Hudson, D. C. (1961), Application of miniaturized electronic devices to the study of tooth contact in complete dentures, *J. Pros. Den.,* 11: 62.

Briskman, B. A. (1963), Telemetry today, *Electronics,* Sept. 6.

Briskier, A. (1959), Heart examination and consultation by radio and radiophototransmission, *J. Amer. Med. Assoc.,* 169: 1981–1983.

Brown, B., Johnston, D., and Duthie, H. L. (1966), Radio-telemetering of alimentary physiological data in man, *J. Physiol.,* 182: 23 pp.

Brown, E. J. (1961), Biomedical instrumentation as applied to the astronauts and primates of project mercury, *Paper ISL 61 ISA* (St. Louis), Jan. 17–19.

Brown, J. S., and Pedico, J. P. (1965), *Micropower miniature temperature telemetry,* presented at 5th Annual Symposium for Biomedical Engineering, June 6–8, San Diego.

Brown, R. L. (1965), Techniques developed and employed in a study of female sharptailed grouse ecology in Montana during 1963, '64, '65, *4th Biennial Meeting, West. States Sage Grouse Committee,* Walden, Col., 4 pp.

Brown, R. L. (1967), The role of social regulation of survival and breeding in male sharptail grouse, *Ann. Conf., Northwest. Sect., The Wildlife Society,* p. 4.

Bruner, H. D. (Ed.) (1960), Methods in medical research, 8, *Year Book Publ.,* Chicago, 368 pp.

Bugrov, B. G., Gorlov, O. G., Petrov, A. V., Serov, A. D., Yugov, Y. M., and Yakovlev, V. I. (1958), Preliminary results of scientific investigations carried out with the aid of the first Soviet artificial earth satellites and rockets, *III: Medico-Biological Invest. Rockets. Sb. Statei,* 1.

Bullis, E. E., and Hinds, M. (1963), *Telemetry report on two horses,* Texas A and M College, July 13.

Burch, G. E., and Gerathewohl, S. J. (1960), Observations on heart rate and cardiodynamics during weightlessness, *Aerospace Med.,* 31: 661–669.

Burger, H., and Minarovjech, V. (1965), Results of complex sportmedical investigation of boxers (in German), *Medizin und Sport,* 5: 69–74.

Bushor, W. E. (1961), Medical electronics, Part I—diagnostic measurements, *Electronics* (Jan. 20), 34: 49–55.

Busser, F. (1963), Telethermometre a modulation de frequence, *Electronique Medicale,* 26: 109–111.

Busser, F. (1963), Radiotransmission des electrocardiogrammes, *Electronique Medicale,* 24: 50–54; 25: 56–80.

Busser, F. (1964), Stand de controle pour telethermometres, *Electronic Medicale,* 29: 81–82.

Busser, J. (1969), Miniature biotelemetry construction project, *BioScience,* 19: 1014–1016.

Butterworth, T. A., and Farkas, D. F. (1969), Temperature profiling of food processing equipment by radio telemetry, *Food Technology,* 23: 81–82.

Byford, G. (1965), Medical radio telemetry, *Proc. Roy. Soc. Med.,* 58: 795–798.

Caceres, C. A., and Cooper, J. K. (1965), Radiotelemetry: A clinical perspective, in C. A. Caceres, *Biomedical telemetry,* pp. 85–105.

Caceres, C. A., Imboden, C. A., and Smith, M. A. (1965), A medical monitoring system, in C. A. Caceres, *Biomedical telemetry,* pp. 107–115.

Caggiano, A. D. (1967), Telemetry tone oscillator consumes microwatts, *Electronics,* 40(8): 89–90.

Cailler, M. B., et al. (1964), Technical and physiological bases for telemetry and its use during biologic experiments in a rocket (in French), *Rev. Corps. Sante Armees,* 5: 557–577.

Callens, E., Colle, J., and Peleman, C. (1965), Technique de telestimulation a modulation de frequence, *J. Physiol.* (Paris), 57: 727.

Cammilli, L., Pszzi, R., Rizzichi, G., and De Saint-Pierre, G. (1964), Radio-frequency pacemaker with receiver coil implanted on the heart, *Ann. N.Y. Acad. Sci.,* 111(3): 1007–1029.

Capanon, A. (1957), Radiotransmission of physiological parameters (in French), *Compt. Rend. Center D'Essair en Vol,* Brittany, France, 344.

Carbery, W. J., Tolles, M. S., and Freiman, A. H. (1960), A system for monitoring the ECG under dynamic conditions, *Aerospace Med.,* (Feb.), 31: 131–137.

Carbery, W. J., Steinberg, C. A., Tolles, W. E., and Freiman, A. H. (1961), Automatic methods for the analysis of physiological data, *Aerospace Med.,* 32(1): 52–59.

Card, W. I. (Ed.) (1961), Modern trends in gastro-enterology, 3, Butterworth, London, 317 pp.

Carlson, L. D. (1959), Requirements for monitoring physiological function in space flight, *IRE Trans. Med. Electron.,* **ME-7.**

Carr, A. (1963), Orientation problems in the high seas travel and terrestrial movements of marine turtles, in L. E. Slater (Ed.), pp. 179–193.

Carr, A. (1962), Orientation problems in the high seas travel and terrestrial movements of marine turtles, *Amer. Sci.,* 50: 359–374.

Carr, A. (1965), The navigation of the green turtle, *Sci. Amer.,* 212: 79–86.

Carr, A. (1967), Caribbean green turtle—imperiled gift of the sea, *Nat. Geogr. Mag.,* 131(6): 876–890.

Cassano, C., Torsoli, A., Ramorino, M. C., and Colagrande, C. (1963), Quelques resultats des recherches sur la motilite intestinale etudiee per les "endoradro sondes" et la cinefluorographic, *World Congr. Gastroenterol.*, 2: 772–774.

Castillo, H. T. (1964), Transducers for physiological measurements during biastronautical research and operations, *Med. Electron. Biol. Eng.*, 2: 109–121.

Cebula, J. J. (1966), *Radio-telemetry as a technique used in greater prairie chicken (Tympanuchos cupido pinnatus) mobility studies,* M.S. Thesis, Kansas State University, 61 pp.

Cerkez, C. T., Steward, G. C., Bacongallo, B., and Manning, G. W. (1964), Telephonocardiography: The transmission of electrocardiograms by telephone, *Can. Med. Assoc. J.*, 91: 727–733.

Cerkez, C. T., Steward, G. C., and Manning, G. W. (1965), Telemetric electrocardiography, *Can. Med. Assoc. J.*, 93: 1187–99.

Chambergs, R., and Nelson, J. G. (1963), Biomedical and psychological instrumentation for monitoring pilot performance during centrifuge simulation of space flight, *Proc. Symp. Bio-Med. Eng.* (San Diego).

Champlin, G. A., and Gerathewohl, S. J. (1959), *Bio-flight project 2 (U)*, Bioastronautics Research Unit of the U.S. Army, Med. Res. and Devl. Command, AOMC, Redstone Arsenal, Huntsville, Ala., Rept. Contr. Symbol CSCRD-16 (Secret); June 21.

Champlin, G. A., and Wilbarger, E. S. (1959), *Bio-flight project 2 B (revision)*, Bioastronautic Research Unit of the AOMC, Redstone Arsenal, Huntsville, Ala., Rept. Contr. Symbol CSCRD-16, July 10.

Charitonov, S. A. (1932), Relation between conditioned reflexes of secretion and motory of free moving animals (in Russian), *Sov. Nevropotol. Psychiatr. Psichogigiena.*, 12: 766–777.

Chatelier, G., and Buser, P. (1965), Etude neurophysiologique effectuee sur un chat lors d'un vol en fusee, *J. Physiol.* (Paris), 57: 787–799.

Chernov, V. N., and Yakovlev, V. I. (1958), Scientific investigations during the flight of an animal on an artificial earth satellite (in Russian), *Iskusstvennye Sputniki Zemli.* (*Artificial Earth Satellites*), Sciences USSR Publishing House, No. 1: 80–94.

Cholvin, N. R. (1962), Development and operation of chronically implanted electronic devices with their effects in experimental animals, *IRE Conv. Record,* Part 9.

Church, F. W. (1965), Development of a personal monitoring instrument for noise, *Amer. Industr. Hyg. Assoc. J.*, 26: 59–63.

Citters, R. L. Van, Kemper, W. S., and Franklin, D. L. (1966), Blood pressure responses of wild giraffes studied by radio telemetry, *Science,* 152: 384–386.

Citters, R. L. Van, Watson, N. W., Franklin, D. L., and Elsner, P. W. (1965), Telemetry of aortic blood pressure and flow in free ranging animals (abstract), *Fedn. Am. Socs. Exp. Biol.*, 24: 525.

Clark, R. T. (1964), *Evaluation on physical fitness levels of children,* Abstracts of papers presented at ICSS, Tokyo, 188.

Clarke, N. P. (1963), Biodynamic response to supersonic ejection, *Aerospace Med.*, 34: 1089–94.

Clegg, B. R., and Schaefer, K. E. (1966), Studies of circadian cycles during prolonged isolation in a constant environment using eight-channel telemetry systems, *Nat. Telemet. Conf. Proc.*, Boston, Mass., May.

Cochran, W. W. (1965), Techniques in aerial telemetric studies of nocturnal migration of birds, *Prog. Rep. 1, NSF GB-3155,* University of Illinois, 16 pp. mimeo.

Cochran, W. W., and Hagen, T. E. (1963), Construction of collar transmitters for deer, *Univ. of Minnesota Mus. Nat. Hist. Tech. Rept. #4,* 1–12.

Cochran, W. W. (1967), 145–160 MHz beacon (tag) transmitter for small animals, *AIBS/BIAC Information Module M15,* 12 pp.

Cochran, W. W., and Lord, Jr., R. D. (1963), A radio-tracking system for wild animals, *J. Wildl. Mgmt.,* 27(1): 9–24.

Cochran, W. W., and Nelson, E. M. (1963), The Model D-11 direction-finding receiver, *Univ. of Minnesota Mus. Nat. Hist. Tech. Rept. #2,* 14 pp. mimeo.

Cochran, W. W., Mech, L. D., and Bellrose, F. C. (1964), A radio-tracking technique for following flying ducks, *Univ. of Minnesota Mus. Nat. Hist. Tech. Rept. #8,* 8 pp. mimeo.

Cochran, W. W., Montgomery, G. G., and Graber, R. R. (1967), Migratory flights of *Hylocichla* thrushes in spring: A radio-telemetry study, *The Living Bird,* 6: 213–225.

Cochran, W. W., Warner, D. W., and Raveling, D. G. (1963), A radio transmitter for tracking geese and other birds, *Univ. of Minnesota Mus. Nat. Hist. Tech. Rept. #1,* 4 pp. mimeo.

Cochran, W. W., Warner, D. W., and Tester, J. R. (1964), The Cedar Creek automatic radio-tracking system, *Univ. of Minnesota Mus. Nat. Hist. Tech. Rept. #7,* 9 pp. mimeo.

Cochran, W. W. (1967), Radio location and tracking of animals—an inexpensive triangulation recorder, *AIBS/BIAC Information Module M14,* 31 pp.

Cockrum, L. (1963), Wildlife telemetry, *Proc. 2nd. Ann. Meeting, New Mexico-ariy. Sect., The Wildlife Society,* 18–20.

Coester, W., Olson, G., and Seubert, J. L. (1960), Miniature radio transmitter development, *South Dakota Dept. Game, Fish, Parks,* pp. 1–6.

Coggshall, J. C., Starr, I. M., and Rader, R. D. (1966), Control of biomedical implants by D.C. magnetic fields, *Proc. 19th Ann. Conf. Eng. Med. Biol.*

Collins, C. C. (1967), Miniature passive pressure transensor for implanting in the eye, *IEEE Trans. Bio-Med. Eng.,* **BME-14,** April, 74–83.

Collins, C. C. (1967), Evoked pressure responses in the rabbit eye, *Science,* 106–108.

Collins, C. C. (1967), Passive telemetry with glass transensors, *Proc. Nat. Telem. Conf.,* 146–151.

Collins, C. C., Bach-y-Rita, P., and Loeb, D. R. (1967), Intraocular pressure variation with oculorotary tension, *Amer. J. Physiol.,* 213: 1039–1043.

Collins, V. P., Hudgins, P. T., and Petrany, Z. (1964), Observations on the physiology of stress in auto racing, *J. Sport. Med. Phys. Fitness,* Torino 4, 3: 188.

Collins, V. P., West, W. D., McTaggard, W. G., and Maxwell, A. R. (1965), Telemetry in a driving safety study, *Proc. Nat. Telemetering Conf.,* 241–243.

Connell, A. M. (1963), Applications of radio telemetering capsules to clinical medicine and research, *Curr. Med. Drugs,* 4: 3–12.

Connell, A. M., et al. (1963), Observations on the clinical use of radio pills, *Brit. Med. J.,* 771–774.

Connell, A. M., and Rowlands, E. N. (1960), Wireless telemetering from the digestive tract, *Gut,* **1.**

Connell, A. M., and Rowlands, E. N. (1960), Clinical applications of radio pills for measuring gastrointestinal pressure, *Inter. Conf. Med. Electron., Procl.,* 3: 126–27.

Connell, A. M., and Waters, T. E. (1964), Assessment of gastric function by pH telemetering capsule, *Lancet,* 11: 227–230.

Cook, G. (1964), *A miniature biopotential telemetry system* by Gordon J. Debo and Thomas B. Fryer, Ames Research Center, NASA TM X-54068, May.

Cook, R. S., White, M., Trainer, D. O., and Glazener, W. C. (1967), Radiotelemetry for fawn mortality studies, *Bull. Wildl. Dis. Assn.,* 3: 160–165.

Cooper, J. K., and Caceres, C. A. (1964), Transmission of electrocardiograms to computers, *Mil. Med.,* 129:457.

Cooper, J. K., and Caceres, C. A. (1965), Telemetry by telephone, in C. A. Caceres, *Biomedical telemetry,* pp. 15–36.

Cooper, O. L. (1962), *Development of transistorized miniature fifteen-channel pulse-time telemeter transmitting set,* Informal Progress Report No. 5 for the Period Oct. 1, 1961 through Dec. 31, 1961, Oklahoma State U. Res. Foundation, Stillwater, Jan., 6 pp.

Cooper, R. (1963), Electrodes, *Am. J. EEG. Technol.,* 3: 91–101.

Cooper, W. N., and Beaupre, M. A. (1961), *Internalized animal telemetry system engineering considerations,* Paper presented 32nd Ann. Meeting, Aerospace Med. Assoc. (Chicago) Apr. 26.

Cope, C. B., and Blackwell, R. (1963), A miniaturized transducer-transmitter for temperature measurements, *Proc. 5 Conf. Int. Electron. Med.,* Liege.

Corson, S. A., Corson, E. O. L., Pasamanick, B., and England, J. M. (1963), The influence of restraint and isolation of physiologic baselines in conditioned reflex studies: The promise of telemetry, in L. E. Slater (Ed.), *Bio-telemetry,* pp. 311–320.

Cox, A. G. (1964), Easier localisation of the crosby capsule in the alimentary tract, *Gut,* 4, 413.

Crabb, R. P., Halonen, C. A., Kuphaldt, R. R., Roger, S., and Sheretz, P. C. (1961), Daisy II, an advanced system for the simultaneous acquisition, recording and interpretation of data from many sensors, *Digest of the 1961 International Conference on Medical Electronics,* New York City, 126.

Craighead, Jr., F. C. (1963), Progress in bio-telemetry studies of grizzly bears, Presented Sept. 9, 1963, at *18th Ann. Instrument Soc. of Amer. Conf. Bio-Telemetry session,* Chicago, 11 pp. mimeo.

Craighead, F. C., Craighead, J. J., and Davies, R. S. (1963), Radiotracking of grizzly bears, in L. E. Slater (Ed.), *Bio-telemetry,* pp. 133–148.

Craighead, F. C., and Craighead, J. J. (1965), *A biotelemetry system for elk,* Yellowstone Field Expedition V, Atmospheric Sciences Research Center, State Univ. of New York, Pub. No. 31: 39–46.

Craighead, J. J., and Craighead, F. C. (1965), *Radio-tracking and telemetering system for large western animals,* Offsite Ecological Research Div. Bio. Med., Terrestrial and Fresh Water, *U.S. Atomic Energy Comm. TID-13358:* 57.

Craighead, F. C., and Craighead, J. J. (1966), Trailing Yellowstone's grizzlies by radio, *Natn. Geogr. Mag.,* 130(2): 252–267.

Crews, Jr., A. H., et al. (1964), Arrhythmias: A study utilizing telemetric monitoring, *J. Newark Beth. Israel Hosp.,* 15: 157–160.

Dadasev, R. S., and Erdman, G. M. (1963), Principles of the construction of a radiotelemetric system for the investigation of atmospheric contamination (in

Russian), in V. V. Parin, *Radiotelemetry in physiology and medicine,* pp. 112–117.

Datnow, B., Sandler, H., and Fryer, T. B. (1967), Telemetered left ventricular pressure in the dog effect of sustained acceleration and exercise, *Digest of 7th Intern. Conf. Med. Biol. Eng.,* 94.

Davey, L. M., Kaada, B. R., and Fulton, J. F. (1949), Effects on gastric secretion of frontal lobe stimulation, *Proc. Assn. Res. Nerv. Ment. Dis.,* 29: 617–627.

Davidoff, D., Davidson, J. M., and Feldman, S. (1962), A subminiature radio transmitter for the EEG, *J. Appl. Physiol.,* 17: 721–3.

Davis, D. A., et al., The clinical applications of telecardiography, *Digest 15th Ann. Conf. Eng. in Med. and Biol.* (Chicago), Nov.

Davis, D. A. (1964), The monitoring of hospital patients, with remarks on radio technique, *Clin. Pharmocol. Ther.,* 5: 546–552.

Davis, D. A. (1965), Radiotelemetry in anesthesia and surgery, *Int. Anesth. Clin.,* 3: 533–45.

Davis, D. A., et al. (1962), The clinical applications of telecardiography, *Digest 15th Ann. Conf. Eng. in Med. and Biol.* (Chicago), Nov.

Davis, D. A., Thornton, W., Grosskreutz, D. C., Sugioka, K., and McKnight (1961), Radio telemetry in patient monitoring, *Anesthesiology,* 22: 1010–1013.

Davydov, M. S., and Sarycev, S. P. (1960), One-channel radiotelemetering system for examination of rowers (in Russian), *Tez.i Ref. Dokl.Itogovoj Konf.,* LNIIFK (1960 g), Leningrad, 13–17.

Davydov, M. S., and Sarycev, S. P. (1962), A multichannel radiotelemetric system for investigation in sports (in Russian), *Tez.i Ref. Dokl.Itogovoj Konf.,* KNIIFK (26–28 dekabrja 1961 g), Leningrad, 16–18.

Deboo, G. J., and Fryer, T. B. (1964), A miniature biopotential telemetry system, *IEEE/ASA 17th Ann. Conf. Eng. Med. Biol.,* Cleveland.

Deboo, G. J., and Fryer, T. B. (1965), Miniature biopotential telemetry system, *Amer. J. Med. Electr.,* 4: 138–142.

Deboo, G. J., and Jenkins, R. S. (1965), A technique for recording a noise-free electrocardiogram from a chicken embryo still in its shell, *Med. Electron. Biol. Eng.,* 3: 443–445.

D'Haens, J. P. M. (1964), The endomotorsonde: A new device for studying the gastrointestinal tract, *Amer. J. Med. Electron.,* 3: 158–161.

Delgado, J. M. R. (1963), Telemetry and telestimulation of the brain, in L. E. Slater (Ed.), pp. 231–249.

Delgado, J. M. R. (1964), Free behavior and brain stimulation, *Intern. Rev. Neurobiol.,* 6: 349–449.

Delgado, J. M. R. (1965), Chronic radiostimulation of the brain in monkey colonies, *Proc. Int. Un. Physiol. Sci.,* 4: 365–371.

Delgado, J. M. R. (1965), Evoking and inhibiting aggressive behavior by radio stimulation in monkey colonies (abstract), *Bull. Ecol. Soc. Amer.,* 46(4): 163.

Delgado, J. M. R. (1966), Aggressive behavior evoked by radio-stimulation in monkey colonies, *Am. Zoolog.,* 6: 669–683.

Deliyiannis, S., Eliakim, M., and Bellet, S. (1962), The electrocardiogram during electroconvulsive therapy as studied by radio electrocardiography, *Am. J. Cardiol.,* 10: 187–192.

Demole, M. (1964), Intestinal pyxigraphy, *Gastroenterologia* (Basel), 101: 185–188.

Dendal, J. (1962), Enregisterement continue et a distance, par linison transistorisee U.H.F., de l'electroardiogramme complet d'un sportif en plein effort, *Electroacoustique* (Buil. Unv. Liege), 4: 27–51; *Rev. Gen. Electron.,* 16: 23–32.

Denisov, V., and Klevtsov, M. (1961), Biotelemetry, *Radio,* 10: 18–19.

Denisov, V. G., Egorov, A. D., Kuzminov, A. P., Silvestrov, M. M., and Sosin, B. A. (1963), Using biotelemetric data for the investigation of the control systems of spacecraft with astronaut's active share (in Russian), in V. V. Parin, *Radiotelemetry in physiology and medicine,* pp. 121–124.

DeVos, A., and Anderka, F. W. (1964), A transmitter-receiver unit for microclimatic data, *Ecology,* 45: 171–172.

Directorate of Bioastronautics Projects, *Discoverer III biomedical data report,* A.F. Ballistic Missile Division, Hq. A.R.D.C., Los Angeles, Calif., WDZP Report No. 2.

Diringshofen, H. Von, and Osypka, R. (1964), Physiologiche und tecnische Erwagungen zur Bioinstrumentierung von Luft- und Raumfahrzeugen, *Electromedizin,* 9: 73–87.

Dodge, W. D., and Church, M. B. (1965), Construction of transmitters for radio tracking hares and mountain beavers, *Northwest Sci.,* 39(3): 118–122.

Doebel, J. H. (1967), Radiotracking Mr. Bushytail, *Virginia Wildl.,* 28(3): 4–5.

Donald, D. E., Milburn, S. E., and Shephard, J. T. (1964), Effect of cardiac denervation on the maximal capacity for exercise in the racing greyhound, *J. Appl. Physiol.,* 19: 849.

Dordick, H. S., Baird, B. C., and Saultz, J. E. (1962), An improved telemetering system for physiological measurements, *ISA Trans.,* 1(4): 380–388.

Dordick, H. S., Balin, H., Kirael, S. L., and Hatke, F. (1961), The determination of ovarian function by means of permanently implanted endoradiosondes, *Inter. Conf. Med. Electron., Proc.* 4: 7.

Douglas, D. W., and Seal, H. R. (1961), Internalized animal telemetry system electronic considerations, *32nd Ann. Meeting Aerospace Med. Assoc.* (Chicago), Apr. 24–27.

Douglas, D. W., Seal, H. R., and Simons, D. D. (1961), Human physiological telemetering system, *Aerospace Med.,* 32: 228–229.

Dow, W. (1960), A telemetering hydrophone, *Deep Sea Res.,* 7: 142–147.

Dracy, A. E., Essler, W. O., and Jahn, J. R. (1963), Recording intrareticular temperatures by radiosonde equipment, *J. Dairy Sci.,* 46: 241–242.

Dracy, A. E., and Essler, W. O. (1964), Some ruminal reticular pressures recorded by radio telemetry, *J. Dairy Sci.,* 47: 1428–1429.

Dracy, A. E., and Jahn, J. R. (1964), Use of electrocardiographic radio telemetry to determine heart rate in ruminants, *J. Dairy Sci.,* 47: 561–63.

Dracy, A. E., and Kurtenbach, A. J. (1964), Radio telemetry—a tool for studying bloat, *South Dakota Farm and Home Res.,* 15: 15–17.

Dracy, A. E., and Kurtenbach, A. J. (1965), Radiosonde for measuring bovine intraruminal reticular pressures, *J. Dairy Sci.,* 48: 1250–1251.

Duche ne-Marullaz, R., Faucon, G., and Delost, P. (1965), Observation par telemetrie de l'electrocardiogramme du chien en rythme nodal, *J. Physiol.* (Paris), 57: 604–605.

Duckworth, J. E., and Shirlaw, D. W. (1955), The development of an apparatus to record the jaw movement of cattle, *Brit. J. Anim. Behav.,* 3: 56–60.

Duckworth, J. E., and Shirlaw, D. W. (1958), The value of animal behaviour records in pasture evaluation studies, *Brit. J. Anim. Behav.,* 6: 139–146.

Duckworth, J. E., and Shirlaw, D. W. (1958), A study of factors affecting feed intake and eating behaviour of cattle, *Brit. J. Anim. Behav.,* 6: 147–154.

Ducote, B. A., and Alexander, H. E. (1965), The measurement of stress in animals by bio-telemetry, *Digest 6. Int. Conf. Med. Electron. Biol. Eng. Tokyo,* 203–204.

Dunn, F. L., and Beenken, H. G. (1959), Short distance radio telemetering of physiological information, *J. Amer. Med. Assoc.,* 169: 1618.

Dunn, F. L., and Rahm, W. E. (1950), Electrocardiography: Modern trends in instrumentation and visual and direct recording electrocardiography, *Ann. Intern. Med.,* 32: 611.

Dutky, S. R., Schechter, M. S., and Sullivan, W. N. (1963), Monitoring electro-physiological locomotive activity of insects to detect biological rhythms, in L. E. Slater (Ed.), *Bio-telemetry,* pp. 273–281.

Dziuk, H. E. (1964), Radiotelemetry for stomach motility in ruminants, *Med. Electron. Biol. Eng.,* 2(3): 281–286.

Eckert, T. (1967), The pH-endoradio transmitter: A method of studying the invivo disintegration of orally applied pharmaceutical preparations, *Arzneimittel-Forsch.,* 17(5): 645–646.

Einthoven, W. (1966), Le Telecardiogramme, *Arch. Intern. Physiol.,* 4: 132.

Eisenberg, L., Mauro, A., and Glenn, W. W. L. (1961), Transistorized pacemaker for remote stimulation of the heart by radio-frequency transmission, *IRE Trans. Bio-Med. Electron.,* BME 8: 253–257.

Eklund, C. R. (1963), Determination of temperatures of incubating eggs of antarctic birds, in L. E. Slater (Ed.), pp. 267–272.

Eliassen, E. (1960), A method for measuring the heart rate and stroke/pulse pressures in birds by normal flight, *Mat. Nat. Sci.,* 12: 1–22.

Eliassen, E. (1963), Telemetric registering of physiological data in birds in normal flight, in L. E. Slater (Ed.), pp. 257–265.

Eliassen, E. (1963), Preliminary results from new methods of investigating the physiology of birds during flight, *Ibis,* 105: 234–237.

Ellis, A. B., McKenzie, J. A., and Stabolspezy, C. A. (1963), *Phase three report of microminiaturized instrumentation package,* FTC-TDR-62-37, Air Force Flight Center, Edwards Air Force Base, Cal., Feb.

Ellis, D., and Schneidermeyer, F. (1962), Multipack, a three-channel system of physiologic/psychophysiologic instrumentation, *Amer. J. Med. Electronics,* 1: 280–286.

Ellis, R. J. (1964), Tracking raccoons by radio, *Wildlife Mag.,* 28: 363–368.

Ellis, J. E., and Lewis, J. B. (1967), Mobility and annual range of wild turkeys in Missouri, *J. Wildl. Mgmt.,* 31(3): 568–581.

Elsner, R. W., Franklin, D. L., and Van Citters, R. L. (1964), Cardiac output during diving in an unrestrained sea lion, *Nature,* 202: 809–810.

Elsner, R. W., and Kenney, D. W. (1966), Muscle blood flow and heart rate in the exercising horse, *Fed. Proc.,* 25: 333.

Elsner, R., Kenney, D. W., and Burgess, K. (1966), Diving bradycardia in the trained dolphin, *Nature,* 212: 408.

Elsner, R. (1966), Diving bradycardia in the unrestrained hippopotamus, *Nature,* 212: 408.

Emanuel, G. R. (1959), *Jupiter missile AM-18, thermal environment analysis unit. Systems report of temperature and pressure control systems, environment temperatures, pressures, and special measurements* (*U*). Evaluation Section, Propulsion and Mechanics Branch, Structures and Mechanics Lab., DOD,

ABMA, Redstone Arsenal, Huntsville, Ala., Rept. No. DSD-TR-27-59 (Secret), July 17.

Emlen, J. T., and Penney, R. L. (1966), The navigation of penguins, *Sci. Amer.*, 215: 104–113.

Epstein, R. J., Haumann, J. R., and Keener, R. B. (1968), An implantable telemetry unit for accurate body temperature measurements, *J. Appl. Physiol.*, **24**, No. 3, Mar.

Esser, A. H., and Etter, T. L. (1966), *Automated location recording on a psychiatric ward: Preliminary notes on continuous monitoring of posture and movement of all individuals in an observation area.* Mimeo of paper presented at Amer. Accos. Advance. Sci. 133rd meeting.

Essler, W. O., *Electrocardiographs and temperatures by radio telemetry,* Elec. Eng. Dept., Univ. of Vermont, Burlington, Vt.

Essler, W. O. (1963), Radiotelemetry of heart rate with subdermally implanted sensors, in L. E. Slater (Ed.), pp. 341–351.

Essler, W. O. (1964), Radiotelemetry of electrocardiograms and body temperatures from surgically implanted transmitters, *State Univ. of Iowa, Studies in Nat. History* (1961), 20: 4.

Essler, W. O., and Folk, G. E. (1960), Monitoring heart rates and body temperatures of unrestrained mammals by radio transmitter, *Anat. Rec.*, 137: 353; *Bull. Ecol. Soc. Amer.*, 41: 91.

Essler, W. O., and Folk, G. E. (1961), Determination of 24-hour physiological rhythms of unrestrained mammals by radiotelemetry, *Nature,* 190: 90–91.

Essler, W. O., and Folk, G. E. (1963), Multi-channel radio telemetry system, *Bull. Wildlife Telemetry,* 2: 4–5.

Essler, W. O., and Folk, G. E. (1962), A method of determining true resting heart rates of unrestrained mammals by radio telemetry, *Animal Behav.*, 10: 168–170.

Essler, W. O., Folk, G. E., and Adamson, G. E. (1961), 24-hour cardiac activity of unrestrained cats, *Fed. Proc.*, New York, 20: 129.

Ettelson, B. L. (1960), Internal animal telemetry—a feasibility test program, *ARS Publ.*, New York, 1426–1460.

Ettleson, B. L., and Pinc, B. W. (1962), Development of an internalized animal telemetry system, *Aerospace Med.*, 33: 75–80.

Evans, D. (1965), Pheasants tune in on space age, *S. Dak. Cons. Dig.*, 32(3): 3–5.

Evans, W. E., and Sutherland, W. W., Studied communication of confined dolphins at Marineland of the Pacific, *Acoustics Lab.*, Lockheed Missiles and Space Co.

Evans, W. E., and Sutherland, W. W. (1963), Potential for telemetry in studies of aquatic animal communication, in L. E. Slater (Ed.), pp. 217–224.

Evrard, E., and Rens, F. (1956), Transmission par radio d'un electrocardiogramme. Realisation d'un appareil pour l'enregistrement en sol de l'electrocardiogramme d'un pilote en vol, *Revue H.F.*, 3: 193–206.

Ewe, K. (1966), Die Beziehung der Mukosamorphologie zur Magensaftabsonderung, *III. Weltkongr. Gastroenterolog.* Tokyo, 18–24, Sept.

Fabris, F., Morea, M., and Todesco, S., et al. (1964), Application of radioelectrocardiography in surgical practice, *Acta. Chir. Ital.*, 20: 511–515.

Falize, J. (1963), Examination of athletes during exercise (in French), *Rev. Educ. Phys.*, 3(1): 19–24.

Farfel, V. S. (1966), Investigation and perfection of sportsman's kinetic function by means of express information, *"S" Beitrage. XVI Weltkongreb fur Sportmedizin,* Hannover No. 125.

Farrar, J. T. (1960), Use of a digital computer in the analysis of intestinal motility records, *IRE Trans. Med. Electron.*, **ME 7**: 259–263.

Farrar, J. T. (1961), Telemetering of gastro-intestinal motility, in W. I. Card (Ed.), *Modern trends in gastro-enterology*, 3, Butterworth, London.

Farrar, J. T., and Bernstein, J. S. (1958), Recording of intraluminal gastrointestinal pressures by a radio-telemetering capsule, *Gastroenterology*, 35: 603.

Farrar, J. T., and Davidson, M. (1960) Measurement of gastrointestinal motility in man, in H. D. Bruner (Ed.), *Methods in medical research*, 8, Year Book Publ., Chicago, pp. 200–221.

Farrar, J. T., and Zworykin, B. (1959), Telemetering of gastrointestinal pressure in man by means of intraluminal capsule energized from an external wireless source, *The Physiologist*, Aug.

Farrar, J. T., Zworykin, B. D., and Berkley, D. (1960), Telemetering of physiological information from the gastrointestinal tract by an externally energized capsule, *Proc. Third Intern. Conf. on Med. Electron.*, July.

Fascenelli, F. W. (1965), Electrocardiography by do-it-yourself radiotelemetry, *New England J. Med.*, 273: 1076.

Fedorov, V. L., Zuravleva, N. V., and Pancenko, E. M. (1965), Analysis of the sequence of a gesture by radio telemetry (in Russian), *Teor.Prakt, Fiz.Kult., Moskva*, 28(8): 28–29.

Fedotov, V. A., and Geller, E. S. (1967), Biologic telemetry. Problems of classification (in Russian), *Izv.Akad.Nauk.SSSR, Ser. Biol.*, 4: 583–594.

Fehr, H., Stavney, L. S., Hamilton, T., Surcus, W., and Smith, A. N. (1966), Hiatal hernia investigated by pH telemetering, *Amer. J. Digestive Diseases*, 11: 747.

Fender, F. A. (1936), A method for prolonged stimulation of the nervous system, *Am. J. Physiol.*, 116: 47.

Fender, F. A. (1937), Prolonged splanchnic stimulation, *Proc. Soc. Exp. Biol. Med.*, 36: 396–398.

Filshie, J. (1965), Radio telemetry of information from the hen's oviduct, *J. Reprod. Fertil.*, 10(2): 302.

Findley, J. D. (1963), Detecting and recording the behavior of animals in a multi-enclosure, programmed environment, in L. E. Slater (Ed.), pp. 213–216.

Fischler, H., and Frei, E. H. (1961), Subminiature electroencephalograph for radio-telemetering, *Digest 1961 Intern. Conf. Med. Electron.*, New York, p. 213.

Fischler, H., and Frei, E. H. (1963), Subminiature apparatus for radio-telemetering of EEG data, *IEEE Trans. Biomed. Electron.*, **BME-10**: 29–36.

Fischler, H., and Rubinstein, M. (1965), Advantages of radio-telemetric nystagmography for clinical examinations, *1965 Proc. Internat'l. Symp. Vestibular and Oculomotor Problems*, Univ. Tokyo, Japan, 207–214.

Fischler, H., Peled, N., and Yerushalmi, S. (1967), FM/FM multiplex radio-telemetry system for handling biological data, *IEEE Trans. Bio-Med. Engineer*, **BME 14**: 30–39.

Fischler, H., Blum, B., Frei, E. H., and Streifler, M. (1959), *Bull. Res. Counc. Israel Sect. E.* 8: 101.

Fischler, H., Blum, B., Frei, E. H., and Streifler, M. (1961), Telemetering of ECG's from unrestrained convulsive cats by a transistorized amplifier-transmitter, *Electroencephalogr. Anim. Neuorphysiol.*, 13: 807–812.

Fischler, H., Blum, B., Frei, E. H., and Streifler, M. (1960), Electrocortigogram transmission by a transistorized subminiature amplifier-transmitter for the normal and epileptic range, *Bull. Res. Coun. Israel*, 8E: 101–104.

Flickinger, D. (1959), *Results of animal investigation in space vehicles to date,* Presented at the 30th Ann. Meeting of the Aeromedical Assoc., Los Angeles, Calif., Apr. 27–29.

Flokin, M. (Ed.) (1965), *Life science and space research,* New York, Wiley, 257 pp.

Flood, M. M. (1963), Experimenting with monkeys to be stimulated by remote control of electrodes planted chronically in the brain, in L. E. Slater (Ed.), pp. 251–254.

Flory, L. E., Hatke, F. L., and Zworykin, V. K. (1963), Telemetering internal biological potentials with passive type capsules, *5th Int. Conf. Med. Electron. Biol. Eng.,* 1: 575.

Flyger, V., and Townsend, Marjorie R. (1968), The migration of polar bears, *Sci. Amer.,* 218(2): 108–116.

Folk, Jr., G. E. (1961), Observations on the daily rhythms of body temperature labile mammals, *Am. New York Acad. Sci.,* 98: 544–969.

Folk, Jr., G. E. (1961), The effect of continuous light upon 24-hour activity of arctic mammals, *Bull. Ecol. Soc. Amer.,* 42: 161.

Folk, Jr., G. E. (1961), 24-hour rhythms in human cardiac function, *Proc. 39th Ross Res. Conf.,* (Univ. Minnesota), 39: 86–88.

Folk, Jr., G. E. (1963), Daily physiological rhythms of arctic carnivors in continuous light, *Amer. Zoologist,* 3: 470.

Folk, Jr., G. E. (1964), The problem of electrodes for use with electrocardiograph radio capsules, *Biomed. Sci. Instrum.,* 2: 235–265.

Folk, Jr., G. E. (1964), Daily physiological rhythms of carnivors exposed to extreme changes in arctic daylight, *Fed. Proc. Symp.,* 23: 1221–1228.

Folk, Jr., G. E. (in press), Day-night physiological rhythms of mammals exposed to extreme changes in arctic daylight, *Fed. Proc. Symp.,* 23.

Folk, Jr., G. E., and Folk, M. A. (1964), Continuous physiological measurements from unrestrained arctic ground squirrels, *Am. Acad. Sci. Fennica,* 71: 157–173.

Folk, Jr., G. E., and Hedge, R. S. (1964), Comparative physiology of heart rate of unrestrained mammals (abstract), *Amer. Zool.,* 4: 297.

Folk, Jr., G. E., and Hedge, R. S. (1965), Further evidence of temperature independence of the mammalian biological clock, *Cryobiology,* 2: 24.

Folk, Jr., G. E., Folk, M. A., and Brewer, M. C. (1966), The daynight (circadian) physiological rhythms of large Arctic carnivors in natural continuous light (summer) and continuous darkness (winter), *Proc. 4th Internat. Biomed. Congress.*

Folk, Jr., G. E., Folk, M. A., and Simmonds, R. C. (1967), Early stages in the onset of hibernation in the Arctic ground squirrel (abstract), *Amer. Zool.,* 7(4): 744.

Folk, Jr., G. E., Schellinger, R. R., and Shyder, D. (1961), Day night changes after exercise in body temperatures and heart rates of hamsters, *Iowa Acad. of Sci.* 68: 594–602.

Folk, Jr., G. E., Simmonds, R. C., and Hedge, R. S. (1965), Telemetered physiological measurements of subarctic bears during natural cold exposure (abstract), *Amer. Zool.,* 5: 239–240.

Folk, Jr., G. E., Folk, M. A., Simmonds, R. C., and Brewer, M. C. (1966), A two-year study of winter lethargy in subarctic bears (abstract), *Amer. Zool.,* 6(4): 583.

Folk, Jr., G. E., Shook, G. L., Hedge, R. S., Brewer, M. C., and Folk, M. A. (1963), Daily (circadian) physiological rhythms of arctic carnivores in continuous light (abstract), *Amer. Zool.,* 3: 470.

Folk, M. A. (1963), *J. Mammal.*, 44: 575.

Forrest, R. H. (1959), A heart-beat detector for remote working with active subjects, *Royal Aircraft Establishment* (Farnborough), Technical Note No. EL 172 (Nov.).

Forstadt, V. M. (1963), Radiotelemetric investigation of fundamental parameters of external respiration (in Russian), in V. V. Parin, *Radiotelemetry in physiology and medicine,* pp. 73–81.

Fox, R. H., and Wolff, H. S. (1961), A high sensitivity, high stability temperature-sensitive radio pill, *Intern. Conf. on Med. Electron.,* July.

Frank, T. H. (1968), Telemetering the ECG of free swimming salnio irideius, *IEEE Trans. Bio. Med. Eng.,* April, 15: 111–114.

Franklin, D. L. (1965), Techniques for measurement of blood flow through intact vessels, *Med. Electron. Biol. Eng.,* 3: 27–37.

Franklin, D. L., Van Citters, R. L., and Watson, N. W. (1965), Applications of telemetry to measurement of blood flow and pressure in unrestrained animals, *Proc. Nat. Telemetering Conf.,* 233–234.

Franklin, D. L., Van Citters, R. L., and Watson, N. W. (1965), The ultrasonic doppler shift blood flowmeter-telemetry system: Application in free roaming experiments and as a flow monitor in humans, *Dig. 6th Int. Conf. Med. Electron. Jap. Soc. Med. Electron. Biol. Eng.,* 205.

Franklin, D. L., Van Citters, R. L., Watson, N. W., and Kemper, W. S. (1965), Telemetry of blood pressure and flow from free-ranging baboons, *Proc. 18th Conf. Eng. Med. Biol.*

Franklin, D. L., Watson, N. W., and Van Citters, R. L. (1964), Blood velocity telemetered from untethered animals, *Nature,* 203: 528–530.

Franklin, D. L., Watson, N. W., Van Citters, R. L., and Smith, O. A. (1964), Blood flow telemetered from dogs and baboons, *Fed. Proc.* 23: 303.

Franklin Institute, Philadelphia, Pa. (1963), Bio-instrumentation for the study of biological rhythms, *Franklin Inst. Lab. Rept.,* July, pp. 1–2.

Freed, M. (1965), Application of telemetry techniques to hard linetransmission of bio-medical information on the 50 foot human centrifuge, *Proc. Nat. Telemetering Conf.,* 56–58.

Freedman, R., and Blockley, W. V. (1959), An instrumentation package for the measurement of physiological response, *Aerospace Med.,* 30: 183–184.

Friauf, Walter S. (1969), Rats: Their comings and goings, *National Telemetering Conference,* p. 34.

Frucht, A. H., and Otto, K. (1958), Kleinstsender sur drahtlosen Übertragung biologischer meßgrössen vom fribeweglichen Menschen oder Tier, *Dtsch. Geswes.,* 13: 1416–1422.

Frucht, A. H., and Otto, K. (1959), Wireless transmission of the ECG of men and animals by a transistorized miniature transmitter (in German), *Pflügers Arch. Ges. Physiol.,* 270: 82.

Frucht, A. H., Liersch, S., and Otto, K. (1960), A multichannel equipment for wireless transmission of biological data for the use in industrial and sports medicine (in German), *Pflügers Arch. Ges. Physiol.,* 272: 84.

Fryer, T. B. (1966), A miniature telemetry system for pressure measurements, *Proc. 19 Ann. Conf. Eng. Med. Biol.*

Fryer, T. B., Deboo, G. J., and Winget, C. M. (1966), Miniature long-life temperature telemetry system, *J. Appl. Physiol.,* 21: 295–299.

Fryer, T. B., Sandler, H., and Datnow, B. (1967), A multi-channel implantable telemetry system, *Digest of 7th Intern. Conf. Med. Biol. Eng.*, p. 86.

Fuchs, G. (1962), Clinical telemetry (in German), *Med.-Markt,* 10(5): 209–212.

Fujie, Z. (1965), Physiological monitoring in flight, *Digest 6 Int. Conf. Med. Electron. Biol. Eng.* Tokyo, 195–196.

Fujii, K., Suda, T., Kawashima, T., and Arai, Z. (1964), Radiotelemetering applied to obstetrical practice, *Jap. J. Med. Electron. Biol. Eng.,* 2: 33–37.

Fullagar, P. J. (1967), The use of radio telemetry in Australian biological research, *Ecol. Soc. Aust.,* Proc. 2: 16–26.

Fuller, J. L., and Gordon, T. M. (1948), The radioinductograph—a device for recording physiological activity in unrestrained animals, *Science,* 108: 287.

Furchner, C. J. (1961), An amplitude modulated radio telemetry for measurement of physiological temperatures, M.A. Thesis, South Dakota State Univ., Brookings, S.D.

Furman, K. I., and Lupu, N. Z. (1963), Cardiac monitoring and telemetering system, *J. Appl. Physiol.,* 18: 840–842.

Galkin, A. M., et al. (1958), Investigations of the vital activity of animals during flights in hermetically sealed cabins to an altitude of 212 km. (in Russian) *III. Sb. Statei,* 1: 1–26.

Galler, S. R. (1961), Biophysical telemetering system, *Electron. Prod.,* July.

Galler, S. R., and Fix, C. E. (1961), Animal tracking gone modern, *Naval Res.,* pp. 11–14.

Ganjuskina, S. M. (1963), Radiotelemetric examination of the pulse rate and the respiration in miners during work (in Russian), in V. V. Parin, *Radiotelemetry in physiology and medicine,* pp. 172–176.

Gardenshire, L. W. (1965), Data redundancy reduction for biomedical telemetry, in C. A. Caceres (Ed.), pp. 255–298.

Gaume, J. G. (1959), Research in space medicine, *Astronautical Sci. Rev.,* 1(2): 20–21.

Gauthreaux, Jr., S. A. (1970), Weather radar quantification of bird migration, *BioScience,* 20(1): 17–20.

Gazenko, O. G., and Baevskij, R. M. (1961), Physiological methods in space medicine (in Russian), *Iskusstvennye Sputniki Zemli,* vyp. 11: 68–77.

Gazenko, O. G. (1962), Some problems of space biology (in Russian), *Vestn.Akad. Nauk SSSR,* 32(1): 30–34.

Gazenko, O. G. (1962), Medical problems of manned space flight, *Space Sci. Rev.,* 1: 369–398.

Gazenko, O. G., and Gjurdzian, A. A. (1965), Medical and biological investigations on the spaceship "Voskhod" (in Russian). *Vestn.Akad.Nauk SSSR,* Aug., No. 8: 19–26.

Geddes, L. A. (1962), A bibliography of biological telemetry, *Amer. J. Med. Electronics,* 1: 294–299.

Geddes, L. A., Hoff, H. E., and Spencer, W. A. (1959), The "Broadcast Demonstration" in the physiology laboratory, *J. Med. Education,* 34: 107–117.

Geddes, L. A., Hoff, H. E., and Spencer, W. A. (1961), Short distance broadcasting of physiological data, *IEEE Trans. Bio-Med. Electron.,* BME-8: 168-172.

Geddes, L. A., Hoff, H. E., and Spencer, W. A. (1965), A closed-circuit data broadcast system, *Health Sci. Bull.,* 2.

Geddes, L. A., Partridge, M., and Hoff, H. E. (1960), An ECG lead for exercising subjects, *J. Appl. Physiol.,* 15.

Geddes, L. A., Hoff, H. E., Hickman, D. M., and Moore, A. G. (1962), The impedance pneumograph, *Aerospace Med.,* 1: 28–33.

Geddes, L. A., Hoff, H. E., Spencer, W. A., and Vallbona, D. (1962), Acquisition of physiologic data at the bedside. Progress report, *Amer. J. Med. Electronics,* 1: 62–69.

Geddes, L. A., Hoff, H. E., Hickman, D. M., Hinds, J., and Baker, L. (1962), Recording respiration and the electrocardiogram with common electrodes, *Aerospace Med.,* 33: 791.

Gengerelli, J. A., and Kallejian, V. (1950), Remote stimulation of the brain of the intact animal, *J. Psychology,* 29: 263–269.

George, J. (1966), The day the gears go to bed, *Readers Digest,* Oct., 137–141.

Gerathewohl, S. J. (1960), Bio-telemetry in nose cones of the U.S. Army Jupiter missiles, *IRE Trans. Med. Electronics,* 4–6: 288–302.

Gerathewohl, S. J. (1960), Space medical experiments with Jupiter rockets (in German), *Raketentechn. Raumfahrtforsch.,* 4(3): 86–92.

Gessamen, J. A., Folk, Jr., G. E., and Brewer, M. C. (1965), Telemetry of heart rate from eight avian species (abstract), *Amer. Zool.,* 5: 696–697.

Gessamen, A. C., Folk, Jr., G. E., and Gesseman, J. A. (1965), The day-night behavior of grizzly bears (abstract), *Bull. Ecol. Soc. Amer.,* 46(4): 196.

Gianascol, A. J., and Yeager, C. L. (1964), Telemetry and child psychiatry: A psychosomatic approach, *Psychosomatics,* 5: 317–321.

Gianascol, A. J., and Yeager, C. L. (1963), Simultaneous study of behavior and brain waves, *Amer. J. Psychiat.,* 81: 279–281.

Gibson, Jr., R. J., Goodman, R. M., and Marmarou, A. (1964), Instrumentation for study of biological rhythms, *FIRL Report F-2029,* Vol. 1.

Gibson, Jr., R. J., Goodman, R. M., and Marmarou, A. (1965), A bio-instrumentation system for circadian rhythm studies, *Digest 6 Int. Conf. Med. Electron. Biol. Eng.* Tokyo, 222–223.

Gibson, T. C., Thornton, W. E., Algary, W. P., and Craige, E. (1963), Telecardiography and the use of simple computers, *New England J. Med.,* 267: 1218–1224.

Gillings, B. R. D. (1964), Jaw movement studies using miniature radio transmitters, *Proc. IREE Australia,* 25(3): 529–533.

Gillings, B. R. D., Kohl, J. T., and Graf, H. (1961), Study of tooth contact patterns with the use of miniature radio transmitter, *Inter. Conf. Med. Electron.,* Proc. 4: 67.

Gillings, B. R. D. (1963), Transmitter fits in a tooth, *Radio electronics,* 34(11): 60, 62, 64.

Gillings, B. R. D., Kohl, J. T., and Zander, H. A. (1963), Contact patterns using miniature radio transmitters, *J.D. Res.,* 42: 177.

Gilson, J. S., and Griffing, R. B. (1961), Continuous electrocardiograms: Electrodes and lead systems, *Amer. J. Cardiol.,* 8: 212–215.

Glasscock, W. R., and Holter, N. J. (1952), Radioelectroencephalography for medical research, *Electronics,* 25: 126–29.

Glatt, R. (1953), L'electrocardiogramme d'un pilote en vol, *Cardiologia,* 22: 238–246.

Glatt, R., Wiesinger, K., and Pircher, L. (1953), The drahtlose Ubertragung des Elektrokardiogramms eines Piolen im Flug, *Helv. Physiol. et Pharmocoal. Acta,* 11: C3–C4.

Gleason, D. M. (1964), Study of micturition with electronic measuring devices, *Ann. New York Acad. Sci.,* 118: 17–25.

Gleason, D. M., and Lattimer, J. K. (1962), A miniature radio transmitter which is inserted into the bladder and which records voiding pressures, *J. Urol.,* 87: 507.

Gleason, D. M., et al. (1964), Pressure telemeter: A method of measuring intravesical pressure without wires, or catheters, or tubes, *Surg. Forum,* 15: 502–503.

Gleason, D. M., Lattimer, J. K., and Bauxbaum, D. (1965), Bladder pressure telemetry, *J. Urol.,* 94: 252–256.

Gleason, W. G. (1960), Bioinstrumentation for space flight, *Proc. Nat. Telemetry Conf.,* St. Monica, Calif.

Glenn, W. G. (1963), The biotelescanner: An instrument for telemetric quantitation of immunodiffusion (antigen-antibody) reactions, in *Biomed. Sci. Instrum.,* 1: 395–399.

Glenn, W. G., et al. (1963), Bioinstrumentation and telemetry for immunochemical analysis, *Proc. 1963 Natl. Tele. Conf.*

Glenn, W. G., and Prather, W. E. (1965), Telus: Telemetric universal sensor for space and terrestrial application, *Proc. Natl. Telemetering Conf.,* 65–67.

Glenn, W. G., Prather, W. E., and Jaeger, H. A. (1965), Telus (telemetric universal sensor), *Technical Report, USAF SAM-TR,* 1–7.

Glenn, W. G., Prather, W. E., and Jaeger, H. A. (1965), Measurement, processing and control of telemetered biological analyses for terrestrial and aerospace applications, *Instrument Society of Amer.,* #1.2-2-65, Defense Documentation Center, Defense Supply Agency.

Glenn, W. G., Prather, W. E., and Jaeger, H. A. (1966), Automatic integration and identification of telemetered biological analyses, *Proc. 19 Ann. Conf. Eng. Med. Biol.*

Glenn, W. W. L., Mauro, A., Longo, E., Levietes, P. H., and Mackay, F. J. (1959), Remote stimulation of the heart by radio-frequency transmission, *New Engl. J. Med.,* 261: 948–951.

Godfrey, G. A. (1967), *The summer and fall movements and behavior of immature ruffed grouse* (*Bonasa umbellus* <L.>), M.A. Thesis, Univ. of Minnesota, Minneapolis, 205 pp.

Goetze, W. (1958), Bioelectrical investigation of healthy and brain diseased people during movement by radiotransmission (in German), *Zbl.Neurochirurgie,* 18 (5/6): 295–303.

Goetze, W., Knudsen, U., and Krokowski, G. (1964), Uber den telemetrishen Nachweis des Einflusses von kreisformigen positiven und negativen Beschleunigungen auf das EEG des Menschen, *Elektromedizin,* 9: 185–188.

Goetze, W., and Kofes, A. (1955), Uber Radio-Elektroencephalographie. (Eine Methode fur Studium des Einflusses korperlicher Belastungen auf die Hirnfunktion), *Z. Ges. Exper. Med.,* 126: 439–443.

Goetze, W., and Kofes, A. (1956), Radio-electroencephalography, *EEG Clin. Neurophysiol.,* 8: 704–705.

Goetze, W., and Kofes, A. (1957), Remote recording of bioelectrical phenomena (in German), *Aerztl. Forschg.,* 11: 11/1–11/3.

Goetze, W., Kofes, A., Kubicki, S. T., and Wolter, M. (1958), EEG- und EMG-untersuchungen an gesunden und himkranken unter arbeitsbelastung nach radio-ubertachung, *Elektro-Med.,* 3: 297–303.

Goetze, W., and Krokowski, G. (1962), Telemetrischer Nachweis von EEG—Potentialen wahrend der Nystagmographie, *Elektromed.,* 7: 254–255.

Goetze, W., Kofes, A., and Stoelzel, R. (1961), Simultaneous recording of several bioelectric phenomena of moving objects by radiotransmission (in German), *Electromedizin,* 6: 42–43.

Goetze, W., Munter, M., and Knudsen, U. (1965), Vergleichende telemetrische EEG—Untersuchungen wahrend korperlicher Belastung und Hyperventilation, *Elektromedizin,* 10: 189–192.

Gogan, P., and Buser, P. (1962), Dispositif simple d'enregistrement electrophysiologique par telemesure, *C.R. Acad. Sci.* (Paris), 255: 2184–2186.

Gold, D. C., and Malcolm, J. C. (1957), Action potentials recorded by radio transmission from the cortex of a non-anaesthetized, unrestrained cat, *J. Physiol.,* 135: 5P, 135.

Gold, D. C., and Perkins, W. J. (1959), A miniature electroencephalograph telemeter system, *Electronic Eng.,* 31: 337.

Goldberg, M. N., and Foster, D. R. (1963), Telemetry of physiological data during parachute research, *Biomed. Sci. Instrum.,* 1: 263–271.

Goldberg, M. N., and Wagoner, E. V. (1961), Development of physiological instrumentation and the evolution toward a logical monitoring system, *Digest 1961 Intern. Conf. Med. Electron.,* New York, p. 120.

Goldberg, M. N., Mills, R. A., and Brockley, W. V. (1960), *Instrumentation package for in-flight studies,* WADC TR 60–83.

Goldberger, E. (1961), Long period continuous electrocardiography of active persons, *Amer. J. Cardiol.,* 8: 603–604.

Goldman, J., and Ross, D. K. (1960), Measuring work performance, *Electronics,* 34: 196.

Goldstein, N., Instrumentation methods for physiological studies, Vol. 1, printed by Operations Services Dept., Univ. of California, Berkeley, Cal.

Goodman, R. M., and Taylor, W. M. (1964), A statement on biotelemetry, *Ann. Biomed. Sci. Instrum. Symp.,* Proc. 2: 213–215.

Goodman, R. M. (1967), Simultaneous multi-parameter data from implantable telemeters *Digest of 7th Intern. Conf. Med. Biol. Eng.,* 87.

Graber, R. R. (1965), Night flight with a thrush, *Audubon,* 67(6): 368–374.

Graber, R. R., and Wunderle, S. L. (1966), Telemetric observations of a robin, *The Auk,* 83(4): 674–677.

Graf, H. (1962), *Occlusal tooth contact patterns in mastication,* Used as M.S. Thesis, University of Rochester, New York.

Graf, H., and Borbely, A. (1966), Radiotelemetrische Korpertemperaturmessung bei Mausen und Ratten, *Experientia,* 22: 339–340.

Graf, H., and Zander, H. A. (1963), Tooth contact patterns in mastication, *J. Pros. Dent.,* 13: 1055–1966.

Graf, H., and Zander, H. A. (1964), Tooth contact patterns and muscle activity in mastication and swallowing (in German), *Schweiz. Mschr. Zahnhk.,* 74: 495–510.

Grandpierre, R., Rozier, J., and Mesnard, E. (1965), Device of miniaturized telestimulation of deep cerebral structures, *C. R Soc. Biol.,* 159: 1965–1968.

Gray Merriam, H. (1960), *Problems in woodchuck population ecology and a plan for telemetric study,* Doctoral Thesis, Cornell Univ.

Gray Merriam, H. (1963), Low frequency telemetric monitoring of woodchuck movements, in L. E. Slater (Ed.), *Bio-telemetry,* pp. 155–171.

Graybiel, A. (1959), An account of experiments in which two monkeys were recovered unharmed after ballistic flight, *Aerospace Med.,* 30: 871.

Graybiel, A., McNinch, J. H., and Holmes, R. H. (1960), Observations on small primates in space flight, *10th Intern. Aeronaut. Conf.,* 1: 394–401.

Graystone, R., McLennan, H., and Plummer, P. M. (1964), A light 2 channel telemetry transmitter for brain activity, *J. Physiol.,* 175: 23–24.

Greatroex, C. A. (1966), A 4-channel radiation radio telemetry system, *Phys. Med. Biol.,* 11: 175.

Green, D., and Shore, J. R. (1969), A two-channel transmitter for the telemetry of cat electrocorticograms, in *J. Physiol.,* 1–2.

Greer, M. A., and Riggle, G. C. (1957), Apparatus for chronic stimulation of the brain of the rat by radiofrequency transmission, *EEG Clin. Neurophysiol.,* 9: 151–155.

Greig, J., and Ritchie, A. (1944), A simple apparatus for remote nerve stimulation in the unanesthetized animal, *J. Physiol.,* 103: 8.

Grieve, D. W., and Humphries, W. J. A. (1957), *A new heart-beat detector and its ancillary recording and telemetry systems,* Ministry of Supply, Directorate of Physiological and Biological Research. Clothing and Stores Experimental Establishment, Report No. 87 (July).

Griffin, D. R. (1963), *Potential uses of telemetry in studies of animal orientation,* Biol. Lab., Harvard Univ., in L. E. Slater (Ed.), pp. 25–31.

Griffith, C., Where DO the deer and antelope play? *S. Dak. Cons. Dig.,* 29(4): 22–25.

Griffith, R. S. (1960), Mouse transmits own temperature, *Radio Electron.,* 101–106.

Grisamore, N. T., Cooper, J. K., and Caceres, C. A. (1965), Evaluating telemetry systems, in C. A. Caceres, pp. 351–375.

Groh, H., Kubeth, A., and Baumann, W. (1964), Zur kinetic und Dynamik schneller menschlicher Korperbewgungen, *Proc. 16 World Congr. Sports. Med. Hannover.*

Grotz, R. C. (1965), Intramuscular R, radio transmitter of muscle potentials, *Arch. Phys. Med.,* 46: 804–808.

Grotz, R. C., Ko, W. H., Long, D., Yon, E., and Greene, I. (1964), Implantable radio transducers for physiological information and implant techniques, *Ann. Biomed. Sci. Instrum. Symp., Proc.* 2: 145–153.

Gumener, P. I., and Poltorak, S. A. (1960), Radiotelemetric methods of investigating thermoregulation in labor physiology and hygiene, *Novyye Fiziol. Metody v Gigiyene* (*New Physiological Methods in Hygiene*), 32–35.

Gumener, P. I., Joffe, V. N., and Poltorak, S. A. (1963), Radiotelemetric recording of bioelectrical activity of muscles under industrial conditions (in Russian), in V. V. Parin, *Radiotelemetry in physiology and medicine,* pp. 94–100.

Gumener, P. I., Poltorak, S. A., Rapoport, K. A., and Raichman, S. P. (1963), Radiotelemetric investigation of the temperature of the skin, the body, and the air (in Russian), in V. V. Parin, *Radiotelemetry in physiology and medicine,* pp. 101–108.

Haahn, F. (1965), Designing for physiological data, *BioScience,* 15(2): 112–115.

Haahn, F., A study of power and frequency requirements in biotelemetry, *BIAC,* 25 pp.

Hagan, W. K. (1965), Telephone applications, in C. A. Caceres, *Bio-Medical Telemetry,* pp. 39–66.

Hagan, W. K., and Larks, S. D. (1962), A fetal electrocardiograph transmission network, *San Diego Symp. Biomed. Eng.,* 66–173.

Hageman, J., Flanigan, S., Harvard, B. M., and Glenn, W. W. L. (1966), Electromicturition by radiofrequency stimulation, *Surg. Gynecol. Obstetrics,* 123: 807.

Halberg, F. (1964), Physiological rhythms and bioastronautics, in K. Schaefer, *Bioastronautics,* Macmillan Co., New York, 181 pp.

Hambrecht, F. T. (1963), A multichannel electroencephalographic telemetering system, *MIT Res. Lab. Electron. Tech. Rep. 413,* Nov.

Hambrecht, F. T. (1965), Multi-channel telemetry systems, *Progr. Brain Res.,* 16: 297–300.

Hambrecht, F. T., Donahne, P. D., and Melzalk, R. (1963), A multi-channel electroencephalogram telemetering-system, *Electroencephalography and Clin. Neurophysiol.,* 15(2): 323–326.

Hanish, H. M. (1959), New techniques in physiological recording under dynamic conditions, WESCON, *Proceedings of Medical Electronics Session,* San Francisco, Aug.

Hanish, H. M. (1959), A subminiature biological telemetry system, *Proc. of the Pilot Clinic on Bio-Instrumentation,* OSU and Foundation for Instrumentation Education and Research.

Hanish, H. M. (1962), *Telemetry of EKG and respiration from the same pair of electrodes,* Rept. from Summer Course, Baylor Med. College.

Hansen, J. T., et al. (1963), Physiological monitoring of animals during space flight, *Biomed. Sci. Instrum.,* 1: 299–307.

Hanson, J. S., and Tabakin, B. S. (1964), Electrocardiographic telemetry in skiers anticipatory and recovery heart rate during competition, *New England J. Med.,* 241: 181–185.

Hanson, M. L., and Preston, R. J. (1963), Bio-telemetry with molecular electronics, *Proc. Symp. Bio-Med. Eng.*

Harris, R. M. (1963), Laying the right lines for electronic monitoring, *Nurs. Outlook,* 11: 573–576.

Harris, C. L., and Siegel, P. B. (1967), An implantable telemeter for determining body temperature and heart rate, *J. Appl. Physiol.,* 22(4): 846–849.

Hart, J. S., and Roy, O. Z. (1965), Respiratory and cardiac responses to flight in pigeons, *Fed. Proc.,* 24: 138.

Hart, J. S., and Roy, O. Z. (1966), Respiratory and cardiac responses to flight in pigeons, *Physiol. Zool.,* 39: 291–307.

Hart, J. S., and Roy, O. Z. (1966), Telemetry of physiological data from birds in flight, *Fed. Proc.,* 25, No. 2.

Harten, G. A., and Koroncai, A. K. (1960), A transistor cardiotachometer for continuous measurements on working persons (in German), *Philips Techn. Rdsch.* (1959/60), 21(10): 293–298. (In English: *Philips Techn. Rev.,* 1960, 21: 304.)

Harten, G. A., and Koroncai, A. K. (1960), Radio transmitter for remote heartbeat measurements, *Electronics* (Dec. 23), 33: 54–55.

Hasler, A. D. (1960), Homing orientation in migrating fish, *Ergebn. Biol.,* 23: 94–115.

Hasler, A. D. (1967), Underwater guideposts for migrating fish, pp. 1–20, in R. M. Storm (Ed.), *Animal orientation and navigation,* Oregon State Univ. Press, Corvallis, 134 pp.

Hasler, A. D., and Henderson, H. F. (1963), Instrumentation problems in the study of homing in fish, in L. E. Slater (Ed.), *Bio-telemetry,* pp. 195–201.

Haupt, G. J., and Birkehad, N. C. (1965), Implantable cardiac pacemaker, in J. H. Lawrence and J. W. Gofman (Eds.), *Advances in biological and medical Physics,* **10**, Academic Press, New York.

Hawthorne, W. E., and Harvey, E. (1961), Telemetering of ventricular circumference in dogs, *J. Appl. Physiol.,* 16: 1124–1125.

Haynes, H. E., and Witchey, A. L. (1957), The radio pill, *Syracuse Scanner* (*IRE*), Dec., 2(4).

Heaf, P. J. (1964), Automation in medicine, *Proc. Roy. Soc. Med.,* 57: 1148–1149.

Heather, M. D. (1965), Beckman defining biomedical tests for Air Force's MOL flight program, *Missiles and Rockets* (Jan. 4), 16: 26–28.

Hebert, C. L. (1960), Physiological monitoring of patients, *Digest Tech. 13th Ann. Conf. Electr. Tech. Med. and Biol.,* 65.

Hedge, R. S., Folk, Jr., G. E., and Brewer, M. C. (1965), Studies on winter lethargy of black and grizzly bears, *Alaskan Sci. Conf.,* AAAS, 16: 31–32.

Heezen, K. L. (1965), *Evaluation of animal-location data obtained by triangulation with observations on winter and spring movements of whitetailed deer in Minnesota,* M.S. Thesis. Univ. Minnesota, Minneapolis. 47 pp.

Heezen, K. L., and Tester, J. R. (1967), Evaluation of radio-tracking by triangulation with special reference to deer movements, *J. Wildl. Mgmt.,* 31(1): 124–141.

Heinkel, K., and Gloor, M. (1966), Untersuchungen zum Alkalitest nach Noller, *Z. Gastroenterol.,* 4: 140.

Helvey, T. C. (1959), Telemetered parameters of primates and humans from space capsules, *IRE Trans. Space Electron. Tele.,* 5: 99–102.

Helvey, W. M., Albright, G. A., and Axelrod, I. (1962), Dynamic biomedical monitoring, *Aerospace Med.,* 33: 338.

Helvey, W. M., Albright, G. A., and Axelrod, I. (1964), A review of biomedical monitoring activities and report on studies on F-105 pilots, *Aerospace Med.,* 35: 23–27.

Hemmati, A. (1965), Uber die Endoradiosonde und ihre praktischen Wert in Vergleich zur fraktionierten Magensaftausheberung, *Dtsch. Arch. Klin. Med.,* 211: 1.

Hemmati, A. (1965), Uber den gastroilealen Reflex, Rontgenkinematographische und elektromanometrische Vergleichsuntersuchungen, *Fortschr. Rontgenol.*

Hemmati, A., Rothe, R., and Werner, H. (1964), Uber eine neue steuerbare Darmkapsel zur Enthahme auf der Gnundlage des von Perrenoud angegebenen Prinzips, *Z. Ges. Exp. Med.,* 138: 366–377.

Hemmati, A., and Werner, H. (1965), Eine neue Hochfrequenz steuerbare kapsel zur Entnahme von Darminhalt. II Mitteilung, *Z. Exp. Med.,* 139(6): 608–620.

Henderson, J. F., Hasler, A. D., and Chipman, G. C. (1966), An ultrasonic transmitter for use in studies of movements of fishes, *Trans. Amer. Fish. Soc.,* 95: 350–356.

Hendler, E., and Santa Maria, L. J. (1961), Response of subjects to some conditions of a simulated orbital flight pattern, *Aerospace Med.,* 32: 126.

Hendrix, C. E. (1960), *U.S. Naval ordnance test station tech. note,* 304–350.

Hendrix, C. E. (1964), Telemetry of dolphins, *Science,* 145: 727.

Henry, J. P. (1952), Animal studies of the subgravity state during rocket flight, *J. Aviat. Med.,* 23(5): 421–432.

Henry, J. P. (1954), Flight above 50,000 feet: A problem in control of the environment, *Astronautics,* 1: 12.

Henry, J. P. (1955), Physiological laboratories in rockets, *Astronautics,* 2: 22–26.

Henry, J. P. (1956), Physiological laboratories in rockets, *Bull. Med. Res.,* 10: 2.

Henry, J. P. (1958), Some correlations between psychologic and physiologic events in aviation biology, *J. Aviation Med.,* 29: 171.

Henry, J. P., et al. (1949), Animal studies of the effects of high negative acceleration, *Rept. 1, Memo Rept. No. MCREXD-695-74N, USAF,* Jan. 20.

Henssge, C. (1966), Application of telemetry in sports (in German), *Informationsdienst WTZ Dresden,* 6(7): 1–3.

Henssge, C., and Otto, K. (1963), Transmitter and receiver for direct biotelemetry of the ECG (in German), *Theorie und Praxis des Leistungssports,* 1: 96–106.

Henssge, C., and Otto, K. (1963), Leading bioelectric potentials under dynamic conditions (in German), *Theorie und Praxis des Leistungssports,* 1: 107–121.

Henssge, C. (1969), Telemetrische Untersuchungsmethoden in der Sportmedizin, *Medizin und Sport,* **5**, pp. 131–139.

Hess, O. W. (1962), Radio-telemetry of fetal heart energy, *Obstet. Gynec.,* 20: 516–521.

Hess, O. W. (1965), A system of telemetering the fetal electrocardiogram and intrauterine pressure, *Digest 6 Int. Conf. Med. Electron. Biol. Eng.* Tokyo, 220–221.

Hess, O. W. (1965), A system of telemetry for fetal electrocardiography, *Proc. Nat. Telemetering Conf.,* 133–135.

Hess, O. W., and Litvenko, E. E. (1963), Telemetering physiological data: Instrumentation of fetal EKG and radiotransmission of intrauterine pressures, *Proc. 1963 Natl. Tele. Conf.*

Hess, O. W., and Litvenko, W. (1965), Radiotelemetry of fetal heart energy, in C. A. Caceres, 117–128.

Hess, O. W., and Litvenko, W. (1967), Radio telemetry of the fetal electrocardiogram, *Digest of 7th Intern. Conf. Med. Biol. Eng.,* 98.

Hessler, W. E. (1968), *Survival of ring-necked pheasants released in selected habitats in Minnesota,* M.S. Thesis, Univ. Minnesota, Minneapolis, 44 pp.

Hester, F. E. (1968), Telemetry from wood ducks in natural environments, *item NCSU-4* in NASA CR-106344.

Hetherington, A. W. (1958), Performance assessing human capability in space, *IRE Wescon Conv. Record,* Part 5.

Hickmann, D. M., et al. (1961), A portable miniature transistorized radio-frequency coupled cardiac pacemaker, *IRE Trans. Bio-Med. Electron.*

Hill, D. W. (1966), Transmitting physiological signals by telephone, *Wld Med. Electron. Instrument.,* 4(4): 108–109.

Himwich, W. A., and Hambrecht, R. I. (1963), Telemetry systems in physiological research, *Proc. 16 Ann. Conf. Eng. Med. and Biol.*

Himwich, W. A., Knamm, F., and Steiner, W. (1965), Electrical activity of the dog's brain: Telemetry and direct wire recording, *Progr. Brain Res.,* 16: 301–317.

Himwich, W. A., and Hambrecht, F. T. (1964), Telemetry systems in physiological and pharmacological research, *Fed. Proc.,* 23: 248.

Hindson, J. C., Schofield, B., Turner, C. V., and Wolff, H. S. (1965), A radio telemetric study of intrauterine pressure in pregnant ewes up to and including parturition, *J. Reprod. Fertil.,* 10(2): 285.

Hindson, J. C., Schofield, B. M., Turner, C. V., and Wolff, H. S. (1965), Parturition in the sheep, *J. Physiol.,* 181: 560–568.

Hinkle, Jr., L. E., Carver, S., Benjamin, B., Christenson, W. N., and Strone, B. W. (1964), Studies in ecology of coronary heart disease. I. Variations in the human electrocardiogram under conditions of daily life (a preliminary report), *Arch. Environ. Health,* 9: 14–24.

Hirsch, C., Kaiser, E., and Petersen, I. (1966), Telemetry of myo-potentials. A preliminary report on telemetering of myo-potentials from implanted microcircuits for servo control of powered prostheses, *Acta Orthop. Scand.,* 37: 156–165.

Hirschberg, H. (1963), Measurements for internal pressure in unrestricted subjects (abstracts), *5 Int. Congr. Med. Electron.*, Liege, p. 29.

Hirschberg, H. (1965), A multichannel radiotelemetry system, *Digest 6 Int. Conf. Med. Electron. Biol. Eng.* Tokyo, 185–187.

Hirschberg, H. (1969), Signal to noise improvement in bio-telemetry systems, *Proc. 8th Int. Conf. Med. and Biol. Eng.*, 30–4.

Hirtzmann, M. (1962), Eine neue Methodik zur Diagnostik und Erforschung des Magendarmkanals insbesondere des Dickdarms (*1 Mittlg.*) *Med. Klin.*, 57: 1002.

Hirtzmann, M., and Reuter, G. (1963), Klinische Erfahrungen mit einer neuen, automatisch gesteurerten Kapsel zur Gewinnung von Darminhalt und bakteriologische Untersuchungen des Inhalts hoherer Darmabschnitte, *Med. Klin.*, 58: 1408.

Hixson, W. C., and Beischer, D. E. (1964), Biotelemetry of the triaxial ballistocardiogram and electrocardiogram in a weightless environment, *Monogr. No. 10, USA Naval School. Aviat. Med.*, 1–112.

Hixson, W. C., Paludan, C. T., and Downs, S. W. (1960), Primate bio-instrumentation for two Jupiter ballistic flights, *IRE Trans. Med. Electron.*, **ME-7**: 318–325.

Hoare, D. W., and Ivison, J. M. (1961), Measuring the heart rate of an active athlete, *Electron. Eng.*, 33: 6.

Hoare, D. W., Ivison, J. M., and Quazi, S. (1969), A multichannel biomedical telemetry system using delta modulation, *Proc. 8th Int. Conf. Med. and Biol. Eng.*, 30–8.

Hochberg, K., Kolig, G., and Struder, M. (1965), Die Endoradiosonde (Heidelberger Kapsel) wurde zur Beurteilung der sauresekretorischen Leistung des Magens vergleichsweise in Parallele zu anderen schlauchlosen Magensaft-Untersuchungsverfahren angewandt und diesen eindeutig uberlegen befunden. Wahrend das eine verglichene Verfahren in 69% der falle versagte, war die Versagerquote Chirurg, 36: 398.

Hochberg, K., Noller, H. G., and Kelly, T. (1964), Magenuntersuchungen mit der "Heidelberger Kapsel," *Munchn. Med. Wschr.*, 106: 789–98.

Holzer, K. H., Binzus, G., and Ritter, U. (1965), Fehler quellen bei der intragastralen pH-Meter um 1 Radiosonden, *Z. Exp. Med.*, 139(6): 589–603.

Hoffman, H., and Reygers, W. (1960), Investigations of the circulatory system of car drivers under various conditions of driving (in German), *Zbl. Verkehrsmedizin,* 6: 131–151.

Hoffmann, H. (1959), An investigation method for the behavior of the heart and the circulation of a car driver (in German), *Munch. Med. Wschr.*, 101: 638–639.

Hoffmann, H. (1961), Experimental investigation of the circulatory system of car drivers under various conditions of driving (in German), *Munch. Med. Wschr.*, 103: 2335–2338, 2385–2389.

Hoffmann, H. (1962), Experimentelle Untersuchungen des Kreislaufverhaltens von Krattfahrzeugfuhrern, *Therapiewoche,* 12(9): 335–337.

Hoffmann, H. (1961), Verkehrsmedizin, *Med. Monatsspiegel* (E. Merck, Darmstadt), 145–49.

Hoitink, N. C., Porter, N. S., and Ratcliffe, C. A. (1966), Animal physiological function telemetry, *Proc. 19 Ann. Conf. Eng. Med. Biol.*

Holden, G. R., Smith, J. R., and Smedal, H. A. (1962), Physiological instrumentation systems for monitoring pilot response to stress at zero and high G, *Aerospace Med.*, 33: 420.

Holmes, Ann C. V., and Sanderson, G. C. (1965), Populations and movements of opossums in east-central Illinois, *J. Wildl. Mgmt.*, 29(2): 287–295.

Holmes, J. R., Alps, B. J., and Darke, P. G. G. (1966), A method of radiotelemetry in equine electrocardiography, *Vet. Record,* 79: 90.

Holt, T. W., and Lamonte, R. J. (1965), Monitoring and recording of physiological data of the manned space flight program, *IEEE Trans. Suppl. to AS-3,* No. 2, June, 341–47. Abstracts: *IEEE Trans. AS-3,* 93–98.

Holter, N. J. (1957), Radioelectrocardiography: A new technique for cardiovascular studies, *Ann. N.Y. Acad. Sci.,* 65: 913–923.

Holter, N. J. (1957), Radioelectrocardiography, *Ann. N.Y. Acad. Sci.,* 65: 913.

Holter, N. J., *Progr. Rept.,* Public Health Service Research grant no. H-2614(C2).

Holter, N. J. (1961), New method for heart studies, *Science,* 134: 1211–1220.

Holter, N. J. (1963), Some perspective on telemetry in the biological sciences, in L. E. Slater (Ed.), pp. 61–64.

Holter, N. J., and Gengerelli, J. A. (1949), Remote recording of physiological data by radio, *Rocky Mt. Med. J.,* 46(9): 749.

Holter, N. J., and Gengerelli, J. A. (1950), A miniature amplifier transmitter for the broadcasting of bioelectric potentials, *117th Ann. Meet. Amer. Assoc. Sci. Bull.*

Honda, M. (1964), Inspection and diagnosis of circulatory function by the telemeter (in Japanese), *Fukuoka Acta Med.,* 55: 78–79.

Honig, W. M. (1962), Microminiature passive telemetry for biological voltages, *Digest 15th Ann. Conf. Eng. Med. and Biol.*

Hoover, G. N., and Ohanson, F. J. (1960), Problems in instrumentation for dynamic subjects, *NEC,* Oct.

Horowitz, L., and Farrar, J. T. (1962), Intraluminal small intestinal pressures in normal patients and in patients with functional gastrointestinal disorders, *Gastroenterology,* 42: 455–464.

Hoshiko, M., and Holloway, G., Radio telemetry for the monitoring of verbal behavior, *Journal of Speech and Hearing Disorders,* 33(1): 48–50.

Hospodar, J. (1964), Radiotelemetric examination of pulse reactions of pilots in flight (in Czech.), *Sborn. Ved. Prac. Lek. Fak. Karlov Univ.,* 7: 505–511.

Howells, T. H. (1965), A radiotelemetric monitor (experience gained from its construction and use), *Anaesthesia,* 20(4): 474–478.

Hoxsie, F., and Robbins, R. (1963), Miniature radio-transmitter development, 1961–1962; 1962–1963, *S. Dak. Dept. Game, Fish and Parks,* Pittman-Robertson Proj., W-75-R-4 and 5, 27 pp. mimeo.

Hudson, P. H., Meindl, J. D., and Foletta, W. D. (1969), An implantable monolithic command receiver, *Proc. 8th Int. Conf. Med. and Biol. Eng.,* 30–1.

Huszar, R. J., and Haloburdo, J. (1969), EKG transmission from emergency vehicles, *National Telemetering Conference,* 211.

Ikai, M. (1963), Some observations on the respiration and circulatory functions during running on a tread-mill, *Symp. Appli. Telem. Sys. Med.,* Tokyo (1962). Abstract in: *Med. Electron. Biol. Eng.* (1963), 1: 578.

Ingram, D. L., and Whittow, G. C. (1961), Measurement of changes in the body temperature of the ox (*Bos taurus*), *Brit. Vet. J.,* 117: 479–484.

Ira, G. H., and Bogdonoff, M. D. (1961), Clinical application of radiotelemetry in man: Continuous recording of heart rate (abstract), *Circulation,* 24: 963–964.

Ira, G. H., Burnette, J. C., Whalen, R. E., and Bogdonoff, M. D. (1962), Heart rate changes in physicians during teaching conference presentations and during cardiac catheterization, *Clin. Research,* 10: 18.

Ira, G. H., and Bogdonoff, M. D. (1962), Application of radiotelemetry in man for continuous recording of heart rate, *J. Amer. Med. Assoc.*, 180: 976–977.

Ira, G. H., et al (1964), Syncope with complete heart block. Differentiation of real and simulated Adams-Stokes seizures by radiotelemetry, *J. Amer. Med. Assoc.*, 188: 707–710.

Ishiko, T. (1964), *Kinesiological study of rowing.* Abstracts of papers presented at at ICSS, Tokyo, 161.

Ishiko, T. (1965), Application of a telemeter to the analysis of rowing movement, *Abstract 23rd Intern. Congr. Physiol. Sci.* Tokyo, 310.

Ishiko, T., and Yamakawa, J. (1963), Application of telemetry to regatta, in *Univ. Tokyo Symp. Appl. Tele. Sys. Med.*, 1962. Abstract: *Med. Electron. Biol. Eng.* (1963), 1: 579.

Iwai, Y. (1963), Electronic circuitry for telemetering systems in medicine, in *Univ. Tokyo Symp. Appl. Tele. Sys. Med.*, 1962. Abstract: *Med. Electron. Biol. Eng.* (1963), 1: 579.

Jacobsmeyer, H. T., and Cooper, T. H. (1962), Telemetry in clinical and experimental cardiovascular physiology, *Digest 15 Ann. Conf. on Eng. Med. Biol.*

Jacobson, B. (1958), Endoradiosondes, *Svenska Lakartidn.*, 55: 587.

Jacobson, B. (1959), Endoradiosonde techniques for telemetering physiological data from the alimentary canal, *Proc. Second Intern. Conf. Med. Electron.*, 300; *2 Intern. Conf. IFME* (Paris).

Jacobson, B. (1960), A critical review of techniques for telemetering physiological data from the alimentary canal, *Proc. 3 Intern. Conf. Med. Electron.*, 108–110.

Jacobson, B. (1961), Servoed antenna tracks radio "pill," *Karolinska Institut Schweden,* 119.

Jacobson, B. (1961), Antenna Servo system for tracking endoradiosonde movements, *Digest Intern. Conf. Med. Electron.*, 95.

Jacobson, B. (1962), Tracking radio pills in the human body, *New Scientist,* 13: 288–290.

Jacobson, B. (1963), Endoradiosondes techniques—a survey, *Med. Electron. Biol. Eng.*, 1: 165.

Jacobson, B., and Lindberg, B. F. M. (1960), FM receiving system for endoradiosonde techniques, *IRE Trans. Med. Electron.*, **ME-7**: 334–339.

Jacobson, B., and Nordberg, L. (1961), Endoradiosonde for pressure telemetering, *IEEE Trans. Bio-Med. Electron.*, **BME-8**: 192–196.

Jahn, J. R. (1964), The use of electrocardiographic radio telemetry to determine heart rate in ruminants, *Diss. Abstracts,* 24: 3396.

Jahn, J. R., and Dracy, A. E. (1962), Use of electrocardiographic radio telemetry, in ruminants, to determine the heart rate before, during and after pertusition, *J. Dairy Sci.*, 45: 667–668.

Jakubaschk, H. (1964), Wireless microphone with tunnel-diode transmitter (in German), *Radio u. Fernsehen,* 13(13): 393–395.

Jazdovskij, V. I., and Baevskij, R. M. (1962), Medico-biological control during space flight (in Russian), *Vestn. An SSSR,* 32(9): 9–15.

Jeffries, R. J. (1963), The frontiers of telemetry technology, in L. E. Slater (Ed.), pp. 65–74.

Jensen, R. E., Siple, W., Benson, V. G., and Squirer, R. D. (1961), *Bioinstrumentation and biotelemetry on Stratolab High-5 balloon flight,* Report No. WADC-MA-6137, Nov., Aviation Medical Acceleration Laboratory, Johnsville, Pa.

Jenssen, G. D., and Mullins, G. L. (1963), A telemetry approach to mother-infant interaction in monkeys, in L. E. Slater (Ed.), pp. 225–229.

Jeter, L. K., and Marchinton, R. K. (1964), Preliminary report of telemetric study of deer movements and behavior on the Eglin Field reservation in Northwestern Florida, *SE Assn. of Game and Fish Commissioners,* Proc. 18: 140–152.

Jeutter, D. C., and Fromm, E. (1969), A micropower subminiature implantable multichannel telemeter, *Proc. 8th Int. Conf. Med. and Biol. Eng.,* 30–10.

Johansen, K., Franklin, D. K., and Van Citters, R. L. (1966), Aortic blood flow in freeswimming elasmobranchs, *Comp. Biochem. Physiol.,* 19: 151.

Johnson, J. H. (1960), Sonic tracking of adult salmon at Bonneville Dam (1957), *Fishery Bull. 176, U.S. Fish and Wildl. Serv.,* 60: 471–485.

Johnston, E. B. (1963), Telemetry of physiological data, *Proc. Natl. Telemetry Conf.*

Johnston, E. B. (1961), Patient remote monitoring system, *Digest Intern. Conf. Med. Electron.,* 121.

Johnston, E. B. (1965), Two way telemetry for hospital use, *Proc. Intern. Telemetering Conf. Washington 1965,* Vol. 1, 215–224.

Johnston, E. B. (1966), Stimulation of the brain by remote telemetry, *Nat. Telemetry Conf. Proc.*

Johnston, E. B., et al. (1963), A multiple patient monitor for intensive care of recovery room, *Biomed. Sci. Instrum.,* 1: 273–285.

Juscenko, A. A., and Cernavkin, L. A. (1932), New radio method in psychophysiology of work (in Russian), *Socialisticeskaja Rekonstrukcija i Nauka,* No. 1: 217–220.

Juscenko, A. A., and Cernavkin, L. A. (1932), A new method for the investigation of unconditioned and conditioned reflexes in free moving animals (in Russian), *Sov. Nevropatol. Psickiatr. i Psichogig.,* No. 8: 327–332.

Kacnelson, B. A., Kedrov, B. D., and Rozenblat, V. V. (1961), Experience in radiotelemetric counting of the frequency of respiration under industrial conditions (in Russian), *Gig. i San.,* 12(11): 61–65.

Kacnelson, B. A., Kedrov, B. D., and Rozenblat, V. V. (1962), Recording of the frequency of respiration in workers under industrial conditions by radio telemetry (in Russian), *Vopr. gigieny, fiziolog. truda, prof. patolog. i toksikolog.,* Sverdlovsk, vyp. 6: 579–583.

Kacnelson, B. A., Rozenblat, V. V., Kedrov, B. D., and Ganjuskina, S. M. (1962), Some results of elements increasing the lung's ventilation in physical work under industrial conditions (in Russian), *Vopr. gig., fiziol. truda, prof. patol. promys. toksikol.,* Sverdlovsk, vyp. 6: 584–599.

Kacnelson, B. A., and Rozenblat, V. V. (1962), Peculiarities of respiratory activity during muscle work performance under industrial conditions (in Russian), *Fiziol. Zurn. SSSR,* 48(10): 1218–1224.

Kacnelson, B. A. (1963), Radiotelemetry and some problems of the physiology of respiration under actual industrial conditions (in Russian), in V. V. Parin, *Radiotelemetry in physiology and medicine,* pp. 183–185.

Kadefors, R., Kaiser, E., and Petersen, I. (1969), Energizing implantable transmitters by means of coupled inductance coils, *IEEE Trans. Bio-Med. Eng.,* July, 16: 177–183.

Kaeburn, L. (1959), *Space Canaries—implicit biological monitoring,* paper presented at American Rocket Society 14th Ann. Meeting, Washington, D.C.

Kahn, A., Ware, R. W., and Siahaya, O. (1963), A digital readout technique for aerospace biomedical monitoring, *Am. J. Med. Electron.,* 2: 152–157.

Kaminir, L. B., *Radioelectronics in biology* (Moskau), 79 pp. English translation: TT-64-11599N.

Kamp, A. (1962), Eight-channel EEG telemetry, *Am. Electroencephalographic Soc. Symp. on the EEG* in relation to Space Travel.

Kamp, A., and Van Leeuwen, S. (1961), A two-channel EEG radio telemetering system, *Electroencephalog. Clin. Neurophysiol.,* 13: 803–806.

Kanatsoulis, A. (1951), Le Tele-electrocardiography, *Presse Med.,* 59:458.

Kantrowitz, A. (1964), Implantable cardiac pacemakers, *Ann. N.Y. Acad. Sci.,* 111(3): 1049–1067.

Kanwisher, J. (1968), Man is not a seagoing animal, *Oceanus,* March, 14(1): 3–7.

Kapitanov, R. A. (1962), The status and prospects of use of means of communication, signalization and telemetry in the therapeutic procedure (in Russian), *Sovet Zdravookhr,* 21(10): 52–58.

Kaplan, M. (1964), A compact laboratory facility for psycho-physiological study of dogs, *ONR Tech. Rept.,* No. 5.

Kaplan, M., and Loesch, R. V. P. (1964), Use of telemetry to compare dog ECTs under routine and test environments, *Naval Research, Creedmor Inst. for Psychobiologic Studies. Tech. Rept.,* No. 6.

Karelina, R. N. (1963), "Radiopulsometry" during training and competition (in Russian), in V. V. Parin, *Radiotelemetry in physiology and medicine,* pp. 152–156.

Kavanagh, D., and Zander, H. (1965), A versatile recording system for studies of mastication, *Med. Electron. Biol. Eng.,* 3: 291–300.

Kavanagh, L. (1967), A subminiature crystal-controlled biological transmitter, *Digest of 7th Intern. Conf. Med. Biol. Eng.,* 96.

Kavanau, J. L. (1963), The study of social interaction between small animals, *Animal Behaviour,* 11: 263–273.

Kavanau, J. L. (1961), Identification of small animals by proximity sensing, *Science,* 134: 1694–1696.

Kavanau, J. S., and Norris, K. S. (1961), Behavior studies by capacitance, sensing, *Science,* 134: 730–732.

Kawamura, M. (1965), Capsule oscillator using esaki diode, *Digest 6 Int. Conf. Med. Electron. Biol. Eng.,* 208–209.

Kawanagh, D. (1963), Development of a miniature telemetry system, *Techn. Docum. Rep. SAM-RDR-63-80,* U.S. Air Force Electron. Sys. Div., 1–12.

Kazakov, M. B., Chodakov, N. M., and Chudorozkov, V. P. (1963), Radiotelemetric results of heart rate of athletes during training and competition (in Russian), in V. V. Parin, *Radiotelemetry in physiology and medicine,* pp. 145–451.

Keeping, G. G. (1967), Implantable instrumentation, *Industrial Research,* Apr., 72–76.

Keever, J. C., Laurda, H. C., and Barborka, C. J. (1963), The endomotorsonde: A microminiatured capsule device for studying gastrointestinal physiology (abstract), *Gastroenterology,* 44: 836.

Kellenyi, L., and Angyan, L. (1964), Electrical recording in freely moving animals (abstract), *Acta Physiol. Acad. Sci. Hung.,* 26: 45 (Suppl.).

Kelly, G. F., and Phipps, C. G. (1962), In-flight bioinstrumentation in a near space operational environment, *USN Missile Cent. Misc. Publ. NMC-MP-62-2.*

Kendall, B., Farell, D., and Kane, H. A. (1962), Fetal radioelectrocardiography: A new method of fetal electrocardiography, *Amer. J. Obstet. Gynecol.,* 83: 1629–1636.

Kenney, D. W., Elsner, R. W., and Franklin, D. L. (1966), Simple surgical approach to iliac arteries of the horse for blood flow measurement, *J. Appl. Physiol.,* 21: 705–706.

Ketz, H. A. (1963), Die Bedeutung der Endoradiosonde fur die Beterinarmedizin, *Monatsh. Vet. Med.,* 18: 645–647.

Kezdi, P., and Naylor, S. (1965), Telemetry system to transmit baroceptor nerve action potentials, *Amer. J. Med. Electron.,* 4: 153.

Kezer, C. F., and Aronson, M. H. (1959), Tiny temperature telemeter is selfblocking oscillator, *Instr. Contr. Sys.,* 32: 724.

Kharitonov, S. A. (1932), Relationship of the secretory and motor conditioned reflexes during the free movement of an animal (in Russian), *Sov. Nevropotal., Psikhiatr. i Psikhogigiyena,* 12: 766–777.

Kijs, V. E., Raudcepp, C. J., and Reeben, V. A. (1962), The "tele-intervallograph" for heart rate recording during exercise (in Russian), *II Vses. Konf. po Primeneniju Radioelektroniki v Biologii i Medicine,* NIITEIR, 71–72.

Kimmich, H. P., and Kreuzer, F. (1967), Telemetry of respiratory oxygen pressure in man during exercise, *Digest of 7th Intern. Conf. Med. Biol. Eng.,* 90.

Kitagawa, K., Nishigori, A., Murata, N., Nishimoto, K., and Takada, H. (1966), A new radio capsule with micro glass electrode for the telemetry of the gastrointestinal pH and its clinical use, *Digest 6 Int. Conf. Med. Electron. Biol. Eng.,* 216–217.

Kitamura, K. (1963), Symposium on new diagnostic procedures in internal medicine 3. Clinical application of the telemetering method (in Japanese), *J. Jap. Soc. Intern. Med.,* 52: 610–616.

Kitamura, K. (1963), The clinical application of telemetering, *Jap. J. Med. Electron. Biol. Eng.,* 1: 290–298.

Kitamura, K., Yamakura, K., Furuya, H., Ezawa, H., Minamitani, K., and Yamanaka, Y. (1963), Clinical use of telemetering (in Japanese), *Naika,* 12: 75–85.

Kinoshita, M. (1965), Telemetering of physiologic information from within the human body by a medical capsule following the method of magnetic transmission, *Digest 6 Int. Conf. Med. Electron. Biol. Eng.,* 210–212.

Klimt, F. (1964), The wireless transmission of the EEG in pediatrics (in German), *Wiss. Arbeitstagung Gesellsch. Exp. Med. DDR,* Berlin, Dec.

Klimt, F. (1964), Drahtlose Herzschlagfrequenzmessung im Kindesalter, *Dtsch. Geswes.,* 19: 1199–1203.

Klimt, F. (1964), Funk-EKG-Registrierung im Kindesalter, *Arztl. Jugendheild.,* 55: 385–392.

Klimt, F. (1964), Atmung und Kreislauf beim tatig Kind., *Med. Klin.,* 59: 1129–1131.

Klimt, F. (1965), Anwendungsmoglichkeiten des Funk-EKG in der Padiatrie, *Padiat.,* Prax. 4: 177–182.

Klimt, F. (1966), Telemetrische Herzschlagfrequenzregistrierungen bei klein Kindern wahrend einer korperlichen tatigkeit, *Dtsch. Geswes.,* 21(13): 599–605.

Knyazev, I. I., and Matov, V. V. (1962), Radiotelelectrocardiographic investigation during the performance of great physical tasks (in Russian), *Teoriya i Prakitika Fizkul'tury,* 25(6): 65–68.

Ko, W. H. (1960), *Electronics,* 33: 93.

Ko, W. H. (1963), Progress in microelectronics biotelemetry, *ISA Conf.*

Ko, W. H. (1961–1964), Microminiature muscle signal transducer development, *Final report summary.*

Ko, W. H. (1963), Micro-electronic technology and its application to bio-telemetry, in L. E. Slater (Ed.), pp. 107–116.

Ko, W. H. (1964), Progress in telemetering muscle potentials, *Bio-Med. Sci. Instr.,* 2: 267–273.

Ko, W. H. (1965), Progress in miniaturized biotelemetry, *Bio. Sci.,* 15: 118–120.

Ko, W. H., Long, C., Grotz, R. C., Yon, E., and Greene, L. (1964), Implantable radio transducers for physiological information and implant techniques, *Biomed. Sci. Instr.,* 2: 145–153.

Ko, W. H., and Newman, M. R. (1967), Implant biotelemetry and microelectronics, *Science,* 156: 351–360.

Ko, W. H., and Reswick, J. B. (1965), Miniature RM implant bio-telemetering transmitters, *Digest 6 Int. Conf. Med. Electron. Biol. Eng.,* 213–214.

Ko, W. H., Slabinski, C. J., and Yon, E. T. (1965), An implant telemetry system for the measurement of internal strain, *18 Ann. Conf. Eng. Med. Biol.*

Ko, W. H., and Slater, L. E. (1965), Bioengineering: A new discipline, *Electronics* (June 14), 111–118.

Ko, W., Thompson, W., and Yon, E. (1964), Temperature compensation of tunnel diode FM transmitters, *Med. Electron. Biol. Eng.,* 2(4): 431–433.

Ko, W. H., Will, H., and Yon, E. (1964), Microminiature signal transducer development, *Report No. EDC 3-64-6 Solid State Electron. Lab. Eng.,* Design Cent., Case Inst. Technol., Cleveland.

Ko, W. H., and Yon, E. (1964), Micro-miniature high impedance FM telemetering transmitter for bio-medical research, *Proc. Nat. Telemetering Conf.,* Session 2-2, 1–8.

Ko, W. H., and Yon, E. (1965), RF induction power supply for implant circuits, *Digest 6 Int. Conf. Med. Electron. Biol. Eng.,* 206–207.

Kobayashi, T., Takeuchi, M., Koro, T., and Tawara, J. (1963), Clinical applications for radio telemetering technique with special reference to patients with coronary artery disease, in *Univ. Tokyo Symp. Appl. Tele. Sys. Med.,* 1962. Abstract: *Med. Electron. Biol. Eng.* (1963), 1: 580.

Kobayashi, T., et al. (1963), The clinical application of telemetering (in Japanese), *J. Jap. Soc. Intern. Med.,* 52: 616–617.

Kobayashi, T., Ito, Y., Takeuchi, M., Koro, T., Tawara, I., Taneichi, Y., Kase, Y., and Nakao, O. (1965), Automatic monitoring instrument for hospitalized patients by telemetering, *Digest 6th Intern. Conf. Med. El. Biol. Eng.,* Tokyo, 368.

Kobriger, G. D. (1968), I wonder where the wild grouse goes, *North Dak. Outdoors,* 30(10): 12–16.

Kofes, A. (1960), Multi channel radio transmission of bioelectric data, *Proc. 3 Intern. Conf. Med. Electron.,* 54.

Kofes, A. (1963), The technique of telemetry of physiological data (in German), *Medizinalmarkt,* 11(10): 424–425.

Kolcum, E. H. (1961), Mercury flight provides severe test for capsule and chimp, *Aviation Week,* 74(6): 26–28.

Kolenosky, M. B., and Johnston, D. H. (1967), Radio-tracking timber wolves in Ontario, *Am. Zoo.,* 7(2): 289–303.

Kolig, G., and Vollmar, J. (1963), Funktionsuntersuchungen mit der Heidelberger pH ludoradisonde bei chirurgischen Erkrankungen des magens und der Speiserohre, *Proc. Welt Congr. Gastroenterol.,* 2: 54–57.

Konecci, E. B., and Shiner, A. J. (1965), Uses of telemetry in space, in C. A. Caceres, pp. 321–348.

Konecci, E. B., and Shiner, A. J. (1965), The developing challenge of biosensor and bioinstrumentation research, in C. A. Caceres, pp. 299–319.

Konikoff, J. J. (1967), A survey of *in vivo* energy sources, *AIBS/BIAC Information Module M9.*

Konikoff, J. J., and Toloat, M. E. (1968), Comments on biological electric power extraction from blood to power cardiac pacemakers, *IEEE Trans. Bio-Med. Eng.,* July, **ME-15**: 232.

Kooyman, G. L. (1964), Techniques used in measuring diving capacities of Weddell seals, *The Polar Record,* 12: 391–394.

Kooyman, G. (1969), The Weddell seal, *Scientific American,* 221(2): 100–107.

Kozar, A. J. (1961), *A study of telemetered heart rate during sports participation of young adult men,* Doctoral dissertation, University of Michigan.

Kozar, A. J. (1963), Telemetered heart rates recorded during gymnastic routines, *Res. Quart.,* 34: 102–106.

Kozar, A. J., and Hunsicker, P. (1963), A study of telemetered heart rate during sports participation of young adult men, *J. Sports Med. Phys. Fitn.,* Torino 3.

Krobath, H., and Bronha, L. (1965), Continuous telemetric recording of cardio respiratory functions, *Fed. Proc.,* 24: 138.

Kryazhev, V. Ya (1932), Experiment in the application of radio procedures to the study of conditioned reflexes in freely-moving animals (in Russian), *Sov. Nevropatol. Psikhiatr. i Psikhogig, Modern Neuropathology, Psychiatry, and Mental Hygiene,* 12: 778–785.

Kryazhev, V. Ya (1945), Experimental neurosis against a background of emotional shock (in Russian), *Fiziol. Zh. SSSR,* 31: 236–259.

Kryzwanek, H. (1963), Untersuchungen uber diurnal Anderungen im Ruminationsverhalten aufgestallter und weidender Schafe., *Diss. Freie. Univ. Berlin.*

Kuck, A., Liebmann, F. M., and Kussick, L. (1963), A miniature transmitter for telemetering muscle potentials, *IEEE Trans. Biol. Med. Electron.,* **BME-10**: 117–119.

Kuck, T. L. (1966), Pheasant radio transmitter study—progress report, South Dakota, 1964–1965, and 1965–1966, *S. Dak. Dept. Game, Fish and Parks,* Pittman-Robertson Proj., W-75-R-7 & 8, 31 pp.

Kuck, T. L. (1966), An improved battery hookup for radio-telemetry studies, *J. Wildl. Mgmt.,* 30(4): 858–859.

Kuck, T. L. (1967), Pheasant radio transmitter study, 1966, South Dakota, *S. Dak. Dept. Game, Fish and Parks,* Pittman-Robertson Proj., W-75-R-9, 11 pp. mimeo.

Kuck, T. L. (1968), *Spring and summer pheasant behavior as determined by radio location,* M.S. Thesis, South Dakota State Univ., Brookings.

Kuechle, L. B. (1967), Batteries for biotelemetry and other applications, *AIBS/BIAC Information Module M10,* 15 pp.

Kuh, E., Noller, H. G., and Stein, W. (1966), Magensaureproduktion bei myotonischer Dystrophie. Untersuchungen mit der Heidelberger Kapsel., *Klin. Wschr.,* 44: 53.

Kuiper, J., Bosman, J., and Boter, J. (1966), Improvement in measuring physical load by wireless transmission of the ECG, *World Med. Electron.,* 4: 304–308.

Kupa, J. J. (1964), An investigation of the motile and action responses of free-ranging white tailed deer using radio telemetry technique, *Progress Report,* Univ. Rhode Is., No. 1.

Kupa, J., and Hill (1964), Use of radio telemetry to study the habitat of deer, *Rhode Island Agric.,* June.

Kurtenbach, A. J., and Dracy, A. E. (1965), The design and application of an FM/AM temperature telemetering system for intact, unrestrained ruminants, *IEEE Trans. Bio-Med. Eng.,* **BME-12**: 187.

Kydd, W. L., and Mullins, G. (1963), A telemetry system for intraoral pressures, *Arch. Oral. Biol.,* 8: 235–236.

Lafferty, J. M., and Farrell, J. J. (1949), A technique for chronic remote nerve stimulation, *Science,* 110: 140–141.

Lamb, L. E. (1966), The influence of manned space flight on cardiovascular function, *Cardiologia,* 48: 118–134.

Lambert, R. K. (1969), A low cost telemetering system for free-range animals, *Med. and Biol. Eng.,* March, 249–252.

Larks, S. D. (1962), Regional FM phone data transmission, *Digest 15 Ann. Conf. Eng. Med. Biol.*

Laughlin, C. P. (1962), In flight measurement, transmission and assessment of astronaut physiologic responses, *Aerospace Eng.*

Lawton, W. (1959), The requirements for biomedical monitoring in space, *Proc. Nat. Telemetry Conf.*

Lebar, J. F., and Ellerbruch, V. G. (1966), A recording system for direct computer analysis of animal temperatures, *Conf. Eng. Med. Biol.,* Proc. 8: 177.

Lebed, A. N. (1964), Teleelectrocardiograph TEK-1 for all mains supply (in Russian), *Med. prom. SSSR,* 18(12): 48–49.

LeMunyan, C. D., White, W., Nybert, E., and Christian, J. J. (1959), Design of a miniature radio transmitter for use in animal studies, *J. Wildl. Mgt.* 23: 107–110.

Lepeschkin, E. (1963), Electrocardiographic instrumentation, *Progr. Cardiov. Assoc. Dis.,* 5: 498–520.

Letunov, S. P. (1950), Electrocardiography in sports medical practice (in Russian), *Moskva-Leningrad.*

Letunov, S. P., and Matov, V. V. (1962), Radio electrocardiographic investigation during high physical performances (in Russian), Mezdunarodnaja naucno-metodieskaja konferenci ja po probleman sportivnoj trenirovki Plenarnoe zasedanie, *Moskva,* 197–206.

Levine, I. M., Jossman, P. B., Tursky, B., Meister, M., and DeAngelis, V. (1964), Telephone telemetry of bioelectric information, *J. Amer. Med. Assoc.,* 188: 794–798.

Levine, I., Jossman, P., and DeAngelis, V. (1966), Monitoring spasticity—home telephone telemetry vs. direct laboratory testing, *Proc. Nat. Telemetry Conf.*

Lewin, J. E., and Perkins, W. J. (1960), Telemetering of the EEG from the conscious animal, *Proc. 3 Intern. Conf. Med. Electron.,* 73.

Liben, W. (1965), Monkeys and microelectronics, *Electronics,* 38(4): 90–93.

Lin, W. C., and Ko, W. H. (1968), A study of microwatt—power pulsed carrier transmitter circuits, *Medical and Biological Engineering,* 6: 309–317.

Lin, W. C., Ko, W. H., and Garg, B. B. (1969), Realization of modular concept in micro-power telemetry systems, *Proc. 8th Int. Conf. Med. and Biol. Eng.,* 30–2.

Lindberg, R. G. (1967), Development of instrumentation and flight hardware to study circadian periodicity in space, *Digest of 7th Intern. Conf. Med. Biol. Eng.,* 89.

Lindsey, J. R. (1965), *Computer utilization of time-line medical data from man in space flight,* US NASA 9, 1–28 (NASA Tech. Note TN D-2695).

Lion, K. S. (1963), Developments in tranducers and sensors for biological events, in L. E. Slater (Ed.), pp. 75–81.

Lion, K. (1967), The organization of instrumentation elements, *AIBS/BIAC Information Module M5.*

Long, F. M. (1962), Biological energy as a power source for a physiological telemetering system, *IEEE Conv. Record 10,* Part 9, 68–73.

Lonsdale, E., Dummire, J. D., and Brown, S. E. (1964), A transistorized self-pulsing oscillator for telemetry, *Proc. Nat. Telemetering Conf.,* Session 2-1, 1–5.

Lonsdale, E. M., Steadman, J. W., and Pancoe, W. L. (1966), A telemetering system for securing data on the motility of internal organs, *IEEE Trans. Bio-Med. Eng.,* **BME-13:** 155–59.

Loon, J. H., van (1964), Continuous pulse rate recording in strenous work, *Proc. 2 Int. Cong. Ergonomics.*

Lord, Jr., R. D., Bellrose, F. C., and Cochran, W. W. (1962), Radiotelemetry of the respiration of a flying duck, *Science,* 137: 39–40.

Lord, R. D., and Cochran, W. W. (1963), in L. E. Slater (Ed.), pp. 149–154.

Lubich, T. (1962), First results of the application of a new method telemetering and telerecording some cardiocirculatory and respiratory parameters during sporting activity (in Italian), *Medic. dello Sport.,* 2: 523–535.

Lupu, N. Z. (1963), A simple FM subcarrier oscillator suitable for physiological telemetry, *Proc. IEEE,* 51: 1621–1622.

Lutherer, L. O., Folk, Jr., G. E., and Essler, W. O. (1962), Daily activity pattern of reindeer in arctic continuous light, *Zoologist,* 2: 536.

Lutherer, et al. (1963), Daily physiological rhythms of arctic carnivores in continuous light, *Zoologist,* 3: 470.

MacInnis, H. F. (1954), The clinical application of radioelectrocardiography, *Med. Assoc. J.* (Canada), 70: 574–576.

Mack, E. (1958), An apparatus for recording heartbeat of a fetus, *IRE Trans. Med. Electron.*

Macek, O. (1963), Miniature radiotelemetric equipment for transmitting biological potentials and other data of free moving subjects (in Czech.), *Cs. Fysiol.,* 12: 367.

Macek, O., and Jilek, L. (1963), Equipment for teletransmission of biological events (in Czech.), *Cs. Fysiol.,* 12: 368.

Mackay, R. S. (1960), Radio telemetering and its application to the alimentary tract, *Proc. 3 Intern. Conf. Med. Electron.,* 115.

Mackay, R. S. (1963), Radio telemetry of physiological information from within the body, *Proc. Natl. Telemetry Conf.,* Session 5-1.

Mackay, R. S. (1963), The potential for telemetry in biological research in the physiology of animals and man, in L. E. Slater (Ed.), pp. 45–58.

Mackay, R. S. (1964), Deep body temperature of untethered dolphin recorded by ingested radio transmitter, *Science,* 144: 864–866.

Mackay, R. S. (1967), Telemetering studies from aquatic animals, *Digest of 7th Intern. Conf. Med. Biol. Eng.,* 85.

Mackay, R. S. (1964), A progress report on radio telemetry from inside the body, *Bio Med. Sci. Instr.,* 2: 275–292.

Mackay, R. S. (1965), Implanted transmitters and body fluid permeability, *IEEE Trans. Bio-Med. Eng.,* **BME-12:** 198.

Mackay, R. S. (1966), Biomedical telemetry and environmental studies, *Arch. Environ. Health,* 13: 683–685.

Mackay, R. S. (1966), Telemetering physiological information from within cetaceans, and the applicability of ultrasound to understanding *in vivo* structure and performance, in K. S. Norris (Ed.), Whales, dolphins, and porpoises, University of California Press, Berkeley, Calif., pp. 445–470.

Mackay, R. S., and Jacobson, B. (1958), Pill telemeters from digestive tract, *Electronics,* 31: 51.

Mackay, R. S., Piasecki, G. J., and Jackson, B. T. (1969), Fetal vectorcardiography, *Proc. 8th Int. Conf. Med. and Biol. Eng.,* 10–31.

Malcom, J. L. (1958), The electrical activity of cortical neurones in relation to behavior as studied with microelectrodes in unrestrained cats, *Ciba Found. Symp. Neurol. Basis Behav.* (Little, Brown, and Co.), 295–301.

Mansberg, H. P., and Hendler, E. (1960), A precision multichannel body temperature measurement system, *Digest 13 Ann. Conf. Electrical Tech. Med. Biol.,* 58.

Marchinton, R. L. (1964), *Activity cycles and mobility of central Florida deer based on telemetric and observational data,* M.S. Thesis, Univ. Florida, 101 pp.

Marcus, J. (1962), Telecommande et telemeasure, *Eyrolles Ed. Press.*

Marko, A. R., McLennan, M. A., and Correll, E. G. (1961), A multi-channel personal telemetry system using pulse position modulation, *ASD Technical Report 61-290,* Aeronautics Systems Division, Wright-Patterson Air Force Base, Ohio, July.

Marko, A. R. (1962), Respiration recording using imped changes of the chest, *Aerospace Med.,* 33: 364.

Marko, A. R. (1961), A multi-channel personal telemetry system using pulse position modulation, Proc. Nat. Telem. Conf. 1961, *Aerospace Med.,* 32: 1019–1022.

Marko, A. R., Monitoring unit for heart and respiration rate, *Aerospace Med. Div. Wright Air Developm. Div. Rep. WAAD RE 50-519.*

Marko, A. R., and McLennan, M. A. (1963), A seven channel personal telemetry system using pulse duration modulation, *Proc. 16th Ann. Con. Eng. Med. Biol.,* Baltimore, Nov. 154.

Marko, A. R. (1965), Research and development work on personal telemetry systems, *Proc. Intern. Telemetering Conf.,* Washington, Vol. 1: 253–257.

Marks, E. (1965), Use of miniature transistor oscillators in dental research, *Nature,* 206: 944–945.

Marmarou, A., Gibson, R. J., and Goodman, R. M. (1966), An automated 54-channel biological data acquisition system—12,000 hours later, *Proc. 19th Ann. Conf. Eng. Med. Biol.*

Marshall, W. H. (1963), Radiotracking of porcupines and ruffed grouse, in L. E. Slater (Ed.), pp. 173–178.

Marshall, W. H., et al. (1960), Development and use of short wave radio transmitters to trace animal movements, *Progr. Rept.,* Univ. Minn., pp. 1–27 mimeo.

Marshall, W. H., et al. (1962), Development and use of short wave radio transmitters to trace animal movements, Aug. 1960–Feb. 1962, *Progr. Rept.,* Univ. Minn., pp. 1–18 mimeo.

Marshall, W. H., et al. (1963), Studies of movements, behavior, and activities of ruffed grouse using radio telemetry techniques, *Progr. Rept.,* Univ. Minn., pp. 1–30 mimeo.

Marshall, W. H., et al. (1964), Studies of movements, behavior and activities of ruffed grouse using radio telemetry techniques, *Progr. Rept.,* Univ. Minn., pp. 1–12 mimeo.

Marshall, W. H., Cullion, G. W., and Schwab, R. G. (1962), Early summer activities of porcupines as determined by radio-positioning techniques, *J. Wildl. Mgmt.,* 26: 75–79.

Marshall, W. H., and Kupa, J. J. (1963), *Development of radio-telemetry techniques for ruffed grouse studies,* Univ. Minn., pp. 1–8 mimeo.

Marshall, W. H., and Kupa, J. J. (1963), Development of radio-telemetry techniques for ruffed grouse studies, *Trans. 28th N. Amer. Wildl., and Natl. Resources Conf.*, 443–456.

Martin, A. R. (1969), An oscillator having no reactive components for use as an integrated circuit biotelemeter, *National Telemetering Conference,* 39.

Master, A. M., and Rosenfeld, I. (1964), Monitored and post-exercise two-step test, *J.A.M.A.*, 190: 494–500.

Masuda, M., and Mihara, T. (1965), *On the arterial blood pressure changes in exercise,* abstracts of papers presented at ICBS, Tokyo, 1964, 63.

Masuda, M., and Mihara, T. (1965), Automatic indirect determination of arterial blood pressure during exchange, *Bull. Physical Fitness Res. Inst.*, Feb., No. 4: 25–33.

Masuda, M. (1965), On the upper limit and transient fluctuations of the capability of human cardiovascular system during exercises, *Abstracts 23rd Intern. Congr. Physiol. Sci.*, Tokyo, 307.

Matov, V. V. (1960), Investigation of remote electrocardiograms in athletes (in Russian), *Novosti Med. Tekhniki (What's New in Medical Technology)*, 3: 42–45.

Matov, V. V. (1959), Some problems of transients in exercise according to electrocardiograms of differently trained sportsmen recorded during sprint exercises (in Russian), *Sportivnaja medicina, Moskva,* 113–114.

Matov, V. V. (1960), Dynamics of the ECG during physical exercises (in Russian), *Diss., Moskva.*

Matov, V. V., and Surkina, I. D. (1964), The electrocardiogram during strong physical performances and ways for its analysis (as recorded by radiotelemetry) (in Russian), *Mteody Issledovanij v Sportivnoj Medicine (red. S. P. Letunov), Moskva,* 5–19.

Matsuda, K., Hoski, T., Amma, T., Yagi, K., Hayaski, H., Kammo, T., Yamagishi, S., Kamiyama, A., and Kato, A. (1963), Telemetering of respiration and cardiac rhythm, *Symp. on Appl. Telemetry Sys. Med.* Tokyo, 1962. Abstract in *Med. Electron. Biol. Eng.*, 1: 580.

Matsui, H. (1963), Measurement of heart rate of track and field athletes with radio-telemeter, *Olimpia,* No. 20.

Matsui, H., Takagi, K., Kobayaski, M., and Iwami, T. (1963), A note on applications of radio telemetering to sports medicine, *Symp. on Appl. of Telemetry Sys. to Med.* Tokyo, 1962. Abstract in *Med. Electron. Biol. Eng.*, 1: 578.

Matsui, H. (1964), The application of radio-telemeter in the study of physical activity, *Res. J. Phys. Educ.*, 7(4): 55–67.

Matsuo, T. (1959), *Experimental electronic equipment for medical telemetry* (in Japanese), National Aeronautics and Space Administration, Washington, D.C., Technical Translation no. F-51, Nov. 1960. Translation of material *No. TM-510001,* Mitsubishi Electric and Manufacturing Co., Tokyo, Nov. 20.

Mattson, R. H., and Ulstadt, M. S. (1960), A system for telemetering physiological data, *Digest 13 Ann. Conf. Electri. Tech. Med. Biol.*, 54.

Maurer, H-Chr. (1964), Telemetrie und schlauchlose Probenentnahme bei Untersuchungen des Intestinaltraktes, *Med. Markt/Acta Medico, Tech.*, 12: 92–95.

Mauro, A. (1948), Technique for *in situ* stimulation, *Yale Biophys. Bull.*, 1: 2.

Mauro, A., Wall, P. D., Davey, L. M., and Scher, A. M. (1950), Central nervous stimulation by implanted high frequency receiver, *Fed. Amer. Soc. Exp. Biol.*, Proc. 9:86.

Mayo-Wells, W. J. (1952), Radio telemetering, pp. 457–459, in C. S. White and O. O. Benson (Eds.), *Physics and medicine of upper atmosphere, a study of the aeropause,* University of New Mexico Press, Albuquerque, 611 pp.

Mazza, V. (1951), High altitude bailouts, *J. Aviation Med.,* 22: 403.

McCall, J. (1960), Receiver systems for use with radio pills, *Proc. 3 Intern. Conf. Med. Electron.,* 124.

McCally, M., and Barnard, G. W. (1962), Endoradiosondes: A state of the art survey, *AMRL-TDR 62-122.*

McDonald, M., and Piper, E. A. (1961), The recording of simultaneous events on persons during normal behaviour, *Digest of the 1961 Intern. Conf. Med. El.,* New York City, 100.

McDonald, M., Perkins, W. J., and Piper, E. A. (1963), The simultaneous recording of biological events on persons during normal behaviour. *Med. Electron. Biol. Eng.,* 1(2): 243–248.

McGinnis, S. M., and Dickson, L. L. (1967), Thermoregulation in the desert iguana *Dipsosaurus dorsalus, Science,* 156(3783): 1757–1758.

McGinnis, S. M., and Southworth, T. P. (1967), Body temperature fluctuations in the northern elephant seal, *J. of Mammalogy* (Aug. 21), 48(3): 484–485.

McLennan, M. A. (1953), Application of radio telemetering to biophysical research, *Proc. Natl. Telemetering Conf.*

McLennan, M. A. (1959), A data system for physiological experiments in satellites, *IRE Nat. Conv. Rec.,* 7(9): 3–9.

McLennan, M. A. (1960), Data versus information—a new look at the physiological data problem in the space situation, *Proc. 3 Intern. Conf. Med. Electron.,* 141.

McLennan, M. A. (1959), Physiological telemetry, *Adv. Astronaut Sci.,* 4: 420–427.

McNitt, J. I., Folk, Jr., G. E., Brewer, M. C., and Ehrenhaft, J. B. (1967), The day-night activity of polar bears exposed to continuous light (abstract), *Amer. Zool.,* 7(4): 807.

Mean, A., and Perrenoud, J. P. (1964), Pratique de la pyxigraphie, *Praxis,* 53: 383.

Mech, D. (1964), Deer 54, where are you? *Nature and Science,* 2(4): 4–6.

Mech, D. (1965), Meanwhile, back at the lab . . . , *Outdoor Life,* 136(5): 16–17, 20, 161.

Mech, L. D. (1965), A critical evaluation of the concept of home range, *Bull. Ecol. Soc. Amer.,* 46(3): 121.

Mech, L. D. (1967), Telemetry as a technique in the study of predation, *J. Wildl. Mgmt.,* 31(3): 492–496.

Mech, L. D., and Tester, J. R. (1965), Biological, behavioral, and physical factors affecting home ranges of snowshoe hares (*Lepus americanus*), raccoons (*Procyon lotos*), and white tailed deer (*Odocoileus virginianus*) under natural conditions, *Univ. Minn., Mus. Nat. Hist. Tech. Rept. #9,* 11 pp. mimeo.

Mech, L. D., Heezen, K. L., and Siniff, D. B. (1966), Onset and cessation of activity in cotton tail rabbits and snowshoe hares in relation to sunset and sunrise, *Anim. Behav.,* 14(4): 410–413.

Mech, L. D., Tester, J. R., and Warner, D. W. (1966), Fall daytime resting habits of raccoons as determined by telemetry, *J. Mammal.,* 47(3): 450–466.

Melvin, J. P. (1964), Telephone telemetry, *J. Mississippi Med. Assoc.,* 5: 84–86.

Menitskii, D. N. (1959), Place of radioelectronics in physiological research, *Fiziol. Zhur. SSSR,* 45.

Merrem, G., and Niebelling, H. G. (1965), Radiotelemetering of human EEG (in German), *Forschungen und Rottschritte,* 39: 321–325.

Merriam, H. G. (1961), Problems in woodchuck population ecology and a plan for telemetric study, *Diss. Abst.*, 22: 684.

Merriam, H. G. (1963), Low frequency telemetric monitoring of woodchuck movements, in L. E. Slater (Ed.), pp. 155–171.

Metzner, A. (1953), Bestimmung von Atmung und Kreislaufgroßen wahrend des Laufes, *Sportartztliche Praxis,* 6: 79–86.

Michael, R. P., Weller, C., and Wolff, H. S. (1965), Telemetry, brain activity and animal behaviour, *Digest 6 Int. Conf. Med. Electron. Biol. Eng.,* 201–202.

Michael, R. P., Weller, C., and Wolff, H. S. (1965), A totally implantable transmitter for telemetering the electrical activity of the brain in cats and monkeys, *J. Physiol.,* 180: 3P–5P.

Michener, M., and Walcott, C. (1965), Airplane tracking of single homing pigeons (abstract), *Bull. Ecol. Soc. Amer.,* 46(4): 204.

Michener, M. C., and Walcott, C. (1966), Navigation of single homing pigeons: Airplane observations by radio tracking, *Science,* 154(3748): 410–413.

Mikiska, A. (1961), Magnetic tape recording and radiofrequency transmission of physiological informations (in Czech, abstracts in English and Russian), *Activ. Nerv. Sup.,* 3: 92–110.

Mikiska, A. (1965), Transmitter for telemetric and tape-recording communication of the heart rhythm (abstract), *Cesk Fysiol.,* 14(5): 384–385.

Mikiska, A. (1967), Short distance radio telemetry of biopotentials in occupational medicine, *Z. Prav.-Med.,* 12(1): 48–62.

Mikiska, A., and Hyska, P. (1967), Radiotelemetry for biopotentials in occupational physiology, *Pracov. Lek.,* 19(5): 211–223.

Mikiska, A., and Hyska, P. (1967), Radio telemetry of biopotentials during industrial work, *Digest of 7th Intern. Conf. Med. Biol. Eng.,* 98.

Miles, G. H. (1962), Telemetering techniques for periodicity studies, *Ann. N.Y. Acad. Sci.,* 98: 858–865.

Miller, B. (1961), System monitors bioastronautic data, *Aviation Week and Space Technol.*

Miller, B. (1961), Simplified bio-instrumentation studies, *Aviation Week,* 74(6): 52–67.

Misin, L. N. (1961), Radio electronics in medicine (in Russian), *Sov. Zdravooshranenie, Moskva,* 20(8): 23–27.

Moen, A. N. (1967), Hypothermia observed in water-chilled deer, *J. Mammal.,* 48(4): 655–656.

Monnier, A. J., Wright, I. S., Lenegre, J., Cameron, D. J., and Coblentz, B. (1965), Ship-to-shore radio transmission of electrocardiograms and X-ray images, *J. Amer. Med. Assoc.,* 193: 1060–1061.

Montana Fish and Game Dept., Evaluation of side effects of experimental insecticides to forest grouse, *Montana Fish and Game Dept., Prog. Rept. State Project No. 32-B-9.*

Moore, A. M., and Kluth, E. O. (1966), A low frequency radiolocation system for monitoring an Armadillo population, *Nat. Telemetry Conf. Proc.*

Moore, M. L., et al. (1963), The use of radio telemetry for electromyography, *J. Amer. Phys. Ther. Assn.,* 43: 787–791.

Morimoto, T. (1965), Continuous recording of blood pH and carbon-dioxide pressure and radiotelemetry of blood pH, *J. Physiol. Soc. Japan,* 27: 15–20.

Morimoto, T. (1965), Carbon dioxide pressure electrode for circulating blood and application of implantable pH carbon dioxide pressure electrode to detection of acidosis and alkaloses, *J. Physiol. Soc. Japan,* 27(1): 21–26.

Morimoto, T. (1965), Construction of implantable pH glass electrode for recording the circulating blood pH, *J. Physiol. Soc. Japan,* 27(1): 15–20.

Morimoto, T. (1965), Radiotelemetry of circulating blood pH, *J. Physiol. Soc. Japan,* 27(1): 27–31.

Morrell, R. M. (1959), Amplitude modulation radio telemetry of nerve action potentials, *Nature,* 184: 1129–1131.

Morrell, R. M. (1959), Radio telemetry of whole nerve action potentials, 12th Ann. Conf. on Electrical techniques in Medicine and Biology, *Digest of Technical Papers,* 32–33.

Morrell, R. M. (1961), Monitoring of astronaut brain function by means of telemetered electroencephalogram with automatic analysis, *Astronautical Sci. Rev.,* 3(4): 13–19.

Morrow, P. E., and Vosteen, R. E. (1953), Pneumotachographic studies in man and dog incorporating a portable wireless transducer, *J. Appl. Physiol.,* 5: 348–360.

Moskalenko, Y. E. (1964), Telemetrical equipment for studying blood circulation of the brain, *J. Physiol.,* 172: 3.

Moulton, D. G. (1967), The measurement of spike activity telemetered from the olfactory bulb and behavioral response to odorants in rats and rabbits, in N. Tanyloac (Ed.), *Odor theories and odor measurement,* NATO Advanced Study Inst. Symp., Istanbul, 1966 Proc.

Moulton, D. G. (1967), Olfaction in mammals, *Amer. Zool.,* 7(3): 421–429.

Moulton, J. C. (1966), *Movement and activity of three-white-tailed deer during the winter of 1964–1965 in east central Minnesota determined by telemetry,* M.S. Thesis, Univ. Minnesota, Minneapolis, 82 pp.

Muller, C. (1965), Cardiotelemetry (in Norwegian), *T. Norsk Laegeforen,* 85: 628–629.

Muller, B. (1965), Medizinische Elektronik in der Luftfahrt und im Spaceprogramm, *Elektromedizin* (1964), 9: 152–155.

Mullin, G. L., Brant, D. H., and Guntheroth, W. G. (1966), Duty cycle telemetry, *Proc. 19 Ann. Conf. Eng. Med. Biol.*

Munter, M., Gotze, W., and Krokowski, G. (1964), Telemetrische EEG-Untersuchungen wahrend roratorischer Bestibularisreizing, *Dtsch. A. Nervenheilkd.,* 186: 137–148.

leMunyan, D. C., White, W., Nyberg, E., and Christian, J. J. (1959), Design of a miniature radio transmitter for use in animal studies, *J. Wildl. Mgmt.,* 23: 107–110.

Murooka, H., Fujita, K., Ouchi, A., and Watanabe, H. (1965), On the detection of ovarian changes of rabbits with radiocapsule in bioassay, *Digest 6 Int. Conf. Med. Electron. Biol. Eng.,* 218–219.

Murray, R. H., Marko, A., Kissen, A. T., and McGuire, D. W. (1968), A new miniaturized multichannel personal radio telemetry system, *J. App. Physiol.,* 24: 588–592.

Murry, W. E., and Salisbury, P. F. (Ed.) (1964), Biomedical sciences instrumentation, Vol. 2, *Proc. 2nd Nat. Biomed. Sci. Instr. Symp., Albuquerque* (New Mexico), Plenum Press, Pitt.

Mussell, L. E., Marcus, R., and Watt, J. (1964), A method of radio-location for use during gastrointestinal intubation, *Physics Med. Biol.,* 9: 73–82.

Myers, G. H., Parsonnet, V., Zucker, I. R., Lotman, H. A., and Asa, M. M. (1963), Biologically-energized cardiac pacemaker, *IRE Trans.,* **BME-10** (2): 83.

Nagumo, J. (1963), Some aspects of the echo-capsule technique, *Univ. Tokyo Symp. Appl. Telem. Syst. Med.,* 1962. Abstract in, *Med. Electron. Biol. Eng.* (1963), 1: 579.

Naraoka, K. (1963), Pattern of ECG determined by telemetering, *Res. J. Physical Educ.*, 8, No. 1.

Naraoka, K. (1963), Motor ability and heart rate in various sports training or test using telemetering, *Res. J. Physical Educ.*, 8, No. 1.

Naraoka, K. (1963), A study on heart rate in endurance exercise with radio-telemetering and determination of the training load, *Res. J. Physical Educ.*, 8, No. 1.

NASA (1965), Medical and biological applications of space telemetry, *NASA*.

Nelson, R. C. (1962), Telemetering and remote control, *Instrum. and Control System* (Feb.), 35: 71–112.

Neufield, H. N., Goor, D., Nathan, D., Fischler, H., and Yerushalmi, S. (1965), Stimulation of the carotid barceptors using a radio-frequency method, *Israel J. Med. Sci.*, 1: 630–633.

Newman, H. W., Fender, F. A., and Saunders, W. (1937), High frequency transmission of stimulating impulses, *Surgery*, 2: 359–362.

Nicholls, T. H., and Warner, D. W. (1966), Biotelemetry—a valuable tool in bird study, *The Passenger Pigeon*, 28(4): 127–131.

Nichols, G. de la M. (1966), Radio transmission of sheep's jaw movements, *N.Z. J. Agric. Res.*, 9(2): 468–473.

Nichols, G. de la M., and O'Reilly, E. D. (1966), Transmission and reception of sheep heart rate, *N.Z. J. Agric. Res.*, 9(2): 460–467.

Nickel, V., and Allen, J. (1961), Conference on external power, *Natl. Acad. Sci., Natl. Res. Council Publ.*, 874.

Nicolai, W., and Gadermann, E. (1956), Eine Methode zur Fernubertragung und Konservierung biologischer Messvorgange (Biophonar-Verfahren), *Kreislauf-forsche.*, 45(7): 293–300.

Nicolai, W. (1957), A method for telemeasurement of audio-frequencies (in German), *Intern. Elektron. Rdsch.*, 11: 8–12.

Nicolai, W. (1960), Zur Frage der Elektronischen Messung und Registrierung physiologischer Vorgange bei Korperlichen Belastung von Versuchspersonen, *Elektronik*, 1: 5–11.

Nicolai, W. (1962), The wireless electrical stimulation of tissues especially of those of the brain, *Dtsch. Gesundheitswesen*, (17): 581–582.

Niess, O. K. (1960), Medicine in the aerospace age, *U.S. Armed Forces Med. J.*, 11(1): 27–37.

Niess, O. K. (1961), The Air Force medical service, *Military Med.*, 126(1): 23–24.

Nixon, W. C. W., and Smyth, C. N. (1960), Instrumentation in obstetric practice, *Inter. Conf. Med. Electron., Proc.* 3(1): 318.

Noble, F. W., T. C. Goldsmith, C. J. Waldsburger, and T. C. Cook (1962), *Digest of 15th Conf. on Eng., Med. and Bio.*, The Conference Committee, Chicago, U.S.A.

Noble, F. W. (1966), *Proc. Inst. Elect. Electron, Eng.*, 54, 1976–1978.

Noller, H. G. (1959), Die Endoradiosondentechnik und ihre Bedutung fur die innere Medizin, *Berh. Ot. Ges. Inn. Med.*, 65: 727–730.

Noller, H. G. (1960), Clinical experience in telemetering gastric condition, *Inter. Conf. Med. Electron.*, Proc. 3: 111–114.

Noller, H. G. (1961), Endoradiosondes for the determination of oxygen content, *Dig. 4th Int. Conf. Med. Electron.*, Proc. 4: 92.

Noller, H. G. (1961), Endoradiosondes for oxygen measurement: Structure, function and practical clinical significance, *Inter. Conf. Med. Electron.*, Proc. 4: 94.

Noller, H. G. (1959), The techniques of measuring by endoradiosonde and their adaptation to pediatrics, *Proc. 2 Int. Conf. Med. Electron.*

Noller, H. G. (1960), Die Endoradiosonde. Zur elektronischen pH-Messung im Magen und ihre klinische Bedeutung, *Dtsche. Med. Wschr.*, 85: 1707.

Noller, H. G. (1960), Die Endoradiosonde ein Mittel zur routinemaßigen, fortlaufenden schlauchlosen Ermittlung des Saurewertes und der Pufferaktivitat im Magen, *Verh. Dtsch. Ges. Inn. Med.*, 66: 496.

Noller, H. G. (1962), Ergebnisse der Magenfunktionsermittlung mit der Endoradiokapsel, einem neuen Hilfsmittel der Magendiagnostik, *Forschr. Med.*, 80: 3310.

Noller, H. G. (1964), Elektronische Hilfsmittel, *Therapiewoche,* 14: 254–260.

Noller, H. G. (1964), Sauresubstituierende Wirkung von Mazur C, intragastral nachgewiesen mit der "Heidelberger Kapsel," *Therapiewoche,* 14: 269.

Noller, H. G. (1964), Der "Heidelberger Kapsel" ein Helfsmittel der Magenfunctions ermittlung und ihr Einsatz zum intragastralen machwiss der antaciden wirkung des Dihydroqaluminiumnatriumkarbonats (compensan), *Med. Will.,* 64: 1203–1206.

Noller, H. G. (1964), Wiretap on gastric acidity, *World Med. News.,* 3: 27.

Noller, H. G. (1964), Hb-Bestimmung auf elektronischem Wege, *Verhandl. der Dtsch. Ges. Med.,* 70: B.212.

Noller, H. G. (1965), "Heidelberger Kapsel" bestatig: Kein "Acid rebound" mit Neutrilac, *Therapiewoche,* 15: 47.

Noller, H. G. (1966), Nit der Endoradiosonde bestatigte Noller die uberlegene sauersubstituierende Wirkung von Citropepsin, *Arztl. Praxis,* 18: 751.

Noller, H. G., and Buchheim, E. (1964), Die antazide Wirkung von Magnesiumaluminathydrat, *Med. Klinik,* 59: 342–345.

Noller, H. G., and Khodabakhsh, G. (1964), Die Sauregildungsleistung des Magens und ihre Individualstrenung, *Fortschritte Med.,* 82: 264–268.

Nomura, S. I. (1962), Telemetry of electrocardiograms of racing horses and jockeys, in *Symp. Appl. Telem. Syst. Med., Tokyo* (1962). Abstract: *Med. Electron. Biol. Eng.* (1963), 1: 578.

Nomura, S. I. (1966), Adaptation of radiotelemetry to equestrian games and horse racing, *Jap. J. Vet. Sci.,* 28: 191–205.

Norland, C. C., and Semler, H. (1964), Angina pectoris and arrhythmias documented by cardiac-telemetry, *J. Amer. Med. Assoc.,* 190: 115–118.

Norris, K. (1963), Preparations for radio-telemetry of the body temperature of large reptiles, in L. E. Slater (Ed.), pp. 283–287.

North American Aviation, Inc. (1963), *Telemetering systems for the Life Sciences—a survey,* Downey, Calif., Mar. 15.

Oka, Y. (1963), Telemetering system in sports medicine (1963), *Univ. Tokyo Symp. Appl. Telem. Syst. Med.* (1962). Abstract: *Med. Electron. Biol. Eng.* (1963), 1: 577.

Oka, Y., Utsuyama, H., Sekine, K., and Noda, U. (1961), A method of the short wave wireless carriage of ECG during muscular exercise, *J. J. Physical Fitness,* 10: 68.

Oka, Y., Utsuyama, H., and Noda, U. (1961), A method of the short wave wireless duplex carriage of ECG and respiratory movement during muscular exercise, *J. Jour. Physical Fitness,* 10: 75.

Oka, Y., Utsuyama, N., Noda, K., and Kimura, M. (1963), Studies of radio telemetering on EKG and respiratory movements during running, jumping and swimming, in *Univ. Tokyo Symp. Appl. Telem. Syst. Med.*, 1962. Abstract: *Med. Electron. Biol. Eng.* (1963), 1: 578–579.

Oka, Y. (1964), *Method of telecontrolled underwater transportation and its application,* Presented at ICSS, Tokyo.

Olbe, L., and Jacobson, L. (1963), Intraluminal pressure waves of the stomach in dogs studied by endoradiosondes, *Gastroenterology,* 44: 787–796.

Oliver, H. (1963), Long distance transmission of electrocardiographical data (in Italian), *Prog. Med.* (Napoli), 19: 672–674.

Olsen, E. R., Collins, C. C., Altenhofen, T. R., Adams, J. E., and Richards, V. (1968), *Amer. J. Surg.*, July, 116: 3–7.

Olsen, E. R., Collins, C. C., Loughvorough, W. F., Richards, V., Adams, J. E., and Pinto, D. W. (1967), Intracranial pressure measurement with miniaturized passive implanted pressures transensor, *Amer. J. Surgery,* June, 113: 727–729.

Ota, H. (1965), We're wiring our chickens for sound! *Broiler Industry,* 28: 39–44.

Ottenjann, R. (1963), Gleichzeitige endogastrale Aziditatsmessung mittels Glaselektrode und Endoradiosonde, *Med. Klinik,* 58: 1999.

Ottenjann, R. (1966), Die endogastrale Aziditatsmessung, *Fortschr. Med.,* 84: 373.

Otto, K. (1960), Wireless transmission of physiological parameters, *Technical Digest,* Feb., 60–63.

Otto, K., and Boernert, D. (1965), Implantable transmitters for transmission of bioloigcal parameters (in German), *Informationsdienst WTZ Dresden,* 5 (Sonderheft 2): 12–15.

Ouchi, A., et al. (1963), "Spurt" (Heart Rate Telemeter model 101) (in Japanese), *NEC,* 61.

Ouchi, A., and Kumano, M. (1963), Endoradiosondes and their technical problems, *Univ. Tokyo Symp. Appl. Telem. Syst. Med.* (1962). Abstract: *Med. Electron. Biol. Eng.* (1963), 1: 580.

Ovcinnikov, E. F., Akulinicev, I. T., Baevskij, R. M., Beljaiv, A. A., and Sacrincev, I. S. (1963), Telemetric systems with successive data transmission (in Russian), in V. V. Parin, *Radiotelemetry in physiology and medicine,* pp. 40–44.

Oviatt, M. C. (1961), Bio-Electronic telemetry system, *13 Southwestern IRE Conf. Electron. Show.*

Oyuki, Y., Seki, M., and Yanagita, N. (1964), *ECG of human body in exercise,* abstracts of papers presented at ICSS, Tokyo, 68.

Pachomov, R. R., Cernavkin, L. A., and Juscenko, A. A. (1932), Equipment for recording of the drop secretion under the conditions of a free roaming and free moving animal (in Russian), *Sov. Nevropatol. Psichiatr. Psichogig.,* No. 12: 764–765.

Palti, Y. (1966), Stimulation of internal organs by means of externally applied electrodes, *J. Appl. Physiol.,* 21: 1619–1623.

Paludan, T. U. (1960), Thermistors, the key to biomedical measuring in space, *Electron. Industries,* 19.

Panin, B. V. (1956), Instrument for the remote recording of the work of the internal organs of sheep at a distance (in Russian), *Karakulevodstvo i Zverovodstvo, Karakul Sheep and Animal Raising,* 5: 62–64.

Panin, B. V. (1958), Radioteleactography—a new method of investigating physiological functions in man and animals (in Russian), *Nauchn. Trudy N.-i In-Ta Karakulevodstva,* scientific works of the Scientific Research Institute of Karakul Sheep Raising, 7: 320–336.

Panin, V. B., and Musatov, G. I. (1959), Radioteleactography as a method of graphical registration of physiological functions by radio (in Russian), *Materialy III (Zonal'noy) Nauchno-Praktich, Konf. po Vrachebn. Kontrolyu i Lechebn. Fizkul'ture* (Materials of the Third (Zonal) Scientific-Practical Conference on Medical Control and Therapeutic Physical Culture), II: 96–101.

Parin, V. V. (1959), Some results and prospects for the use of radioelectronics in medicine and biology (in Russian), *Vestn. Acad. Med. Plauk SSSR,* 5: 27–39.

Parin, V. V. (1960), Use of radioelectronics in medicine and biology (in Russian), *Elektronika v Meditsine* (A. I. Berg, Ed.), Moscow-Leningrad, 19–37.

Parin, V. V. (Ed.) (1963), *Radiotelemetry in physiology and medicine* (in Russian), Materialy vtorogo simpoziuma 9–11 dekabrja 1963 g, Sverdlovsk.

Parin, V. V. (1963), Problems of biological telemetry. Abstract: *Aerospace Med.,* 607.

Parin, V. V. (1965), Electronics in space biology and medicine, Suppl. II to *Digest 6th Intern. Conf. Med. El. Biol. Eng.,* Tokyo, 1–9.

Parin, V. V., and Babskii, B. (1964), The main trends and immediate prospects of application of modern technical equipment in medicine, *Vestnik. Akad. Med. Nauk SSSR,* 19: 2.

Parin, V. V., and Baevskij, R. M. (1963), Classification of biotelemetric system (in Russian), in V. V. Parin, *Radiotelemetry in physiology and medicine,* pp. 5–9.

Parin, V. V., and Jazdovskij, V. I. (1961), Advances in space physiology in the Soviet Union (in Russian), *Fiziol. Zurn. SSSR,* 47(10): 1217–1226.

Parin, V. V., and Rozenblat, V. V. (1964), Second symposium on biotelemetry, held in Sverdlovsk (in Russian), *Fiziol. Zurn. SSSR* (Sept.), 50: 1191–1193.

Parin, V. V., and Jazdovskij, V. I. (1961), Means of soviet space physiology (in Russian), *Fiziol. Ah. SSSR,* 47(10): 1217–1726.

Parin, V. V., Baevskij, R. M., and Gazenko, O. G. (1965), Heart and circulation under space conditions, *Acta Cardiol., Brux.* 7(2): 105–129.

Parker, C. S., Breakell, C. C., and Christopherson, F. (1953), The radioelectrophysiologogram. Radiotransmission of electrophysiological data from the ambulant and active patient, *Lancet,* 1:1285.

Parrot, J. L., Dinanian, J., Billot, Y., and Thouvenot, J. (1956), Inscription de la frequence cardiaque instantanee, Possibilité de transmission a distance par ondes courtes, *J. Physiol.* (*Paris*) 48:758–59.

Parsonnet, V., Myers, G., Zucker, I. R., and Lotman, H. (1964), The potentiality of the use of biologic energy as a power source for implantable pacemakers, *Ann. N.Y. Acad. Sci.,* 111(3): 915–921.

Parsonnet, V. (1965), Stimulation parameters for carotid sinus pacemakers, *Proc. 18 Ann. Conf. Eng. Med. Biol.,* 1965, 210.

Patric, E. F., Longacre, A., and Doan, R. J. (1965), A modified approach to animal position finding, *Huntington Wildl. Forest, State U. College Forestry, Syracuse U.,* New York, 4 pp. mimeo.

Payne, L. C. (1960), Simple telemetering system for signaling high rumen pressures, *Science,* 131: 611–612.

Peiss, C. N., McCook, R. D., Rovick, A. A., and Randall, W. C. (1960), Electronic instrumentation in a medical physiology laboratory, *J. Med. Educ.,* 35: 660–663.

Peleska, B. (1964), Biotelemetric (C_2), *Cesk. Fysiol.,* 13: 146–164.

Penney, R. L. (1965), Some practical aspects of penguin navigation-orientation studies, *BioScience,* 15(2): 268–270.

Penney, R. L., and Emlen, J. T. (1967), Further experiments on distance navigation in the adelie penguin, *Pysogcelis adeliae, ibis,* 109: 99–109.

Perrenoud, J. P., and Possel, J., and Wagner, J. P. (1964), Caracteristiques techniques de la pyxigraphie, *Praxis,* 53: 379.

Pessar, T., Krobath, H., and Yanover, R. R. (1962), The application of telemetry to industrial medicine, *Amer. J. Med. Electron.,* 1: 287–293.

Petrovick, M. L. (1962), Detection of glossal motion patterns, *15 Ann. Conf. Eng. Med. Biol.*

Petrovick, M. L., and Wolfson, A. (1966), A doppler head movement sensor for migratory birds, *Proc. 19th Ann. Conf. Eng. Med. Biol.*

Phibbs, B. (1965), Monitored exercise electrocardiograms, *Amer. J. Cardio.,* 15: 738–740.

Pienkowski, E. C. (1962), *The development of radio equipment to be used for tracking small animals,* Thesis, Ohio State Univ., vii–79.

Pienkowski, E. C. (1962), Miniature 6 meter transmitter, *QST,* 58.

Pinc, B. W., and Ettleson, B. L. (1961), Development of an internalized animal telemetry system, *Aerospace Med.,* 32: 244.

Pinc, B. W., and Foster, D. L. (1967), Detection of oxygen consumption rhythms in the potato with a closed environment control system, *Digest of 7th Intern. Conf. Med. Biol. Eng.,* 84.

Pircher, L. (1964), An instrument for registration of heart frequency at the working place during sport activity, *Helv. Physiol. Pharmacol. Acta,* 22: C4–6.

Plas, F., and Talbot, P. (1962), Electrocardiography in sport medicine, *J. Sports Med. Phys. Fitness,* 2(3): 141–151.

Plisnier, H. (1962), Clinical study of the action of antitussive agents by means of an acoustic telewriter (in French), *Bruxelles Med.,* 42: 1409–1417.

Pocztarski-Bielecka, J. (1964), The application of the telemetric method to the physiologic studies of physical efforts (in Polish), *Wychowanie Fizyczne i Sport,* 8(2): 185–190.

Podar, G. K. (1963), Remote recording of periods of employed legs during skating (in Russian), in V. V. Parin, *Radiotelemetry in physiology and medicine,* pp. 109–111.

Pokorovski (1956), *Study of the vital activity of animals during rocket flights into the upper atmosphere,* International Congress of Rockets and Guided Missiles, Paris.

Poszner, W. (1965), Some remarks on telemetric problems in industrial hygiene (in German), *Z. Aerztl. Fortbildung,* 59: 1084–1087.

Potocko, R. J. (1963), *Selected bibliography on biotelemetry,* Bibliography no. Q497, Man Machine Information Center Documentation, Inc.

Potor, G., and Marko, A. R. (1960), Physiological instrumentation in high altitude bailout (Project Excelsior), *American Rocket Society Publication,* 1427–60, New York.

Potter, A. G., and McMechan, J. D. (1962), A pulse-rate modulation system for biomedical radiotelemetry, *Digest 15 Ann. Conf. Eng. Med. and Biol.*

Potter, A. G., and McMechan, J. D. (1963), Designing a simple telemeter for medical research, *Electronics,* 36(4): 47–49.

Powell, J. T. (1959), Biomedical measuring circuitry, *Proc. Natl. Telem. Conf.,* 308–320.

Powell, R. N. (1963), *Tooth contact during sleep,* Thesis, Univ. of Rochester.

Pravosudov, V. P. (1960), Remote recording of the frequency of heart contractions in bicyclists (in Russian), Tez. Dokl. Nauch. Conf. GDOIFK im. Lesgafta po Itogam Raboty za 1959 (Chair of Physiology, Sports, Medicine, and Hygiene) 13–14 Leningrad 1960, Tez. Dokl. Konf. po Vopr. Fiziologii Sporta (summaries of reports of the conference on problems of sports physiology), 163–164 Tbilisi.

Proud, J. C. (1968), *Wild turkey studies in New York by radio telemetry,* M.S. Thesis, Cornell Univ., Ithaca, 88 pp.

Proud, J. C. (1966), How and where the turkeys roam, *The Conservationist,* 20(5): 6–8.

Purpuro, C., Kilduff, J., Welkowitz, W. M., and Traite, M. (1959), Physiological transducers for measurements in space vehicles, *Proc. Nat. Telem. Conf.,* 301.

Pyrah, D. (1967), Ecological effects of chemical and mechanical sagebrush control, *Montana Dept. Fish and Game,* Pittman-Robertson Proj., W-105-R-1, 15 pp. mimeo.

Quinn, Jr., F. B., and Brechner, V. O. (1961), Selection of physiologic measurements, *Planetary and Space Science,* 7.

Rader, R., and Griswold, K. (1965), Carrier and subcarrier oscillators for implanted transmitters, *18th Ann. Conf. Eng. Med. Biol.*

Rader, R., and Meehan, J. P. (1966), Multiple channel physiological telemetry data acquisition system for restrained and mobile subjects, *Nat. Telem. Conf. Proc.*

Rader, R., Homs, E., Meehan, J. P., Henriksen, J., and Griswold, K. (1966), A miniature EEG transmitter, *Proc. 19th Ann. Conf. Eng. Med. Biol.*

Raffel, I. (1961), Ein neuer Intertinal empfanger, *Nachrichtentechnik,* 11: 441–444.

Ragland, Jr., S., (1964), *A discussion of medical monitoring in relation to safety in centrifuge operations,* NADC-ML-6410, U.S. Naval Aviat. Med. Accel. Lab, 1–8.

Rahm, E. L. (1961), The design and construction of a radio telemetry unit for the study of ruminal pressures, M.S. Thesis South Dakota S.C.

Ramorino, M. L., and Lepri, F. (1961), Studies on gastrointestinal motility; radio-telemetered recordings combined with synchronized cinefluorography, *Intern. Conf. Med. Electron.*

Ramorino, M. L., et al. (1964), Intestinal motility: Preliminary studies with telemetering capsules and synchronized fluorocinematography, *Amer. J. Dig. Dis.,* 9: 64–71.

Ramseth, D., Yon, E. T., and Ko, W. H. (1969), A multiple channel integrated circuit biomedical telemetry system, *Proc. 8th Int. Conf. Med. and Biol. Eng.,* 30–7.

Rauterkus, T., Feltz, J. F., and Fickes, J. W. (1966), Frequency analysis of korotkov blood pressures sounds using the fourier transform, Technical Report SAM-TR-66-8, *U.S. Air Force School Aerospace Med.,* 1–46.

Rawson, K. S., and Hartline, P. H. (1964), Telemetry of homing behaviour by the deer, mouse, permyscus, *Science,* 146: 1596–1597.

Rawson, R. O., Stolwijk, J. A. J., Graichen, H., and Abrams, R. (1965), Continuous radiotelemetry of hypothalmic temperatures from unrestrained animals, *J. Appl. Physiol.,* 20: 321–325.

Ray, C. D., and Bickford, R. G. (1963), Electroencephalographic and response-averaged bioelectric data transmitted via telephone, *Link. Proc. Natl. Telem. Conf.*

Ray, C. D., Bickford, R. G., Walter, W. G., and Remond, A. (1965), Experiences with telemetry of biomedical data by telephone, cable and satellite: Domestic and international, *Med. Electron. Biol. Eng.,* 3: 169–177.

Rebiffe, R., and Tarriere, C. (1964), Telecompteur de frequence cardia avec inscription numerique, *Proc. 2 Int. Cong. Ergonomics.*

Reinhardt, Microminiaturized electronics in biology and medicine, *Am. Biophy. Res. Lab. Monogr. Ser. No. 1.*

Reswick, J. B., and Ko, W. H. (1964), Micro-miniature muscle signal transducer development, *Final Report Summary,* Case Institute of Technology, Cleveland, Apr.

Rimskich, E. I. (1963), Some results on recording receivers for biotelemetry and ways for constructing a portable test device (in Russian), in V. V. Parin, *Radiotelemetry in physiology and medicine,* pp. 82–88.

Rimskich, E. I. (1963), A radiotelemetric receiver based on a commercial one (in Russian), in V. V. Parin, *Radiotelemetry in physiology and medicine,* pp. 89–93.

Robbins, K. E., and Marko, A. (1962), An improved method of registering respiration rate, *Digest 15 Ann. Conf. Eng. Med. Biol.*

Robbins, R., and Lee, O. B. (1964), Chess game with a muley, *S. Dak. Conserv. Dig.,* 31: 2–3.

Robel, R. J. (1965), *Use of radio-telemetry in movement studies of greater prairie chickens in Kansas,* abstract of paper presented on March 9, 1965 at the Radio-Telemetry Session of the 30th N. Am. Wildl. and Natural Resources Conf., Wash., D.C.

Robel, R. J. (1965), Function of territory size in mating success of booming *Tympanachus cupido pinnatus.* Abstract, *Bull. Ecol. Soc. Amer.,* 46(4): 175.

Robinson, B. W., Warner, H., and Rosvold, H. E. (1964), A head-mounted remote-controlled brain stimulator for use on rhesus monkeys, *Electroenceph. Clin. Neurophysiol.,* 17: 200–203.

Robinson, B. W., Warner, H., and Rosvold, H. E. (1965), Brain telestimulator with solar cell power supply, *Science,* 148: 1111–1113.

Robinson, S. G. (1964), A temperature telemetering system with constant accuracy, *Med. Electron. Biol. Eng.,* 2: 81–83.

Robinson, S. G. (1964), Temperature telemetry for small pigs, *J. Physiol.,* 175: 28 pp.

Robrock, II, R. B. (1965), *A six channel physiological radio telemetry system,* M.S. Thesis, Case Inst. Technology, Cleveland, Ohio.

Robrock, II, R. B., and Ko, W. H. (1967), A six channel physiological telemetering system, *IEEE Trans. Bio-Med. Eng.,* **BME-14**: 40–46.

Rokushima, H. (1969), A multichannel PWM/FM radio-telemetry system for EEG, *Proc. 8th Int. Conf. Med. and Biol. Eng.,* 30–9.

Roman, J. A., War, R. W., Adams, R. M., Warrne, B. H., and Kahm, A. R. (1962), School of Aerospace Medicine physiological studies in high performance aircraft, *Aerospace Med.,* 33: 412–419.

Roman, J. A., and Lamb, L. E. (1962), Electrocardiography in flight, *Aerospace Med.,* 33: 527–544.

Roman, J. A. (1963), Cardiorespiratory functioning in-flight, *Aerospace Med.,* 34: 322–337.

Romonino, M. L., and Colagrande, C. (1964), Intestinal motility: Preliminary studies with telemetering capsules and synchronized fluorocinematography, *Amer. J. Dig. Dis.,* 9: 64.

Rose, K. D., et al. (1964), Telemeter electrocardiography: A study of heart functions in athletes, *Nebraska Med. J.,* 49: 447–456.

Rose, K. D. (1964), Physiology of running studied by use of radiotelemetry, *J. Sport Med. Phys. Fitness,* Torino 4, 3: 188.

Rose, K. D., and Dunn, F. L. (1964), The heart of the spectator sportsman, *Med. Times,* 92: 945–951.

Rose, K. D., and Dunn, F. L. (1964), A study of heart function in athletes by telemetered electrocardiography, *Proc. 5th Ann. Nat. Conf. Medical Aspects of Sports, Amer. Med. Assn.,* Chicago, 30–37.

Rose, K. D. (1965), Telemetering physiological data from athletes, *Proc. Intern. Telemetering Conf. Washington,* Vol. 1: 225–241.

Rosenfeld, I., Master, A. M., and Rosenfeld, C. (1964), Recording the electrocardiogram during the performance of the Master two-step test: part, *Circulation,* 29(2): 204–211.

Ross, B., Watson, B. W., and Kay, A. W. (1963), Studies on the effect of vagotomy on small intestinal motility using the radio-telemetering capsule, *Gut,* 4: 77–81.

Rothfeld, E. L., Bernstein, A., and Crews, Jr., A. H., et al. (1965), Telemetric monitoring of arrhythmias in acute myocardial infarction, *Amer. J. Cardiol.,* 15: 38–44.

Rowen, B. (1959), Human factors support of the X-15 program, *Aerospace Med.,* 30: 816–820.

Rowen, B. (1959), Aeromedical support of the X-15 program (in English, French, Italian), *II Congresso mondiale e IV Europeo di medicina aeronautic e spaziale,* Roma.

Rowen, B. (1960), Physiological telemetry in research aircraft, *J. Appl. Nutrition,* 13(4): 189–192.

Rowen, B. (1961), *Biomedical monitoring of the X-15 program,* AFFTC-TN-61-4, Air Force Flight Test Center, Edwards Air Force Base, Cal., May.

Rowen, B. (1961), Aeromedical support of the X-15 program, *Aerospace Med.,* 32: 246.

Rowen, B. (1961), *Bioastronautics support of the X-15 program,* FTC-TDR-61-6, Air Force Flight Test Center, Edwards Air Force Base, Cal., Dec.

Rowen, B., and Roman, J. A. (1962), *Optimization of biomedical monitoring capabilities,* Presented at Meeting of Institute of Aerospace Sciences, New York City, Jan.

Rowen, B. (1962), Bioastronautics support of the X-15, *Aerospace Med.,* 33: 350.

Rowlands, E. N. (1960), Application of electronic techniques to the study of diseases of the gastrointestinal tract, *Inter. Conf. Med. Electron.,* Proc. 3: 106–107.

Rowlands, E. N., and Wolff, H. S. (1960), The radio pill. Telemetering from the digestive tract, *Br. Commun. Electron.,* 7:598–601.

Roy, O. Z. (1966), Telemetry on a shoe-string, *Med. Biol. Eng.,* 4: 599–602.

Roy, O. Z., and Hart, J. S. (1963), Transmitter for telemetry of biological data from birds in flight, *IEEE Trans. Bio-Med. Electron.,* **BME-10**: 114–116.

Roy, O. Z., and Hart, J. S. (1966), A multi-channel transmitter for the physiological study of birds in flight, *Med. Biol. Eng.,* 4: 457–466.

Rozenblat, V. V. (1959), Modern state of the problem of the use of radiotelemetry for obtaining physiological information (in Russian), *Materialy III (Zonal'noy) Nauchno-Praktich. Konf. po Vrachebn. Kontrolyu i Lechebn. Fizkul'ture,* II: 73–81.

Rozenblat, V. V. (1960), Radio observation of certain functions of the athlete during the performance of physical exercises (in Russian), *Tez. Dokl. Konf. po Vopr. Fiziologii Sporta,* summaries of reports at the conference on Problems of Sport Physiology, 173–76.

Rozenblat, V. V. (1960), On the value of the radiotelemetric procedure in investigations in the field of labor hygiene (in Russian), *Tez. Dokl. III Nauchn. Konf. po Vopr. Fiziologii Truda,* summaries of reports of the Third Scientific Conference on Problems of Labor Physiology, Moscow, 129–130.

Rozenblat, V. V. (1961), The dynamical radiotelemetry in the medicine of sport, *Digest Int. Conf. Med. Electron.,* 127.

Rozenblat, V. V. (1961), The problem of fatigue (in Russian), *Medgiz.*

Rozenblat, V. V. (1962), The state and means of development of dynamic radiotelemetry (in Russian), *Materialy Konf. po Metodam. Fiziol. Issledovaniy Cheloveka,* Materials of the Conference on Methods of Physiological Investigations of Man, 156–159.

Rozenblat, V. V. (1962), On the frequency of heart contractions under natural conditions of muscular activity of man (according to the data of dynamic radiotelemetry) (in Russian), *Fiziol. Zh. SSSR,* 48(12): 1454–1465.

Rozenblat, V. V. (1963), Dynamic biotelemetry (in Russian), *Usp. Sovrem. Biol.,* 56: 341–364.

Rozenblat, V. V. (1963), Some principles in constructing radiotelemetric equipment (in Russian), in V. V. Parin, *Radiotelemetry in physiology and medicine,* pp. 35–39.

Rozenblat, V. V. (1963), The use of radiotelemetry in industrial and sports physiology and some results of the Sverdlovsk biotelemetric team (in Russian), in V. V. Parin, *Radiotelemetry in physiology and medicine,* pp. 14–23.

Rozenblat, V. V. (1964), Dynamic biotelemetry, *U.S. Dept. Commerce,* Mar. 27.

Rozenblat, V. V. (1964), On the use of radiotelemetry in physiological research on work and sport (in Russian), *Vestn. Akad. Med. Nauk SSSR,* 19: 66–71.

Rozenblat, V. V. (1966), *Radiotelemetric researches in sports physiology and sports medicine* (in Russian), abstracts of papers presented at ICSS, Tokyo 1964, 88–89.

Rozenblat, V. V. (1966), Radiotelemetric researches in sports physiology and sports medicine (in Russian), *Moskva, Medicina.*

Rozenblat, V. V., and Dombrovskij, L. S. (1957), On means of studying the functional state of an athlete during the performance of physical exercises (with a demonstration of the radiopulsophone—an instrument of radiorecording of the frequency of heart contractions in a freely-moving person) (in Russian), *Materialy II Nauchno-Praktich. Konf. po Vopr. Vrachebn. Kontrolya i Lechebn. Fizkul'tury,* 112–123.

Rozenblat, V. V., and Dombrovskij, L. S. (1959), On the recording of heart rate of sportsmen during training by means of the "Radiopulsophon" (in Russian), *Sportivnaja Medicina.* Trudy XII jubilejnogo mezdunarodnogo kongressa sportivnof mediciny, Moskva 28 maja–4 ijunja 1958 g., Moskva, Medgiz, 576–577.

Rozenblat, V. V., and Dombrovskij, L. S. (1959), Registration by radio of the frequency of heart contractions in a freely-moving person (in Russian), *Fiziol. Zh. SSSR,* 45(6): 718–724.

Rozenblat, V. V., and Dombrovskij, L. S. (1962), Methods of studying the functional condition of sportsmen during exercise (with demonstration of the radiopulsophon —an equipment for wireless recording of heart rate of free-moving persons) (in German), *Theorie und Praxis der Korperkultur,* II: 628–633.

Rozenblat, V. V., and Vorobev, A. T. (1961), On a procedure for drawing off the heart biocurrents in man during dynamic radiotelemetry (in Russian), Byull,

Eksperim. Biol. i Med., *Bulletin of Experimental Biology and Medicine,* 52(10): 119–122.

Rozenblat, V. V., and Vorobev, A. T. (1962), A laboratory for radio electronics serving sports medicine (in Russian), Bjull, ucenogo medicinskogo soveta ministerstva zdravoochranija RSFSR, *Moskva,* No. 4: 24–27.

Rozenblat, V. V., Vorobev, A. T., and Sanachev, L. M. (1961), Materials of radiotelemetric investigations of the pulse frequency in sport physiology (in Russian), *Sb Dokl. II Nauchn. Konf. Fiziologiv, Biokhimikov i Farmakologiv Zap Sibiri, Posvyashch. XXII Syezdu KPSS.* Collection of reports at the Second Scientific Conference of Physiologists, Biochemists, and Pharmacologists of Western Siberia (Dedicated to the 22nd Party Congress), 123–124 Tomsk.

Rozenblat, V. V., Vorobev, A. T., and Unzhin, R. V. (1962), First experiment in radioelectrocardiography in skaters during competitions (in Russian), Teoriya i Praktika Fiz. Kultury, *Theory and Practice of Physical Culture,* 25(10): 62–68.

Rozenblat, V. V., Kazakov, M. B., and Vorobev, A. T. (1966), Biotelemetric observations in sports medicine, "S" Beitrage XVI. *Weltkongress fur Spormedizin,* Hanover, Nr. 179.

Rozenblat, V. V., Dombrovskij, L. S., Unzhin, R. V., and Vorobev, A. T. (1962), Apparatus for dynamic radiotelemetry in investigations in labor and sport physiology (in Russian), *Tez. Dokl. II Vses, Konf. po Primeneniyu Radioelektroniki v Biologii i Meditsine.,* NIITEIR, 65–67.

Rozenblat, V. V., Dombrovskij, L. S., Unzhin, R. V., Vorobev, A. T., Karmanov, G. L., Rimskich, E. I., and Forstadt, V. M. (1963), Diagrams of equipment for dynamic radiotelemetric measurements in industrial and sports physiology (in Russian), *Biol. i Med. Elektronika,* 2: 34.

Rubenstein, L. (1962), Continuous radio telemetry of human activity, *Nature,* 193: 849–850.

Russ, R. F., and Wolff, H. E. (1960), Constructional aspects of radio pills suitable for mass production, *Proc. Third Int. Conf. Med. Electron.,* 122.

Saito, M. (1963), Telemetry systems (in Japanese), *Medicine,* 20: 261–264.

Sakamoto, T. (1963), Technical problems in telemetering systems for medical purposes, *Univ. Tokyo Symp. Appl. Tele. Sys. Med.* (1962). Abstract: *Med. Electron. Biol. Eng.* (1963), 1: 579.

Sakamoto, T., Saito, M., and Sakai, T. (1963), A comparison of multiplexing systems for telemeters in medicine, *Univ. Tokyo Symp. Appl. Tele. Sys. Med.* (1962). Abstract: *Med. Electron. Biol. Eng.* (1963), 1: 579.

Sakamoto, T., Saito, M., and Hayashi, S. (1965), Telephone transmission of ECG data, *Digest 6 Int. Conf. Med. Electron. Biol. Eng.,* 197–198.

Salmon, H., and Shapiro, M. (1965), An electronic method for telemetering cough, *18 Ann. Conf. Eng. Med. Biol.*

Salmons, S. (1969), Implantable devices for stimulation and telemetering tension from skeletal muscles, *Proc. 8th Int. Conf. Med. and Biol. Eng.,* 30–5.

Samojlov, G. V., Peskov, E., and Mjazdrikov, V. A. (1960), A method for recording the main physiological parameters of man over a distance by means of radiotelemetry (in Russian), *Voenno-med. Zurn.,* 2: 70–73.

Sanderson, G. C., and Sanderson, B. C. (1964), Radio-tracking rats in Malaysia—a preliminary study, *J. Wildl. Mgmt.,* 28: 752–768.

Sanderson, G. C. (1963), *Radio-tracking rats in Malaysia—a preliminary study,* Armed Serv. Tech. Inform. Agency, Arlington Hall Station, Arlington, Va., 17 pp.

Sanderson, G. C. (1965), Studies of white-tailed deer in spread of zoonoses, *Annual Progress Rept.*, Project CC 00052-01 (CDC), Illinois Nat. His. Survey, Urbana.

Sanderson, G. C. (1966), The study of mammal movements—a review, *J. Wildl. Mgmt.*, 30(1): 215–235.

Sargeant, A. B., Forbes, J. E., and Warner, D. W. (1965), Accuracy of data obtained through the Cedar Creek automatic radio tracking system, *Univ. of Minn. Museum Nat. Hist. Tech. Rept. #10,* 20 pp. mimeo.

Sarychev, S. P. (1958), Electromyographic investigations during the process of training by the remote recording of biocurrents (in Russian), *XII Yubil. Mezhdunar. Kongress Sport Meditsiny* (Ref. Soobshch), 12 Anniversary International Congress of Sport Medicine, abstracts of reports, Moscow, 38–39.

Sarychev, S. P. (1959), Radioelectromyographic investigations in the process of athletic training (in Russian), *Sportivnaya Meditsina.* Trudy XII Yubil. Mezhdunar. Kongressa Sport. Meditsiny Sports Medicine. Proceedings of the 12 Anniversary International Congress of Sports Medicine, Moscow, 133–36.

Sarychev, S. P. (1959), Radiotelemetric investigation on oarsmen at changing rowing speeds in training and competition (in Russian), *Tez. i Ref. ZNIIFK,* Leningrad.

Sarychev, S. P. (1959), Radiotelemetry in sports investigations (in Russian), *Materialy III (Zonal'noy) Naucho-Praktich. Konf. po Brachebn. Kontrolyu i Lechebn. Fizkul'ture,* II: 91–95.

Sarychev, S. P. (1962), Radiotelemetric methods for investigation in sports physiology (in Russian), *Mezdunarodn. nauchno-methodich. knof. po probl. sport. trenirovki, plenarnoe zasedanie,* Moskva, 183–197.

Sarychev, S. P. (1960), Radiotelemetric investigations of rowers during variable rates of work (in Russian), *Tez. Dokl. Konf. po Vopr. Fiziologii Sporta.,* 186–188.

Sarychev, S. P. (1962), Radiotelemetric method of examining athletes (in Russian), *Materialy Konf. po Metodam Fiziol. Issledovaniy Cheloveka,* 163–164.

Sarychev, S. P., Davydov, M. S., Lazareva, A. M., and Tichvinskij, S. B. (1960), Radiotelemetric examination of rowers of the USSR combined team (Kayak, Canoe) (in Russian), *Tez. i Ref. Dokl. Itogovoj Konf. LNIIFK,* 50–53.

Sarychev, S. P., Davydov, M. S., Lazareva, A. M., Tichvinskij, S. B., and Sabkov, J. T. (1962), On physiological characteristics of distances at academic rowing (radiotelemetric investigation) (in Russian), *Tez. i Ref. Dokl. Itogovoj Konf. LNIIFK* (26–28 dekabrja 1961 g.), Leningrad, 57–60.

Sarychev, S. P., et al. (1963), A six-channel radio telemetry system for investigation in academic rowing (in Russian), *Tez. i Ref. ZNIIFK,* Leningrad.

Sarychev, S. P., Lazareva, A. M., and Shapkov, U. T. (1964), *Biotelemetric investigations in sports medicine,* abstracts of papers presented at ICSS, Tokyo, 86–87.

Sasamoto, H., Hosono, K., Nakamura, Y., Shimada, H., Ogino, T., and Nagoshi, H. (1965), Wired tele-electrocardiography, *Digest 6 Int. Conf. Med. Electron. Biol. Eng.,* 199–200.

Savage, J. C. (1966), Telemetry and automatic data acquisition systems, pp. 69–98, in K. E. F. Watt (Ed.), Systems analysis in ecology, Academic Press, New York, 276 pp.

Savulescu, V., Popescu, D. W., and Teodorescu, P. (1962), Microradiosound in the investigation of the motoricity and pH of the gastrointestinal tract (preliminary note) (in Roumanian), *Med. Intern. (Buchur),* 14: 1009–1016.

Schaff, G., and Schieber, J. P. (1960), Testing the physical fitness by continuous recording of heart rate and respiration (in French), *Med. Educ. Phys. et Sport,* 34(4): 255–259.

Schaffer, K. E. (Ed.) (1964), *Bioastronautics,* Collier-Macmillan, New York, London.

Scharer, P., and Stallare, R. E. (1965), The use of multiple radio transmitters in studies of tooth contact patterns, *Peridontology,* 5: 5–9.

Schevill, William E., and Watkins, William A. (1966), Radio-tagging of whales, *Woods Hole Oceanographic Institution,* Tech. Rept., Reference No. 66-17, 14 pp.

Schicketanz, W. (1964), Der Ablouf der Magenverdaung m ach plotzlichen Futterumstellungen bei Schwein und seine Ermittlung und den pH-Intestinalsender (abstract), *Monatsh. Vet. Med.,* 19 (Sonderheft 48).

Schicketanz, W. (1964), Der Alkalitest zur Megenfunktionsprufung beim Hund mit Hilfe des pH-Intestinalsenders, *Monasch. Vet. Med.,* 19: 466–469.

Schladweiler, Philip (1965), *Movements and activities of ruffed grouse (Bonasa umbellus (L.)) during the summer period,* M.S. Thesis, Univ. Minnesota, Minneapolis, 106 pp.

Schladweiler, Philip (1968), Feeding behavior of incubating ruffed grouse females, *J. Wildl. Mgmt.,* 32(2): 426–428.

Schladweiler, R., and Mussehl, T. W. (1967), Use of radio-telemetry during forest grouse-pesticide studies, *Ann. Conf., Northwest Section, The Wildlife Society,* p. 5.

Schlotthauser, B., and Noller, H. G. (1964), Ergebnisverfalschung bei der Magenuntersuchung, bedingt durch die Schlauchtechnik, *Munch. Med. Wschr.,* 106: 785.

Schmal, R. L., Goodson, C. M., Pfunke, P. C., Tucker, C. H., and Zollner, M. J. (1965), Telephone systems, in C. A. Caceres, pp. 67–82.

Schmidt-Koenig, K. (1963), The problems of distant tracking in experiments in bird orientation, in L. E. Slater (Ed.), *Bio-telemetry,* pp. 119–124.

Schmidt, F. L. (1962), The heart rate measurement by wireless transmission during exercise on fields and in a rowing-basin (in German), *Sportarzt,* 13: 294–301, 338–340.

Schmitt, O. H. (1965), A beginning in biotelemetry, Ch. 1, in C. A. Caceres, *Biomedical telemetry,* Academic Press, New York, 392 pp.

Schmitt, O. H., and Caceres, C. A. (1964), *Electronic and computer-assisted studies of bio-medical problems,* Charles C Thomas, Springfield, Ill.

Schneider, D. G., Mech, L. D., and Tester, J. R. (1966), An eight month radio-tracking study of racoon families (abstract), *Bull. Ecol. Soc. Am.,* 47(3): 149–150.

Schock, G. J. D., and Castillo, H. T. (1960), A multi-channel physiological acquisition system for bio-astronautical research, *Digest 13 Ann. Conf. Electr. Techn. Med. Biol.,* 48.

Scholl, H. (1959), Microsender und Empfangsgerat zur fortlaufenden Pulszahregistrierung, *Internat. Z. Angew. Physiol.,* 17(6): 485–489.

Schuchentanz, W., and Bornert, D. (1963), Die Registrierung der wasserstoffionenkonsentration in Verdaurungs kanal bei Hund und Schwein mit Hilfe der Intestinalsendermethode, *Monatsh. Vet. Med.,* 18: 653–657.

Schuder, J., et al. (1962), Theoretical analysis of micromodule pacemaker receiver, *Digest 15 Ann. Conf. on Eng. Med. Biol.*

Schuder, J. C., and Stephenson, Jr., H. E. (1965), Energy transport to a coil which circumscribes a ferrite core and is implanted within the body, *IEEE Trans. Bio-Med. Eng.,* **BME-12**: 154–163.

Schwartz, B. (1961), Remote recording of electrocardiograms from patients in motion, *Inter. Conf. Med. Electron.,* Proc. 4: 122.

Schwartz, B. (1963), Telemetry, *Amer. J. Med. Electron.,* 2: 181.

Scott, I. (1966), A six-channel intraoral transmitter for measuring occlusal forces, *J. Prosth. Dent.,* 16: 56–61.

Seliger, V. (1950), Tepova frekvence u bezechych vykonu a ve startovnim stavu (in Czech.), *Ceskoslov. Fisiol.,* 1: 264–271.

Seliger, V. (1966), Metabolism of volleyball, soccer, basketball and ice-hockey players, "S" Beitrage XVI, *Weltkongress fur Sportmedizin,* Hannover, Nr. 192.

Seliger, V., and Hrdlicka, J. (1965), Wireless transmission of pulse frequency, *Prac. Lek.,* 17: 109–111.

Sem-Jacobsen, C. W. (1959), Airborne EEG recording as a means of studying and selecting pilots and crews in high performance aircraft and space vehicles, *Aerospace Med.,* 30: 202.

Sem-Jacobsen, C. W., et al. (1964), Collection of biological information during prolonged flight missions with "yes and no" data reduction analysis, *Aerospace Med.,* 35: 880–883.

Semler, H. J. (1965), Radiotelemetry during cardiac exercise tests (Ch. 8), in C. A. Caceres, *Biomedical telemetry.*

Semler, H. J., and Gustafson, R. H. (1963), Postural effects on the radioelectrocardiogram simulating myocardial ischemia during exercise (abstract), *Circulation,* 28: 802.

Shillito, E. E. (1966), A method for recording the effect of drugs on the activity of small mammals over long periods of time, *Brit. J. Pharmacol.,* 26: 248–256.

Shinner, M., Water, T. E., and Gray, J. D. A. (1963), Culture studies of the gastrointestinal tract with a newly devised capsule, *Gastroenterology,* 45: 625.

Shipp, L. M. (1965), Electronics and medicine, *J. Occup. Med.,* 7: 423–430.

Shipton, H. W. (1960), A simple telemetering system for telemetering electrophysiological data, *Electroencephalog. Clin. Neurophysiol.,* 12: 922.

Shiraishi, N. (1965), Studies on the circulatory and respiratory changes in a long distance running by radio telemetering of CTG, ECG and respiratory movement, *Abstracts 23 Intern. Cong. Physiol. Sci.,* Tokyo, 307.

Shiraishi, N. (1965), *Studies on the circulatory and respiratory changes in a long distance running by radio telemetering of CTG, ECG and respiratory movement,* Department of Hygiene, Nihon University School of Medicine, Tokyo, Sept.

Shiraishi, N., and Shirahata, A. (1964), *Studies on the radio telemetering system of biological information and its application in sports medicine,* Abstracts of papers presented at ICSS, Tokyo, 132.

Shiraishi, N., and Shirahata, A. (1964), *Studies on the radio telemetering system of biological information and its applications in sports medicine,* Department of Hygiene, Nihon University School of Medicine, Tokyo, Oct.

Shiraishi, N., Shirahata, A., Nakamura, K., and Shiraishi, Y. (1967), Telemetering system for measuring minute volume, tidal volume and respiratory frequency, *Digest of 7th Intern. Conf. Med. Biol. Eng.,* 93.

Shook, G. L., and Folk, G. E. (1965), Body moisture and the operating life of implantable heartrate transmitters, *IEEE Trans. Bio-med., Eng.,* **BME-12** (1): 44–46.

Shuvator, L. R. (1958), Application of miniature radio-telemetering equipment for registration of certain physiological functions, *Proc. Conf. Appl. Radioelectronics Med. Biol.* (Moscow), p. 75.

Shuvatov, L. P. (1959), Radiotelemetric apparatus for investigating the pulse, respiration, and biocurrents of the muscles in children during physical exercises (in Russian), *Materialy III (Zonal'noy) Nauchno-Praktich. Konf. po Vrachebn. Kontrolyu i Lechebn. Fizkul'ture,* II: 82–86.

Shuvatov, L. P. (1959), Midroapparatura dlya Registratsii po Radio Nekotorykh Fiziologicheskikh Funktsiy (microapparatus for the recording of certain physiological functions by radio), *Medgiz. Moskow.*

Shuvatov, L. P. (1960), Miniature Radiotelemetric apparatus for the recording of physiological functions (in Russian), *Elektronika v Meditsine,* 177–182.

Sidorowicz, W. (1964), The pulse rate during running (in Polish), *Lekka Atlet.,* Warszawa, 5: 11–12.

Sidorowicz, W. (1965), Tele-electrocardiographic investigations—warming up and evaluation of the physical fitness (in Polish), *Lekka Atlet.,* Warszawa, 5: 12–14.

Sildmjae, C. J. (1962), The investigation of the circulatory system during skiing (in Russian), *Materialy 7-j Naucn. Konf. po Vopr. Morfol., Fiziol. i Biochim. Myshechn. dejateln, Moskva,* 258–259.

Sildmjae, C. J. (1963), Radiotelemetric recording of the dynamic changes of heart rate during skiing (in Russian), in V. V. Parin, *Radiotelemetry in physiology and medicine,* pp. 164–167.

Simmons, K. R., Dracy, A. E., and Essler, W. O. (1965), Recording uterine activity by radio telemetry techniques, *J. Dairy Sci.,* 48: 1126–1128.

Simmons, K. R., Dracy, A. E., and Essler, W. O. (1965), Diurnal temperature patterns in unrestrained cows, *J. Dairy Sci.,* 48: 1490–1493.

Simons, D. G. (1958), *Psychophysiological aspects of the manhigh experiment,* Sealed Cabins Session 13th Ann. Meeting ARS.

Simons, D. G. (1959), *Manhigh I,* Technical Report AFMDC-TR-59-24, Holloman Air Force Base, New Mexico.

Simons, D. G., *Manhigh II,* AFMDC Technical Report 59-28, Holloman Air Force Base, New Mexico.

Simons, D. G. (1963), Psychophysiological approach to the study of fatigue in space flight, *Proc. 12th Intern. Astronaut. Congr.,* Washington 1961, Wien-New York-London, 581–583.

Simons, D. G. (1962), Use of personalized radio telemetry techniques for physiological monitoring in aerospace flight, *J. Miss. St. Med. Assn.,* Illiniois, 413–420.

Simons, D. G. (1962), A personalized radiotelemetry system for studying CNS function in aerospace flight, *Digest 14 Ann. Conf. Eng. Med. Biol.*

Simons, D. G., and Burch, N. R. (1963), EEG telemetry and automatic analysis under simulated flight conditions: A comparation study of analytic techniques, *Amer. Meeting EEG Soc.* (1962). Abstract in: *Electroenceph. Clin. Neuroph.* (1963), 15: 165–166.

Simons, D. G., Flinn, D. W., and Hartman, B., *Psychophysiology of high altitude experiences. Unusual environments and human behavior,* N. M. Burns and R. M. Chambers (Eds.), School of Aerospace Med., Brooks AFB, Tex.

Simons, D. G., and Johnson, R. L. (1965), Heart rate patterns observed in medical monitoring, *Aerospace Med.,* 36: 504–513.

Simons, D. G., and Prather, W. (1964), A personalized radio telemetry system for monitoring central nervous system arousal in aerospace flight, *IEEE Trans. Bio-Med. Eng.* (1964), **BME-11:** 40–51.

Simons, D. G., and Prather, W. E. (1965), An operational portable biomedical monitoring system, *Proc. Nat. Telemetering Conf.,* 59–64.

Simons, D. G., Prather, W., and Coombs, F. K. (1965), The personalized telemetry medical monitoring and performance data-gathering system for the 1962 SAM-

MATS fatigue study, *Tech. Rept. SAM-TR-65-17,* 1965 USAF School of Aerospace Medicine, Aerospace Med. Div., Brooks Air Force Base, Tex., 1–29.

Simpson, D. C. (1962), A patient-monitor, *Lancet,* 1: 759–760.

Simpson, D. C., and Greening, J. R. (1965), Patient monitoring, *Phys. Med. Biol.,* 10: 1–16.

Singer, A. (1963), Some solutions to the problems of homing, in L. E. Slater (Ed.), pp. 125–132.

Singer, A., Thomas, R., and Spector, B. (1964), An automatic system for the measurement and recording of basal temperature in the human female, *Fertility Sterility,* 15: 44–51.

Siniff, D. B. (1966), Computer programs for analyzing radio tracking data, *Univ. of Minn. Mus. Nat. Hist. Tech. Rept. #12,* 22 pp. mimeo.

Siniff, D. B. (1967), *A simulation model of animal movement patterns,* Ph.D. Thesis, University of Minnesota, Minneapolis, 170 pp.

Siniff, D. B., and Tester, J. R. (1965), Computer analysis of animal-movement data obtained by telemetry, *BioScience,* 15(2): 104–108.

Siniff, D. B., Tester, J. R., and Kuechle, L. B. (1969), Population studies of Weddell seals at McMurdo Station, *Antarctic Journal,* IV(4): 120–121.

Sipple, W. C., Squires, R. D., Jensen, R. E., and Gordon, J. J. (1961), Miniaturized physiological telemetry systems, *Aerospace Med.,* 32: 247.

Sisakyan, N. M., and Yazdovskiy, V. I. (1962), Pervyye Kosmicheskiye Polety Cheloveka (in Russian), *Nauchnye Rezul'taty Mediko-Biologicheskikh Issledovaniy Provedennykh vo Vremya Orbital'nykh Poletov Korablev-Sputnikov "Vostok-1" i "Vostok-2"* (First space flights of Man. Scientific results of medical-biological investigations conducted during orbital flights of the spaceships "Vostok-1" and "Vostok-2"), Moscow.

Sisakyan, N. M. (Ed.), Problems of space biology (collected reports) (in Russian), *Moskva "Nauka,"* Vols. 1–4, 1962–1965.

Sisakyan, N. M. (Ed.), Problems of space biology, *NASA Technical Translation F-174,* Washington, D.C.

Skubic, V., and Hilgendorf, J. (1964), Anticipatory, exercise, and recovery heart rates of girls as affected by four running events, *J. Appl. Physiol.,* 19: 853–856.

Skutt, H. R., Beschle, R. G., Moulton, D. G., and Koella, W. P. (1967), New sub-miniature amplifier transmitters for telemetering biopotentials, *Electroenceph. Clin. Neurophysiol.,* 22: 275–277.

Slade, N. A., Cebula, J. J., and Robel, R. J. (1965), Accuracy and reliability of biotelemetric instruments used in animal movement studies in prairie grasslands of Kansas, *Trans. Kansas Acad. Sci.,* 68(1): 173–179.

Sladen, W. J. L., Boyd, J. C., and Pedersen, J. M. (1966), Biotelemetry studies on penguin body temperatures, *Antarctic Journal,* I: 142–143.

Slagle, A. K. (1963), *An electronic system for the study of the movements of wild animals,* M.S. Thesis, University of Missouri, Columbia, 107 pp.

Slagle, A. K. (1965), Designing systems for the field, *BioScience,* 15(2): 109–112.

Slater, L. E. (1963), The future of micro-miniaturization in medicine, *5 Int. Cong. Med. Electron.,* Liege, abstract: p. 26.

Slater, L. E. (1963), The future of micro-miniaturization in biological sensing systems, *Proc. 5 Int. Conf. Med. Electron.,* 16 pp.

Slater, L. E. (1964), What's ahead in biomedical measurements?, *ISA Journal,* Feb., 55–60.

Slater, L. E. (1964), Some facts about bio-telemetry, *Case Inst. of Tech.*, Cleveland, 23 pp. mimeo.

Slater, L. E. (1965), A broad-brush survey of biomedical telemetric progress, in C. A. Caceres, pp. 377–382.

Slater, L. E. (1965), Biotelemetry and the physician, *N.Y. State J. Med.*, 65: 2893–2901.

Slater, L. E. (1965), The role of bio-telemetry in predictive medicine, Preprint Nr. 4, *20 Ann. Instrum. Soc. Amer. Conf.*

Slater, L. E. (1965), Introduction, as guest editor (volume on Bio-telemetry), *Bio-Science,* 15: 81–82.

Slater, L. E., The challenge of bio-instrumentation, *ISA J.*

Slater, L. E. (1966), Telemetering from inside the body, *Proc. Nat. Telem. Conf.*

Slater, L. E. (1966), A sampling of some current projects in implant bio-telemetering, *Bioinstrumentation Advisory Council,* Washington, D.C.

Slater, L. E., Kilpatrick, D. G., and Bellet, S. (1965), Telemetering of the human electrocardiogram underwater, *18 Ann. Conf. Eng. Med. Biol.*

Slater, A., Bellet, S., and Kilpatrick, D. G. (1969), Instrumentation for telemetering the ECG data analysis, Apr. **BME-16**(2): 148–151.

Sloan, I. A. (1967), An improved tele-thermometer, *Canad. Anaesth. Soc. J.*, 14(1): 62–63.

Smirnov, G. V., Tichvinskij, S. B., and Vasilev, N. L. (1963), A method for recording the ventilation volume by teletransmission (in Russian), *Materialy 4-j Nauchn. Konf. po Vopr. Fiziol. Truda,* Leningrad, 304–305.

Smith, A. M., and Ridgway, M. (1961), The management of recording pressure change in the gastrointestinal tract using a telemetering capsule, *J. R. Coll. Surg. Edinb.*, 6:192–198.

Smith, A. M., and Ridgway, M. (1962), The use of telemetering capsules in disorders of the alimentary tract. 1. A technique for recording pressure changes, *Gut,* 3: 366–371. 2. The application to the study of human gastrointestinal motility, *Gut,* 3: 372–376.

Smith, A. M., and Ridgway, M. (1963), Recording pressures in the gastrointestinal tract using a telemetering capsule, *Bull. Soc. Int. Chir.*, 22: 97–102.

Smith, D. D. (1961), Telemetering physiological data, *Naval Res. Rev.*, Oct.

Smith, G., and Lamb, L. (1960), Vectorcardiography in aerospace flight, *Amer. J. Cardiol.*, 6: 63–69.

Smith, Jr., P. E. (1963), Instrumentation for work physiology studies, in L. E. Slater (Ed.), pp. 363–369.

Smyth, C. N. (Ed.) (1960), Medical electronics, *Proceedings of the Second International Conference on Medical Electronics,* Paris. Iliffe: London, 614 pp.

Sokol, K. (1962), The wireless transmission of fetal heart sounds (in German), *Wien. Klin. Wschr.*, 74: 738.

Sokol, K. (1963), A new equipment for wireless transmission of fetal heart sounds (in German), *Medizinalmarkt,* 11: 66.

Solonin, J. G. (1963), Using the radiopulsometry under the condition of industrial production during heavy physical work and under the conditions of micro-climate getting warm (in Russian), in V. V. Parin, *Radiotelemetry in physiology and medicine,* pp. 177–182.

Sorin, A. M. (1962), Radiotelemetric system for investigating the human gastro-intestinal tract (in Russian), *Tez. Dokl. II Vese. Konf. po Primeneniyu Radio-elektroniki v Biologii Meditsine,* NIITEIR, 62–64.

Southern, W. E. (1963), The application of radio telemetry in bald eagle studies, *Prog. Rept. 1,* North. Illinois Univ.

Southern, W. E. (1963), *The application of bio-telemetry in avian navigation research,* presented Sept. 9, 1963, at 18th Ann. Instru. Soc. of Amer. Conf., Bio-telemetry Session, 6 pp. mimeo.

Southern, W. E. (1964), *Equipment and techniques for using radio-telemetry in wild-life studies,* North. Illinois Univ., Dept. Biol. Sci., Rept. No. 3, 33 pp. mimeo.

Southern, W. E. (1964), Additional observations on winter bald eagle populations: Including remarks on biotelemetry techniques and immature plumages, *The Wilson Bulletin,* 76(2): 121–137.

Southern, W. E. (1965), Avian navigation, *BioScience,* 15(2): 87–88.

Southern, W. E. (1965), Biotelemetry: A new technique for wildlife research, *The Living Bird,* 4: 45–58.

Specht, D. (1965), Telemetische Untersuchungen der Atumfrequenzen von Restpferden in allen gangarhn und beim springen, *Vet. Diss.,* Berlin.

Specht, D. F., and Drapkin, P. E. (1965), Biomedical data compression, *Proc. Nat. Telemetering Conf.,* 68–74.

Spencer, W. A., and Vallbona, C. (1965), Application of computers in clinical practice, *J. Amer. Med. Assoc.,* 191: 917–921.

Spencer, W. A., White, S. C., Geddes, L. A., and Vogt, F. B. (1964), The impact of electronics on medicine: Part 2, *Postgraduate Med.,* 36: 516–523.

Squires, R. D., Gordon, J. J., Jensen, R. E., and Sipple, W. (1961), The remote monitoring of physiological data from personnel in flight, *Aerospace Med.,* 32: 248.

Sperry, C. S., Gadsden, C. P., Rodriguez, C., and Bach, L. M. N. (1961), Miniature subcutaneous frequency modulated transmitter for brain potentials, *Science,* 123: 1423–1424.

Sprung, H. B. (1958), Der verschluckbaren Intestinalsender und seine ersten Messergebnisse, *Arch. Klin. Chir.,* 289: 605.

Sprung, H. B. (1959), Uber Messergebnisse mit dem verschluckbaren Intestinalsender in Magen-Darmkanal, *Arch. Klin. Chir.,* 291: 80–97.

Sprung, H. B. (1959), Uber die Registrierung der normalen und pathologischen Dunndarmmotilitat mit dem Intestinalsender, *Arch. Klin. Chir.,* 292: 537.

Sprung, H. B. (1959), Zur Problematik des verschluckbaren Intestinalsender Festschrift zur 5 Jahrfeier der Medizinischen akademie "Carl Gustav Carus" Dresden, *Schriften der Medizinischen Akademie Dresden,* 1: 93.

Sprung, H. B. (1959), Uber mit dem verschluckbaren Intestinalsender gweonnene Messungen im Magen-Darm-Kanal, *Medsche Bild,* 2:97.

Sprung, H. B. (1960), Clinical results of the application of the intestinal transmitter, in Smyth, C. N. (Ed.), *Medical Electronics* (Iliffe: London), 281–295.

Sprung, H. B. (1960), Uber Messungen der Motilitat im Oesophagus vermittels des Verschluckbaren Intestinalsenders, *Proc. 3rd. Int. Conf. Med. Electron.,* London, 1: 128–131.

Sprung, H. B., and Roisch, R. (1960), Uber die Registrierung der Dunndarmotilitat mit dem Intestinalsenders, *Gastroenterologia,* 93: 145–157.

Stander, R. W., et al. (1963), Telemetry of physiological data during parturition, *Proc. Natl. Telem. Conf.*

Stander, R. W., Hagan, W. K., and Bolner, M. H. (1965), A system for analog-to-digital conversion, transmission and storage of physiological data, *Digest 6 Int. Conf. Med. Electron. Biol. Eng.,* 190.

Stattelman, A., and Buck, W. (1965), A transmitter for telemetering electrophysiological data, *Proc. Soc. Exp. Biol. Med.,* 119: 352–356.

Stattelman, A., and Cook, H. (1966), A transducer for motion study by radio telemetry, *Proc. Soc. Exp. Biol. Med.,* 121: 505–508.

Staubitz, W., Cheng, S., Gillen, H., Holmquist, B., Zurlo, P., and Greatbatch, W. (1966), Management of neurogenic bladder in paraplegic dogs by electrical stimulation of peloric nerves, *Invest. Urol.,* 4: 20.

Stavney, L. S., Hamilton, T., Sircus, W., and Smith, A. N. (1966), Evaluation of the pH-sensitive telemetering capsule in the estimation of gastric secretory capacity, *Amer. J. Digest. Dis.,* 11: 753.

Stebbins, R. C., and Barwick, R. W. (1968), Radiotelemetric study of thermoregulation in a lace monitor, *Copeia,* Aug. 31, pp. 541–547.

Steinberg, W. H., Mina, F. A., Pick, P. G., and Frey, G. H. (1966), In Vitro evaluation of a new instrument for measuring intragastic pH, *J. Phyrm. Sci.,* 54: 772–776.

Steinberg, C. A., Sullivan, W. E., and Farrar, J. T. (1960), A physiological telemetry and data conversion system for gastrointestinal pressures, *SWIRECO,* 20–22.

Sterup, K. (1965), Gastric acid secretion test by means of a radio transmitter, *Danish Med. Bull.,* 12: 189–192.

Stewart, W. K. (1960), Physiological requirements for monitoring man in space, *Proc. 3 Int. Conf. Med. Electronics,* 133.

Stewart, W. W., and Ryan, C. S. (1965), Advanced bio-electronics. The value of the RKG 100 radio-cardiogram in industry, *Industr. Med. Surgery,* 34: 788.

Stolbun, B. M., and Forstadt, V. M. (1963), The first experiment of radio telemetric recording of the pulse wave velocity under dynamic conditions (in Russian), in V. V. Parin, *Radiotelemetry in physiology and medicine,* pp. 125–132.

Storer, E. H., Dodd, D. T., Snyder, P. A., and Eddlemon, C. O. (1961), A telemetering capsule for gastrointestinal pH measurement, *Amer. J. Med. Assoc.,* 178: 830.

Storm, G. L. (1965), Movements and activities of foxes as determined by radio-tracking, *J. Wildl. Mgmt.,* 29(1): 1–13.

Storm, G. L., and Verts, B. J. (1966), Movements of a striped skunk infected with rabies, *J. Mammal.,* 47: 705–708.

Storm, R. M. (Ed.), *Animal orientation and navigation,* Oregon State Univ. Press, Corvallis, 134 pp.

Storm van Leuwen, W., Kamp, A., Kok, M. L., and Zaal, J. (1963), Monitoring states of alertness by telemetering the EEG, *EEG Clin. Neurophysiol.,* 15: 164.

Straneo, G., Taccola, A., and Andreuzzi, P. (1963), Radio-electrocardiography during exertion for the functional exploration of the heart (Italian), *Minerva Cardioangiol.,* 11: 436–440.

Stuart, J. (1963), Extraterrestrial biological instrumentation problems, *Proc. Symp. Bio-Med. Eng.*

Sturkie, P. D. (1962), Radio telemetry measures heart rhythm in chickens, *New Jersey Agric.,* 44: 12–13.

Sturkie, P. D. (1963), Heart rate of chickens determined by radio telemetry during light and dark periods, *Poultry Sci.,* 42: 797–798.

Sullivan, G. H., and Bredon, A. D. (1961), Biotel-human biological telemetry system, *Proc. Natl. Tele. Conf.*

Sullivan, G. H., Hoefener, C., and Bolie, V. (1963), Electronic system for biological telemetry, in L. E. Slater (Ed.), pp. 83–106.

Sullivan, G. H., Schulkins, T. A., and Freedman, T. (1961), Internalized animal telemetry biomedical-surgical considerations, 32 Ann. Meeting Aerospace Med. Assoc., *Aerospace Med.,* 32: 249.

Sullivan, W. E., Farr, J. T., and Steinberg, C. A. (1960), A physiological telemetry and data conversion system, *IRE Natl. Conv.*

Suma, K. (1963), Study on medical application of ultramicro radiotransmitters, *Jap. J. Med. Electron. Biol. Eng.,* 1: 129–142.

Summers, G. D., and Temps, A. J. (1969), A biomedical telemetry system for clinical applications, *National Telemetering Conference,* 215.

Sunquist, M. E. (1967), Effects of fire on raccoon behavior, *J. Mammal.,* 48(4): 673–674.

Takagi, K. (1963), Study of ECG in physical exercise study of running using radio telemeter 1-pulse and ECG in interval running, *Res. J. Physical Educ.,* No. 1.

Takagi, K., Nagasaka, T., Ueda, K., and Wakita, S. (1965), A new method for indirect measurement of human blood pressure, *Digest 6th Intern. Conf. Med. El. Biol. Eng. Tokyo,* 38.

Takagi, S. (1963), University of Tokyo Symposium on applications of telemetry systems to medicine, *Med. Electron. Biol. Eng.,* 1: 577–581.

Takahashi, T., Watanabe, H., Watanuki, T., and Niwa, N. (1967), Endoradiocapsule for enzyme activity measurement, *Digest of 7th Intern. Conf. Med. Biol. Eng.,* 95.

Tarriere, C., Rebiffe, R., and Pittier, M. (1964), Application of a telecounter for human heart rate in (in French), *Trav. Humain,* 27 (3/4): 239–258.

Tepper, M. (1959), *Fundamentals of radio telemetry,* J. F. Rider Publ., New York, 116 pp.

Ter-Stepanyan, S. M. (1961), Laboratory apparatus for the simultaneous and complex investigation of physiological changes in the organism of the worker during the work process (in Russian), *Materialy Nauchn. Konf. In-Ta Gigiyeny Truda i Prof. Zapol. ArmSSR* (Materials of the Scientific Conference of the Institute of Labor Hygiene and Occupational Diseases, Armenian SSR), Yerevan, 19.

Tester, J. R. (1963), Radio tracking ducks, deer, and toads, *Univ. of Minnesota Mus. Nat. Hist. Tech. Rept. #6,* 9 pp. mimeo.

Tester, J. R. (1965), Status and value of biotelemetry in studies of vertebrate ecology, *Trans. Congress Intl. Union Game Biol., Beograd, Yugoslavia,* 7: 195–199.

Tester, J. R. (1966), Potentials of radio-tracking in studies of vertebrate behaviour (abstract), *Conf. Engr. Med. Biol.,* Proc. 8: 155.

Tester, J. R., and Heezen, K. L. (1965), Deer response to a drive census determined by radio tracking, *BioScience,* 15(2): 100–104.

Tester, J. R., and Moulton, J. C. (1965), Movements and activity of pen-raised white-tailed deer (*Odocoileus virginianus*) after release in unfamiliar habitat (abstract), *Bull. Ecol. Soc. Amer.,* 46(3): 121.

Tester, J. R., Warner, D. W., and Cochran, W. W. (1964), A radio-tracking system for studying movements of deer, *J. Wildl. Mgmt.,* 28(1): 42–45.

Tester, J. R., Warner, D. W., and Cochran, W. W. (1963), A radio-tracking system for studying movements of deer with observations on movements of penned deer after release, *Univ. of Minnesota Mus. Nat. Hist. Tech. Rept. #13,* 24 pp. mimeo.

Tharp, G. D., and Folk, Jr., G. (1965), Rhythmic changes in rate of the mammalian heart and heart cells during prolonged isolation, *Comp. Biochem. Physiol.,* 14: 255–273.

Thompson, N. P. (1964), Implanted perialuial pressure transducers, *Med. Electron. Biol. Eng.*, 2: 387–391.

Thompson, R. D. (1964), Design and implantation of sensors and transducers for physiological measurements in wild birds, *Biomed. Sci. Instrum.*, 2: 123–130.

Thompson, R. D., Grant, C. V., Pearson, E. W., and Corner, G. W. (1966), Physiological response of starlings to auditory stimuli, *Proc. 19 Ann. Conf. Eng. Med. Biol.*

Thompson, R. D., and Rukberg, D. L. (1967), A miniaturized four-channel radio transmitter with receiving system for obtaining physiological data from birds, *Med. Biol. Eng.*, 5(5): 495–504.

Thornton, E. W., and Davis, D. A. (1964), Some practical aspects—radio telemetry in clinical medicine, *Proc. Nat. Telemetering Conf., Session 2–4,* 1–3.

Tichvinskij, S. B., et al. (1963), The radiotelespirograph (in Russian), *Tez. i Ref. ZNIIFK,* Leningrad.

Timakov, V. A. (1965), Use of radiotelemetry for studying the functional state of the stomach in man (in Russian), *Eksp. Khir. Anesteziolgiya,* 10(5): 26–30.

Timofeyeva, T. Ye, and Anteselevich, V. A. (1960), Apparatus for recording human electrocardiograms at a distance (in Russian), *Novosti Med. Tekhniki,* 3: 27–41.

Timofeyeva, T. Ye (1961), Device for tele-electrocardiography (in Russian), Med. Prom-St' SSSR, *USSR Medical Industry,* 7: 46–50.

Timofeyeva, T. Ye (1962), Tel-electrocardiograph and the development of a multi-channel system for the recording of the ECG and human respiration (in Russian), *Tez. Dokl. II Vses. Konf. po Primeneniyu Radioelektronikii v Biologii i Meditsine,* NIITEIR, 64–65.

Timofeyeva, T. Ye, Smoljak, L. I., Kljukachev, V. A., and Bodrjagin, G. I. (1963), Radiotelemetric two-channel electrocardiospirograph EKS-1 (in Russian), *Trudy Instituta VNIIMIO, Moskva,* 3: 134–145.

Tobach, E. (1963), The potential for telemetry in the study of the social behavior of laboratory animals, in L. E. Slater (Ed.), pp. 33–44.

Todesco, S., Gambari, P., and Vincenci, M. (1965), The radioelectrocardiographic examination during heavy exertion in healthy young persons (in Italian), *Folia Cardio.* (Milano), 24: 88–98.

Tolles, W. E. (1963), In medical research and patient care, *Prog. Cardiov. Dis.,* 5: 595–609.

Tolles, W. E. (1963), Short range telemetry of ingested or implanted sensors, in L. E. Slater (Ed.), pp. 339–340.

Tolles, W. E., and Carbery, W. J. (1959), *A system for monitoring the electrocardiogram during body movement,* Aero Med. Lab., Wright Development Center, Air Res. and Development Command, USA Air Force, Wright-Patterson Air Force Base, Ohio WADC, Technic. Rep. 58–453, ASTIA Doc. No. AD 215538.

Toloat, M. E., Kraft, J. H., Cowley, R. A., and Khazei, A. H. (1967), *Biological Electrical Power extraction from blood to power cardiac pacemakers,* Oct., BME-14(4): 263–265.

Tomlinson, J. T. (1957), Pigeon wing beats synchronized with breathing, *Condor,* 59: 401.

Traite, M. (1959), Environmental testing of future spacemen, *Electronics,* 32(42): 65–69.

Trummer, J. R., and Reining, W. N. (1969), A four channel telemetry system for intensive care, *Proc. 8th Int. Conf. Med. and Biol. Eng.,* 30–6.

Turkowski, F. J., and Mech, L. D. (1966), An analysis of the movements of a young male raccoon, *Univ. of Minnesota Mus. Nat. Hist. Tech. Rept. #13,* 24 pp. mimeo.

Tursky, B., et al. (1962), Problems encountered in telemetering of physiological activity of schizophrenics in an experimental psychiatric ward, *Digest 15 Ann. Conf. Eng. Med. Biol.*

Uchiyama, A., Morimoto, T., and Yoshimura, H. (1964), Continuous recording of blood pH and its radiotelemetry, *Jap. J. Physio.,* 14: 630–637.

Uherka, M., and Kistler, W. (1958), *Army medical service research and development command special experiment for Jupiter missile AM-13 (U),* Missile Systems Engineering Branch, Structures and Mechanics Lab., DOD, ABMA, Redstone Arsenal, Huntsville, Ala., Rept. No. DSL-TN-28-58 (secret); October 27.

Unzhin, R. V. (1963), On the time dividing of the channels in two-channel transmission of physiological information (in Russian), in V. V. Parin, *Radiotelemetry in physiology and medicine,* pp. 45–47.

Unzhin, R. V. (1963), On the medico-technical requirements and the general application of semiconductor amplifiers in radioelectrocardiography (in Russian), in V. V. Parin, *Radiotelemetry in physiology and medicine,* pp. 64–72.

Unzhin, R. V. (1963), On the transducer in sets for wireless transmission of physiological data (in Russian), in V. V. Parin, *Radiotelemetry in physiology and medicine,* pp. 58–63.

Unzhin, R. V., and Rozenblat, V. V. (1959), On the recording by radio of respiratory motions in a freely-moving man using a semiconductor radiopneumograph (in Russian), *Materialy III (Zonal'noy) Nauchno Praktich. Konf. po Vrachebn. Kontrolyu i Lechebn. Fizkul'ture,* Sverdlovsk, II: 102–109.

Unzhin, R. V., and Rozenblat, V. V. (1962), Combined radiotelemetric instrument for physiological investigations (in Russian), *Tez. Dokl. II Vses, Konf. Po Primeneniyu radioelektronikii v Biologii i Meditsine,* NIITEIR, 67–68.

Unzhin, R. V., and Rozenblat, V. V. (1963), On the combined radioelectronic equipment KRP-3 (in Russian), *Biol. i Med. Elektronika,* #2.

Unzhin, R. V., and Rozenblat, V. V. (1964), A transistor device for remote recording of heart rate, respiration and movement, *Bull. Exper. Biology and Medicine* (Feb.), 57: 237–241.

Unzhin, R. V., Rozenblat, V. V., and Forstadt, V. M. (1962), The KRP-2 combined radiotelemetric instrument for physiological investigations (in Russian), *Materialy IV Obyed. Ural'skoy Konf. Fiziologiv, Farmakologov, i Biokhimikov* (Materials of the fourth joint Ural Conference of Physiologists, Pharmacologists, and Biochemists), Chelyabinsk, 247–249.

Upson, J. D., King, F. A., and Roberts, L. (1962), A constant-amplitude transistorized unit for remote brain stimulation, *Electroenceph. Clin. Neurophysiol.,* 14: 928–930.

Uvarova, Z. S. (1959), Study of the influence of physical exercises on the cardiovascular and respiratory function in young children using the radiotelemetric procedure (in Russian), *Materialy III (Zonal'noy) Nauchno-Praktich. Konf. po Vrachebn. Kontrolyu i Lechebn. Fizkul'ture,* Sverdlovsk, II, 87–90.

Vallbona, C., Spencer, W. A., Geddes, L. A., Blose, W. F., and Cansoneri, J. (1966), Experience with on-line monitoring in critical illness, *Proc. Natl. Telem. Conf. Proc.*

Van Citters, R. L., and Franklin, D. L. (1965), Blood pressure and flow telemetered from free ranging baboons, *Abstracts 23 Intern. Congr. Physiol Sci.,* Tokyo, 141.

Van Citters, R. L., Evonuk, E., and Franklin, D. (1966), Blood flow distribution in Alaska sled dogs during extended exercise (abstract), *The Physiologist,* 9(3): 130.

Van Citters, R. L., Watson, N. W., Franklin, D. L., and Elsner, R. W. (1965), Telemetry of aortic blood pressure and flow in free ranging animals, *Fed. Proc.,* 24: 525.

Van Citters, R. L., Franklin, D. L., Smith, O. A., Watson, N. W., and Elsner, R. W. (1965), Cardiovascular adaptations to driving in the northern elephant seal murounga angustirosters, *Comp. Biochem. Physiol.,* 16: 267.

Van Citters, R. L., Smith, Jr., O. S., Franklin, D. L., Kemper, W. S., and Watson, N. W., Radio telemetry of blood flow and blood pressure in fetal baboons: A preliminary report, in H. Vagtborg (Ed.), *The Baboon in Medical Research, Vol. II.*

Vanderval, F. L., and Young, W. D. (1958), *A preliminary experiment with recoverable biological payloads in ballistic rockets,* Project MIA, Space Technology Laboratories, Los Angeles, Cal.

Van Heeckeren, D. W., and Glenn, W. W. L. (1966), Electrophrenic respiration by radiofrequency induction, *J. Thorac. Cardiovasc. Surg.,* 52: 655–666.

Van Patten, R. E. (1963), An annotated bibliography on biomedical (animal) satellite vehicles, life support systems and related technology, Techn. Docum. Rept. AMRL-TDR-63-126, *U.S. Air Force 6570, Aerospace Med. Res. Lab,* 1–25.

Van Trappen, G., D'Haens, J., Verbeke, S., and Van Denbroucke, J. (1962), The endomotorsonde (in Dutch), *T. Gastroent.,* 5: 578–585.

Vardishvili, I. A., Zandukeli, D. N., and Levin, E. M. (1965), An equipment for radio connection between the coach and the sportsman in exercise (in Russian), *Teor. Prakt. Fiz. Kult., Moskva,* 28(6): 70–73.

Varney, R. J., et al. (1963), *Special tracking project final report,* Philco Corp. Western Development Labs, 49 pp.

Vasil'yeva, V. V., Gracheva, R. P., Yelshina, L. V., Kozlov, I., and Kossovskaya, E. B. (1960), Telemetric investigations of the frequency of heart contractions during the running of various distances (in Russian), *Tez. Dokl. Nauchn. Konf. GDOIFK im. Lesgafta po Itogam Raboty za 1960,* summaries of reports at the Scientific Conference of the GDOIFK imeni Lesgaft on the results of work for 1960, VI: 9–10.

Vasil'yeva, V. V., Gracheva, R. P., Yelshina, L. V., Kozlov, I. M., and Kossovskaya, E. B. (1961), Telemetric investigations of the frequency of heart contractions during the running of various distances (in Russian), *Teoriya i Praktika Fiz. Kul'tury* (Theory and Practice of Physical Culture), 24(3): 188–192.

Vasil'yeva, V. V., Gracheva, R. P., Yelshina, L. V., Kozlov, I. M., and Kossovskaya, E. B. (1963), Telemetric investigation of heart rate during running various distances (in French), *Rev. Educ. Phys.,* 3: 25–30.

Verts, B. J. (1963), Equipment and technique for radio-tracking striped skunks, *Wildl. Mgmt.,* 27: 325–339.

Verzeano, M., and French, J. D. (1953), Transistor circuits in remote stimulation, *EEG Clin. Neurophysiol.,* 5: 613–615.

Verzeano, M., Webb, R. C., and Kelly, M. (1958), Radio control of ventricular contraction in experimental heart block, *Science,* 128: 1003.

Viru, E. A., Oya, S. M., Sil'dmyae, Kh. Yu., and Viru, A. A. (1962), On changes in the frequency of heart contractions and the arterial pressure during sports exercises (in Russian), *Materialy 7-y Nauchn. Konf. po Vopr. Morfologii,*

Fiziologii i Biokhimii Myshechn. Deyatel'n (Materials of the 7th Scientific Conference on the Problems of Morphology, Physiology, and Biochemistry of Muscular Activity), Moscow, 51–52.

Viru, A. A., and Sildmjae, C. J. (1963), On the possibility of using radio telemetry for determination of the functional abilities of sportsmen's organisms (in Russian), in V. V. Parin, *Radiotelemetry in physiology and medicine,* pp. 168–171.

Vodolazskij, L. A. (1957), Recording of the electrocardiogram in workers during labor in industry (in Russian), *Bjull. Eksper. Biol. Med., Prilozenie k Zurnalu No. 1,* 43: 59–61.

Vodolazskij, L. A. (1959), Problems in the method of recording the EMG and the ECG of industrial workers during labor (in Russian), *Avtoref. Diss. Kand. M.*

Vodolazskij, L. A. (1960), Electronic device for routine electrocardiography (in Russian), in Berg, *Electronics in medicine,* pp. 368–373.

Vodolazskij, L. A., Podoba, E. V., and Solovyeva, V. P. (1961), Further development of the method of recording of electrocardiograms and electromyograms of a worker during his work in the factory, *Digest. Intern. Conf. Med. Electron.,* 128.

Vodolazskij, L. A., Podoba, E. V., and Solovyeva, V. P. (1963), Application of the teleelectrocardiogram TEK-1 in measurements of industrial physiology (in Russian), *Trudy VNIIMIO Moskva,* 3: 146–147.

Volyukin, Yu. M., Gazenko, O. G., Agadzhanyan, G. A., and Bayevskiy, R. M. (1962), Some results of a medicobiological study of space flights (in Russian), *Voyenno-med. Zh.* (*Military Medical Journal*), 9: 3–9.

Von Ardenne, M. (1958), Der verschiuckbare intestinalsender ein beispiel fur den einsatz moderner tecknik in der medizin, *Tecknik,* 12: 614–616.

Von Ardenne, M. (1960), Some techniques of the swallowable intestinal transmitter, in C. N. Smyth (Ed.), *Medical electronics* (*Iliffe*), 614.

Von Ardenne, M., and Sprung, H. B. (1958), Uber die gleichzeitige registrierung von druckanderugen und lageanderungen bei dem verschluckbaren intestinalsender, *Z. Ges. Inn. Med.,* 13: 596–601.

Von Ardenne, M., and Sprung, H. B. (1958), Uber versuche mit einem verschluckbaren intestinalsender, *Naturwissenschaften,* 45: 154–155.

Vorobev, A. T., and Stolbun, V. M. (1963), The evaluation of some types of precordial leads used in radio electrocardiography under dynamic conditions (in Russian), in V. V. Parin, *Radiotelemetry in physiology and medicine,* pp. 48–57.

Vorobev, A. T. (1963), The application of radio electrocardiography during training and competition in athletics (in Russian), in V. V. Parin, *Radiotelemetry in physiology and medicine,* pp. 133–44.

Voronkov, G. L., and Shekhtman, M. L. (1964), Tele-electrocardiographic study of convulsive seizures caused by electroshock therapy (in Russian), *Zh. Hevropatol Psikhiat im SS Korsakova,* 64(12): 1845–1851.

Vos, A. De., and Anderka, F. W. (1964), A transmitter-receiver unit for microclimatic data, *Ecology,* 45: 171–172.

Vreeland, R. W., Williams, L. A., Yeager, C. L., and Henderson, J. (1958), Unit telemeters scalp voltages, *Electronics,* 31: 86.

Vreeland, R. W., Collins, C., Williams, I., Yeager, C., Gianascol, A., and Henderson, J. (1963), A subminiature radio electroencephalogram telemeter for studies of disturbed children, *Electroenceph. and Clin. Neurophysiol.,* 15(2): 32.

Vreeland, R. W., and Yeager, C. L. (1967), A four channel integrated circuit telemeter for seizure monitoring, *Digest of 7th Intern. Conf. Med. Biol. Eng.*, 93.

Vrinciawu, R., Arsenescu, G., Repta, V., Bobic, D., and Broskamu, E. (1963), Radio recording of biological parameters in diesel engine drivers, *Rumanian Med. Rev.*, 7: 19–23.

Walker, A. E., and Burton, C. V. (1966), Radiofrequency telethermocoagulation, *J. Amer. Med. Assoc.*, 197: 700.

Wallace, V., Dracy, A. E., and Oines, R. K. (1959), A radiosonde for measuring ruminal pressure, *Proc. South Dakota Acad. Sci.*, 38: 146.

Wangel, A. G., and Deller, D. J. (1965), Intestinal motility in man, 3. Mechanisms of constipation and diarrhea with particular reference to irritable colon syndrome, *Gastroenterology,* 48: 69–84.

Ware, R. W. (1960), A method for in-flight measurement of systolic and diastolic blood pressure pulse rate and rate and depth of respiration, *Pan. Amer. Med. Assn. Sci. Meet.*

Ware, R. W., and Kahn, A. R. (1963), Automatic indirect blood pressure determination in flight, *J. Appl. Physiol.,* 18: 210.

Warner, D. W. (1963), Space tracks, *Nat. Hist.,* New York, 62: 8–15.

Warner, D. W. (1963), Fundamental problems in the use of telemetry in ecological studies, in L. E. Slater (Ed.), pp. 15–24.

Warner, D. W., Tester, J. R., Jacobson, S. L., Lindmeier, J. P., Hartnett, J. P., and Birkebak, R. C. (1960), A study of the mobile responses of animals to radiation fields and to other physical and biotic factors in the natural environment, *Mus. Nat. His., Univ. Minnesota,* 1–15.

Warner, D. W., Tester, J. R., Jacobson, S. L., Lindmeier, J. P., Hartnett, J. P., and Birkebak, R. C. (1961), A study of the mobile responses of animals to radiation fields and to other physical and biotic factors in the natural environment, *Mus. Nat. His., Univ. Minnesota,* 2(i–vii): 65.

Warner, D. W. (1967), Space Tracks, in I. W. Knoblock, *Readings in biological science* (2nd ed.), pp. 221–228.

Warner, H. A. (1962), A remote control brain stimulator system, *Digest 15 Ann. Conf. Eng. Med. Biol.*

Warner, H. A., Robinson, B. W., Rosvold, H. E., Wechsler, L. D., and Zampini, J. J. (1968), A remote control brain telestimulator with solar cell power supply, Apr. **BME-15**(2).

Watanuki, T., Hori, G., Atsumi, K., Suma, K., Toyoda, T., Saknrai, Y., and Omoto, R. (1963), Clinical uses of "echo-capsule" for telemetering temperature within the body, in *Symp. on Appl. Telem. Syst. to Med. Tokyo* (1962). Abstract: *Med. Electron. Biol. Eng.* (1963), 1: 580.

Watanuki, T., Niwa, H., Kawai, K., Watanabe, H., and Ouchi, A. (1965), Study on radio capsule, *Digest 6 Int. Conf. Med. Electron. Biol. Eng.,* 215.

Watanuki, T., et al. (1964), Gastrointestinal diagnosis by capsules (in Japanese), *Clin. All. Round* (Osaka), 13: 1033–1038.

Watanuki, T. (1966), III. Weltkongress fur Gastroenterologie Tokyo, Sept., 18–24.

Watson, B. W., Ross, B., and Kay, A. W. (1962), Telemetering from within the body using a pressure sensitive radio pill, *Gut,* 3: 181–186.

Watson, W. C., and Paton, E. (1965), Studies on intestinal pH by radiotelemetering, *Gut,* 6: 606–613.

Watson, W. C., Watt, J. K., Paton, E., Glen, A., and Lewis, G. J. T. (1966), Radiotelemetering studies of jejunal pH before and after vagotomy and gastroenterostomy, *Gut,* 7: 700–705.

Watson, B. W., Riddle, H. C., and Currie, J. C. M. (1967), The measurement of intracranial pressure in man using radio telemetry, *Digest of 7th Intern. Conf. Med. Biol. Eng.*, 91.

Webb, G. N. (1963), The display of physiological data from extended periods of time, in L. E. Slater (Ed.), pp. 353–361.

Webb, J. C., Campbell, L. E., and Hartsock, J. G. (1958), Electrocardiograph telemetering (Radio), *IRE Nat. Conven. Record Med. Electron.*, Part A, 43–49.

Wegner, M. A., Henderson, E. G., and Dunning, J. S. (1957), Magnetometer method for recording gastric motility, *Science*, 125: 990.

Wehner, A. P., and Hahn, F. J. (1963), Implantation of a miniature multi-channel physiological data acquisition system in a canine, *5 Int. Cong. Med. Electron.*, Liege, Abstr., p. 27.

Weimann, G., and Schweckendiek, R. (1965), Was leistet die "Heidelberger Kapsel"? Vergleichende Untersuchungen mit der Schlauchsonde, *Med. Welt*, 628–634.

Weisler, A. C., and Haygood, C. C. (1959), *Biological experimental acceleration environment Jupiter missile AM-18 (U)*, Flight Evaluation Branch, Aeroballistics Lab., DOD, ABMA, Redstone Arsenal, Huntsville, Ala., Rept. No. DA-TM-87-59 (secret), July 10.

Welford, N. T. (1961), The setar (special event timer and recorder) and its uses for recording physiological and behavioural data, *Digest Int. Conf. Med. Electron.*, 101.

Werner, H., and Hemmati, A. (1964), Klinisch-bakteriologische Erfahrungen mit einer neuen steuerbaren Kapsel zur Entnahme von Carminhalt, *Internat. Mikrookolog Sympos.*

West, J. K., and Merrick, A. W. (1966), A three-channel EEG telemetry system for large animals, *Med. Biol. Eng.*, 4: 273–281.

White, P. D., and Matthews, S. W. (1956), Hunting the heartbeat of a whale, *Nat. Geograph. Mag.*, 110(1): 49–64.

Widmann, W. D., Glenn, W. W. L., Eisenberg, L., and Mauro, A. (1964), Radiofrequency cardiac pacemaker, *Ann. N.Y. Acad. Sci.*, 111: 992.

Wilbarger, E. S. (1959), *Biomedical measurement in ballistic missiles*, U.S. Army Med. and Res. and Develop. Command, Bioastronautics Res. Office, AOMC, Huntsville, Ala. Special Rept. CSCRD-16-3, Sept. 20.

Wilbarger, Jr., E. S., et al. (1964), Indirect heart rate measuring device, *Amer. J. Med. Electronics*, 3: 199–200.

Williams, J. M., Berneski, J. P., and McDermott, J. F. (1965), A signal conditioner for telemetered fetal electrocardiograms, *18 Ann. Conf. Eng. Med. Biol.*

Williams, M. W., Baldwin, H. A., and Williams, C. S. (1968), Body temperature observations on a caged Mexican wolf (*Canis lupus Baileyi*) by remote telemetry, *J. Mamm.*, 49(2): 329–331.

Wilton-Davies, C. C. (1965), Telemetry in pressure chambers, *World Med. Electronics*, 3(6): 227.

Wilton-Davies, C. C. (1965), Telemetry in pressure chambers, *World Med. Electronics*, 3(9): 329–330.

Wilz, H., and Noller, G. (1964), Experimentelle und Klinische Prufung eines neuen Magentherapeutikums, *Med. Klin.*, 59: 663–666.

Winders, S. R. (1921), Diagnosis by wireless, *Scientific American*, 124: 465.

Winget, C. M., Averkin, E. G., and Fryer, T. B. (1965), Quantitative measurement by telemetry of ovulation and oviposition in fowl, *Amer. J. Physiol.*, 209: 853.

Winget, C. M., and Fryer, T. B. (1966), Telemetry system for the acquisition of circadian rhythm data, *Aerospace Med.*, 37: 800.

Winsor, T., Sibley, W. A., and Fisher, E. K. (1961), Electrocardiograms by telemetry, *Calif. Med.*, 94: 284–286.

Wodzicka-Tomaszewska, M., and Walmsley, R. R. (1966), A note on the testing of a sheep heart-rate transmitter, *New Zeal. J. Agr. Res.*, 9: 455–459.

Wojcieszak, I., Burkhard, K., and Kisiecki, A. (1965), Analysis of the comparison between the Harvard step test and the running test, both used as a criterion of the physical fitness (in Polish), *Wychow. Fiz. Sport,* Warszawa, 9(4): 389–405.

Wolff, H. S. (1960), Discussion on the swallowable intestinal transmitter, in C. N. Smyth, *Medical electronics,* p. 342.

Wolff, H. S., and Russ, R. F. (1960), Construction aspects of radio pills suitable for mass production, *3 Int. Conf. Med. Electron.*

Wright, L. N. (1963), *Internal four-channel physiological telemetry system prototype development,* Final Report, North American Aviation, Inc., Downey, Cal., Space and Information Systems Division, Los Angeles, AFSC, July, 66 pp.

Wu, L. S. (1963), A proposed medical instrumentation system for early manned research spacecraft and space laboratory, *Proc. Symp. on Bio-Med. Eng.*

Yamakawa, K., Kitamura, K., Kato, T., Nauba, T., Kawaguchi, S., Bando, J., Yamakura, K., Uesugi, M., Watabe, T., Takagi, K., Furuya, H., Ezawa, H., Yamanaka, Y., Minamitani, K., and Morita, J. (1963), The clinical application of telemetering (abstract), *Jap. Circulat. J.*, 27: 716.

Yamakawa, K., Kitamura, K., Yamakura, K., and Furuya, H. (1963), Clinical evaluation of exercise tests using the radio-electrocardiograph, *Univ. Tokyo. Symp. Appl. Telem. Syst. Med.* (1962). Abstract: *Med. Electron. Biol. Eng.* (1963), 1: 581.

Yamakawa, K. (1964), Various problems with the telemeter circuit (in Japanese), *Clin. All. Round* (Osaka), 13: 980–987.

Yamamura, H., Kiyohara, H., Saito, M., and Sakai, T. (1965), The optimum design of telemetering system and its application in medicine, *Digest 6 Int. Conf. Med. Electron. Biol. Eng.*, 191–192.

Yokobori, S. (1964), *Analysis of the interval training of hockey athletes by telemetering,* Abstracts of papers presented at ICSS, Tokyo, 261.

Yoshida, T., and Tani, S. (1963), A miniaturized telemeter set for medical uses, abstract: *Med. Electronics Biol. Eng.*, 1: 579.

Yoshimoto, C. (1955), Transducers in medical and biological fields, *Bull. Res. Inst. Appl. Elect. Hokkaido Univ. Sapporo,* Japan, 175–186.

Young, I. F., and Naylor, W. S. (1964), Implanted two way telemetry in laboratory animals, Digest 15 Ann. Conf. Eng. Med. Biol. 1962, *Amer. J. Med. Electron.* (1964), 3: 28–33.

Yushchenko, A. A., and Chernavkin, L. A. (1932), New method of studying unconditioned and conditioned reflexes of freely moving animals (in Russian), *Sov. Nevropatol. Psikhiatr. i Psikhogig.*, 8: 327–332.

Yushchenko, A. A., and Chernavkin, L. A. (1932), New radio procedures in labor psychophysiology (in Russian), Sots Rekonstruktsiya i Nauka, *Socialist Reconstruction and Science,* 1: 217–220.

Zakrzewski, H. (1964), Telemetric methods for studying the physiological parameters in man during exercise (in Polish), *Wychow. Fiz. Sport,* Warszawa, 8(2): 179–184.

Zemlyakov, K., Ovanov, D., and Fedorov, T. (1938), A teleradio device, recording the work of the heart (in Russian), Voyenno-san. Delo, *Military Sanitation,* 2: 75–78.

Ziecheck, L. N. (1969), The general design considerations for a wireless telemetry system in the medical diagnosis of humans, *National Telemetering Conference,* 201.

Zorcher, G., and Heinkel, K. (1966), Funkfermessung der Wasserstoffionenkonsentration, *Z. Gastroenterol.,* 4: 130–139.

Zweizig, J. R., Kuho, R. T., Hanley, J., and Adey, W. R. (1967), The design and use of AM/FM radiotelemetry system for multichannel recording of biological data, Oct., **BME-14**(4): 263–265.

Zweizig, J. R., Adey, W. R., and Hanley, J. (1969), Clinical monitoring using a circularly polarized R. F. Link, *Proc. 8th Int. Conf. Med. and Biol. Eng.,* 30–3.

Zweizig, J., Hanley, J., Cockett, A., Hahn, P., Adey, W. R., and Ruspini, E. (1969), EEG monitoring during the treatment of decompression illness ("bends"), *National Telemetering Conference,* 195.

Zworykin, V. K. (1959), Recent advances in medical electronics, *IRE Nat. Conv. Record. Med. Electron.,* 42–46.

Zworykin, V. K. (1960), Prospects in medical electronics, *Proc. 3 Int. Conf. Med. Electron.,* 6.

Zworykin, V. K., Farrar, J. T., Bostrom, R. C., Hatke, F. L., and Deboo, G. L. (1961), The measurement of internal physiological phenomena using the passive type telemetering capsule, *IRE Conv. Part 9,* 141–144.

Zworykin, V. K., and Hatke, F. L. (1961), A miniaturized hospital telemetering system, *Digest Int. Conf. Med. Electron.,* 125.

Zworykin, V. K., et al. (1962), New advances in passive telemetering implants, *Digest 15 Ann. Conf. Eng. Med. Biol.*

Index